“I like redheads; their n
of strawberry jam i

ROGER STERLI

“Once in his life,
every man is entitled
to fall **MADLY IN LOVE**
with a gorgeous redhead.”

LUCILLE BALL

“It was like autumn,
looking at her. It was
like driving up north
to see the colors.”

JEFFREY EUGENIDES,
MIDDLESEX

“You’d find it easier
to be bad than good if
you had red hair.”

L. M. MONTGOMERY,
ANNE OF GREEN GABLES

“Red hair, sir,
in my opinion,
is **DANGEROUS**.”

P. G. WODEHOUSE,
VERY GOOD, JEEVES!

“I’m a ginger, and there’s not much more fun you can get as a ginger. And I think that’s why there’s a lot of resentment toward the ginger community. **WE’RE VIKINGS, ESSENTIALLY**.”

MICHAEL FASSBENDER

“We redheads are a minority,
WE TEND TO NOTICE EACH OTHER—
you know, and notice our identity.” JULIANNE MOORE

RED

A *History* of the *Redhead*

RED

A *History* of the *Redhead*

JACKY COLLISS HARVEY

Black Dog
& Leventhal
Publishers
New York

Black Dog & Leventhal Publishers
Hachette Book Group
1290 Avenue of the Americas
New York, NY 10104

www.blackdogandleventhal.com

Printed in the United States of America

Cover design by Nicole Caputo
Interior design by Cindy Joy

WOR

First Edition: June 2015

10 9 8 7 6 5 4 3 2 1

Black Dog & Leventhal Publishers is an imprint of Hachette Books, a division of Hachette Book Group. The Black Dog & Leventhal Publishers name and logo are trademarks of Hachette Book Group, Inc.

Library of Congress Control Number: 2015936960

This one is for Mark.

CONTENTS

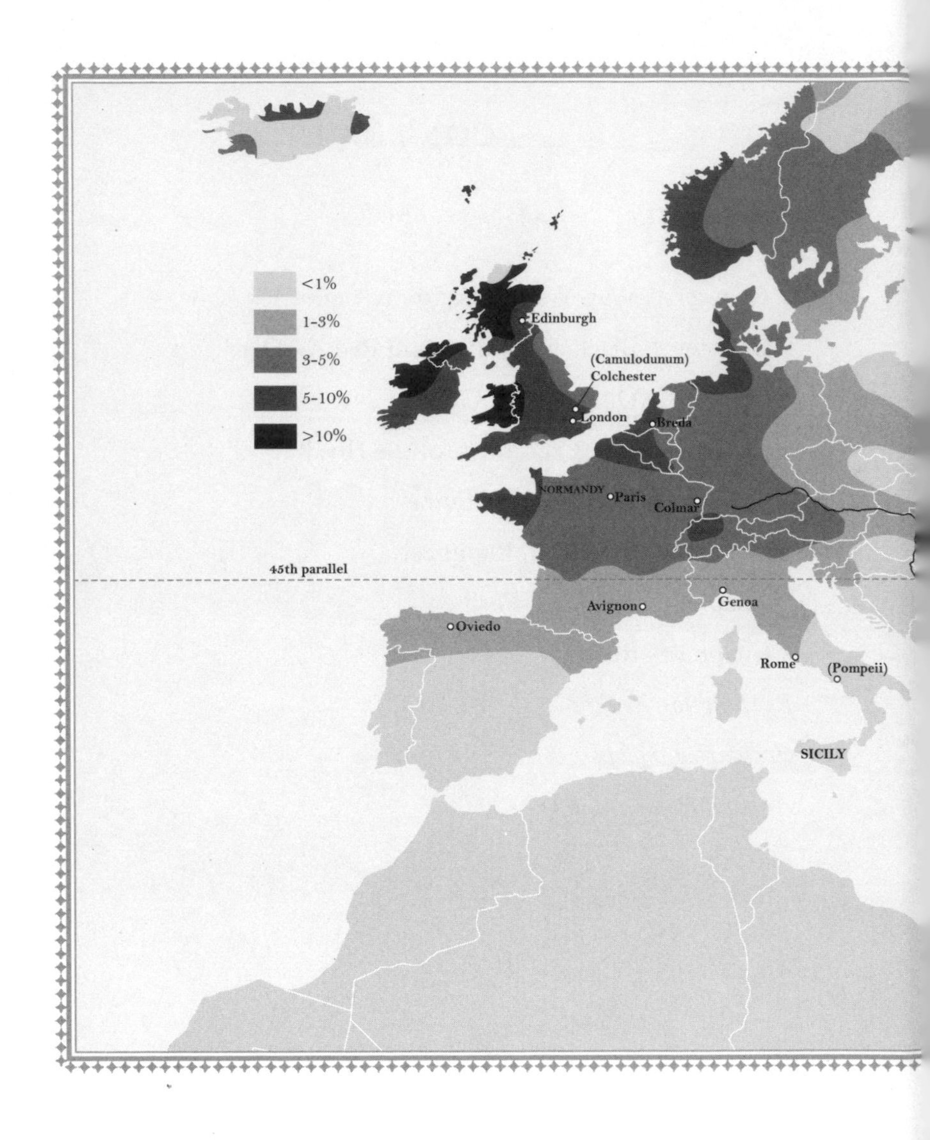

<1%
1-3%
3-5%
5-10%
>10%
Edinburgh
(Camulodunum)
Colchester
London
Breda
NORMANDY
Paris
Colmar
45th parallel
Avignon
Genoa
Oviedo
Rome
(Pompeii)
SICILY

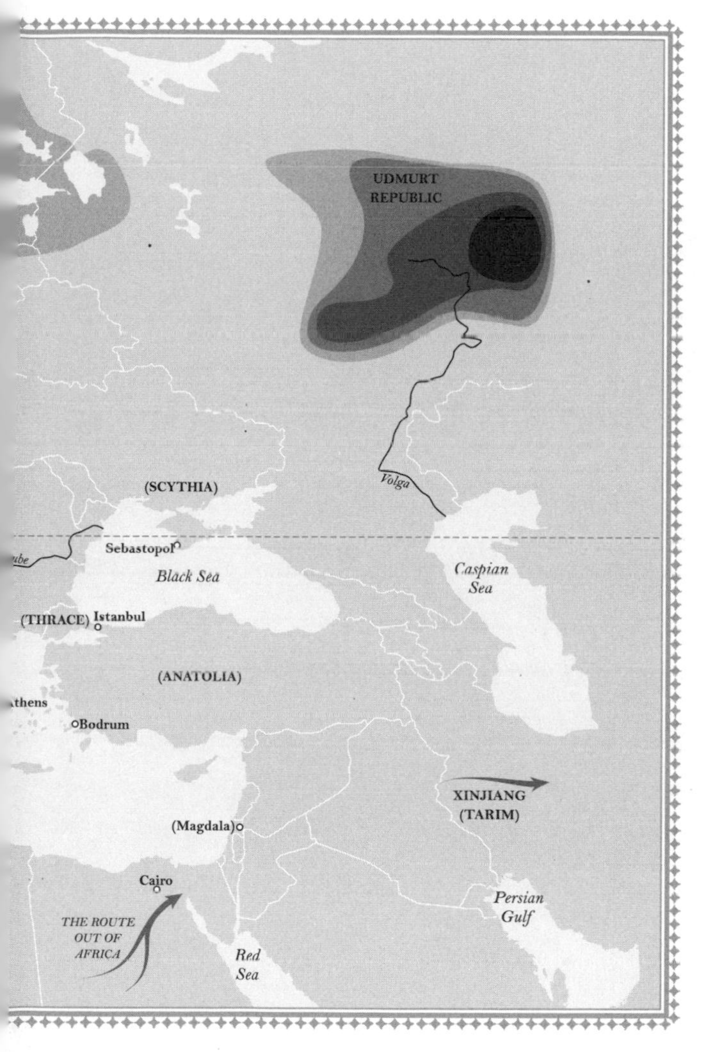

The Redhead Map of Europe.

There is a good deal of controversy over the accuracy of such maps, as there is indeed over so many issues associated with red hair, but what it shows very clearly is the hotspot in Russia of the Udmurt population on the River Volga and the increasing frequency of red hair the farther north and west you go, whether in Scandinavia, Iceland, the British Isles, or Ireland.

INTRODUCTION

The study of hair, I found out, does not *take you to the superficial edge of our society, the place where everything silly and insubstantial must dwell. It takes you, instead, to the centre of things.*

GRANT McCRACKEN, *BIG HAIR*, 1995

I am the only redhead in my family, a situation with which many a redhead will be familiar. My mother, now gray-haired, was blonde (was still a blonde, well into her seventies). My father's hair was dark brown. My brother is also blond. My brother's kids have hair that shades from brownish to blondish to positively Aryan. Yet mine is red. When I was little, it was the same orange color as the label on a bottle of Worcestershire sauce; with age it has toned down, to a proper copper. It is not carrots, nor ginger, nor the astonishing fuzz of paprika I remember on the head of a girl at school, a child with skin so white it was almost luminous. I'm not quite at that end of the spectrum, but I am red. It is, with me, as with many other redheads, the single most significant characteristic of my life. If that sounds a little extreme to you, well, you're obviously not a redhead, are you?

Red hair is a recessive gene, and it's rare. Worldwide, it occurs in only 2 percent of the population, although it is slightly more common (2 to 6 percent) in northern and western Europe, or in those with that ancestry (see the map on pages viii-ix).[1] In the great genetic card game, the shuffling of the deck that has made us all, red hair is the two of clubs. It is trumped by every other card in the pack. Therefore, for a red-haired child to result, both parents have to carry the gene, which, blond- or brown-haired as they may very well be, they can be carrying completely unaware. So when a baby appears with that telltale tint to its peach fuzz, expect many jokes and much hilarity. For all my toddlerhood, my mother would blithely ascribe my red hair to either her craving for tomato juice during her pregnancy or to a mysterious redheaded milkman. My grandmother, meanwhile, was fond of quoting the wise old saying that "God gives a woman red hair for the same reason He gives a wasp stripes." But then she was a native of Hampshire, a West Country girl, where redheads were once also known, charmingly, as Dane's bastards, so really, that was letting me off lightly.

I was five before I realized there might be more to being a redhead than incomprehensible teasing by adults. My village school in Suffolk was terrorized by a kindergarten Caligula, a bully from day one, whom we'll call Brian. The rest of us five-year-olds watched in disbelieving horror as Brian roamed about the playground, dispensing armlocks, yanking out hair by the roots, and knocking down birds' nests and laughing as he stamped on the eggs, or fledglings, inside. His genius was to find the thing most precious to you and

1 By comparison, somewhere between 16 percent and 17 percent of the population of the planet has blue eyes and 10 to 12 percent of the population is left-handed. Roughly 1 in 10 Caucasian men are born with some degree of color blindness, while 1.1 percent of all births worldwide are of twins. The incidence of albinoism, worldwide, is roughly 0.006 percent.

destroy it. One afternoon at the end of a school day he came up behind my friend Karen, who was sporting a new woolly hat of a pale and pretty blue with a large fluffy bobble on the top. Brian seized the hat from Karen's head, ripped off the bobble, and threw it to the ground.

I can still summon up the extraordinary feeling of liberation as the red mists descended. I wound up my right arm like Popeye and punched Brian in the face.

It was a fantastic blow. Brian was knocked flat. As he made to get to his feet, his eye was already swelling shut. Most incredible of all, Brian was in tears. Only then did I realize that my David-and-Goliath moment had been witnessed by all the mothers arriving at the school gate to collect their children, my mother included.

One did not punch. I knew this from the number of times I'd been told off for fighting with my own younger brother. I imagined my punishment. I awaited my mum's reaction, and the reactions of the other mothers at the gate. I was proudly unrepentant, but I knew I was also in any amount of trouble.

The punishment never came. Someone—one of the teachers, I think—picked Brian up and brushed him down. There was laughter. There was an air, astonishingly, of adult approval. My mum, who seemed embarrassed, took my hand and began to hustle me down the road. "Well, what did he expect?" one of her friends remarked, above my head. "She's a redhead!"

She's a redhead. I was five years old, and I had just learned two very important lessons. One, that the world has expectations of redheads, and two, that those expectations give you a license not granted to blondes or brunettes. I was expected to lose my temper.

I dutifully produced appalling tantrums as a child. I was meant to be confident, assertive, and, if I wished, slightly kooky. I could be a screwball. I could be fiery. As I grew older, the list of things I was allowed to do, simply because of the color of my hair, increased. I was allowed to be impulsive. I was allowed to be hot-blooded and passionate (once I reached the age for boyfriends and relationships, it seemed I was almost required so to be). The assumptions and expectations the world made about me and my fellow redheads were endless. I must be Irish. Or Scottish. I must be artistic. I must be spiritual. Was I by any chance psychic? And I must be good in bed. There's a point where all those "musts" start taking on the tone of a command. *She's a redhead.* That was all the world need know, apparently, to know me.

I grew up, and the world got bigger, too. I taught English to a brother and sister from Sicily who were even redder haired and paler skinned and bluer eyed than I am. How did that happen? I traveled farther. I discovered new attitudes toward my red hair, not the same at all as those I had grown up with. Yet the common denominator in every reaction I experienced was this: redheads were viewed as being different. And there has, of course, to be a point when you start asking yourself *why*. Why these assumptions? What's their basis? Do they even have one? Why do they differ from one country to another? Why have they changed, or why have they not, from one century to the next? Where do redheads come from, anyway?

The term "redde-headed" as a synonym for red hair can be tracked back at least to 1565, when it appears in *Thomas Cooper's Thesaurus Linguae Romanae et Britannicae*, otherwise known as *Cooper's Dictionary.*[2] This mighty achievement, admired by no less a redhead than Elizabeth I, who made its author Dean of Christ Church Oxford as reward for his labors, is a building block of the English language and is believed to have been one of the most significant resources used by that great word-smelter William Shakespeare. But the specific chromosome responsible for red hair was only identified in 1995, by Professor Jonathan Rees of the Department of Dermatology at the University of Edinburgh.[3] So for almost the entirety of its 50,000-year existence on this planet, red hair, across every society where it has appeared, has been wrestled with as an unaccountable mystery. In the search for an explanation for it, it has been hailed as a sign of divinity; damned as the awful consequence of breaking one of the oldest sexual taboos; ostracized and persecuted as a marker of religion or race; vilified or celebrated as an indicator of character; and proclaimed as a result of the influence of the stars. It is, unsurprisingly, none of these things, yet at the same time, society's—any society's—responses to red hair have become so inseparable from the thing itself that it has become all of them. And there is as much

2 Bishop Cooper's great work almost never saw the light of day at all. When it was half completed, Cooper's wife ("a shrew," according t o that indefatigable chronicler John Aubrey), "irreconcileably angrie" with him for neglecting her in favor of his studies, broke into his study and threw his papers on the fire. Aubrey does not record if she was a redhead, too.

3 No one could write a book on redheads without a debt to Professor Rees and his research. I am very happy to record my gratitude to him for his early assistance and generous advice.

mistaken nonsense written about it now as there was a hundred, or two hundred, or, for that matter, five hundred years ago.

Let me illustrate what I mean here by way of a famous red-headed tale, which functions almost as a parable. In 1891 Sir Arthur Conan Doyle published the classic Sherlock Holmes short story "The Adventure of the Red-Headed League." The redhead at the center of the tale is a pawnbroker, Jabez Wilson. Owing to the particular tint of his rare red hair, Jabez is selected by the mysterious League and lured from his pawnbroker's shop to an empty office, where he is paid to spend his time pointlessly copying out chunks of text from the *Encyclopedia Britannica*—a task that, according to the League, he and only he, with his unique flame-red hair, is fit to do (you can see why this wouldn't have worked as a ruse were he garden-variety blond- or dark-haired). Sherlock Holmes, of course, spots at once that the pawnbroker's shop sits next to a bank and that the "work" offered to Wilson by the Red-Headed League is no more than a trick to get him out of the way. The League is simply the cover for a gang of robbers who plan to break into the bank through the pawnshop's basement; and Jabez has been selected not because of his hair but because of the location of his shop. In other words, it is a story whose final explanation is completely different from the one you might expect. In exploring the history of red hair, such will very often prove to be the case.

We live in this extraordinary age in which a butterfly flapping its wings on one side of the planet truly can create a tornado on the other—but only if the butterfly then sets up a website. There is an entire alternate solar system of knowledge and its opposite circling around up there. This is often quite miraculously wonderful—I can fly across the wastes of the Taklamakan Desert in western China, ancient home of the mysterious blond and redheaded Tarim mummies, like Luke Skywalker in his landspeeder, then order up the definitive account of the mummies' discovery by simply dipping my pinkie. What would once have been completely beyond human imagination is now as quotidian as a grocery list, and it seems we are all putting them together. In this universe of information and of *un*formation, some are born redheaded, some achieve it, and some poor souls simply have it thrust upon them. There are endless lists of supposed historical redheads out there, a tangle that links one site to another like binary mycelia; layers of space junk that typify redheads as impulsive, irrational, quick-tempered, passionate, and iconoclastic; great drifting rafts of internet factoids (currently the most notorious: the notion that redheads are facing extinction), with repetition alone creating a kind of false positive, a sort of virtual truth by citation. This, I discovered to my delight, is known as a "woozle," after the mythical and perpetually multiplying beasts hunted in the Hundred Acre Wood by Winnie-the-Pooh and Piglet, way back in 1926. It takes Christopher Robin to point out that the pair are simply following their own ever-increasing footprints in a circle around a tree. This book will endeavor not to add to the population of woozles.

Red is a color that has exceptional resonance for our species. There's an argument that it may have been the first color early primates learned to distinguish, in order to be able to select ripe fruits from unripe, and it still seems to speak to something primal in the human brain today: those suffering temporary color blindness as a result of brain damage are able to perceive red before any other color. And it is full of contradictions. It is the color of love, but also that of war; we see red when furiously angry, yet send our love a red, red rose; it is the color of blood, and can thus symbolize both life *and* death, and in the form of red ochre or other natural pigments scattered over the dead, it has played a part in the funerary rites of civilizations from the Minoans to the Mayans. Our worst sins are scarlet, according to the prophet Isaiah, and it is the color of Satan in much Western art, but it is the color of luck and prosperity in the East. It is universally recognized as the color of warning, in red for danger; it is the color of sex in red-light districts across the planet. The symbolism and associations of red hair embody all these opposites and more.

Red hair has always been seen as "other," but fascinatingly and most unusually, it is a white-skinned other. In the aristocracy of skin, as the historian Noel Ignatiev has described it, and in the Western world of the twenty-first century, discrimination is rarely overtly practiced *against* those with white skin. Yet people still express biases against red hair in language and in attitudes of unthinking mistrust that they would no longer dream of espousing or of exposing if the subject were skin color, or religion, or sexual orientation. And these expressions of prejudice slip under the radar precisely because by and large there is almost no difference in appearance (aside from the hair) between those discriminating against it and those being discriminated against. It is as if in these circumstances, prejudice doesn't count. Attitudes toward red hair are also extraordinarily gendered, something we'll encounter very frequently in the following pages. In brief, red hair in men equals bad, in women equals good, or at least sexually interesting. But even within this simplistic categorization there is a glaring contradiction, since culturally it seems we can get our heads around red-haired men as both psychopathically violent (Viking berserkers; or in the UK, the drunk swaying down the street with a can of super-strength lager in one fist and on his head a comically oversize tartan beret, complete with fuzz of fake ginger hair; or even Animal from *The Muppets*), and as denatured, unmasculine, and wimpish (Napoleon Dynamite, for example, or Rod and Todd Flanders from *The Simpsons*). Stereotypes of redheaded children reflect these opposites, with red hair being used to characterize both the bullied and as the bully—Scut Farkus in the movie *A Christmas Story* being a memorable example of the latter—unless the children in question are girls, in which case they are generally perky (Anne Shirley in *Anne of Green Gables*) and plucky (Princess Merida in Disney's *Brave*) and winningly cute (Little Orphan Annie). Redheaded women are supposedly the least desired by their peers of the opposite sex among American college students, yet (and I have to say, this has been my own experience) the popular construct of the female redhead is often profoundly eroticized and escapes the rules and morality applied to the rest of female society.[4] For this far from godly point of view, we have in large part to thank the medieval church.

---

4 Saul Feinman and George W. Gill, "Sex Differences in Physical Attractiveness Preferences" *The Journal of Social Psychology* 105, no. 1 (June 1978): 43–52.

This happens with red hair, time and again. Its presence, and attitudes toward it—the cultural stereotyping, cultural usage, cultural development—link one historical period, one civilization, to another, sometimes in the most surprising ways and very often flying in the face of all logic and common sense as well. One of the most intriguing aspects of the history of red hair is the way these links run on through time. You begin by investigating the impact of redheaded Thracian slaves in Athens more than 2,000 years ago and end at Ronald McDonald. You examine the workings of recessive characteristics and genetic drift in isolated populations and come to a stop at the Wildlings in *Game of Thrones.* You explore depictions of Mary Magdalene and find yourself at Christina Hendricks.

Why is the Magdalene so often depicted as a redhead? What possible reason can there be for that? If there ever was a specific individual of this name (which is a big assumption—the Magdalene as the Western church created her is a conflation of a number of different Biblical characters), her name suggests that she could have been a native of Magdala, on the western shore of the Sea of Galilee, well below the forty-fifth parallel. Beneath this latitude, redheads, while not unknown, are vanishingly rare, so the Magdalene's coloring in Western art and literature is unlikely to recall some Biblical truth. What, then, for so many artists, from the medieval period onward, is the explanation for showing Mary Magdalene with red hair? What message did that convey to an audience five hundred years ago? What might it tell us about that audience? And what might it illuminate about our own attitudes toward red hair today?

To begin with, from my own experience in my student days of working as an artist's model, I think artists simply enjoy depicting red hair. They like the turning shades and tints, they relish the glint and gleam of light upon it and the way that light bounces off the pale skin that so often goes with it. But the meaning of the red hair of the Magdalene takes one somewhere else altogether. It reflects the fact that the version of Mary Magdalene that the Western church has always found most fascinating is that of a reformed prostitute, a penitent whore, and culturally, for centuries, red hair in women has been linked with carnality and with prostitution. It still is today. What color is the hair of Helena Bonham Carter's character, Red Harrington, in Disney's 2013 film *The Lone Ranger* (and yes, there is a clue in her character's name)? As flaming a red as Piero di Cosimo's poised, calm, intellectual *Magdalene* of *c.* 1500, sat at her window, reading her book. Two entirely different women, centuries apart, yet linked by their societies' identical responses to their hair color. And this leads back to one of the stereotypes that began this discussion, and back to one of the greatest contradictions in the cultural history of the redhead: the centuries-long linkage between red-haired women and sexual desirability, and the fact that (despite, at the time of this writing, the naturally red-haired *Sherlock* actor, Benedict Cumberbatch, being voted sexiest actor on the planet) the exact opposite seems to hold true for redheaded men.[5]

This book is a synoptic overview of red hair and redheadedness: scientifically, historically, culturally, and artistically. It will use examples from art, from literature, and, as we come up to the present day, from film and advertising, too. It will discuss red hair not

5 *Metro* newspaper, October 3, 2013.

just as a physiological but as a cultural phenomenon, both as it has been in the past and as it is now. Redheaded women and sex and the gendering of red hair is a subject to which it will return in detail, but this book will also journey through the science of redheadedness, its history and the emerging genetic inheritance that is starting to be understood by modern medicine today. It will examine the many conflicting attitudes toward the redhead, male and female, good and bad, West and East. It is a study of other, and as always, what we say about "other" is far less interesting than what that says about us. But if you are going to ask what and who and how and why, the place to start is where and when.

WAY, WAY BACK, MANY CENTURIES AGO

In solving a problem of this sort,
the grand thing is to be able to reason backwards.

ARTHUR CONAN DOYLE,
A STUDY IN SCARLET, 1887

The ur-redhead, the first carrier among early modern humans of the gene for red hair and thus the genetic grandparent of the vast majority of redheads now alive, appeared on this planet some time around 50,000 years ago.

The world, at this point, was a very different place from how it appears today. Those parts now dry and arid, such as the Sahara, were green and pleasant; areas we think of as temperate, including most of western Europe, were either tundra or under an ice sheet. Stomping or slinking across that ice sheet went the fantastic mega-fauna of the Later Stone Age, or Upper Paleolithic period—woolly mammoth, giant elks, two-hundred-pound hyenas, saber-toothed cats. Trailing after them went Europe's resident population of Neanderthals, who had lived as hunter-gatherers in this landscape for 200,000 years; and creeping cautiously along as a distant and possibly rather puny-looking third came the first early modern humans.

These early humans had left Africa some 10,000 years before. They had already created populations in the Middle East and Central Asia; they were to explore around the coastlines of the Indian subcontinent; reach as far across the Pacific as Australia and as high as Arctic Russia; find toeholds in the Far East; and at some point cross the land bridge into what is now North America. Their expansion was driven (along with, one might suspect, hunger or greed) by an event referred to by paleontologists as the Upper Paleolithic Revolution. What this term encompasses is a step-change in toolmaking, from basic stone implements to highly specialized artifacts of bone or flint that range from needles to spearheads; evidence for the first purposeful engagement in fishing; figurative art, such as cave-painting, along with self-adornment and bead-making; long-distance trade or bartering between different communities; game-playing; music; cooking and seasoning food; burial rituals; and in all probability at this date, the emergence of language.

There is no one reason why this evolutionary jump happened when it did, nor even a consensus as to when or where it began. It may have been driven by changes in climate, as the ice sheets receded or grew, causing these early modern humans to create new technologies and survival strategies or perish. It may have been a very gradual process, but simply without a large-enough surviving debris field of earlier artifacts and evidence for us to be able now to judge how gradual; it might have been triggered by some sudden and singular genetic anomaly; or any possibility between these two. We simply don't know; too much evidence has been lost to us. Melting glaciation and rising sea levels have drowned the evidence of the earliest coastal settlements. All trace of those who may have battled out a nomadic existence on the ice sheets has disappeared. What we do know is that some 40,000 to 35,000 years ago, those people who had settled the grasslands of Central Asia began to explore outward, west and north, working their way from Iran to the Black Sea, up the valley of the Danube, and into Russia and the rest of Europe. With them, along with their new technologies and beliefs and emerging ethnicities, they carried the gene for red hair.

This may come as a surprise. The gene for red hair, for pale skin, for freckles (on both of which more later), did not originate in Scotland, nor in Ireland, despite the fact that in both those places you will now find the highest proportion of redheads anywhere on Earth (Scotland leads the way with 13 percent of the population being redheads, and maybe 40 percent carrying the gene for red hair, while 10 percent of the people of Ireland have red hair and up to 46 percent carry the gene). Logically, the place with the greatest incidence of any particular characteristic would, one might think, be the place where it first came into being, but not in the case of red hair. The gene emerged at some point in time between the migration from Africa and the settling of those grasslands of Central Asia.

We're able to make this assertion because of a superbly elegant hypothesis known to scientists as the molecular clock. This piece of evolutionary calculus uses the fossil record and rates of minute molecular change to estimate the length of time since two species diverged, charting this over thousands upon thousands of years, if need be. It enables us to calculate not only how long populations have been separate but how separate they are, with each change in an amino acid, or a DNA sequence, being a tick of the clock. Calibrated to the fossil record, the molecular clock makes it possible to estimate the point in geological history when new genetic traits first came into being. Such as, for example, red hair, or the gene for red hair at least, since at this point that is all we are talking about, the gene, rather than its expression, the appearance of red hair itself.[6] And the reason for this lies in the fact that the number of individuals making these migrations seems to have been quite astonishingly low.

It's been estimated that at the time of that final successful migration out of Africa 60,000 years ago, there may have been no more than 5,000 individuals in the entire African continent. The number estimated to have crossed from Africa over the then-shallow mouth of the Red Sea and into the Middle East, whose footprints now lie fathoms deep and whose descendants would become *Homo sapiens*, may have been no more than 1,000 and could have been as few as 150. There had been other migrations before this one—the fossilized remains of Java Man and Peking Man, or the 1.8-million-year-old remains of *Homo erectus* recently discovered in Georgia represent earlier offshoots. The Neanderthal population of Europe is thought to have descended from another even earlier common ancestor, shared with early modern humans, maybe 300,000 years before. There may have been many other earlier migrations out of Africa that failed—just as there were many Roanokes before there was a single Pilgrim Father—or if they did not, their genetic traces are still too deeply hidden within us for science to be able as yet

---

6 Accuracy over eons is of course a rather different matter from setting your alarm clock in the morning. Some paleontologists would push the appearance of the gene for red hair back to between 100,000 and 50,000 years ago.

to make them out. But even without ice storms and blizzards of a ferocity that would make headlines in Siberia today, even without enormous carnivores, for this nascent population of early modern humans, the world was not only a staggeringly dangerous place to be, but one with precious few others of the species to share it with.[7]

And how do we know this? Because, despite the many variations in skin, hair, and eye color that loom so large for us, despite the many rich and challenging social and cultural differences across our planet, genetically we are so very *un*diverse.

Consider, for example, the heather, or *Ericaceae*. It colors the hillsides about my brother's house on the Isle of Skye, fills cranberry bogs in Canada, and romps across the foothills of the Himalayas. Heather, cranberry, and rhododendron are all *Ericaceae*. There are 4,000 different species in total. You need many, many ancestors to create that much genetic divergence. Consider the dog, *Canis lupis familiaris*. There are nine separate breeds of dog, from wolves and jackals to man's best friend, snoozing at your feet. The domesticated breed itself encompasses a spectrum of variation from dachshunds to bulldogs to Great Danes. Think how much genetic diversity that requires. Compared to the differences between a Chihuahua and a St. Bernard, what do we offer? Our limbs are always pretty much in the same proportion to the rest of our bodies, as are our facial features. Our skulls do not radically change shape; we do not come some with

7 As an example of both the danger and the isolation faced by our ancestors, the Toba Catastrophe, a supervolcanic eruption that occurred in Indonesia between 77,000 and 69,000 years ago, created a ten-year winter that could have reduced the entire population of the planet to something between 3,000 and 10,000 individuals and may have been one of the catalysts for that ancient migration from Africa in the first place.

snouts and some not; our ears do not stand up or flop, or trail along the ground. Yet we have been breeding and interbreeding for millennia. The reason why there is so little differentiation between any one of us and every other one of us, compared to so many other species of fauna and flora, is because the number of genetically unique individuals the process began with was so mind-bogglingly small.[8] And the one certainty about red hair is that for a red-haired baby to result, both mother and father have to be carrying the gene, and what is more, both have then to donate that specific recessive gene in sperm and egg. So if the gene for red hair was present in the early human population in the same percentage in which it is found today, with such tiny numbers of people spread out over such a huge area, it might have existed unseen, unsuspected, and unexpressed for generation after generation without any such propitious meeting taking place.

It must also be remembered that in discussing our ancestors of all those many millennia ago, almost nothing can be stated as irrefutable fact. There are a half dozen equally valid theories for every tiny piece of ancient evidence. Do we know, for example, that *Homo sapiens* was responsible for the extinction of all those giant animals and the disappearance of the Neanderthals? No, we don't. We know that the arrival of one *coincided* with the disappearance of the other; we know, for example, that the last Neanderthals in Europe, who were physically far stronger and had bigger brains than the incomers who displaced them, had completely disappeared by 24,000 years ago, the last of them dying

8 To make matters worse, we are also the only version of our species, *Hominidae*, to exist. No other version of us has survived.

in remote caves facing the sea on the coast of Gibraltar. Looking at our more recent actions on this planet we may conclude that we're a depressingly good candidate for the prime suspect; but the extinction of the cave hyenas and saber-toothed cats (*circa* 11,000 years ago), giant elk (7,000 years ago), the mammoth (the last herd lived on Wrangel Island, off Siberia, as recently as 1,650 years ago), and of the Neanderthals themselves could equally well have been caused by climate change or by disease as by the effects of our aggression or over-predation. Or it might have been caused by a combination of all these things. We simply don't know.

All the same. We are an ambitious, covetous, exploitative, and destructive species, and there has rarely been much good in our engagement with anything we perceive as "other." We interpret difference as threat rather than potential. And sadly, our race is not alone in that.

The El Sidrón caves are in northern Spain, inland from the coastline of the Bay of Biscay. The nearest large town is Oviedo. During the Spanish Civil War, the caves were used as hideouts by Republican fighters. They have always attracted the curious and intrepid, and when in 1994 what appeared to be two human jawbones were discovered in the gravel and mud on the floor of one of the caves, it was assumed that as the remains were in such good condition they were the tragic relics of some misadventure from the conflict of 1936–39—the victims of some forgotten Civil War atrocity.

The bones were indeed evidence of an atrocity, but of one much, much longer ago. The remains of twelve individuals—three men, three women, three teenage boys, and three children, including an infant—were those of an extended family group of Neanderthals who had presumably been ambushed outside the cave, killed, dismembered, and then cannibalized, the flesh removed from their bones with sharp flint knives, the long bones split to get at the marrow. It's presumed they must have strayed into the hunting territory of another, rival group of Neanderthals and paid this dreadful price. Or perhaps conditions were so harsh it was a simple case of them or us. After their deaths, some event, perhaps a storm and flash flood, caused the roof of the cave in which their bones were found to collapse, washing their remains into the caverns, and thus they were immured together for another 50,000 years.

Any discovery of so much material in one place is of course of immense importance to archaeologists, but what makes the El Sidrón family so significant a find is just that: these individuals all seem to have been related to one another. They share no more than three groups of mitochondrial DNA, the type that is passed unchanged from mothers to children. In fact all three of the adult men have the same type. And so well preserved were the remains that forensic science could not only extract readable amounts of fragmentary DNA from them but could even fit individual teeth back into jaws. There are similar idiosyncrasies in dentition between the teeth and jawbones of different members of the group, and two of the men shared the same gene variant, which, it is thought, would have given them pale skin, freckles, and red hair.[9]

9 See http://humanorigins.si.edu/evidence/genetics/ancient-dna-and-neanderthals/neanderthal-genes-red-hair-and-more.

We all possess some DNA (maybe 1–4 percent) in common with Neanderthals, and presumably with that original common ancestor, all those many thousands of years ago. If you reach back in time, the percentage of Neanderthal DNA in modern humans seems to increase. Ötzi the Iceman, whose mummified body, frozen into a glacier, was found in 1991 in the mountains of the Austrian-Italian border and who died in about 3,300 BC, had more Neanderthal DNA than today's modern humans. Not a huge amount (it's been estimated at 5.5 percent), but a statistically significant one.[10]

It would be very neat, therefore, to assume that the gene for red hair as it is found in most redheads today is a Neanderthal characteristic, and that when the two species, Neanderthals and early modern humans, met and interbred (scientific thought goes back and forth on this point, but let's assume usual things happen most usually), the gene was transferred over. It would be very neat; and given the redhead's reputation for violent bad temper, it's a theory that has given amusement and much satisfaction to many non-redheads since the first days of anthropological science back in the nineteenth century. It is still encountered time and again in discussions of red hair today, but it is also completely wrong. The genetic mutation that produced the red-headed men from El Sidrón is different from that found in redheads today. It is instead an example of a phenomenon where what appears to be the same result stems from very different causes—something else we will be encountering again. In any case, the far more fundamental point of interest with the

10 See http://johnhawks.net/weblog/reviews/neandertals/neandertal_dna/neandertal-ancestry-iced-2012.html.

red-headed Neanderthals of El Sidrón is not the hair—it's the freckles. It's the skin.

❁

The gene that today results in the red hair of almost every redhead on the planet sits on chromosome 16, and if you have red hair, it's because the version you have of that gene is not working as well as it might. Working perfectly, that MC1R gene, or melanocortin 1 receptor, to give it its full name, would give you brown eyes, dark skin, and an ability to withstand strong sunlight without developing sunburn, sunstroke, or worse. It would do this by stimulating the production of a substance called eumelanin, which colors dark skin, dark eyes, and dark hair. However, MC1R is fritzy. Like a bad internet provider, it flips in and out. If you have red hair, it is almost certainly (there are rare medical conditions that produce red hair, and in the Solomon Islands in the Pacific, an entirely different genetic mutation gives some islanders the most striking gingery-blond afros) *almost* certainly, therefore, down to the fact that you carry two copies of a specific recessive variant of the MC1R gene that dials eumelanin production (with all its protective benefits under strong sun) right down, replacing it with yellow or red phaeomelanin, in an extraordinarily complex set of variants that determines the color of individual hairs on your head and of individual cells in your skin.

MC1R, however, is not alone in this process. There is another gene, HCL2, on chromosome 4 (rather unpoetically, HCL2's full name is simply "hair color 2 [red]"), that also contributes to red hair. Moreover (I did warn you this was complex), while red hair is indeed caused by a recessive gene, there are many possible variants, from recessive red to fully dominant brown or black, with an equal number of manifestations of so-called "codominance" in between.

Thus, many a blonde or brunette has freckles. Many brunettes also have a red tint to their hair, or rather, hair cells that are individually and one by one a mix of eumelanin- and phaeomelanin-producers. A man can have brown or blond hair on his head, yet a beard that grows in red (and very annoying some of them seem to find it, too). This pairing of, say, brown hair, freckles, and red beard is known as "mosaicism," which describes it pretty much perfectly. You can have red hair in any shade from palest strawberry blond to deepest chestnut. It can come hand in hand, as it were, with blue/green eyes and pale skin and freckles, as it has with me, or with amber eyes, or hazel, or dark brown. A recent survey undertaken by Redhead Days, who run the largest annual festival of redheads in the world, uncovered the fact that up to a third of those who responded classified their eye color as hazel or brown. And (and this is where things get really interesting) it can come with skin dark enough to protect you from sun that would send many another redhead running for sun hat, sunglasses, and sunblock. In general, redheads and strong sun most emphatically do not mix. If you look back at the Redhead Map of Europe, at the beginning of this book, you will see how suddenly the incidence of red hair drops off below that forty-fifth line of latitude. But Afghanistan, Morocco, Algeria, Iran, northern India and Pakistan, and the province of Xinjiang in China all have ancient, native populations of redheads. Shah Ismail I (1487–1524), commander in chief of the Qizilbash and founder of the Safavid dynasty in Iran, was described by the sixteenth-century Italian chronicler Giosafat Barbaro as having reddish hair, and indeed his portrait in the Uffizi, by an unknown Venetian artist, shows a man with an aquiline nose, red beard, and red mustache. The sixteenth-century *Shahnama* of his Safavid successor, Shah Tamasp, now in the Metropolitan Museum of Art, shows a red-bearded hero, Rustam. In the ancient world Alexander the Great's Roxana, who was born in Bactria, now northern Afghanistan, and who the Napoleon of the ancient world married in 327 BC, was reputedly a redhead; the present-day Princess Lalla Salma of Morocco perhaps gives an idea of quite how beautiful she might have been. Then there is that different shuffle of the genetic pack in the Solomon Islands in the Pacific—which, again, comes with skin dark enough to withstand in this case tropical sun. (Fig. 29). In the history of red hair, what you might call its typical Celtic manifestation—pale skinned, blue eyed, suited for cool and rainy climes and cloudy skies—may not originally have been typical at all. In fact pale skin in early modern humans has been estimated as having appeared as recently as 20,000 years ago.[11]

Most genetic mutations die out rather than become more common, and if they are as unadvantageous as pale skin under fiercely sunny African skies, they take their unfortunate carriers with them.

11 Other studies give the date as being between 12,000 and 6,000 years ago. One estimate is as recent as 100 generations, or a mere 2,500 years ago, although this is rather unconvincing. Had pale skin suddenly begun appearing among the native populations of the countries they had conquered, surely one of those indefatigable chroniclers of the Roman world (many of whom we will meet in the next chapter) would have made specific mention of it. Thanks to my colleague Kate Owen for this point.

Blue eye color, meanwhile, or rather the switching-off of the gene for brown eye color (although the genetics of eye color are so astonishingly complex as to admit almost any possibility) is thought to have come into being in what is now modern Romania, perhaps 18,000, perhaps 10,000 to 6,000, years ago.

It may be going too far to characterize Mother Nature as smart, but left to herself she does have a heartlessly effective way of cleaning out the gene pool. If you're not fit to survive, you won't. But if a genetic quirk confers a benefit upon those carrying that gene, it, and they, will flourish. And pale skin under Northern skies does exactly that. If your eumelanin production is dialed back, if you have pale skin rather than dark, your body will be much more effective at synthesizing vitamin D, using whatever sunlight is available, than if your skin were darker. And as the ice sheets retreated, and that growing population of early modern humans moved from Russia into Scandinavia, and eventually into the whole of northern Europe, the absence of strong sun in these climes allowed the MC1R gene to mutate into what geneticists term "dysfunctional variants" without these variants proving fatal to those carrying them. In fact the farther north this population went, the more advantageous pale skin became. If you have enough vitamin D, your skeleton develops as it should. If you do not, your bones are soft and stunted, and as you learn to walk, your legs bow under your weight. This is osteomalacia, or rickets. In adults, it causes calcium to leach away from your bones; in children, it cripples you. Women of child-bearing age who suffered from rickets when they themselves were children have distorted pelvises that make carrying a pregnancy to full term difficult and childbirth hazardous, if not fatal (Fig. 3).

Communities that eat most of their protein in the form of meat, as with the early hunter-gatherers, rarely suffer from vitamin D deficiency. But the tendency in the early modern human population was to settle; to become farmers, to grow and eat grain. In these circumstances, pale skin helped keep you strong and healthy. In

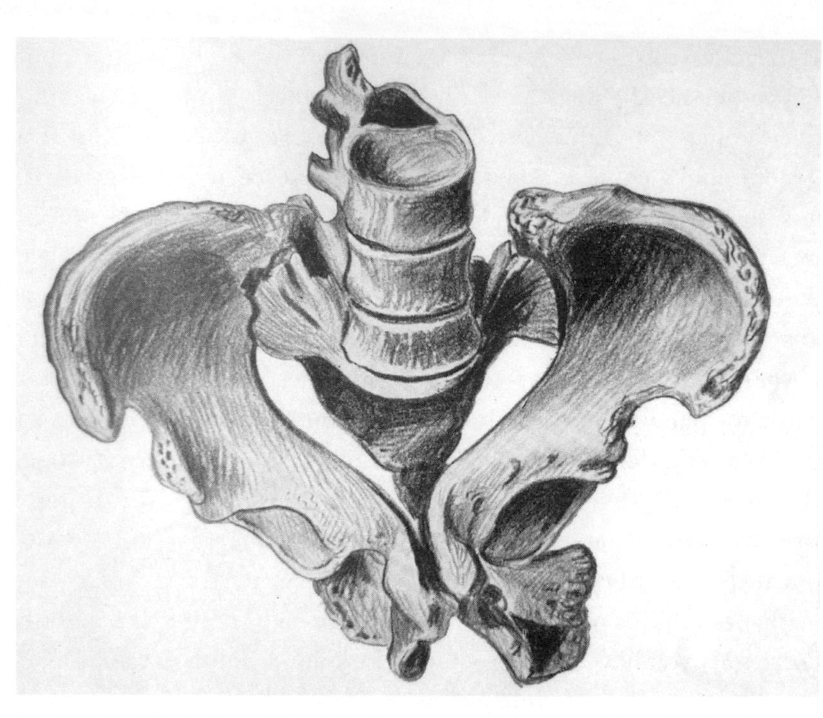

Fig. 3 Medical illustration of the pelvis of a woman suffering from rickets, showing the narrowed and distorted birth canal.

particular, it gave a significant advantage to women during pregnancy and breast-feeding, when their bodies' demand for vitamin D was at an all-time high, which along with all the other ancient and instinctive associations of the color red (fire, blood, passion, ripeness) does rather open the question as to whether the often highly sexualized image of female redheads might not start here, with the simple fact that choosing a redhead as a mate meant you bred successfully, and that your pale-skinned children, themselves now carrying the gene for red hair, did the same.[12]

This is also where the random mysteries of "genetic drift" come into play. Genetic drift is the term used to describe random changes in the frequency of a genetic variant, or "allele," as it is also known within a given population. There is some essential scientific terminology to become familiar with here, and perhaps the simplest way to conceptualize it is if you cast your mind back to your first school photograph. You, seated there cross-legged on the ground, squinting at the sun, are an individual allele, an individual genetic variant. The front row of you and all your classmates make what is known as a haplotype—a cluster of linked alleles, all of whom are likely to be inherited together. Your haplogroup would be everyone in the school.

If a population is large, it can take a considerable time before random changes in the frequency of an allele become noticeable, and indeed as they are governed only by chance, and their frequency can alter one way or the other from one generation to the next, some may never result in any noticeable change at all. But in a small population, the fixation of a particular allele within that group, such as (for example) red hair, can happen very rapidly indeed. In small populations, in borderlands, in any community set apart from the great genetic ebb and flow of the human ocean, in places such as Ireland or on the west coast of Scotland, its effect can be established within just a few generations. Add to this the not-random-at-all mysteries

12 In light of the association between red hair and successful childbirth, it's intriguing that C. G. Leland records in his *Gypsy Sorcery and Fortune Telling* (1891) the belief that for an easy birth, red hair should be sewn into a bag and worn next to the skin of the belly during pregnancy.

of sexual selection, and the fixing of red hair among these liminal populations, and among liminal populations in the Levant, the Caucasus, and the Atlas Mountains, becomes a phenomenon that exists logically and obviously. As of course it also does among the famous Wildlings in *Game of Thrones.* Redheads: guaranteed throughout history to crop up in the last place you expect them.

BLACK AND WHITE AND RED ALL OVER

Men make gods in their own image; those of the Ethiopians are black and snub-nosed, those of the Thracians have blue eyes and red hair.

XENOPHANES

The house of Marcus Fabius Rufus, on the so-called Vico del Farmacista, is the largest domestic dwelling so far excavated in Pompeii. It has four separate levels and appears to have been continuously occupied from the time of the Roman conquest of the then–port of Pompeii, in 80 BC, up to the destruction of the city in AD 79. M. Fabius Rufus (whose body may very well be one of the four found within the house, and whose name was preserved for us in a piece of scurrilous below-stairs graffiti) was but its final owner; over its 160-year history the house had many others, all of whom contributed in some way to the layout and decoration of its rooms—opening up doorways here, closing them there, enlarging this room, repainting that. Then came the eruption of Vesuvius, with its earthquakes and choking blizzards of ash; and when archaeologists finally opened up the house of M. Fabius Rufus, they found the earthquakes of the eruption had partially demolished a wall in Room 71 and revealed behind it another, the style of whose fresco decoration dates it to the century before.

The fresco shows a woman standing between two partially opened doors. She has large eyes, a delicately rounded mouth, and a fashionably piled-up hairstyle; she holds a child against her shoulder (he's maybe a year old, an elongated Roman *putto*, held against his mother with his back to us and his naked rump bared endearingly to posterity); she wears what has been interpreted as a royal diadem, and her hair is a warm, or one might say a reddish, brown. Recent research suggests that the woman is intended to be Cleopatra, in the guise of Venus Genetrix, the mother-goddess of the Roman world, and that the child should be identified as her son Caesarion, born in 47 BC, whose father, so his mother claimed, was Julius Caesar.[13]

The identification of the woman is based upon her likeness to two marble busts of Cleopatra, one now in the Vatican (which may also once have been completed by the figure of a child), the other in Berlin; the presence of the diadem; and the fact that the two half-open doors and the rest of the frescoed scene around her seem to allude to the appearance of the temple to Venus Genetrix, set up by Caesar in the Forum Julium in Rome in 46 BC. The temple was graced with a gilded statue of the goddess, widely and scandalously reputed to have been modeled on Cleopatra herself, who was in Rome from 46 to 44 BC. The thinking is that on Caesar's assassination, in 44 BC, and the accession of his official heir, his great nephew Octavian, as the Emperor Augustus, the then-owner of the house hid the

13 Susan Walker, "Cleopatra in Pompeii?," *Papers of the British School at Rome* 76, (2008): 35-46.

fresco behind a wall to cover up his own Julian sympathies.[14] Thus, intriguingly, the history of red hair and the world of Roman realpolitik cross paths. But is any of this proof, as has been claimed, that Cleopatra was a redhead?

Or to put it another way, *what,* and *who,* is red? One person's unmistakable red is another's vaguely chestnut, and even that authoritative-sounding quote from Xenophanes is not as undisputable as it appears: Xenophanes wrote in ancient Greek, and some authorities translate his "red" as "fair." Moreover, his writings have come down to us only via their use in the writings of others. The quote was preserved, five centuries after Xenophanes (*c.* 570–*c.* 475 BC) worded the original thought, in the work of Clement of Alexandria, a founding father of the Christian church. That so-called portrait of Cleopatra in the house of M. Fabius Rufus is the result of a similar process of transmutation. Judging by her official likeness on the coinage of her rule, Cleopatra herself had neither large eyes nor a rounded mouth. The coins show instead a woman with a long nose, almost hooked, and a sharp, knowing smile. Also, Cleopatra was a native of Egypt, obviously enough, a country that lies well below the forty-fifth parallel, which makes the possibility of her being a natural redhead unlikely. Then one has to remember the Egyptians used wigs, some of which, recovered by archaeologists, have proved to be made from the reddish fibers of the date palm. They also dyed their hair. In fact they seem to have applied as many colors, gels, and waxes to their hair as we use today. The mummy now in the Egyptian Museum in Cairo of the great

14 This would have been only smart. Caesarion himself was killed, probably by strangulation, on Augustus's orders in 30 BC at the age of seventeen—eleven days after his mother's famous suicide.

pharaoh Ramesses II, who ruled Egypt some 1,200 years before Cleopatra, has dyed red hair. Ramesses died in 1213 BC, at the age of ninety, and his own hair by this point was white, unsurprisingly, but either before death, or as part of the embalming process it was dyed with henna.[15] One might speculate that this emulated its color in life, but this would be speculation only. In fact red hair seems to have had something of a conflicted history in Egyptian culture. Red was the symbolic color of Set, the god of violence and disorder and lord of the hostile desert (there is supposedly an Egyptian prayer to Isis, begging for deliverance from "all things evil and red"[16]). Before Ramesses's family, who were followers of Set, came to power, supposedly every year a redheaded male was burned alive as a sacrifice, or so the Roman chronicler Diodorus Siculus, writing more than a millennium later between 60 and 30 BC, informs us. These commentators and chroniclers of the ancient world are the only speaking witnesses we have, but with all of them, their words come to us as light from a dead star, distorted as echoes, sometimes as little more than static, down a long, long line from far away. We cannot know all of the details or sometimes indeed any of the circumstances in which they were writing; we can only guess, and to use these sources to reconstruct that ancient world is to be a detective—almost an archaeologist—yourself. None of this is as black and white as it may first appear. •

---

15 For the arguments that red was Ramesses's hair color in life, see http://www.lorealdiscovery.com, and Bob Brier, *Egyptian Mummies: Unraveling the Secrets of an Ancient Art* (New York: William Morrow & Co., 1994), 153. The question is still open, however: recent research by Silvana Tridico suggests that hair decays after death and its color can alter as a result of fungal or bacterial growth upon it. See http://rspb.royalsocietypublishing.org/content/281/1796/20141755.

16 Quoted in Eva Heller, *Psychologie de la couleur: Effets et symboliques* (Paris: Editions Pyramyd, 2009) 45.

But back to Cleopatra. What the fresco in Pompeii records, therefore, is not the appearance of an actual woman but the appearance of a statue. It is a sizable and significant step away from the living, breathing, original. Rome was 150 miles from Pompeii, not too far a journey, by any means, for a painter on a commission. Statues in the ancient world were rarely left in that blanched state of seashell whiteness in which we see them today. Once carved, they were often painted, and the grandest were gilded, too. So the coloring of the figure in the fresco may record the original coloring of the now-lost statue, or it may simply bear testimony to the paints available to the painter. The palette of the ancient world was limited to earth colors, mineral pigments, or vegetable dyes, but no matter what the painter intended, the iconic force of his subject would still overpower its actual appearance. The eye of the beholder is all. With so many of those names on website lists of historic redheads, the one thing that they all have in common (aside from the fact that we are most unlikely ever to be able to reach an indisputable conclusion as to their hair color) is that they all behaved as if their hair *should* have been red. We look at this Pompeian fresco, and we see not a woman with reddish-brown hair, we see Cleopatra, Caesar's mistress, lover of Mark Antony, the queen who hazarded a kingdom and chose death over conquest, an archetype to which this book will return over and over again: the flame-haired seductress, exotic, sensual, impulsive, passionate. We see her hair as red because we want to do so. What other color would it be?

So who were the redheads of the ancient world? How real, or not, were they?

The kingdom of Thrace, if such a set of tribal territories can be so referred to, existed from roughly 1000 BC to the final dissipation of the Roman Empire some 1,700 years later. It sat across the western side of the Black Sea and stretched down to the Aegean, over an area that now includes most of Bulgaria and parts of Turkey and Greece. The Thracians were horsemen and warriors, and early contact with them, as both Greece and later Rome would discover, tended to be very bloody indeed. Even their war dances were violent enough to leave the odd participant lifeless. According to the Greek historian Herodotus, the Thracians believed that "to live by war and plunder is of all things the most glorious."[17] When as a Thracian you weren't occupying yourself with that, you might perhaps be indulging in a popular drinking game, consisting of standing on a rock, with your head in a noose. One of your friends kicked the rock away and then the trick was to be quick enough with your Thracian short-sword to slice through the rope before you throttled—a sort of Thracian Russian roulette.[18]

The Greeks were recruiting Thracian mercenaries into their armies as early as 600 BC, by which time they had a strip settlement of Greek trading posts along the Thracian coasts. Alexander the Great, three hundred years later, would do the same, both fighting the Thracians and signing them up. To this day, a common explana-

17 Herodotus, *The Histories*. The full text is available on Project Gutenberg. See https://www.gutenberg.org.

18 Lionel Casson, "The Thracians," *The Metropolitan Museum of Art Bulletin* XXXV, no. 1 (Summer 1977) 3–6.

tion for the unexpected appearance of green eyes or red hair in a child in Afghanistan or Kashmir is the onetime presence of Alexander's troops in those regions more than two thousand years ago, and who is to say such hand-me-down folklore doesn't still preserve some kernel of genetic truth? The Thracians were as prized as troops as they were feared as enemies. In 73 BC Rome found itself facing an internal revolt led by the gladiator Spartacus, a Thracian from the border tribe of the Maedi with a Roman military background (the Roman chronicler Plutarch gives us a wealth of this familial detail on him but, maddeningly, fails to record the color of his hair). In addition to their military skills, the Thracians were superb metalworkers in bronze and gold, and they also had a highly evolved belief system covering the underworld and afterlife, as evinced by their elaborately decorated tombs. They lent much of this mythology to the Greeks in turn, although the gods of Thrace seem to have been even darker and less tractable than those of Olympus. The Thracians were also, notoriously, within the ancient world, "barbarian." Not only did they disdain speaking Greek, they also refused to give up the traditional structure of their society, rejecting the whole idea of creating and living in cities, and remained in their small tribal communities. "If they had one head or were agreed among themselves," Herodotus observes sadly, "it is my belief that . . . they would far surpass all other nations. But such union is impossible for them."

To the north of Thrace, at the top of the Black Sea, were the lands of the Scythians, equestrian tribespeople with an origin as far back east as Iran. There is also Biblical mention of the Scythians, in Colossians 3:11: "Here there is no Gentile or Jew, circumcised or uncircumcised, barbarian, Scythian, slave or free," which sounds as if the Scythians are being used as an even more extreme example of barbarism. They too were noted warriors, feared as archers in particular, and flourished from around the seventh century BC to the fourth century AD. Not losing sight of the fact that ancient writers used the term "Scythian" pretty broadly (as, perforce, do archaeologists today), Herodotus is our guide here as well. Writing of a city, Gelonus, in the northern part of Scythia, he describes its people, the Budini, as "a large and powerful nation: they have all deep blue eyes [or gray, depending upon translation] and bright red hair." It's thought that the Scythians, specifically the Budini, might have been the ancestors of the Udmurts of the republic of Udmurtia in Russia, on the Volga River.[19] Why is this thought? Because ever since the anthropologists of the nineteenth century encountered them, the Udmurts have been celebrated as among the most redheaded people on earth, with almost as high a percentage of redheads in their population as the Irish and the Scots. They still are. The Udmurts and the area around Udmurtia on the Volga are the hotspot on the map of European redheads, north of the Caspian Sea.

This is intriguing enough. Even more so is the possibility that the backstory of the Scythians, as it were, might reach even farther eastward, as far as Tibet, Mongolia, and the border of modern China, and that the ancestors of the Scythians might be linked with the civilization of the Tarim Basin, and the Tarim mummies.

The history of red hair is tied to the history of human migration, of one people encountering another, or even "an other," and each of

---

19 The site of Gelonus has been sought by archaeologists for some time; various ancient settlements have been put forward as candidates in the Ukraine or along the Volga River.

these encounters adds another layer to the cultural response to red hair, right down to the present day. Red hair is an unmistakable and very convenient marker of these encounters and is tied in particular to four great human diasporas. Those of the Celts, the Vikings, and the Jews are to come. But let's begin with the first of these four key diasporas: that of the tribes who made the journey across the Middle East to the shores of the Black Sea, and then settled the valley of the Danube (which may itself have acquired its name from a Scythian loanword). If you were to stretch the history of these people eastward, rather than west, and back in time, back beyond the Thracians and Scythians, back even beyond the reign of Ramesses the Great, you would reach the grasslands of central Asia where, it is believed, the history of red hair began. And if, thousands of years ago, rather than trekking west, you had turned east, you would eventually have reached what is now the Taklamakan Desert, in the Tarim Basin.

Almost everything about the civilization of the Tarim Basin and its discovery by Western archaeologists sounds as if it should have come straight out of *Indiana Jones*. There are the stories of the first European explorers who reached the area, to begin with: Nikolai Przhevalsky, who gave his name to Przewalski's Horse, and whom internet mythology would have as the father of Joseph Stalin; or Albert von Le Coq, a German beer and wine magnate who began studying archaeology at the age of forty, whose expeditions were financed by none other than the German emperor Wilhelm II, and who ended by shipping more than seven hundredweight of artifacts back to Berlin, convinced that the presence of red-haired, blue-eyed figures in the frescoes he chipped, carved, and sawed out of caves in northwest China meant he had discovered a new Aryan heartland;

or Sir Aurel Stein, who owed his knighthood as much to the role he played as a spy in the "Great Game," the battling-out between Britain and Russia for influence over Central Asia, as he did to his archaeological discoveries. These early archaeologists discovered the ruins of settlements, orchards, and oases that had once been shaded by poplar and tamarisk trees and watered by rivers that had run dry centuries before, all now buried under the sand dunes of the Taklamakan Desert (the name can be translated as "You go in, but you don't come out"). And, at various graveyard sites around the rim of the Tarim Basin, they also discovered the Tarim mummies themselves, hundreds of them at least, almost perfectly preserved by the cold, dry climate, a climate so perfectly suited to mummification that it has preserved even the bodies of ancient mice in the remains of ancient granaries. What these Tarim mummies reveal is the fact that, in what is now western China, in the province of Xinjiang, bordered on one side by the 'Stans, and on the other by Mongolia, from at least 2000 BC to roughly AD 200 there lived a people of almost modern height with fair skin and blond hair, and in a couple of cases at least, as reported in the authoritative history of these discoveries, with actual red hair. They had angular, Caucasian features and light-colored eyes set in very un-Asian recessed eye sockets. And, they wove, wore, and maybe traded textiles that link them to the tribes of Celtic Europe.[20] Basically, 4,000 years ago, there were people living in western China who looked as European as the tribes living

20 The magisterial study on the Tarim mummies is by J. P. Mallory and Victor Mair, *The Tarim Mummies* (London: Thames and Hudson, 2008). Their hair color is variously described as blond, fair, or red, but as far as I am aware, none of the mummies has as yet been specifically tested to see if they carry the MC1R gene.

at that time around the Seine or the Thames. The Roman writer Pliny the Elder (AD 23–78) who, like M. Fabius Rufus, also perished in the eruption of Vesuvius (Pliny the Elder was both corpulent and asthmatic, and couldn't make it to safety through the pumice fall), included a description in his *Natural History* of these people of western China, given to him in turn by a diplomat from Taprobane (modern-day Sri Lanka), who was visiting the Emperor Claudius:

> These people, they said, exceeded the ordinary human height, had flaxen hair, and blue eyes, and made an uncouth sort of noise by way of talking, having no language of their own for the purpose of communicating their thoughts.

The "uncouth sort of noise" may have been a very early Indo-European language, which scholars now refer to as Tocharian. The people speaking it are given the name "Seres" in Pliny's account, or "people of the land of silk." This is a very big clue as to how and why contact might have existed between the people of the Tarim Basin and those of the Black Sea and even farther into Europe: Tarim was on the Silk Road, one of the most important global trading routes that has ever existed.

We think of our planet as being divided into continents and countries, each of which is the place of origin for people of a specific appearance: African, Eurasian, Caucasian, and so on, but this is simply how the world looks to us at our moment in time. Thousands of years ago, these boundaries might not have been the same. Our ancestors were intrepid and tireless explorers; it may have been 2,500 miles from Cornwall to the shores of Phoenicia, but trade existed between the two. It may have taken seven months to travel overland from the Hindu Kush, between Pakistan and Afghanistan, to China, but the journey was made. (Clearly, so was that—more than 7,500 miles—between Sri Lanka and Rome.) And if goods were being traded, it would be very strange if tribal alliances and marriages weren't taking place as well.

The Scythians left tomb-mounds, known as *kurgans*, across the whole of Eurasia, from the Ukraine to the Altai mountain plateaus between Mongolia and Siberia. Many of these have proved to be rich in the most exquisite examples of metalwork, in gold and bronze. They have also yielded up human remains, which in turn have revealed that the Scythians belonged to a haplogroup, R-M17, that is much more closely related to people living now in eastern Europe than it is to those of central Asia, and which would have given them fair skin, blue or green eyes, and light-colored hair.[21] Hence the link suggested between the ancestors of the Scythians and the ancestors of the fair-skinned, Caucasian-featured mummies from Tarim. This also raises at least the possibility that those unexpected red-haired, green-eyed children of Kashmir and Afghanistan record not the later passing of Alexander's troops in the 320s BC but are perhaps reminders of an even more ancient lineage of some of the earliest redheads on the planet. What the tombs of the mobile, nomadic Scythians, who kept their art small and portable, have not so far rendered up is much in the way of paintings that shows us how these people depicted themselves. The tombs of the Thracians, however, have.

---

21 Some of these remains were of women, but buried with the accoutrements of warriors, leading to the suggestion that Greek stories of Amazons came from their neighbors the Scythians.

Some hundred miles east of Sofia, almost exactly in the center of modern Bulgaria, there is a valley containing no fewer than three hundred Thracian tumuli, or tomb-mounds. One of these is the so-called Ostrusha tomb, dating to 330–310 BC. Like most such tombs, its funerary bed is now empty (the reputation of the Thracians as metalworkers and the lure of Thracian gold meant many tombs were robbed from a very early date), although in this case its original occupant might not have been buried alone. When the tomb was opened, the skeleton of a horse was found inside it, with a knife rusting on its chest, suggesting that the poor beast was led in there and killed by being stabbed through the heart, in order to accompany its master to the Thracian heaven depicted above the bed, on the ceiling.[22]

The coffered ceiling of the Ostrusha tomb is extraordinary. Carved out of solid rock, it is divided into square fields, deeply inset, with both the framing borders and the central square being appropriately decorated. The coffers are painted with scenes of mourning (one, for example, shows the goddess Thetis mourning her son the Trojan hero Achilles—also a redhead, according to some accounts) and of the journey into the afterlife, and in coffer 32, there is the head and shoulders of a young woman [Fig. 4]. Her head is tilted to

22 Julia Valeva, *The Painted Coffers of the Ostrusha Tomb* (Sofia, Bulgaria: Bulgarski Houdozhnik, 2005). With grateful thanks to the author for providing me much information and a copy of her superbly detailed book.

the left, as if looking down on the funerary bed, and she is extremely fair of face, even with the damage 2,300 years have inflicted on the fresco. She has skin like a rose petal, an air of gentle, clear-eyed calm that still catches the heart, and she has red hair. It is possible to conjecture that in this setting she represents Demeter or her daughter, Persephone. Both goddesses had powers within the cycle of life, death, and rebirth and were intimately connected with the notion of the turning of the year, and it is natural to see the young woman's red hair as symbolic of fire, of the low winter sun, of sunsets and sunrise, of returning life, of winter and spring.

One hundred miles or so farther east of Ostrusha, toward the coast of the Black Sea and close by the village of Alexandrovo, is another tomb. This too dates to the fourth century BC. Its configuration differs from that in Ostrusha—there is no coffered ceiling; instead in cross-section it resembles an igloo, with a tumulus of earth mounded up over it. You enter through a low, narrow tunnel, as did the treasure hunters who discovered the site as recently as December 2000, and in the main chamber of the tomb, as you stand upright, you see above your head the frescoed decoration of a whole year's worth of hunting scenes. The quarry are boars and deer, depicted as the huntsman thrusts his spear down their gullet or the hounds leap upon their backs. The hunters are shown wearing short tunics, or warmer trousers and boots (one reason for seeing these scenes as taking place in different seasons).[23] One hunter is on

23 G. Kitov, "New Discoveries in the Thracian Tomb with Frescoes by Alexandrovo No 1," *Archaeologia Bulgarica*, 9 (2005): 15-28. One should note that even after his death in 2008 Georgi Kitov is a controversial figure among the archaeologists and scholars of Thrace. A reconstruction of the tomb and its frescoes can be seen at http://www.aleksandrovo.com/en.

horseback, and it has been conjectured that he is the "hero" figure, who might have been buried here. In that case, who are the others? One scene shows a chubby man with a prominent belly, stark naked, enraged and armed with an axe, charging toward a deer twice his size. Another hunter, heavier muscled—he has been described as soldierly—stands with spear raised, ready to deliver the coup de grace to a boar. This hunter's legs are bare and might be interpreted as sunburned; his hair is dark, but when it came to the beard, the artist deliberately changed his palette and painted the beard red.

You can go badly astray in trying to read the ancient world as if it were our own. Archaeologically, it is clear that the Alexandrovo tomb must have been opened at least twice; it also contains the remains of what might have been a stone couch, or might have been a table. The decreasing height of the passage into the central chamber, with the frescoed hunt eternally circling its ceiling, might suggest that the chamber itself was used as a temple, and it and the couch/table inside it were for ceremonies and rituals lost to us. But the scenes and details are so specific—the portly naked man in such furious pursuit of that deer, and the red-bearded "soldier"—that they tempt one to another explanation: that they and the man on horseback hunted together, that they would have recognized themselves and the events in these scenes, and that on the rider's death, for a period, his friends gathered here, and drank, and feasted, and remembered. And this is what they looked like. And one of them had a red beard.

Herodotus is the man to provide some context here. Herodotus was born *c.* 484 BC (so a little after Xenophanes's long life came to its end), in what is now Bodrum in southern Turkey. Thrace would have been no more remote and its customs no less known to him than those of Canada are to the United States. Around 450 BC Herodotus was in Thrace, and in his *Histories* he describes the various habits of the various Thracian tribes. Some, he says, practiced polygamy, with the most favored wife following her husband to the grave and, indeed, fighting for the privilege of doing so. He details with relish how the Thracians kept "no watch" on their maidens "but leave them altogether free," and how unwanted children—possibly, one imagines, the resulting unwanted children—would be sold to slave traders. He also describes their funeral rites and their worship of Dionysus. Thrace was reputedly the birthplace of Orpheus, the musician whose songs were so sweet they could divert rivers, coax trees to dance, and almost freed his dead wife, Eurydice, from the underworld. It was also the site of his death, torn limb from limb by Thracian women supposedly in the throes of Dionysian ecstasy.

You can also, of course, go wildly astray by taking the words of Herodotus at face value, as was noted by writers from Plutarch to Voltaire and as is still debated today. Where Thrace is concerned, though, he seems to have known what he was writing about. And for the Greeks, with their city-states and hierarchy of gods and government, and in particular the Athenians, with their strict notions of order and their rigid seclusion of women to the home, the Orpheus story must have been the perfect Thracian myth. It is inescapably violent and full of both transgressive females and the terrifying mysteries of the afterlife. Thrace both appalled and entranced the Athenians. And it was their single greatest source of slaves.

Besides their depictions of themselves in their tombs, the other means by which we know what the Thracians looked like is from their appearance in Greek art. Thracian women (when not pursuing Orpheus) shuffle slipshod and disconsolate around many a Greek vase. Their hair is shorn short, and their limbs are decorated with tattoos. To the Greeks, who punished escaped slaves with tattoos or branding, this was a sure sign of servitude, but to the Thracians themselves, ironically, these rosettes and dotted lines, these whorls and stylized animals (twiggy-antlered stags seem to have been a particular favorite) were signs of noble birth.[24] Thracian men are depicted as warriors, sometimes fallen, sometimes not. They sport pointed beards, they wear cloaks decorated with bands of geometric patterns and caps that Herodotus informs us were made of fox-skin. There is obvious potential here for confusion between the color of a fox-skin cap and that of the hair of the head, similar to the confusion engendered between the Qizilbash warriors of thirteenth-century Anatolia and *their* crimson headwear. In the case of the Thracians, it may also reflect a connection made in the ancient world between the behavior of the animal—supposedly cunning, sly and untrustworthy—and the character ascribed to redheaded barbarians of whatever tribal identity. As the third-century *Physiognomonica* would have us believe, "The reddish are of bad character.

---

24 Bodies found in Scythian *kurgans* also sport elaborate tattoos, for example that of the famous Siberian Ice Maiden, discovered in 1993. See the article by Anna Liesowska in *The Siberian Times*, October 14, 2014: http://siberiantimes.com/culture/others/features/siberian-princess-reveals-her-2500-year-old-tattoos/.

Witness the foxes." (Of course now it is used of redheaded women in the sense of the 1967 Hendrix song "Foxy Lady," and is thus part of another of those historical associations of red hair with a shelf life of millennia.) But the Thracians are also sometimes shown, unmistakably, as redheads, as with King Rhesos of Thrace, a supporter of the Trojans in the *Iliad.* King Rhesos was late getting to Troy (a little local trouble with those pesky Scythians); then before he had so much as set foot on the battlefield, he was done to death in his tent by Diomedes and Odysseus, who stole his famous horses, too. His death is shown on a black-figure vase now in the Getty Museum in Los Angeles (Fig. 5). Rather less elevated is a little terracotta figure of a runaway slave from the British Museum in London (Fig. 6). It's thought to have been made in Athens, at about 350–325 BC, so the same century as the Alexandrovo tomb. It's no more than five inches, or about thirteen centimeters, high, and it shows a chubby little man, this time sitting on an altar. His shoulders slouch, his left hand grasps his left knee, and his right is raised to his ear, as if he is hard of hearing, or perhaps (given the open wail of his mouth), someone has just clipped him around the head, and here he is, scrambled atop the altar, claiming sanctuary, wailing and bemoaning his lot.

These little terracotta sculptures were traded all across the Classical world, from North Africa to southern Russia. They were the Toby-jugs (or maybe today the bobbleheads) of ancient Greece: cheap, coarse, and dispensable. Scenes of tragedy, such as the death of King Rhesos, were reserved for pottery of the grander sort—the black-figure vases, for example, which were intended for those who could appreciate them. Figures such as this runaway slave derived from comedy and were intended to appeal to the common sort.

And they reflect exactly the "Three Stooges" slapstick aesthetic of Greek comic drama. On stage, our runaway slave would have had a red leather phallus bouncing between his legs under his miniskirt-length tunic (it makes this little figure even more poignant, that his manhood has been broken off). The grotesque expression of the mask he wears would have spoken to his audience of the primitive and ungovernable nature of his emotions. And his hair, or rather the wig attached to his mask (and even now on the terracotta, traces of pigment remain), his hair would have been red.

The Greeks liked lists. They liked the world ordered and subdivided. In the second century AD, the Greek scholar and rhetorician Julius Pollux compiled one of the world's first thesauruses: a dictionary arranged not alphabetically but by subject matter. Within this, the *Onomasticon*, he includes descriptions of the seven different slave types in Greek drama, and four of the seven have red hair. On stage, the names of their characters repeat the same message: Pyrrhias, a slave in Menander's comedy *Dyskolos*, ("The Grouch"), whose name means "Fiery." In the comedies of Aristophanes no fewer than five different slave characters in five separate dramas share the same name: Xanthias, meaning "Goldy," or "Red."

It seems unlikely, to say the least, that every slave in every Greek household was a Thracian redhead. Red hair is still recessive; it would have been at least as much of a rarity in the ancient world as it is today. But in a phenomenon that repeats throughout history, this clearly made no difference. Where "other" is concerned, we focus upon its epitome to the exclusion of every other detail. One thing comes to stand as a symbol for all, and this one characteristic, red hair, came to stand for an entire class, if not in fact an entire nationality.

The association was created; at some subconscious societal level it was accepted, and it stuck. Red hair, in the ancient world, equaled "barbarian"; then via the Greek stage and figures such as our little terracotta runaway, it began to equal "clown." You can trace a line of development from these primitive slave characters of the Greek stage to the white-faced, red-haired clown of the circus big top to Ronald McDonald, unnerving children across the planet (and originally incarnated by Buttons, Ringling Brothers' red-wigged clown); to the rather more endearing Obelix, Asterix the Gaul's red-pigtailed bosom buddy in the cartoon books by Goscinny and Uderzo. It is as if we are watching two redhead archetypes, the ungovernable savage and the comic buffoon, coming into being before our eyes. In fact, it is not even as if. We *are.*

All freeborn people, as the historian Sandra Joshel puts it, are defined by their physical integrity. So to insult their appearance is to insult both their social standing and personal identity. But slaves had no personal identity. They were identified only by the work they might be fit for, like so many differently sized tools in a tool kit: this one has a singing voice. This one would make a good plowman. This one could be a wet nurse. This one could work with vines. And they had no ethnicity either, so red hair no longer marked you out as Thracian; it marked you out instead as disempowered, subservient. The good slave was one who accepted this and subsumed their own identity; the "bad" slave, and the one who got the most laughs on stage, was the one who insisted on retaining a human personality—always a bad one.

The pseudoscience of physiognomy, in which personality is inferred from physical characteristics, also held great appeal to the

Greeks. Here was another way of getting the world to measure up. Aside from its views on foxes, the *Physiognomonica* (a treatise now fittingly attributed to an author known as the "pseudo-Aristotle") relates how those with very fiery hair (*agan purroi*) are rascals (*panourgi*), while very white skin (*agan leukoi*) was a sign of cowardice (is this a third stereotype, of the redheaded man as wimp, coming into being?). But then nothing about red hair in the Greek world was good: in *The Clouds* Aristophanes has his Chorus grumble that the state is now in the hands of "men of base metal—foreigners and redheads," and Aristophanes created more sympathetic slave characters, such as Xanthias in *The Frogs*, written *c.* 405 BC, than most. Otherwise slaves in Greek drama were grumbling, self-pitying, oversexed (masturbation in the Greek world was a vice of slaves and foreigners), uncouth, dim-witted, clownish, lazy, dishonest, and petulant. And, on stage at least, they had red hair.

The Roman encounter with the redheaded world tended to be at the sword's point, rather than within the *theatron.*

Red hair is liminal, as has been said (this of course only contributes to its status as "other" to begin with). It's out on the edge, geographically as well as genetically, with an undisturbed gene pool giving it, as a recessive gene, the best chance of coming up. It's like blackjack: if you're playing the same cards over and over, sooner or later, you'll get a natural twenty-one. If there truly were unusually large numbers of redheads in Thrace, this might be why;

all those separate tribes unable to agree, or presumably to intermarry, between themselves. You find redheads throughout Europe, but as a rule of thumb you find them in greater numbers the farther north you go. By far the longest-lasting of the tribal civilizations encountered by the Romans was that of the Celts, whose lands, at their greatest extent (depending on where you place the borders of the Celtic language, and where you place the borders of what was, and was not, Celtic in the first place) can be said to have reached from Ireland to Poland, as far north as the Hebrides, as far south as what would become Genoa. As the Roman legions marched and conquered and pushed the borders of the Empire ever up into the Celtic heartlands of Germania and Gaul, they encountered more and more redheads among the peoples they subjugated. Caesar's Gallic campaigns of 58–51 BC alone are estimated to have added a million slaves to the Roman Empire. And although there is a notion that blond- or red-haired Celtic slaves were prized, and although they may have been so, as novelties, there is little evidence that the individual slaves themselves were regarded as being of any extra worth at all. Rather the reverse, in fact. The philosopher Cicero (106–43 BC) writes of British slaves, "I think you would not expect any of them to be learned in literature or music." You can almost hear the sniff of disdain.

But contact with these Northern tribes did lead to the recognition that people who looked so similar must be connected in some way. Here is Tacitus (AD 61–after 117) on the tribes of Britain, and he's worth quoting at length because here, finally, history has allowed the historian to catch up, and Tacitus is reflecting upon events within his own living memory. The original invasion of Britain by Julius Caesar in 55 BC led to conquest under the Emperor Claudius in AD 43 and was followed by decades more of on-and-off campaigning. Indeed Hadrian's Wall, which marked the outermost northern border of the Roman world, was begun only in AD 122. In his life of the Emperor Agricola, Tacitus writes:

> Who were the original inhabitants of Britain, whether they were indigenous or foreign, is as usual among barbarians, little known. Their physical characteristics are various, and from these conclusions may be drawn. The red hair and large limbs of the inhabitants of Caledonia point clearly to a German origin. . . . Those who are nearest to the Gauls are also like them, either from the permanent influence of original descent, or, because in countries which run out so far to meet each other, climate has produced similar physical qualities. But a general survey inclines me to believe that the Gauls established themselves in an island so near to them. Their religious belief may be traced in the strongly marked British superstition. The language differs but little; there is the same boldness in challenging danger, and, when it is near, the same timidity in shrinking from it. The Britons, however, exhibit more spirit, as being a people whom a long peace has not yet enervated. Indeed we have understood that even the Gauls were once renowned in war; but, after a while, sloth following on ease crept over them, and they lost their courage along with their freedom. This too has happened to the long-conquered tribes of Britain; the rest are still what the Gauls once were.

Not quite as long-conquered as Rome might have thought.

The counties of Suffolk, Norfolk, and Essex, in the east of England, exist under what Tacitus describes as a "sky obscured by continual rain and cloud." The shoulders of the fields are dotted with Bronze or Iron Age tumuli that predate even the Romans, the churches are mostly medieval, the towns are small, the villages are tiny. The landscape dwarfs the people. It feels old. The skies go on forever, and dwarf everything. And as every East-Anglian schoolchild knows, in Roman times these were the homelands of the Celtic tribe of the Iceni, and of their queen, Boudicca. To such children (I was one myself) Boudicca's chariots, with the terrifying scythes mounted Persian-fashion on their wheels, ready to cut down her enemies, are every bit as real and realizable as the tractor in the next field, grumbling home.

It was Rome's policy to move from subjugation to colonization to integration, and there were native Britons, especially in the south and east of the country, who accepted the conquest, the changed state of their world, recognized the emperor in Rome, and prospered and grew rich on Roman loans. Boudicca's husband, Prasutagus, was one of these. At the same time, again according to Tacitus, all the tensions one would expect in a newly conquered territory were present: there were Roman veterans living around the colony of Camulodunum, now Colchester in Essex, who had taken over native lands and homesteads and were especially hated, while even Roman slaves found in the native population people they could patronize and insult. Then Prasutagus died, the loans were called in, his lands taken away from his family, his wife was scourged, his daughters raped, and in AD 60 the Iceni rose in revolt. And thus Boudicca has come down to us, in the words not of Tacitus this time but of a later historian, Cassius Dio. He describes her as "possessed of greater intelligence than often belongs to women. . . . In stature she was very tall, in appearance most terrifying, in the glance of her eye most fierce, and her voice was harsh; a great mass of red hair fell to her hips; around her neck was a large golden necklace; and she wore a tunic of divers colors over which a thick mantle was fastened with a brooch."

Again, the evidence has to be sifted. Cassius Dio lived AD 155–235, so his account is certainly not as contemporaneous as Tacitus, and shows it, in its elaboration. What both sources agree upon is that the Iceni and their neighbors the Trinovantes joined forces, and with Boudicca at their head swept down on Camulodunum, then on the new trading post of Londinium, and then Verulanium, present day St. Albans, too, and razed them to the ground. Eighty thousand Roman citizens are thought have perished before the rebellion was put down, and in fact both Cassius Dio and Tacitus speak of the distinct possibility of the island of Britain being lost to Rome altogether.

Cassius Dio, like Xenophanes, wrote in Greek, and that one word, "red," in his description of Boudicca is also translatable as tawny, or as reddish-brown. Nor is there any evidence that scythe-bearing chariots were used by the Britons. Yet in the sculpture set up in 1902 on the corner of Westminster Bridge and the Victoria Embankment in London, there Boudicca is, in all her stern-browed Victorian glory, with knives a yard long protruding from the hub of her chariot's wheels. And, right up to Marvel Comics' *Red Sonja*, red hair has been indispensable to the image of the indomitable, ferocious, and usually voluptuous female barbarian. Because, again, what other color could Boudicca's hair be? In her case it speaks of her unvanquished determination, her patriotic courage (one reason why she was so popular with the Victorians), her non-Roman-ness. Red hair meant Celt, it meant Gaul, it meant Thracian. Sometimes, indiscriminately, it meant all three.

The Capitoline Museum in Rome contains a sculpture now known as *The Dying Galatian*, but for centuries after it was unearthed in Rome in 1623, it was known as *The Dying Gladiator*, or *The Dying*

Gaul. It's a Roman marble copy (that process of historical transmutation again) of a lost Hellenistic original, thought to have been cast in bronze and commissioned by Attalus I, king of Pergamon in Turkey from 241 to 197 BC. It shows a naked warrior, half-lying on his shield. Scattered about him are his sword, his belt, his war trumpet. Around his neck there is a twisted, Celtic-looking torque, and his hair is in the short, punky, lime-washed spikes that must, one imagines, be just as it would have appeared and exactly as the Greek historian Diodorus Siculus, who wrote between 60 and 30 BC, describes it.[25] The man's head is bent; he props himself up on one arm. There is a wound, a sword-thrust, bleeding, under his right breast, and the sculptor somehow managed to catch the one still, central moment, when the man knows death is upon him but has not yet tumbled beneath it. Lord Byron saw the sculpture and was moved to include it in *Childe Harold's Pilgrimage*:

I see before me the Gladiator lie:
He leans upon his hand—his manly brow
Consents to death, but conquers agony . . .

. . . his eyes
Were with his heart, and that was far away;
He reck'd not of the life he lost nor prize,

25 In his vast *Bibliotheca Historica* Diodorus Siculus wrote, "The Gauls are tall of body, with rippling muscles, and white of skin. Their hair is blond, and not only naturally so, but they also make it their practice by artificial means to increase the color which nature has given it, for they are always washing their hair in lime-water, and they pull it back from the forehead and back to the nape of the neck, with the result that their appearance is like that of satyrs and Pans, since the treatment of their hair makes it so heavy and coarse that it differs in no respect from a horse's mane." In fact the hair of *The Dying Galatian* was recarved, perhaps with this description as a guide, in the seventeenth century.

But where his rude hut by the Danube lay,
THERE were his young barbarians all at play,
THERE was their Dacian mother—he, their sire,
Butchered to make a Roman holiday—
All this rush'd with his blood—Shall he expire,
And unavenged?—Arise! ye Goths, and glut your ire!

But the dying warrior is not a gladiator, nor a Goth, nor a Gaul. He's a Thracian, and we are back where we began.

The commission from Attalus I for the original bronze is thought to have been in celebration of his victory over a force of marauding Thracians who had settled in Galatia, in the highlands of Anatolia, and from there created such terror that they were able to extort tribute from as far afield, apparently, as the kingdom of Syria. Livy—the final Roman chronicler in this chapter—describes them thus:

> Their tall stature, their long red hair, their huge shields, their extraordinarily long swords; still more, their songs as they enter into battle, their war-whoops and dances, and the horrible clash of arms as they shake their shields in the way their fathers did before them—all these things are intended to terrify and appal.

In fact he puts these words into Attalus's mouth, as he exhorts his troops to action against the "Gauls." By the time Livy was writing, in the first century BC, it didn't matter if you were Gaulish, Celt, or Thracian. You had red hair, you were barbarian—that was all.

In 2014 workmen digging in the center of Colchester, down through the layered past, down to the level of the two-foot-thick seam of blackened rubble that still marks where Roman Camulodunum stood and burned, unearthed a horde of gold jewelry—the treasured possession of some Roman woman, hidden under the floor of her home as Boudicca's warriors descended on the town.[26] The floor of the room in which it had been hidden was strewn with the remains of food, carbonized by the heat of the fire that had destroyed the building; and mixed in with this were fragments of human bone—a piece of jaw; a piece of shin. The reality of Boudicca's attack on Camulodunum is grotesquely at odds, as these things always are, with an image of fierce and indomitable courage, with Byron's notion of innate nobility, refusing to bow beneath the oppressor's yoke. The actions of Boudicca's troops were as savage as anything perpetrated today, with particular brutality meted out to the Roman women she took captive: dragged to ancient sacred groves outside the town and mutilated with a vindictiveness that it seems almost impossible to countenance any one woman inflicting on another (purportedly the Roman women's breasts were cut off and sewn to their mouths), and then they were killed. We don't know the fate of the woman who once lived in this building in Roman Colchester, but she would have been no more an equal being to those she hid her jewelry from than they would have been to her. And she did not live to retrieve it.

---

26 See http://www.telegraph.co.uk/history/11074055/Unearthed-a-golden-Roman-hoard-hidden-from-Boadiceas-army.html.

DIFFERENT FOR GIRLS

Thus human beings judge of one another, superficially, casually, throwing contempt on one another, with but little reason, and no charity.

BARONESS ORCZY, *THE SCARLET PIMPERNEL*, 1905

The nineteenth-century anthropologist John Munro, who published his *The Story of the British Race* in 1899, makes much of a supposed difference in hair color between the Vikings and the Saxons. "The Danes were distinguished by their red hair and fiery temper," he writes, "from the more phlegmatic Anglo-Saxons with light brown or flaxen hair and blue eyes." This idea of some atavistic race-memory accounting for the suspicion in which red hair was held crops up very frequently in the nineteenth century (and down in Hampshire, was clearly still going strong in my grandmother's day), but given the mixing and mingling that had gone on between the tribes of northern Europe for all those centuries before, it seems unlikely that the average Saxon and the average Dane would have varied in appearance by much at all. The Vikings were no less flaxen or any the more redheaded than those they terrorized (and never mind Hagar the Horrible and his bristling red beard). It was the new and sudden nature of these seafarers' raids that made them so feared.

The first Viking raids upon the Christian West came in the summer of AD 793, on the monastery community of Lindisfarne, an island off the coast of the north of England, and in one form or another Viking raiders alternately terrorized or settled (or both) coastal and river communities throughout Europe for at least the next three hundred years. The Vikings settled in Iceland, hence its appearance on the Redhead Map of Europe. In Greenland, the first community was founded by Erik the Red, whose name almost certainly commemorates the color of his hair, while the story of the slow extinction of one such Viking settlement there, Herjolfsnes, provides a fascinating if appalling example of what happens to an isolated community as its inner stores of Vitamin D are steadily depleted.

By the eleventh century, the Vikings had reached Newfoundland, which they regarded as the last land before one's Viking longboat tipped into the abyss, in an expedition led by Erik the Red's son, Leif Ericson. They certainly sent emissaries to Byzantium, and there is archaeological evidence that Viking traders reached Baghdad (did they add their redhead genes to those already in the local population?). There are also intriguing legends among the Paiute people of Nevada of their tribe's encounter, long ago, with a redheaded enemy around the Lovelock Caves, which have led some, rather rashly, to assume that Vikings must have penetrated far into the North American continent as well. The Vikings of Sweden raided and traded down the Volga, no doubt meeting many a remnant Scythian or emergent Udmurt population as they did so, and perhaps giving what we know as Russia its name, from a Slavic version of theirs: "Rus," or "the men who row." Norwegian Vikings took captives from the coastal settlements of Ireland and the west of Scotland, adding these Celtic genes to their own, and no doubt left descendants in these places, too. They were only finally defeated in Ireland by the great Brian Boru, at the Battle of Clontarf in 1014. But to present some race-memory of Viking raiders as the reason for the dislike and distrust of red hair is to ignore far more active and pertinent prejudices that still exist today, just as they did in John Munro's time, and which in medieval Europe were even more virulent and deeply rooted.

Moreover in England at least, Viking raiders from Denmark reached an accommodation with the indigenous population, establishing a kingdom for themselves in the northeast that stretched from York almost down to London—a third of the country, contained by the Danelaw, and subject to the treaty signed with Alfred the Great in about 880. They seeded their own language, place names, and culture, along no doubt with a few supplementary redheads, throughout northern Britain (my own sandy-haired, blue-eyed grandfather was a Cougill, from Lancashire), a region where William the Conqueror would encounter seething resistance for years after his conquest of England in 1066, and which he would viciously repress. And, of course, in the tenth century they would colonize Normandy. William's own great-great-great-grandfather was Rollo, or Hrólfr (*c.* 846–*c.* 931), who was either Danish or Norwegian, depending upon which source you believe.

The Bayeux Tapestry, marking William's victory over the English at the Battle of Hastings, and which could be contemporary evidence for the color of his hair, is not much help here, identifying various figures as *WILLEM* or *WILLELMO DUCI*, who are both sandy and brown-haired. But William's third son and the heir to the English throne, the blessedly short-lived William Rufus, seems to have owed his name either to his hair or to the florid complexion long regarded as accompanying a boorish temper.[27] The associations created in the Classical world between red hair and a suspect character were clearly still going strong, as were those linking red hair, especially where it came with a ruddy, weather-beaten skin, with low birth or as a signifier of vulgarity—if not still slave, then certainly serf. The historian Ruth Mellinkoff cites an example from the ninth-century *Life of Charlemagne* in the cautionary tale of an ill-mannered peasant who refuses to uncover his head in church. When his cap is finally dragged from his ears and his head is exposed, the priest thunders from the pulpit in a final inculpatory denunciation, "Lo and behold, all ye people, *the boor is redheaded.*" The same association would be exploited by Chaucer for the character of Robin, the drunken loudmouth miller in the *Canterbury Tales* (1380s–90s), with his spade-shaped beard as red as "anye sowe or fox," and the wart on his nose, sprouting bristles as red as those of a sow's ears. Even in Scandinavia, golden-blond Odin, Mellinkoff suggests, was the god of the nobility; redheaded Thor with his hammer was the god of the laboring man. You could rule a kingdom, in fact, and your red hair would still be used to denigrate you.

Roger I, who had taken Sicily from the Muslims by 1091, was but one of many generations of Norman mercenaries fighting in Italy, and his descendants would rule Sicily for the next century (hence the

27 William Rufus, "hated by almost all his people and abhorrent to God," according to the *Anglo-Saxon Chronicle*, died on August 2, 1100, while hunting in the New Forest in Hampshire. A stag crossed his path, he shouted to his companion to shoot—and the man did. William's body was left where it fell; a peasant found it and brought it to Winchester in his cart.

blue-eyed, red-haired, gloriously freckled Sicilian brother and sister I encountered in a language school in Cambridge). Frederick II, Roger I's great-grandson, extended that rule across Italy and Germany and even as far as Jerusalem. Frederick spoke six languages, launched two crusades, and was elected Holy Roman Emperor, a role in which he clashed with the papacy so frequently that he was excommunicated four times over and once denounced as the Antichrist. He kept a menagerie and, according to his enemies, a harem. He was an empiricist in matters of science, investigating the process of human digestion by practicing what was basically vivisection on his dinner guests; and a skeptic in matters of religion; and was proclaimed by his contemporaries as the wonder of the world—"stupor mundi," which somehow sounds even more impressive. Then you read the Syrian chronicler Sibt-ibn al-Jawzi's description of him: "The Emperor was covered with red hair and was bald and myopic. Had he been a slave he would not have fetched 200 dirhams at market," which seems a little harsh, however true. Frederick spoke Arabic and was one of the few European rulers of the time to approach the Muslim kingdoms of the Middle East with interest and respect. He also stood up for Sicily's community of Jews.

There had been Jewish communities living around the Mediterranean as part of the Roman Empire, and the first Jewish communities in France and Germany may date back to the same early period—indeed, possibly even predate it. There is circumstantial evidence that Jewish settlers in Spain were trading with the Phoenicians, which would have been centuries before the creation of that small, insignificant Roman settlement on a gravelly ford of the Tiber in 753 BC. By the seventh century AD there were Jewish settlements spread as far afield as China, very likely of travelers and traders who had followed the Silk Road. In the eighth century the kingdom of the Khazars represented for a century or so a Jewish power base between the Caspian and Black Seas and was a major crossroads between Russia, Byzantium, and traders from the Middle East. The first documented communities of Jews in England are perhaps the only people in all its flotsam-and-jetsam population who might genuinely claim to have come over with the Conqueror, or at least to have been brought over, from Rouen, under William's watchful eye, in 1070. William had a new kingdom to subdue, and a mighty program of castle-, or rather fortress-, building to fund. The man who could come up with the concept of the Domesday Book could put a price on anything, but William needed tribute rendered in and an economy based upon coin, not Saxon barter. Bring on the first money men of Europe, the Jews enlisted by William I, to kick-start the transformation.

In 1079 another group, five French Norman Jews (and plucky souls they must have been, too), crossed the Irish Sea to set foot on what was then the extreme western edge of the known world, in a trade delegation to the king of Munster, Toirdhealbhach Ua Briain, or so the *Annals of Inisfallen* tell us. Toirdhealbhach was seventy by this date, and had spent the previous thirty years fighting, exiling, or murdering his rivals. By 1079 he was effective ruler of half Ireland. His Jewish visitors clearly knew who the Big Fellow was, but Toirdhealbhach seems to have been less certain of them. They came with gifts, but the annals record only that these visitors "were sent back again over the sea." The event does, however, open up the pleasing possibility of this grandson of Brian Boru, this aging Celtic khan, redheaded as we may surely imagine him, greeting visitors who might have been as red-haired and red-bearded as he. Were there Jewish redheads? There both were and are.

Red hair, as has been said, survives best out of the great ebb and flow of a changing population. Those circumstances are also found in communities that are endogamous—that is, that marry within their own specific ethnic group, something the Jewish population has done for centuries. There are many, many redheaded Jews, and as these communities moved into western Europe, it was a mighty cultural mischance that they brought with them a characteristic already aligned in European culture with bad character at best and barbarity at worst. One that was already perceived as apart from the norm, was already picked out and commented upon; that was, in other words, already "racialized." Or as Eleanor Anderson describes it in her thesis on discrimination against redheads today, and in the language of contemporary psychology, with the Jews of medieval Europe we see all the hallmarks of a situation where "an individual who might have been received easily in ordinary social intercourse possesses a trait that can obtrude itself upon attention and turn those who he meets away from him, breaking the claim that his other attributes have on us."[28]

---

28 Eleanor Anderson, "There Are Some Things in Life You Can't Choose . . . : An Investigation into Discrimination Against People with Red Hair," *Sociology Working Papers* 28 (2002). Anderson's thesis does a superb job of exploring prejudice against red hair as part of the overall landscape of discrimination and cultural stereotyping.

We in our communities today know that stereotyping or stigmatizing individuals on the grounds of their difference is as destructive of those who think that way as it is of those victimized by their thinking. We may not all always understand this, but we all know it, and in the First World, at least, many if not most modes of living and of belief that would have reduced our ancestors to a blood-flecked mob do so no longer. But in medieval Europe, Jews were Christ-killers and the abductors of Christian children. They were known as usurers, moneylenders, and the financiers of the state (which very often turned on them as well). And some of them had red hair. And since Judas was a Jew, in a noticeable number of examples in European art, particularly in Germany, he is also depicted as a redhead. Even the redhead's freckles were not spared. In medieval Germany, one term for freckles was "Judasdreck."[29] As the scholar Paul Franklin Baum put it in 1922, "There can be little doubt that this tradition is simply the application of the old belief—much older than Judas Iscariot—that red-haired men are treacherous and dangerous, to the arch-traitor, some time during the early Middle Ages."[30]

You read with sinking heart the first edicts forcing Jews to identify themselves by wearing specific badges; those moving them into ghettoes; the first instances of persecution, the first massacres, the first expulsions. Jewish communities were expelled from France in 1182; recalled in 1198 (the royal coffers were running low); expelled again in 1306 and their property confiscated for exactly the

29 Which basically translates as "Judas-shit." See Ruth Mellinkoff, *op. cit.*, p. 168.

30 Paul Franklin Baum, "Judas's Red Hair," *Journal of English and German Philology* 21, no. 3 (July 1922): 520–29.

same motive. There were no Jewish communities in England after 1290; and in 1492, in Sicily, which was under the control of Ferdinand and Isabella of Spain, those same Jewish communities that had been defended from the zeal of the Crusaders by Frederick II were driven out of the island altogether. In that year the redheaded Genoese sea captain Christopher Columbus followed Leif Ericson across the Atlantic in an expedition funded by Ferdinand and Isabella, and the world changed once again. Those monarchs, newly in control of a kingdom from which any religion other than Catholicism was to be extirpated, also, notoriously, put into motion the diktats of the Spanish Inquisition.

Historians have argued back and forth over the number and nature of the victims of the Inquisition in Spain and its territories in the New World, but however many and whoever died because of it, their suffering was inflicted by a bureaucracy of terror that used symbols of "other" as a means of identifying its victims. Trials were secret; punishment was public and vindictive enough to extend even to the burning of corpses. The Inquisition was not directed solely against those of faiths other than Catholicism, either; it went after homosexuals, those in possession of prohibited texts, and those suspected of witchcraft. In the nineteenth century it was hunting down those "suspected" of freemasonry, as well, which rather starts to sound as if there was no one else left. Right from the start it was suspected that those persecuted were being chosen because they were wealthy, and the state could confiscate their property. The processes of the Inquisition were as absurd as they were appalling, as such systems always are. But if red hair meant you were Jewish (it might equally have meant you were Protestant, or merely un-Spanish in some way), red hair might also mean you were a backsliding *converso*, especially given all those other qualities ascribed to redheads concerning their treacherous and untrustworthy nature. You end up going around in a circle here, where the prejudice justifies the racism and the racism strengthens the prejudice. Attitudes toward red hair in Spain only reflected the fears and prejudices of Catholic Europe as a whole. If you want to look for reasons for the continuing and increasing antipathy toward redheads, in particular redheaded men, in medieval Europe, look no further than its anti-Semitism. And if you want to place the point at which attitudes toward red hair in men and women begin so radically to diverge, likewise.

To understand what a people thought or believed, start by looking at what they were looking at. In this case, two panel paintings, one German, one French, created less than ten years apart, demonstrate the polar opposites of attitudes toward red hair in the medieval world. The German work shows Christ on the Mount of Olives, or *The Agony in the Garden*, as the scene is perhaps better known. It dates from about 1444–45 and is now in the Bayerisches Nationalmuseum in Munich (Fig. 7). For many years—centuries, in fact—it was unattributed to any particular artist; another of those anonymous works that in their lost history speak with quiet eloquence of the unknown, uncountable faithful who knelt in front of them. Its creator was identified only as late as the 1980s as Gabriel Angler the Elder, a Munich artist whose dates are thought to have been *c.* 1405–62.[31] Previously his works were simply grouped together under the title of the most famous of them, and he was known as the "Master of the Tegernsee Altar."

In his painting, we are in a garden of sorts, although parts of it seem very rocky and untended. It's bounded by a wicker fence, which might bring to mind images of that first garden, Eden, or of unicorns and virgins, or indeed of any place private, secluded, and reserved for contemplation or for prayer. And it's night. There are four figures in the garden, three of them huddled in sleep, and although one should be aware that the background was repainted in the late seventeenth century, the re-painter, in his creation of a sky literally dark with foreboding, did a good and sympathetic job. This seems to be one of those nights where heaven has come very near to earth—the sky is framed by delicate Gothic tracery, and in this realm of the divine, top left, there hovers an angel bearing a scroll. You can almost hear the stillness—the nighttime insects, the lazy flapping of the scroll as it unfurls from the angel's hand, the whispering of the one man awake, kneeling at the rocky outcrop as at a prie-dieu.

Enter, stage right, as from some other world beyond the painting and through the flimsiest-looking portal, a half dozen soldiers in full fifteenth-century armor; helmeted, with swords at hip, and spear and halberd glinting above them—a pre-Reformation SWAT team. They are being led through the fragile doorway into the garden by a

31 V. Liedke, "Die Münchner Tafelmalerei und Schnitzkunst der Spätgotik, II: Vom Pestjahr 1430 bis zum Tod Ulrich Neunhausers 1472", *Ars Bavarica*, xxix/xxx (1982): 1–34.

man in a long, Biblical, yellowish robe, as if he has summoned them from the artist's time into his. One of his hands is raised, a finger pointing in admonition—*shush!*—but of course also, ironically, up to the divine; while with the other he supports the weight of the heavy purse hung around his neck. This is Judas. And his hair, and his beard, and even the skin of his cheeks, are red.

Redheaded men in medieval Europe were at just as much of a disadvantage as they would have been in classical Greece or Rome (and it should be remembered, throughout this period, men's hair was much more publicly visible than women's). Those who wished to justify their prejudice might point to the Old Testament example of boorish Esau, born "ruddy, all over like a hairy mantle," and dim-witted enough to lose his birthright to his younger, more intellectual brother, Jacob. Cross such men (trick them out of their rights as firstborn over a bowl of lentil soup, for example, as Jacob did), and the dark flush of their ruddy visages betokened violence, loss of temper, retribution that would be both swift and unthinking.[32] Even the Biblical King David might be so judged: "And when Samuel saw that David was ruddy he was smitten with fear, thinking he might also be a murderer." And if King David's temper might be suspected on the basis of his coloring, how much more likely, and easier, was it to project one's fears onto one's foreign, heretical, idolatrous, synagogue-going neighbor, with his strange dietary observances and his red hair?

It's ridiculous to assume that in reality red hair was any more

32 Ruth Mellinkoff *(op. cit.)* points out that the tiny figure of Cain, a marginal illumination in the Mosan Park Bible of *c.* 1148, also has red hair. BL Add MS 14788, folio 6, verso.

common then than it is today. It was and is still a minority characteristic, among Jews just as it is among the Irish and the Scots, and as it no doubt was with the Thracians, too. But once again it comes to stand for an entire ethnicity of other: once viewed, apparently blinding those who saw it to any other characteristic.[33] In medieval art red hair in men, in particular in combination with a ruddy skin, is like the black hat on the baddie in a Western. It's visual shorthand for a brutal character of a particular unthinking sort—animalistic, unintellectual, unreachable by reason, and all the more frightening for that. It's a means of signifying not simply villein, or serf, but villain, too.[34] It's the color of the hair of Christ's (Jewish) tormentors, with their whips and goads and what in so many medieval paintings is the artist's invention of ridiculously, sadistically, overelaborated means of pressing the crown of thorns down on Christ's forehead (why not simply have his torturers wear gloves?). It's noticeable that it's in this combination with ruddy, rough, or weather-beaten skin that red hair appears in some of the most nastily caricatured depictions of Judas. And nakedly exposed, time and again, as Mellinkoff points out, it is the coloring of the impenitent thief at Christ's crucifixion (Fig. 8). Distorted, damned and writhing in agony, possessed by devils, abandoned by God and man (and the Catholic church), this is the uncouth barbarian again. The coloring of the impenitent thief,

---

33 Or as Bishop Jocelin puts it in William Golding's *The Spire* (1964), on seeing Goody Pangall's red hair, "It was as if the red hair, sprung so unexpectedly from the decent covering of the wimple, had wounded all the time before, or erased it."

34 And as such, well into the nineteenth century, it stigmatizes Quasimodo in Victor Hugo's *The Hunchback of Notre Dame* (1831): "a huge head bristling with red hair . . . that little left eye obstructed with a red, bushy, bristling eyebrow." Sure enough, Quasimodo's looks are taken as proof that he is "as wicked as he is ugly." He even remains a redhead in the Disney version of 1996.

by contrast to the pale, serene, penitent thief, heading for Paradise, is used both to evoke his freakishness and his inhumanity, the fact that his fate doesn't matter, and to demonstrate, with the artist as our proxy, as it were, society's and our Christian revenge.

And oh, how that prejudice linking Judas and Jewishness and red hair persisted. In *As You Like It*, written in 1599, a century and a half after the completion of Angler's painting, Shakespeare has Celia describe Orlando's hair, defensively, as "something browner than Judas's." In the 1690s, in the midst of a bitter falling-out with his publisher Jacob Tonson (a businessman canny enough to have put the first copyright on Shakespeare's plays), the poet John Dryden anathematized Tonson as having "two left legs and Judas-colored hair." The association between red hair and Jewishness was so strong that Shylock was still being played in a red wig until Edmund Kean performed the part at Drury Lane in 1814.[35] Even Charles Dickens, who could make as impassioned a case for the dispossessed and the minoritized as any writer before or since, in his 1838 novel *Oliver Twist* gave the world the character of Fagin, "a very old, shriveled Jew, whose villainous-looking and repulsive face was obscured by a quantity of matted red hair." Fagin's character very likely contributed at least as much to the negative associations of red hair as the prejudices that gave us red-haired Judas in the first place.[36]

---

35 John Gross, *Shylock: A Legend and Its Legacy* (New York: Simon & Schuster, 1994).

36 In his defense, Dickens did tone down the anti-Semitism in later editions of the book, and in his public readings of the part, but by then, one might argue, the harm was done. It's also intriguing if again deeply depressing to speculate how much the belief in the Jews as the exploiters of Gentile children might have influenced the creation of Fagin's "gang" of child thieves and pickpockets, too.

And how it still persists. The notorious "Ginger Kids" episode of *South Park*, broadcast in 2005, with Cartman proclaiming that "Gingers have no souls," and can't go out in sunlight (much to the fury of Kyle, possessor of a little-seen but magnificent red "Jewfro"), was only mining a much older tradition that Judas became a vampire and roamed the world as one of the undead. It's pretty much a perfect example of how such folklore accretes, one layer around another, with our own age merely adding the most recent. In 1887 the French geographer Élisée Reclus recorded a Romanian belief that "if the deceased has red hair . . . he would come back in the form of dog, frog, flea or bedbug, and . . . enter into houses at night to suck the blood of beautiful young girls." Vampires, so the thinking goes, are bloody; red is the color of blood, therefore redness must predispose one to vampirism.[37] Indeed, until John Polidori, Lord Byron's physician (and uncle, incidentally, of the pre-Raphaelite artist Dante Gabriel Rossetti), published his genre-defining short story "The Vampyre" in 1819, rather than being the pale-skinned, dead-eyed aristo embodied by the mysterious Lord Ruthven, vampires were given away by the ruddiness of their countenance, and the way in which it suggested that they were gorged with blood.

In fact the association between red hair and vampires noted by Reclus might hark back to even older folklore, as far back as Byron's Dacians, in fact, who were another Indo-European tribe in the region west of the Black Sea, and who may have been Thracian, originally, but with a greater influence from Scythia and the

37 For this and further detail on the evolution in European folklore of the undead, see Paul Barber, *Vampires, Burial, and Death: Folklore and Reality* (New Haven, CT: Yale University Press, 1988).

Celts. Or they may not. It's as well to take all this with a pinch of salt, especially as one of the earliest writers to interest himself in this area was one Montague Summers (1880–1948). Summers's personal *shtick* was to present himself as an academic witch-hunter, as opposed to the insane necromancer role inhabited by his notorious contemporary Aleister Crowley. (The two were acquainted, naturally enough—Summers was very possibly a Satanist himself, and very likely a pedophile as well.) Summers traces a Greek belief that those who commit crimes in life return as vampires to Slavic mythology, and he states that people with red hair and gray eyes are regarded as vampires in Serbia.[38] There are other associations as well that help explain this particular bit of mythology. Judas's putative wanderings as a vampire, rejected by both heaven and hell, mimic those of the Wandering Jew, while the vampire's invulnerability might suggest the Old Testament narrative of the Mark of Cain. The silver bullet that supposedly downs witches, werewolves, and everything else that goes bump in the night could suggest the thirty pieces of silver paid to Judas for betraying Christ. It all goes to show that you need neither the internet nor a butterfly to create a woozle.

The *South Park* "Ginger Kids" episode was the catalyst at a school in Yorkshire (more Northern redheads) for a seriously unpleasant "Kick a Ginger Day." So if this story has a moral, it's this: don't show sophisticated satire on the stupidity of racism to the stupid. Or as Ruth Mellinkoff puts it, "Red hair is a minority feature, and this fact is sufficient to explain why it is used in the visual arts as a

38 Montague Summers, *The Vampire in Europe*, reprinted Aquarian Press, 1980.

negative attribute and why it is still widely treated with suspicion. Antipathy to red hair, a red beard, and ruddy skin, is as simple, and as complicated as that."

Quite.

The French panel-painting is different from the German work in every way. It shows *The Coronation of the Virgin* and was commissioned for the Carthusian monastery of Chartreuse du Val de Bénédiction, Villeneuve-lés-Avignon, where it remains to this day, and where we can assume it has been since September 29, 1454, the delivery date specified in the contract for the work, drawn up between the clergyman who paid for it, one Jean de Montagny, and the painter, Enguerrand Quarton (Fig. 9).

The survival of Enguerrand Quarton's contract for his *Coronation*, and the astonishing detail this document goes into about the commissioning of the work, is one reason why the painting is so well known today—among students and scholars of medieval art, that is. (The other and much better reason is of course the painting's peerless, nigh-on flawless, beauty.) It is unusual for any document of this period to survive. Scholars of medieval art spend years scouring archives and parish records for a single mention of the artist they are researching, and the discovery of even the most abbreviated reference will generate years' more speculation and analysis. In this context, the commissioning document for Quarton's *Coronation* has the same kind of importance as the discovery of the El Sidrón family of Neanderthals. Among other details that the commission specifies are these: the overarching composition of the painting (there should be, it says, with unconscious poetry, "the form of Paradise"), the specific identities of the major saints to be included in that Paradise, that the Mass of St. Gregory was to be shown in the cutaway church lower left, and that in the representation of the Holy Trinity "there should not be any difference between the Father and the Son." It specifies the payment stages and how long the artist would have to complete the work. It even records where the document was drawn up and signed: in the spice shop of one Jean de Bria. It describes what the Virgin Mary should be shown wearing ("white damask"), and that the Holy Trinity should be surrounded by cherubim and seraphim, as indeed in the finished work they are. Yet in all these specifications it leaves the details of the appearance of the Virgin, the central figure of the painting, to the artist, "as it will seem best to Master Enguerrand."

In fact no fewer than six separate contracts for Enguerrand Quarton have been found, although only two of the works they describe have survived. We know or can extrapolate from the documents that he must have been born around 1420, that he was in Aix-en-Provence in 1444, Arles in 1446, and rented a house in Avignon in 1447. He seems (unusually, for an artist of this or perhaps any period) to have delivered the works commissioned from him without default. As far as we can judge, he moves seamlessly from locale to locale and patron to patron, from one work to a larger work, and from one fee to a greater fee. His first commission in Avignon was for a *Madonna of Mercy* for the Celestine convent there (the painting is now in the Musée Condé in Chantilly); the *Coronation* followed five or six years later.[39]

Everything about this painting is special. There is its glorious condition, to begin with. There is Quarton's palette—the fiery reds and deep cool blues for heaven, the pastels for our world beneath. There are the saints and priests and bishops, the holy martyrs, each one in this cast of hundreds differentiated from the next, each face as real and fully realized as any you might encounter in Avignon today. There are God the Son and God the Father, an almost identical pair of bearded brothers, clothed in red silk lined with green. And between them there is the Virgin, shown as seemed best to the artist. And how did she seem best to him? Not in white damask, for sure. Despite the stipulation in the contract her robe is cloth-of-gold, shot with a bold pattern of stylized blooms. And her hair is red.

There she kneels, on a kind of ermine of fluffy cloud and blue cherubs with her aristocratic, long-fingered hands crossed over her breast. Her head is tilted beneath the fluttering wings of the Holy Spirit and the weight of the crown being placed on her head by God the Father to one side and her son to the other. Her eyes are half lowered and her bright red hair is falling over her shoulders. In fact such a redhead is she that the color of her hair and the ivory of her skin can hold their own against even the background of burning seraphim. Nor as a redhead is she alone in Quarton's works. The Virgin of his *Madonna of Mercy* also has red hair visible under the cloak with which she shelters the faithful. She also has the same oval face, point

39 The last occurrence of his name is in 1466. The plague came to Provence in that year; Quarton, it is thought, may have been one of its victims.

to her chin, long, straight nose, fine eyebrows, and narrow, slightly Asian-looking eyes. Was she simply Quarton's ideal? Looking at the face of the Virgin in the Coronation, some might think they can see personality, challenge even, in the gaze with which she meets the viewer from under those half-lowered lids, a vivid comprehension of her role. Might that hair have been encountered for real in the streets of Avignon, six centuries ago?

Artists are among the most fervent admirers of red hair, and of the pale skin that so often goes with it. Hans Memling's triptych of *The Last Judgement* of *c.* 1467–73 has two, if not more, proudly nude female redheads, having been greeted by St. Peter, in the crowd dignifiedly walking up the crystal stairs of the painting's left wing to Paradise (Fig. 10).[40] They have their backs to us, and their rippling red hair reaches to their hips. The Archangel Michael in the same work, dividing the saved from the damned, has red hair, too. Red hair on women, on angels, is a thing of beauty, so these medieval artists seem to be telling us, except of course (they add) when it's not. So in these medieval paintings we have on the one hand the Queen of Heaven, in all her beauty and divinity, depicted as a redhead, and on the other hand we have Judas Iscariot, also portrayed as a redhead. Clearly in the one case red is good, and in

---

40 This work has perhaps one of the most adventurous life stories of any medieval painting. Commissioned by the Medicis' banker in Bruges, Angeleo Tani, it was being shipped to Italy by his successor, Tommaso Portinari (who may be the little figure in the left-hand bowl of St. Michael's dreadful scales), when the vessel carrying it had the bad fortune to run into the hands of a Polish privateer, Paul Beneke. Clearly a man of taste, Beneke presented the painting to the Church of St. Mary in Gdansk, and there, despite lawsuits demanding its return, it remained until looted by Napoleon in 1807. After Napoleon's fall, it found itself in 1817 in Berlin, from where St. Mary's Church was able after some effort to reclaim it; after World War II it turned up in Leningrad, and was finally returned to Gdańsk in 1956. It is now in the National Museum.

the other it is about as bad as bad can be, and the question begging to be answered is *why? Why* is red hair so gendered? Why is it so different for girls?

Here is another painting: this time in the Musée Unterlinden in Colmar, and that enjambment of French and German in the museum's name should hint at a good deal of Colmar's history. Founded in the ninth century, the city nudges up against the Franco-German border, which has swung back and forth over Colmar like a jump rope. It was a part of the Holy Roman Empire until 1673, when it was conquered by Louis XIV. It came back under German rule in 1871, as a result of the Franco-Prussian War, returned to France in 1919, was annexed by Nazi Germany in 1940, and returned to France again in 1945. With so much war being waged around it for so long, you might think very little of old Colmar would survive, but you'd be wrong. The city is as pretty as a film set, its elaborate timber-framed architecture, with witch's hat roofs and bold carved sandstone doorways, suggesting quite how much money might be made out of being a border trading post in the fifteenth and sixteenth centuries.[41] The wines are delicious, the cuisine heart-stoppingly good, the people pragmatically bilingual. In 1834 it was the birthplace of Frédéric Auguste Bartholdi, designer of the Statue

41 The Maison Pfister in Colmar was in fact in a movie—it is the inspiration for *Howl's Moving Castle* in the eponymous animation from Studio Ghibli (2004).

of Liberty, and in 1450 or thereabouts, the birthplace of Martin Schongauer, whom one may speak of as the foremost German artist of his generation, and whose workshop (every successful artist of the period created a workshop of assistants and apprentices) in *c.* 1480 created the panel painting shown in Fig. 11. It too blurs boundaries.

Once again, we are in a garden, but a very different one from that of Gabriel Angler's *Betrayal.* The wattle fence is solid, the gate is of knotty two-by-four, painted with such solid realism you could build a copy of it today. Like the lych-gate to a church, it has a little roof. The sky is neither dark with night nor overpaint, but gold—this is a holy space—and the ground is green with grass as lush as fur. It might be a latter-day Garden of Eden. There is a rose bush, blooming, what may be a tree peony. There are birds in the branches of the trees. One looks like a wagtail and the other a chaffinch, in which case, because no detail in paintings of this period is accidental, the wagtail may symbolize earthly love and beauty (the bird was linked in Classical mythology to Aphrodite), while the chaffinch was a symbol of celibacy. They create a counterpoint to the figures in the foreground, the Adam and Eve of this place: a woman kneeling on the grass, her hair in thick coppery ringlets, and a man, looking back at her even as he walks away. The man is Christ, carrying the banner of the Church and dressed in a triumphant scarlet toga, arranged to display the sacrificial wound in his side (the same wound, incidentally, as the *Dying Galatian*—how these things repeat themselves); the woman is Mary Magdalene. This is the *Noli me tangere*, the moment after the Magdalene's recognition of the risen Christ outside the tomb. For my money it is one of the most psychologically complex moments in all Western art, and the figure of the redheaded

Magdalene, as Western art and literature has created her, one of the most multilayered and compelling—and, I would argue, the single most important reason why Western attitudes toward redheaded men and redheaded women diverge so thoroughly. Accidents of history have given redheaded men the stereotypes of barbarian warriors, clowns, milksops, arch-traitors, or violent God-damned brutes. It has given redheaded women Mary Magdalene.

The Magdalene as she has come down to us today is a conflation of at least four figures from the Biblical story. There was Mary Magdalene herself, present at both the Crucifixion and the Resurrection, and the first to whom the risen Christ revealed himself. There was Mary of Bethany, sister of Martha and Lazarus, who according to the Gospel of St. John had anointed Christ's feet with perfume and wiped them with her hair. And there is the unnamed woman who in Luke's gospel approaches Christ in the house of the Pharisee and does the same, and who is described only as "a sinner." There was another Mary, whom Christ cured of demonic possession in the Gospel of St. Mark. Finally, Mary Magdalene's medieval biography, in its best-known telling, by the thirteenth-century churchman Jacobus de Voragine, seems to owe something to the story of Mary of Egypt, another prostitute-saint, who wandered the desert clothed only in her hair.

The problem with the Bible, certainly for the early fathers of the Christian church, who saw truth as having to be one narrative or as being nothing at all, is that so often the sources they were teaching and preaching from were various and contradictory. Nor was this a problem confined to Christianity. The Talmudic *Alphabet of Ben Sira*, of the eighth to tenth century, conjures into being a first wife for Adam, pre-Eve, to explain the fact that in the Book of Genesis a "wife" is mentioned twice: first at Genesis 1:27:

> So God created man in his own image, in the image of God created he him; male and female created he them.

and then at 2:22:

> and the rib, which the LORD God had taken from the man, made he a woman, and brought her unto the man.

Out of this minor inconsistency, and a good deal of padding from earlier Jewish and even Babylonian mythology, was created the legend of Lilith, the woman who, seeing herself as her husband's equal (unlike docile Eve, borne of his rib), refuses to "lie beneath," quarrels with Adam, and strikes out into the Babylonian wilderness on her own. Disputatious and disobedient therefore, and yes, to this day, very often depicted with red hair.

Such painstaking elaboration to explain what was basically even less than a typo might strike us as hardly worthy of the effort, but to the first Catholic popes of the post-Classical world, pressed by the Goths on one side and the power of Byzantium on the other, yes, it was. In 591 Pope Gregory the Great (he of the Gregorian chant) decreed in his Homily 33 that the unnamed sinner of the Gospel of St. Luke, Mary of Bethany, the woman cured of seven devils in the Gospel of St. Mark, and the Mary Magdalene of Christ's earliest followers were one and the same.

> She whom Luke calls the sinful woman, whom John calls Mary, we believe to be the Mary from whom seven devils were ejected according to Mark . . .

> . . . It is clear, brothers, that the woman previously used the unguent to perfume her flesh in forbidden acts. What she therefore displayed more scandalously, she was now offering to God in a more praiseworthy manner. She had coveted with earthly eyes, but now through penitence these are consumed with tears. She displayed her hair to set off her face, but now her hair dries her tears. She had spoken proud things with her mouth, but in kissing the Lord's feet, she now planted her mouth on the Redeemer's feet. For every delight, therefore, she had had in herself, she now immolated herself. She turned the mass of her crimes to virtues, in order to serve God entirely in penance.

And what a winning combination this sinful woman, this "peccable female," was to prove to be.

According to Jacobus de Voragine in the *Golden Legend*, his best-selling compilation of the lives of the saints, Mary Magdalene was the daughter of wealthy parents, and "shone in beauty," as de Voragine puts it (or rather as William Caxton translated him, in 1483). Her sins were her pleasure in her looks and her riches, and her giving up of her body to "delight," rather than to straightforward prostitution. Female beauty, female wealth, and, above all, female delight were deeply suspect to the medieval Christian church (there are times when it feels they are hardly less so today), and sure enough, they are Mary Magdalene's downfall. Inspired by the Holy Ghost, she is filled with remorse for her sins, renounces her wealth, buys a pot of the costliest ointment (it's the object looking like a miniature fire hydrant by her knee in the Colmar panel) and, penitent and weeping, approaches Christ in the house of the Pharisee. There follows the well-known story of the tears, the feet, the ointment, the hair. After the Resurrection, in a crackdown by the authorities, Mary Magdalene and companions, including St. Maximin of Aix, are cast into a rudderless boat and set adrift, washing ashore at Marseille, where she begins to preach. There is the conversion of the local prince, there are miracles and visions, and once the prince has destroyed his heathen temples, Mary Magdalene retires to "a right sharp desert," as Caxton feelingly describes it, where she spends the next thirty years in solitary contemplation, sustained only by her faith and by a daily angelic "assumption," where choirs of angels sing to her and feed her. She dies after receiving a last communion from St. Maximin, is herself anointed "with divers precious ointments," and buried. By the tenth century her relics were to be found many hundreds of miles inland, at Vézelay in Burgundy.

But holy relics were big business in the Middle Ages, and in 1279 Charles II, Prince of Salerno, announced that he had been told in a dream that the true relics of Mary Magdalene still lay within the chapel of St. Maximin, near Aix-en-Provence. Charles was involved for most of his life in the tripartite power plays between the kingdoms of Aragon, Naples, and Sicily for control of the latter (a centuries-long dispute that can be traced back to the death of Frederick II), but he was also Count of Anjou, whose territories included Aix, and the details of his dream, were they to be proved true, would give him a valuable degree of both divine and papal endorsement. So a tomb in the crypt of St. Maximin was opened, the relics were indeed discovered (unperished and exquisitely perfumed), and their approval by the pope, Boniface VIII, was sought and granted. Charles founded a massive pilgrim's basilica on the site, the saint's remains were enclosed in a gold reliquary, complete with rippling beaten gold hair, and the cult of the saint began to spread, first via Naples into Italy and then throughout Europe.

So much for the history. Let's talk about the woman.

❋

How does one begin to unpack the symbolism around Mary Magdalene, to analyze what it was that made her so significant, so ubiquitous and so beloved a figure in the medieval mind? Where do you start? There is her narrative to begin with—the wealth, the privilege, come to naught, rejected for something greater.[42] This is a storytelling arc to which something very deep and very human in us still responds today—that combination of schadenfreude evolving into empathy and then even into admiration drives most reality TV, for example. There is her role, even her posture, as the tearful penitent, the seeker of reconciliation, the rebellious female brought to heel, contrite and come to make up, as if to a lover—on her knees at Christ's feet, wiping them with her hair, or in its counterpart, kneeling in the garden, reaching up toward him—and the fact that this scene takes place in a garden makes her also, of course, the antidote to the first peccable female, Eve. Christ touches her on her forehead, which might be taken as an antidote to the Mark of Cain, as well. Then there is her own bittersweet agony of "*Noli me tangere*," the role reversal of look-but-don't-touch, the way in which, although it is the Magdalene kneeling, thus apparently in the subservient posture, her reaching out toward Christ might be seen as placing him

42 St. Francis of Assisi, who is sometimes cited as the male version of Mary Magdalene, shares the same kind of narrative arc to his life, and is another saint with now more than pan-European appeal.

in the role normally reserved in the medieval world for the unattainable female object of adoration or desire (you see what I mean about the Magdalene's compelling nature?). And then there are her tears. The Magdalene weeps, she feels, she has human emotions and artists depict her as giving them vent—gathering her tresses around her in Titian's *Penitent Magdalene* of 1533, now in the Palazzo Pitti in Florence, brimming eyes turned up to heaven. Or, a hundred years earlier, in Jan van Eyck's *Crucifixion* (Fig. 12, *c.* 1428, Metropolitan Museum of Art). Here the Virgin slumps bottom left, face hidden, literally shrouded in grief. Mary Magdalene is beside her with her back to the viewer (and yes, again, the fall of rippling red hair) raising her arms as if opening her body to the full agony of the scene, knitting her hands together, imploring heaven to intercede.

And then there's her sensuality and human passion.

Red hair in your female mate might be a relative guarantee of successful childbirth and healthy heirs, but there is surely more at play here than that. Red is the color of blood. One of the most ancient slurs thrown at redheads is that they are the product of sex during menstruation, in itself one of the oldest of sexual taboos. And if red is the color of blood, it is thus also the color of passion—whether of rage or of erotic arousal. And it is the color of fire. "Do not," thunders St. Jerome, in AD 403 in a letter to one Laeta, concerning her daughter Paula, "do not dye her hair red, and thereby presage for her the fires of hell," thus neatly linking all three. There seems to be no specific historic connection between red hair and prostitution, but there is a connection with the color and the profession, going back to the Old Testament, the Book of Joshua, and Rahab, the whore with a heart of gold, another peccable female who redeemed herself by hiding Joshua's spies in the roof thatch of her house in Jericho (more hair symbolism, one wonders?), and when Joshua's troops sacked the city, they hung a red cord out of her window to identify her house, and thus she and her family were spared. It's suggested that Rahab's cord may be the derivation of the red-light district, as if the color and its other associations were not already sufficient explanation. Red stands out. If as a streetwalker you wish to be noticed, it's a choice, and for many centuries it was also one of the hair colors easily achievable through natural dyestuffs, in this case henna. And it is the Magdalene's color. Rogier van der Weyden, painting Mary Magdalene in the guise of a nun in *c.* 1445, in the left-hand panel of his *Crucifixion*, now in the Kunsthistorisches Museum in Vienna, clothes her in black and hides her hair under a wimple, but nonetheless gives her a flash of red underskirt. You might speculate that artists also used the association to contrast the Magdalene with the blue robes of the Virgin. Equally it's been argued that the color red in the Middle Ages symbolized *caritas*, or the love of God (this at the same time, mark you, as it was symbolizing the complete opposite for the figure of Judas). Then there are very definite associations between long or loosened hair and sex—medieval virgins wore their hair loose, wives (whether Gentile or Jewish) wore it bundled away—indeed, one means of shaming a suspected adulteress in Jewish society was by exposing and unloosing her hair. We all know the cliché of the dowdy secretary, relieved of her glasses and with her hair out of its chignon, before whom her (male) boss stands amazed, exclaiming "My God, Miss Peabody, you're beautiful!" Is this where it began?[43] It is as if by unloosing the hair, one was setting free the sexual nature, too, and prostitutes in this period did indeed wear their hair loose on the street, as no respectable, sexually mature woman would have done. (And there were a lot of them—in the sixteenth century the diarist Marin Sanudo estimated that in Venice there were 11,000 prostitutes working in a city with a total population of some 100,000 souls.) Prostitutes were tolerated in cities such as Venice as representing a lesser vice than that of sodomy, in an age when economic circumstance forced men to marry late, and sex outside marriage all too often resulted in children born outside marriage, too. This was also the period when syphilis first appeared in Europe, very probably brought back from the New World by Christopher Columbus's crewmen, and it has also been suggested that the enormous popularity of Mary Magdalene as a saint mirrors the rise in the disease and in the number of prostitutes working the streets and at risk from it.

Yet connect the streetwalker's attribute of long loosened hair with Mary Magdalene, make it red, and its meaning is turned upside down. Instantly it becomes not only how we recognize her but socially acceptable and even desirable. Piero di Cosimo's *Magdalene* of *c.* 1500 would simply be a portrait of a woman reading at her window (here's another attribute of Mary Magdalene that I like, the presentation of her as a thinking creature, a woman with

43 There is a very similar meme in Indian culture, where the unloosing of the woman's knot of hair presages the unloosening of the sari, too.

Fig. 1: This portrait of a redheaded Afghan girl was taken in 2004 by REZA in the Pashtun tribal zone in Afghanistan.

Fig. 2: Uyghur girl photographed in Kashgar, in China's Xinjiang Uyghur Autonomous Region. The ethnicity of the Uyghur is Turkic, rather than Han Chinese. They see themselves as an occupied people, and identify both ethnically and culturally with the non-Chinese civilization of the Tarim mummies rather than with that of Bejing, referring to their region as "East Turkestan." Recently there has been ethnic violence in the region, bombings and knife attacks, indiscriminate as these events always are in the victims they have claimed.

Fig. 4: The red-haired divinity from coffer 32 in the ceiling of the Ostrusha tomb in central Bulgaria. Dating back to 330–310 BC, it is one of the earliest representations of a redhead in Western art.

Fig. 5: King Rhesos and a very rude awakening. This Chalkidian black-figure amphora was made in southern Italy about 540 BC. The artist, known to us as the Inscription Painter, uses a red glaze for Rhesos's hair and beard. As the story was recounted in Homer's *Iliad*, Odysseus and Diomedes infiltrated the Thracian camp outside the walls of Troy, hoping to steal their fine horses, one of which can be seen being led away to the left.

Fig. 6: This tiny terracotta figure, just 13 centimeters high, of an actor playing the part of a runaway slave, was made in Athens between 350 and 325 BC. It came to the British Museum from the collection of Eugene Piot (1812–90). Piot squandered a fortune collecting treasures from across the Classical world. Microscopic remains of red paint still adhere to the figure's terracotta hair.

Fig. 7: Gabriel Angler the Elder, *The Agony in the Garden*, 1444–45, Bayerisches Nationalmuseum, Munich. The altarpiece this panel comes from was made for Tegernsee Abbey in Bavaria, where, four centuries before, in the eleventh century, an unknown monk wrote the *Ruodlieb*, one of the earliest knightly epics, containing one of the first warnings against men with red hair: "A red beard rarely hides a good nature. . . . "

Fig. 8: Antonello da Messina, *Calvary*, 1475. The well-tended landscape, the figures of the Virgin and St. John the Evangelist, even the central figure of Christ on the Cross seem oblivious of the agonies going on around them to left and right. Note, however, the point being made by the artist in showing the impenitent thief's red hair. Royal Museum of Fine Arts, Antwerp.

Fig. 9: Enguerrand Quarton, *The Coronation of the Virgin*, 1454, Monastery of Chartreuse du Val de Bénédiction, Villeneuve-lès-Avignon. The Carthusian order is closed and silent, with monks undertaking a life of prayer and contemplation, each within his own cell. Each cell, however, has a garden, and once a week the community walks in the countryside. An atmosphere of calm and quiet, and of love of the world while being removed from it, permeates Quarton's painting.

Fig. 10: Hans Memling's triptych of *The Last Judgement*, *c.*1467-73. Memling was around forty years old when he painted this work. Born in Germany, he lived in Brussels and Bruges, and his paintings, which represent one of the summits of Early Netherlandish art, feature redheaded Virgins as well as angels and here, two souls of the saved, walking up the crystal stairs to Paradise.

Fig. 11: Studio of Martin Schongauer, *Noli Me Tangere*, *c.* 1480. Schongauer, like Memling, was a follower of Rogier van der Weyden. His works, like Memling's, are prized for their use of color and the quality of their execution. And like Memling, Schongauer created redheaded Virgins, saints, and angels, as well as Magdalenes.

a contemplative, hungry mind), were it not for the jar of ointment on the windowsill with her book, and her beautifully arranged and lushly painted long red hair, twined here with pearls. If the Magdalene is the subject, even her nudity is acceptable; and far from this being frowned on by society, is welcomed, is celebrated, putting it too strongly? Titian's Magdalene is supposedly oblivious to her naked breasts and the nipples so teasingly presented through her protective gathering up of her hair, yet we, as the viewer, can hardly be imagined so to be. We are being encouraged by the artist to gaze to our satisfaction and enjoy. This is the Magdalene as *Venus Pudica*, drawing our attention to the very thing she pretends to hide.[44] And so an age-old connection—one that we will return to again— between red hair in women and sexual desirability is pushed a little further.

And just to show how far it could be pushed, here is one final example of a painting of the Magdalene. In this she's lost her jewels, her pot of precious nard, every stitch of clothing, all habitation, every connection to the world, but if she has her red hair, we recognize her, and her original audience (almost unbelievably) found her acceptable still. In 1876 Jules Joseph Lefebvre paints her alone and naked lying at the entrance to a cave (Fig. 13). Or rather, she is his excuse for painting as erotic a nude as a salon artist in the nineteenth century

44 There is also a winningly eccentric painting of the *Elevation of Mary Magdalene* by Jan Polack (*c.* 1435/50–1519), supposedly showing her in the desert, clothed in her hair, but instead employing and drastically misunderstanding the "feathered tights" convention of costume in medieval drama to display her instead in a little furry cat-suit, with peekaboo cutouts for her breasts. The Magdalene's hands are curved about them, in a gesture that pulls the eye toward the very thing that it pretends to hide. See http://commons.wikimedia.org/wiki/File:Mary_Magdalene_01.jpg.

could expect to get away with, and still be invited to those salons, that is. Let's be blunt: he paints a naked woman laid out across a rock. One of her legs is drawn up; if her audience (they should be so lucky) could have moved around her even by another step she would be as totally exposed as the model in Gustave Courbet's infamous *L'Origine du Monde* (a painting which also has a unique place in the history of the redhead in art). She has a delectable body: slim-legged, round hipped, high-breasted, and her pose is calculated to display it at its most alluring. Her skin is pale, pale, pale; and Lefebvre has so arranged her that she doesn't even have a face to trouble her audience with. She's lost even that—one arm is raised to hide almost all her features but her forehead. So she's not even a whole nude; she's simply a body, save for two things: the stem of thorns that trails across her calf, and the immense spill, down to her waist, of her shimmering copper-colored hair.[45]

In the run of history redheaded women may have reason to be more grateful to Mary Magdalene than not, but we're a long way from the Queen of Heaven, that's for sure.

45 Lefebvre liked redheaded models, so far as one can judge. His other redheaded nudes include *Diana* (1879), *Undine* (1881), *Pandora* (1882), and the undated *Fleurs des Champs*. In 1870 he painted *La Verité*, a dark-haired model posed nude with one arm raised, who seems to have influenced Bartholdi's pose of the Statue of Liberty. Jean-Jacques Henner (1829–1905) is another French painter, more or less contemporary with Lefebvre, who favors redheaded models, but depicting them with his characteristic chiaroscuro and with rather more dignity.

THE EXCREMENT OF THE HEAD

Each people has its own barbarians.
HERODOTUS, *HISTORIES*

London, Christmas 1578, and the translator and editor Ralph Holinshed is learning just how difficult it can be to please all of the people all of the time.

In the best tradition of projects from hell, the one now causing him such concern shouldn't even have been his. More than twenty-five years before, Holinshed, then a stripling of twenty or so, had been taken on as an assistant by one Reyner Wolfe, a Dutchman who had arrived in London in 1533 and set himself up (with some success) as a bookseller and printer. The publishing world was as hungry for the next big thing then as it is now, and in 1548, a year after the death of Henry VIII, and one year into the reign of his son, the boy-king Edward VI, Wolfe had hit upon a plan for a publication that would leave his rivals in the shade.

What Wolfe had come up with was the concept of a "Universal Cosmographie"—a kind of Tudor Wikipedia. This mighty work would contain in two volumes a complete and newly written account of the history of every nation then known, from the Biblical

Flood up to Wolfe's own present day. With what one can only view as heart-warming Anglophilia, Wolfe had decreed that volume one would be devoted solely to the history of Britain, while volume two would deal with the rest of the planet. Now, two and a half decades later, the project has finally been published. In the intervening years, King Edward VI has died at the age of fifteen and been succeeded by first his cousin Lady Jane Grey, the pitiful "Nine Days Queen," then by his half sister Princess Mary, or "Bloody Mary," as she became known. In 1558 Mary in turn had been followed to the throne by her half sister, Elizabeth I. And volume one of the "Cosmographie" had grown. The description and history of England, Scotland, and Ireland were now the entire work. And Wolfe too had died, in 1573, and his backers and his widow had handed the entire project, plus its contributors, over to Holinshed (with, one imagines, a mighty sigh of relief) to bring to completion. And astonishingly, twenty-five years after Reyner Wolfe first came up with the idea, Holinshed's *Chronicle*, as it has ever since been known—one of the largest books that has ever been printed in England—is a best-seller.

Just as well, too. Best guess is that the work took its printer a year and a half simply to typeset. We can imagine many a quill-penned cost sheet being argued over, many a candle being burned low into the small hours, but nonetheless demand and interest were enough for its backers to press on. By 1584 it was clear that the original edition would sell out (a highly desirable state of affairs for any publication, then or now), and the *Chronicle*'s backers would set about creating a new and even more extensive revised version, which Shakespeare would famously use as a source for *Macbeth*, *King Lear*, and for his history plays. But all that is in the future. Right now, the *Chronicle* and all those involved in it are looking at a significant problem. On December 5, 1578, the Queen's Privy Council bans all further sales of the book, and Richard Stanyhurst, one of Holinshed's writers on the *Chronicle*, is summoned to appear before them. Plainly, something among the book's two-and-a-half-million words has offended someone.

This was not quite such a heart-stopping prospect as it would have been during the reigns of Elizabeth's father or her sister, Mary, when such a summons was all too often the start of a short walk to imprisonment and the gallows, but it would nonetheless have been enough to give any publisher a sleepless night, with nightmares of bonfires of books on Tyburn Hill, and the publisher himself facing a hefty fine or even worse. The consequences could be ruinous.

And as Holinshed must have been all too aware, the *Chronicle* could not but, in places, sail close to the wind. It came right up to the (then) present day; thus it not only had to deal with the execution of Elizabeth's mother, Anne Boleyn, in 1536, it also had to reach some kind of a summation of the reign of her father (not to mention those of her half brother and half sister), when the dust was still settling from the social and religious upheaval of the Reformation, and with Protestant England still defining its position on the edge of a largely Catholic and more or less hostile Europe. With this the temper of the times, it's unsurprising that the lines around who was your friend and who were your enemies were drawn in thick and black. This is one reason why Holinshed's *Chronicle* is so much a part of our story here: its recording of Elizabethan England's attitudes toward its "other," in the shape of the Irish and the Scots. But before that there is its place in the iconography of one of the most famous redheads in history: Elizabeth herself.

One of the innovations of the 1577 *Chronicle* is that it includes illustrations—woodcuts—integrated at relevant points into the text. Thus we have Macbeth meeting the witches in the account of his reign, or in the "Description of Ireland," an Irish chieftain feasting his boisterous retinue out-of-doors. Many of the illustrations are charmingly anachronistic (Macbeth's witches wear a sort of Elizabethan masque costume; Macbeth himself is in ribbon breeches and a fashionable beaver hat), a fact that seems not to have troubled Holinshed at all. The images are there, so far as we can judge, simply to help the reader visualize what is described in the text, not to represent it. They also appear to have been pretty crudely made; in many cases the printer simply reused existing blocks. The artist or artists of those newly made for the work hasn't been identified, although Marcus Gheeraerts the Elder (*c.* 1520–*c.* 1590) has been suggested as their designer by some. Gheeraerts was noted as an etcher and printmaker and if he really did have a hand in the *Chronicle*'s woodcuts, there must have been a considerable distance between design and execution, because the illustrations in the *Chronicle* are definitely at the cheap and cheerful end of Tudor art. But then that is in the nature of woodcuts: reused, recut, and much repeated, they are the artistic equivalent of a game of telephone. Details get exaggerated, others are lost altogether. You can never be quite sure if what you see is what the artist originally intended.

For example, one of the most well-known portraits of Elizabeth, even among those for whom a copy of the *Chronicle* was out of reach, would have been the frontispiece to the *Queen's Prayer*

Fig. 14 The frontispiece to the Protestant *Queen's Prayer Book* of 1569, showing Elizabeth I at prayer, in a woodcut supposedly by Levina Teerlinc. These images took the place in the book of traditional illuminations. Their presence is telling evidence of how the Anglican Church was still feeling its way toward a public image.

Book of 1569, showing Elizabeth herself at prayer. This was supposedly cut by one Levina Teerlinc. Elizabeth, unlikeher father, had no official artist (you can't help but suspect this was because she wanted no one in charge of the queen's image but the queen), but at this date Teerlinc was perhaps the closest thing to it, in which case the clumsiness of this image of the queen is all the more extraordinary Elizabeth, the Virgin Queen, facing right, and neatly takingthe place that would otherwise have been reserved for the Virgin Mary in what has been called a Protestant Book of Hours, is shown in profile, with trademark nose, and a forehead as high and bald as an Aztec priest (Fig. 14). It was fashionable to pluck the hairline, yes, but this is absurd. Yet whoever cut the image of Boudicca addressing her troops for the *Chronicle* had seen it, giving Boudicca the same profile (Fig. 15). They give her the loose, flowing hair of Herodotus's description of the queen of the Iceni, but also that of Elizabeth's own "Coronation portrait," where her loosened hair signifies her youth and virginal status, and they place a knopped crown, with open arches and very similar to the one Elizabeth wears in that painting, on Boudicca's head.[46] There is no attempt to represent Boudicca in Celtic robes, any more than the men-at-arms she is addressing are made to look like Celtic warriors. Boudicca wears what looks like a sumptuous embroidered Elizabethan gown, while her soldiers are armored as would have been any crack regiment in Elizabeth's own army. It's difficult to imagine any Elizabethan looking at this image and not seeing in it a reference to their queen and their armored men of war, defying not the might of Imperial Rome but that of Hapsburg Spain. England was as terrified of invasion in 1578 as it would be in 1940. And if it wasn't the Scots, threatening to let the French in down the chimney, it would be the Irish, letting the Spaniards in through the back door.[47]

46 The "Coronation portrait" of the queen can be seen today at the National Portrait Gallery in London, or on their excellent website, as NPG 5175. The painting is now thought to be a copy, made *c.* 1600, of a lost original of *c.* 1559. The cloth of gold Elizabeth wears in the portrait had also been worn by her half sister, Mary I. It's not so far removed from the brocade in which Enguerrand Quarton dresses his Queen of Heaven for her Coronation.

47 The whole of Europe knew how the Spanish treated those they saw as heretics, and while it's most unlikely anyone in England cared one way or the other how Spain dealt with its Jewish or Moorish *conversos*, let an honest Englishman find himself caught in the thralls of the Inquisition, as happened to one Robert Tomson in 1556, and the outrage was national. Tomson's account (another Tudor best-seller) was republished in G. R. G Conway's *An Englishman and the Mexican Inquisition, 1556-1560*, Sidewinder Studies in History and Sociology, 1997.

Fig. 15 Boudicca addressing her troops from Holinshed's *Chronicles* of 1577. Note the very Elizabethan-looking armor on her men-at-arms. The hare is being used as a means of prognostication; the little scene in the tent in the background seems to record the treatment of the queen and her daughters at the hands of the Romans. But to the Irish of this period, the invaders were the English themselves.

By 1578 Elizabeth I, also known as "Gloriana" and "the Virgin Queen," the archetype of a Protestant monarch, let alone of a Protestant female monarch, was in the twentieth year of her reign and the forty-fifth year of her life. This was way past one's prime as a woman in the sixteenth century, yet to look at Elizabeth's portraits, you would never know it. It may have been quietly accepted by her subjects that their queen would never marry, that she was beyond the age when there would be any Tudor heirs, just as it seems to have been apparent to all by this point that Protestant England was heading slowly and steadily into war with Europe's superpower, Catholic Spain. According to Elizabeth's image and her image-makers, however, none of this had happened. Time had stood still. The queen was still as young and wrinkle-free, as limber, as on her accession day. Only the magnificence, if one may put it that way, of her presence had increased as the years had passed.

One of the earliest portraits of Elizabeth was a gift, it is thought, from her to her brother, Edward, and shows her as a girl of sixteen or so. Psychologically it is extremely astute: Elizabeth looks both tightly wound and incapable of being fooled by anything.[48] It shows her with smooth, gingerish hair, parted in the center (although more or less hidden under her Tudor gable-style headdress), and with her mother's dark brown eyes—sadly wary in their expression, in her daughter's case. Anne Boleyn had been a glossy brunette but obviously must have been carrying that redheaded gene; Elizabeth's red-haired father, Henry VIII, was, as Holinshed describes him, "in his latter days, somewhat gross, or as we terme it, *bourly*." Burly, that would be, but with a hint of other meanings to it, too: "boorish" being one, and boar-like another—short-tempered, unpredictable, highly dangerous, and like Chaucer's miller, with his inner nature announced by his red hair and bristles. Given that he had both beheaded her mother and declared her illegitimate, you might have thought Elizabeth would have been tempted to distance herself visually from her father, but no. Elizabeth's public image, once she became queen, is a celebration both of redheadedness and of the pale skin that so often goes with it, and it is an elective one. Once the gable headdress was out of fashion, Elizabeth's hair is as public as the rest of her—curled and tight to the head, almost boyish, when she first became queen, in keeping with the somewhat masculine

48 The portrait is part of the Royal Collection and hangs at Windsor Castle. See http://www.royalcollection.org.uk/collection/404444/elizabeth-i-when-a-princess.

fashion for tight doublets and high collars in the early Elizabethan period. Then, as Gloriana's wardrobe became ever more elaborate, those tight, tight curls ascend to heights that are almost Pharaonic (Fig. 16). Ordinary people did not have hair like this, any more than ordinary people wore foot-high collars to set off their head with the spread of gauzy wings. Of course they did not; they were not the queen. Unsurprisingly, most of these later hairstyles, glinting with diamonds and pearls, are wigs. Unless Elizabeth wished to spend half the day under the hands of her hairdresser, wigs were the only way to achieve such a coiffure, and in any case, how would natural hair support the weight of those jewels?[49]

Bewigged, therefore (she is said to have owned eighty), Elizabeth might have sported hair of any color she wished, yet she chose red—rarely a popular choice, and most likely more popular in England in her day than at any time up until our own. Red hair and pale skin were the Elizabethan brand, if you like, and those courtiers not blessed genetically, and all those many more wishing to copy the fashions of the ruling class, might dye their beards red if men (there is a splendid portrait by Marcus Gheeraerts the Younger of Elizabeth's favorite, the Earl of Essex, with a fashionably square-shaped, fox-red beard), or if women, change the color of their hair in order to emulate the queen with the use of such folksy tinctures as rhubarb juice, or the rather less attractive-sounding oil of vitriol (that would

49 Elizabeth's use of wigs has prompted speculation that she lost her hair when she fell ill with smallpox in 1562. This seems to be based on an errant bit of historical supposition started by F. C. Chamberlain in 1922. Elizabeth did however go gray, if the evidence of the lock of her hair preserved at Wilton House is trusted. Redheads "go gray" just as much as any other hair color, but if you're lucky, with red hair the unpigmented hairs are to an extent disguised, and simply look like fairer hair among the red.

be sulphuric acid to the rest of us). Elizabeth is even, at one time, said to have dyed the tails of her horses orange.[50] For the pale skin there was of course white lead, splendid for giving the skin a satiny white finish, and horribly injurious to health in any degree of contact whatsoever. We can hardly assume the Elizabethan court was ignorant of this, as the hair fell out and the skin withered, and the headaches and tremors took possession of the first ladies-in-waiting to succumb to lead poisoning. But for Elizabeth, the use of such cosmetics may have been a necessity. She may not have naturally enjoyed the ethereal, moonlike pallor with which she is depicted, and which so often accompanies red hair. In 1557 the Venetian ambassador Giovanni Michieli described the queen-in-waiting as having "good skin, although swarthy." Another Italian diplomat, Francesco Gradenigo, describes her in 1596 as "ruddy in complexion." Possibly Elizabeth was sensitive about her apparent high color; Sherrow's *Encyclopedia of Hair* records her as having asked whether her hair was superior to that of her cousin and rival, Mary, Queen of Scots (in 1578, languishing in the eleventh year of captivity in England), and if her skin was fairer. In her happier youth, Mary, Queen of Scots, had been married to Francis II of France, and there is a memorable portrait of her painted by the French artist Francois Clouet in 1560, showing her in all-white mourning following the deaths of her own mother and of her father-in-law, the French king Henry II. It shows

50 Victoria Sherrow, *Encyclopedia of Hair: A Cultural History* (Westport, CT: Greenwood, 2006). It didn't do to emulate the queen too successfully, however. Lettice Knollys had married another of Elizabeth's favorites, Robert Dudley, in September 1578, without royal permission. To judge from her portrait of *c.* 1585 by George Gower, Lettice was as head-turning a redhead as Elizabeth (portraits of the two can be extremely difficult to tell apart), and she was ten years younger. Lettice was banished from court, never to return.

her with skin almost as pale as her mourning veil and hair of a very dark red. However on her execution in 1587 at the age of forty-five, it was discovered that Mary too had resorted to wearing a wig, and that her natural hair was by then a close-cropped gray—"as gray as one of threescore and ten years old," in the words of Robert Wynkfielde, a witness to her death. You wonder how acutely aware these two women, cousins and once sister queens of sister kingdoms, might have been of each other's looks.

But why would Elizabeth create an image so frankly outlandish and dangerous to maintain? Part of the explanation might lie with Elizabeth's own complex psychology. She had been declared illegitimate by her father; therefore to parade his red hair so prominently was one way of giving the lie to that. Her mother had been beheaded and Elizabeth herself, during the reign of her sister, Mary, came close to the same fate. In Elizabethan portraiture there is a great deal of attention devoted to the head and the surroundings and the accoutrements of the head—jewels in the hair, earrings, the framing element of the famous Elizabethan ruffs of the 1580s, which put the head almost on a platter, John the Baptist–style; and then the enormous stand-away, look-at-me collars of the 1590s.

Another part of the reason may lie within the lines of the "Hymn to Astraea," a grandiloquently awful poem written for Elizabeth in 1599 by Sir John Davies, to celebrate the anniversary of her accession.[51] Davies was an Elizabethan heavyweight in every sense, and his flattery of the queen is not so much laid on as ladled:

But here are colours, red and white
Each line and each proportion right;
These lines, this red and whitenesse
Have wanting yet a life and light
A Majestie and brightnesse . . .

And so it goes on. There are twenty-six of these hymns, and the first letters of each line in each of them do indeed form an acrostic of the queen's name. But the colors Davies singles out as "being" Elizabeth—red and white—are also the colors not only of the Tudor rose, the imprimatur of the Tudor dynasty, but equally of the flag of St. George, the patron saint of England (as he is to this day) and the only saint whose banner remained in use in England after the break with Rome. The poet Edmund Spenser's epic *The Faerie Queene*, of 1590, would use the same badge of a "bloudie crosse" to distinguish its hero, the Redcrosse Knight, from the various personifications of foreign villainy that pop up throughout the poem. It is, admittedly, hard to read this work today without Monty Python's "Knights who say *Ni!*" creeping into your thoughts, but it still does a pretty good job of conjuring up Elizabethan England's xenophobia. And its queen's royal branding, in red and white (remembering also that white was the color of virginity, for a Virgin Queen), and

51 The portly Sir John died of apoplexy in December 1626, after what must have been an extremely good supper-party celebrating his appointment as Lord Chief Justice. He had been lobbying for the role for years. His wife, Dame Eleanor Davys, believed herself to be a prophetess, giving rise to the anagram *Never so mad a ladye.* But she foresaw the date of her husband's death and wore mourning for the three years before it came to pass.

the magnificence of her image-making, created an aligning of sovereign, symbol, and state of a different order to anything that had gone before. Ironically, in later portraits of Elizabeth, with a few rare exceptions, the splendid clothes and the jewels may be the only elements painted from life—the ivory face and crown of red hair are an icon, mass-produced, and instantly recognizable even today. The "Elizabeth I" brand is by that measure one of the most long-lived and successful in history.

Only toward the end of her life, when she was poignantly described by Sir Walter Raleigh as "a lady whom time has surprised," does Elizabeth seem to have varied the formula, with wigs just as high but paler, fairer in color: "her hair," so the German lawyer Paul Hentzner describes her, at Greenwich in 1598, "an auburn color, but false." In this case Hentzner seems to have been using the word "auburn" in its original sense of brownish-white. The term that would later become so handy and socially acceptable a catch-all for hair with any reddish tint to it at all first entered the English language around 1430, and comes from the Latin *alburnus*, or white. Only in the seventeenth century did its sense change, and the notion of it meaning a color more brown than pale become common. Ever since then it has been red hair's aristocratic first cousin, but the color Hentzner meant is probably that shown in Robert Peake's *Procession Picture* of the queen of 1601 (Fig. 17). Perhaps the pallor of seventy-year-old skin prompted the change, when the contrast with a red wig began to look too harsh. Elizabeth understood her coloring. Even today, any redheads wishing to know which colors they should favor in their wardrobe need look no further than her portraits.

The *Chronicle*, then, was created at a time when there was a sense of Englishness as being more than nationality, of it standing for some nebulous but estimable set of qualities that governed not merely a man's (or woman's) tongue but his or her beliefs and attitudes and values. It contains something of everything Elizabethan England knew or believed of itself. It is, if you like, a stethoscope, laid against the Tudor heart. Let the pope excommunicate us all (Elizabeth herself had been excommunicated in 1570); we knew better—not for nothing was the so-called Act of Supremacy so called.[52] There was no more Mary Magdalene to bring your woes to, no more pope, no more Church of Rome. Not for nothing, either, had a work that was to have recorded the whole world become one that devoted itself instead to describing one rainy island for the delectation of its subjects. Here too Holinshed caught the temper of the times: the narrative arc of the *Chronicle* is upward to the (then) present day, the "perfect monarchie," the happy ending represented by Elizabeth I—if, that is, this halcyon moment, this little earthly paradise, could only be preserved from its enemies, both abroad and those rather nearer home.

Holinshed's writer for the Scottish chapters of his *Chronicle* was one William Harrison. Harrison's life epitomizes the seesawing nature

52 The 1558 Act of Supremacy finally established the Protestant faith as the official religion of the state and made the monarch, instead of the pope, head of the Church of England.

of what it was to be English during this period. Born in 1534, he had been raised a Protestant; converted to Catholicism while a student at Oxford during the reign of Mary I; converted back again slightly before her death, and by 1559 was establishment enough to have been granted his first living as a clergyman. Harrison writes rather daintily, as if he were in conversation with his readers. For him, where Scotland was concerned, it was a question of united we stand:

> If the kingdoms of Britain had such grace given them from above as they might once live in unity, or by any means be brought under the subjection of one Prince, they should ere long feel such a favour in this amity that they would not only live frankly on their own without any foreign purchase of things, but also resist all outward invasion, with small travail and less damage. . . .[53]

He sounds like Herodotus bemoaning the state of Thrace, and it shouldn't be surprising that he does. Scotland and Ireland were as much England's "other" as Thrace had been to Greece. And there is that fear of invasion again. Sadly for Harrison, there was not much to be expected of England's closest neighbor; for him, the abiding characteristic of the Scots is that they are drunks. Although he describes them as otherwise "courageous and hardy. . . . They cannot refrain from the immoderate use of wine," with the result that "if you knew them when they be children and young men, you shall hardly remember them when they be old and aged . . . but rather suppose them to be changelings and monsters." These are the lowland Scots.

53 The Holinshed Project gives the text of both editions of this work, and just about every tool a researcher could ask for to search through them. See http://www.english.ox.ac.uk/holinshed.

Highlanders fare rather better, being described as "less delicate and not so much corrupted by strange blood and alliance" (that strange alliance would be the one with the French). They are "more hard of constitution . . . watch better and abstain long . . . bold, nimble and more skillful in wars"—all qualities, we may reflect, that would be sorely needed by the Scots and sorely tested in the following centuries of English rule. The one thing Harrison does not do, however, which may seem unexpected, is castigate the Scots on the basis of their hair color.

In fact there is little in the way of description of the appearance of peoples or individuals of any sort in the *Chronicle.* William Rufus is described as "William the Red," with some pointed reflections on his character. King Henry II of England, invader of Ireland in 1171, has his "rednesse" recorded for posterity (red hair and ruddy skin; very suitable for such a warmonger of a king), but beyond those two, not much. There might have been more in the description of Boudicca, but the *Chronicle*, while leaving one in no doubt of the impression a queen might create upon her subjects—"Her mightie tall personage, comely shape, severe countenance and sharp voice . . . her brave and gorgeous apparel also caused the people to have her in great reverence"—has her as a blonde, with "her long and yellow tresses of hair reaching down to her thighs." A matter of translation, perhaps, or you might wonder whether this wasn't simply being politic. With an actual red-haired queen on the throne, it would hardly increase your success as a publisher to describe her hair color as being a characteristic of the treacherous barbarian, or even worse, of a queen who lost her life to an invading force. However, there is a third possibility. The evidence is scanty and very circumstantial, but it is intriguing. It might just be that red hair in the sixteenth century was less common in Scotland than it is today.

John Munro, he of those opinions on the coloring of the Vikings, also writes in 1899 of the Scots as being "of all complexions, from very dark to very fair, with a dash of red hair, about 4–5 percent more or less, in different localities, that is to say rather more than in England." That is to say, significantly less than the 13 percent of the population estimated to sport red hair in Scotland today. This would rather give the lie to all those tedious reports of redhead extinction. While in 1911 the Irish journalist T. W. Rolleston, in his *Myths and Legends of the Celtic Race*, would write that "the prevalence of red hair among the Celtic-speaking people is, it seems to me, a most striking characteristic . . . eleven men out of every hundred whose hair is absolutely red," suggesting that the present proportion of redheads in Ireland hasn't reduced by a single percentage point in the past hundred years, either.

Perhaps the number of redheads in Ireland in the sixteenth century was also lower than it is today. Richard Stanyhurst (in December 1578 awaiting that nerve-racking interview with Elizabeth's Privy Council) describes the Irish thus. First he accuses them of having the Spanish as their "mightiest ancestors," therefore being pretty much born traitors. Then, he says, they are uncouth, they live like animals, and they corrupt the English language—making a "mingle-mangle" of it, in Stanyhurst's own memorable phrase—and without constant vigilance do the same to the manners and morals of any of the English sent to dwell among them.[54] (The poet Edmund Spenser was even less pleasant, warning against the use of Irish wet nurses, as if barbarism was something that might be sucked in with a mother's milk.) The description goes on; the Irish are "religious, frank, amorous, irefull . . . very glorious, many sorcerors, excellent horsemen, delighted with wars . . . the men clean of skin and hew, of stature tall, the women well-favored" and "proud of their long, crisped bushes of hair." But there is not a mention of that hair's color as being red. And Stanyhurst should have known what he was talking about; he, like Edmund Spenser, was Anglo-Irish, one of those many whose ancestors had been shipped in from England or Scotland to occupy lands around Dublin, in a piece of thickheaded imperialism that unworked in Ireland in the same way as it would on the West Bank four hundred years later. English foreign policy in Ireland was a disaster, had been for centuries. No wonder the Privy Council was so sensitive about it. No wonder they thundered on about "report of matters that . . . are not meet to be published in such sort" being put out there for discussion in a book that "falsely recorded events." One suspects rather that Stanyhurst unwittingly recorded events all too well.

Or we might simply need to adjust our vision and read the *Chronicle* as an Elizabethan would have done. Red hair might not be made much of in its pages, but the Scythians and their "red haire" certainly were. This is the description from the *Chronicle* of the first

54 Stanyhurst could give James Joyce a run for his money. Among other gems in the *Chronicle*, he comes up with the phrase "idle benchwhistlers" for the lazy and describes those who would take credit for another's work as flies who fall in another man's soup. Rather less likeably, he also describes the Irish language as "a ringworm."

inhabitants of Britain:

> The people called Picts invaded this land, who are judged to be descended of the nation of the Scythians, near kinsmen to the Goths, both by country and manners, a cruel kind of men and much given to the wars. This people . . . entering the Ocean sea after the manner of rovers, arrived on the coasts of Ireland, where they required of the Scots new seats to inhabit in: for the Scots which (as some think) were also descended of the Scythians did as then inhabit in Ireland. . . .

And this is what an Elizabethan understood of the Scythians. This is from *King Lear,* which, with *Macbeth,* could claim to be the *Chronicle*'s greatest legacy:

> *The barbarous Scythian*
> *Or he that makes his generation messes*
> *To gorge his appetite, shall to my bosom*
> *Be as well neighbour'd, pitied, and relieved,*
> *As thou my sometime daughter.*

And here is Edmund Spenser, again going even further: "I Suppose," he says, the Irish "to be Scithians." They are "the most barbarous Nation in Christendom," Spanish blood (the ancestors of the Irish, remember, along with the Scythians) "is the most mingled, most uncertain, and most bastardly," and as for Scotland, it and the Irish are "one and the same."[55] It is deeply depressing to read the same piece of nonsense linking the Irish, the Scots, and the Scyth-

55 *A View of the Present State of Ireland,* 1596. This pamphlet is so unpleasant and so incendiary it was kept secret during Spenser's lifetime. Spenser bought lands and an estate in Ireland; he was also present at the Smerwick Massacre of 1580, when a ridiculously small force of maybe five hundred Spanish and Italian troops first surrendered to and were then killed by English soldiers.

ians being put forward by John Munro three hundred years later. Tying himself up in knots to account for all those random redheads, Munro decides that immigrants from Scythia first populate Ireland, then leave it, after exactly 216 years, to populate Thrace, then return to Ireland and after that people Scotland, too. Perhaps one shouldn't put too much trust in his math, either.

Might there have been fewer redheads in Scotland and Ireland then than there are now? Given the workings of genetic drift, it's certainly possible. But whether red hair, rather than general barbarity, was associated with either nationality at this date, the reviling and denigrating of the Scots and Irish by the English is beyond doubt. Edmund Spenser would have cleared the Irish from Ireland entirely—language, culture, customs, and people. The *Chronicle* played its part in enshrining their outsider status, and the red hair of the Celts inherited and suffered under these same attitudes as well. One red-haired queen did not redress the balance.

"A conquest," in Stanyhurst's view, "ought to draw with it three things, to wit, law, apparel, and language." Neither conquest nor union with Ireland or Scotland drew with it from England any such things. What Scotland and Ireland, misprized for centuries, drew from their relationship with England goes on to become a part of the history of red hair in the New World as much as in the Old. As for Stanyhurst, he survived his interview with the Privy Council, but on emerging from it he promptly left the country, never to return. Some years later he was working in the alchemical laboratory at the Escorial of Philip II of Spain, the king who was to launch the Armada against England in 1588. One wonders if Stanyhurst shared with the king his theories on Spain's links to those Irish barbarians. He ended his days as chaplain to the Catholic Archduke Albert of Austria in the Netherlands.

Holinshed, meanwhile, the poor soul, having patiently excised all those pages from his *Chronicle* that had so provoked the wrath of the Privy Council, retired to the country and died two years later.

Sometime after Shakespeare's death in 1616, *Macbeth* was revised by Shakespeare's near contemporary, the playwright Thomas Middleton, and among his additions was a song, "Black Spirits." This seems to have been lifted from Middleton's own play *The Witch* of 1615. By now there was a Scottish king, James I and VI (son of Mary, Queen of Scots), on England's throne, who was known for his fascination with witchcraft.[56] Black magic was, well, the new black.

The lines Middleton inserted into Shakespeare's *Macbeth* include the usual Addams Family list of ingredients: blood of a bat, libbard's (leopard's) bane, juice of a toad, oil of an adder, and "three ounces of the red-haired wench." Scholars have speculated that the red hair is to be read as an allusion to lechery, or as an allusion to Jewishness and the Jews' anti-Christian rites, or as an allusion to poisonous substances, or as all three.[57] But this is *Macbeth*, the Scottish play. It might simply be a rare touch of local color. It might also be playing with

56 James inherited the Scottish throne, as the sixth king of that name, from his mother, Mary, Queen of Scots. He was also a great-grandson of Henry VIII's sister Margaret and inherited the English throne, as England's first King James, from Elizabeth on her death.

57 Jeffrey Kahan, "Red Hair as a Sign of Jewry in Middleton's Additions to *Macbeth*," *English Language Notes* 40, no. 1 (September 2002).

another ancient prejudice directed at red hair—that it is somehow connected to witchcraft and the supernatural.

The figure of red-haired Lilith today has abandoned her Babylonian wastes for a shabby hinterland somewhere between mythology and pornography, but originally conflated with the *striges*—vampirelike demons of Ancient Greece—she killed children in their cots and seduced men in their sleep. Wet dreams were Lilith's doing, and she might also leave her victims impotent or even cause their penises to disappear entirely (still an accusation thrown at those suspected of witchcraft in Africa today). Woman as sexual predator has always terrified and aroused in equal measure, and witches have always been bewitching, in art and popular culture at least. The reality was rather different.

The handbook in the early modern age for witchery and witchfinders both was the *Malleus Maleficarum*, written in 1486 by one Heinrich Kramer, aide to the Archbishop of Salzburg, with another German clergyman, Jacob Sprenger, as a kind of publicist and PR man for the work. The pair of them were charlatans to a degree unrivaled even by Aleister Crowley and Montague Summers in the twentieth century. Unsurprisingly, Summers was the work's translator into English. Thanks to him, to this day you will find the *Malleus Maleficarum*, or *Hammer of Witches* as it is known, cited as an authority for tales of young, nubile, redheaded, green-eyed women being dragged off to horrible deaths at the stake. But if you read his translation, or as much of it as you can bear, what you find is a work of unrelenting misogyny that held all women in equal contempt, whatever their hair color might have been. Witches, it says, are driven by lust, and those most likely to be witches are adulteresses, fornicators, and concubines; and those most likely to be victims of the unwanted attentions of the devil are "women and girls with beautiful hair; either because they devote themselves too much to the care and adornment of their hair, or because they are boastfully vain about it." It certainly doesn't have to be red.

In reality it was pretty much certain to be gray or even white. While desirable young women may be depicted in the art of the period as witches, as in the works of Hans Baldung Grien in particular, and even with the *Chronique de France* of 1492 recording the Frankish kings as burning red-haired women as witches seven hundred years before, in the great witch hunts in Europe of the sixteenth and seventeenth centuries, those going to the stake or the gallows on a charge of witchcraft were almost bound to be poor, elderly, widowed, and unprotected.[58] This was recognized even at the time. Reginald Scot (*c.* 1538–99) writes in his *Discoverie of Witches*, a work that bravely set out to prove there was no such thing, that "one sort of such as are said to be witches, are women which are commonly old, lame, bleare-eyed, pale, foul, and full of wrinkles, poor, sullen, superstitious or papists, or such as know no religion, in whose drowsy minds the devil hath gotten a fine seat."

The beauteous redheaded victims of European witch hunts exist in our imaginations only, and in so imagining them, we are being seduced ourselves by an association between otherness and otherworldliness, between red hair and supernatural forces, and between red hair and erotic circumstance that simply refuses to quit: there

---

58 Brian P. Levack, *The Witch-Hunt in Early Modern Europe* (Oxford: Routledge, 2006).

was red-haired Malachai in Stephen King's *Children of the Corn* in 1984; there is the sorceress Melisandre, the Red Woman of *Game of Thrones*; and in 2014 the casting director of *The Last Witch Hunter* was advertising for a redhead with a pale complexion as the movie went into production.[59] How very confusing all this can be.

Or as Obadiah Walker put it, "Each man disparageth his fellow-creature, and gratifies his haughty humor in the derision of his brother." Obadiah Walker was Master of University College, Oxford, from 1676 to 1688, where supposedly his melancholy ghost still walks. He lost his post for refusing to abandon his Catholic faith, so knew more about the workings of discrimination than many another. In his *Periamma epidemion, or, Vulgar errours in practice censured* of 1659, Obadiah writes of "a common yet causeless calumniation: viz. the vilifying of red-hair'd men, the putting of disesteem upon persons, merely because of the native color of the excrement of the head." He means the hair, but you do begin to think sometimes that the phrase might as justly be applied to the historic bigotry and prejudice in our thinking. "I could wish," he says "that men would not hoodwink themselves with their own prejudice." But they do so still. As far as that as a signifier of barbarity goes, it's global.

And yet, and yet. . . . If "other" repels, it also fascinates. There

59 In 1887, in the *Ancient Legends, Mystic Charms, and Superstitions of Ireland*, you might read this: "Red hair is supposed to have a most malign influence, and has even passed into a proverb—'Let not the eye of a red-haired woman rest upon you.'" Who is the writer? Some chauvinist Englishman? Some witch-hunting mittel-European cleric? No, it is Lady Francesca Speranza Wilde, Irish mother of Oscar. More recently, in her thesis *Sirens and Scapegoats: The Gendered Rhetoric of Red Hair*, Emily Cameron Walker draws attention to the number of times the disgraced CEO of News International Rebekah Brooks was referred to in the English press as a red-haired witch during her trial in 2014. See http://ecameronwalker.blogspot.com/2012/09/thesis.html.

is always that contradictory desire within us to stand out. Even in the eighteenth century, when any hair color other than gray was as unfashionable as could be, when every head wore a wig, and every wig looked like powdered topiary, there was a moment in 1782, recorded by the diarist John Crozier, when despite "much aversion as people in general have to red hair, the appearance thereof was so much admired that it became the fashion, for all the Beaus and Belles wore red powder."[60] Nor is the fashion for red hair restricted to the West. Something very similar to the craze for red hair in London in the 1780s happened in Japan at the end of the twentieth century, and the Japanese had otherwise reviled red hair for centuries.

When the first Westerners set foot in Japan in 1543 (the same year, incidentally, that the nine-month-old Mary Stuart was crowned Queen of Scots) the Japanese were appalled by them. These crude, semi-civilized creatures, who ate with their fingers and had all the self-control of children, were immediately compared by the genteel and sophisticated Japanese to monkeys and monsters, to legends of primitive wildmen, covered in fur, and all such, without any reference to their actual appearance, had been lumped together under the label of "red-haired barbarians." Body hair and hair on the head in any color other than Japanese black became the dominant symbol of otherness in Japan for centuries.[61] Yet in a society where the term *ang-mo*, or "red-haired ape," is still bandied around as an insult to

60 Quoted in C. Willet Cunnington and P. E. Cunnington, *Handbook of English Costume in the Eighteenth Century* (London: Faber, 1957): 1952–9.

61 Alf Hiltebeitel and Barbara D. Miller, *Hair: Its Power and Meaning in Asian Cultures* (New York: SUNY Press, 1998).

foreigners, in the 1990s Japanese teenagers began dying their hair all shades of red and brown.

This fashion for *chapatsu*, or "tea-color" hair, became a national controversy. Questions were asked in Parliament. Schools created "hair police," and even now students with naturally brown or curly hair can be asked to prove it should not be naturally black and straight.[62] Japan is a conservative and very homogenous society, and some of its teenage fashions consequently can seem over the top (an accompanying fad was for glittery stickers of fake tears pasted to many tea-haired teens' cheeks). But we are all barbarians to someone. In the 1930s, *Japanese* hairiness became a major feature of the anti-Japanese propaganda coming out of China. As Professor Alf Hiltebeitel puts it, "Nothing is ordinary about hair. It gets into everything, but whatever it gets into, it never seems to be explained in the same way; rather it always seems to be used differently to explain something else." And queen or commoner, we all want to shake a brighter tailfeather than the rest.

---

62 See http://www.japantimes.co.jp/community/2013/07/29/issues/prove-youre-japanese-when-being-bicultural-can-be-a-burden.

SINNERS AND STUNNERS

The truth about red hair, like many other truths, lies enclosed in a nutshell, generally a hard one, and people are often very short of crackers.

THE PHILOSOPHY OF RED HAIR, 1890

The area around St. Paul's Cathedral in London is one of the few parts of the city where you can still summon up its past. Close your eyes and ignore the traffic; imagine instead of honking taxis the shouts of irate draymen, the creak and squeak of wooden wheels, the clop of hooves, the endless music of an endless press of people—probably very similar to those who throng the same streets today, if rather less well soaped and washed. The street names still record their presence, these hordes of ghostly Londoners, and their doings here: Pilgrim Lane, Ironmongers Lane, Limeburner Lane, Old Jewry. Watling Street, leading to St. Paul's, was trodden by Roman legionaries and then by Boudicca's rampaging army; close by, off Cheapside, were once to be found the colorfully named Pissing Alley and, even better, Gropecunt Lane, until times changed and renaming became inevitable (the rather more acceptable Love Lane, where no doubt exactly the same activity took place, still exists). Paternoster Square, Amen Corner, and Ave

Maria Lane celebrate the permanence of human faith and worship in this area, no matter what religion, and as Ave Maria Lane runs north to Newgate Street its name changes, to Warwick Lane. And in a house on Warwick Lane in 1865 there lived a girl of about sixteen whose name was Alice Wilding. She earned her living (or rather contributed to the household, which included her grandmother, two uncles, and at least one infant) as a dressmaker. And secretly, like so many girls of any age or class, she dreamed of a career on stage. Alice, however, was no head-in-the-clouds innocent: when, one evening early in the year, she found herself being first stared at and then followed up the Strand by a short, tubby, balding man of middle age and clearly a higher social class than hers, she seems to have taken the experience in her stride. Very possibly, this was not the first time such a thing had happened to her. Alice had a face both strong and feminine, feline eyes, milky skin, and the most wonderful head of red hair, something between copper and the color of a marigold. The man introduced himself as an artist, Dante Gabriel Rossetti, and in the account of the meeting recorded by his studio assistant, went on to explain that he was "painting a picture and her face was the very one that he required for the subject he was at work on." The man begged Alice to come to his studio at Cheyne Walk in Chelsea the next day, and to sit for him, promising her that she would be paid. Once satisfied that she understood, and had agreed, he went on his way. The next day, "Rossetti made every preparation to receive her and make a study of her head for *The Blessed Damozel.* His palette was set, the canvas on the easel and everything in readiness. . . . "

Alice stood him up. Of course she did. Go sit in an "artist's studio," for an artist she had very possibly never heard of, a man, on her own? Was he mad?

Fortunately for us, however, that wasted day in the studio is not the end of the story.

One significant development in artistic life in the nineteenth century is that we start to learn the names and in some cases details of the biographies of the artists' most significant models. Sir Frederic Leighton, president of the Royal Academy from 1878 to 1896, and arguably England's premier and most successful painter of the period, had an entire family of sisters, the Pullans, who sat to him and his circle. Rossetti had first Lizzie Siddal and later Fanny Cornforth, with Alice Wilding modeling for him as well. Most of these women were working class, and their relationship with "their" artists was one of social and economic dependency as well as collaboration: outside the artist's studio many, in fact all those who posed nude, would be regarded as little better than prostitutes (Fanny Cornforth, who became Rossetti's live-in housekeeper and whose robust humor and working-class attitudes alarmed his family and friends all his life, may very likely at one time have been a streetwalker for real.) Various ruses were adopted by these young women to cover for the true nature of what they did in the artist's studio. Ada Pullan listed herself on the 1881 census as an "art student," for example. It was just about feasible by the latter half of the nineteenth century for a woman to have a career in the arts, to enjoy some measure of financial freedom without having been born to riches, to hold some control over her own life in a way that would have been exceptional in previous centuries, and to negotiate a place for herself within an artist's circle that was neither wholly sexual nor without respect. One who managed to do so was Joanna Hiffernan, an Irishwoman who met the American artist James Abbott McNeill Whistler in a studio in Rathbone Place, London, in 1860, when she too would have been about sixteen. She went on to have a six-year relationship with him, acted in loco parentis to his illegitimate son even after their relationship as lovers had come to an end, and was the inspiration for some of Whistler's most sophisticated and innovative works, including his *Symphony in White, No. 1* of 1862 (Fig. 18), and *Symphony in White, No. 2* of 1864–5. Both paintings make wonderful play of her pale skin, soulful eyes, and almost oversize features, and her dark red hair, hair that Whistler would describe ecstatically as "a red not golden but copper—as Venetian as a dream."[63] Whistler's biographer Joseph Pennell described the woman herself as being not only beautiful but intelligent and sympathetic. Joanna was also someone who steered her own very independent course through the world. She was unconventional (in her morality) and daring (in her professional life), both qualities the world ascribes all too willingly to redheads, and she has a unique role in the history of art in that her most famous supposed "portrait," a work with which she has been intimately associated for decades, is one that as a redhead she simply cannot have sat for at all.

The story begins with a trip to Paris, where while modeling for the second *Symphony in White*, Jo, as she was known, met the French artist Gustave Courbet. In 1865–6 Courbet painted her portrait as *La Belle Irlandaise*, showing her in close-up, before a mirror, combing out a tangle in her hair and creating his own slab-sided, meaty

63 Quoted in Margaret F. MacDonald and Patricia de Montfort, *An American in London: Whistler and the Thames* (London: Philip Wilson Publishers Ltd), 2013, which contains much useful information on Whistler and his relationship to Johanna Hiffernan or Heffernan, as she was also known.

version of the ethereal beauty who had captivated Whistler. In 1866 she began an affair with Courbet and posed as one of the two women in his *The Sleepers* (Fig. 19). This painting, which was the subject of a police report on what seem to have been the only occasion in the nineteenth century when it was publicly exhibited, in 1872, shows two women, one dark-haired, one Jo, curled about each other naked in bed, supposedly sleeping after making love. It has sometimes been hailed as a ground-breaking depiction of lesbianism, but it is surely rather a male heterosexual fantasy about female lovemaking. The poses in which the two women are supposed to have fallen asleep are unnatural (and, speaking as an ex–life model, look excruciatingly uncomfortable, too). Jo's head seems oddly unsupported, and the expression on her face suggests extreme concentration in holding a difficult pose rather than languorous afterglow. Like much erotica, perhaps it hasn't aged very well. But it created an association between the Bohemian, free-spirited Jo and Courbet's erotic paintings that is running to this day.

Courbet's other notorious erotic work, also of 1866, is known simply as *L'Origine du Monde*. For those who don't know it, Courbet's "origin of the world" is, predictably, a close-up of the view the artist would have had if he had set up his easel at the foot of his model's couch and asked his model to raise her shift above her breast and to open her legs. There's a piece of Anglo-Saxon, no doubt familiar to many an historic wanderer in London's St. Paul's, that would describe the subject of the painting perfectly. Its framing excludes everything else, including head, arms, and lower legs. It was commissioned by one Halil Bey, an Ottoman diplomat who also owned *The Sleepers*, and who can therefore claim front rank among the grand dirty old men of erotic art, and the painting still takes one aback, even today. But there is simply no possibility of it being a portrait of Jo, untrammeled by society's mores as she may well have been, and saddled with all the sexual baggage of being a redhead as she undoubtedly is. The pubic hair of the woman in *L'Origine* is so dark as to be almost black. The pubic hair of a redhead is, unsurprisingly, red. In fact you can see a tiny suggestion of Jo's own pubic hair in *The Sleepers*, a minute triangle of gold above the leg of her bedmate. This second woman, an unnamed brunette whose dark hair is spread out across the pillow beside Jo's coppery curls, may have her sharp features echoed in another painting, recently discovered, of a woman's head, mouth open, dark hair thrown back and purporting to show *L'Origine*'s missing upper half. Or, the latter may have nothing to do with the former at all. But neither is Joanna Hiffernan.

Back to the Strand in 1865, and Rossetti's studio assistant, Henry Treffry Dunn, continues the story of Rossetti's vanishing redhead:

> Days and weeks went by, and he [Rossetti] had given up all hope of seeing the young lady again and had even abandoned the picture, when one afternoon in company with [the sometime art dealer Charles Augustus] Howell in the same part of the Strand, he again caught sight of her. He was then in a cab, [&] telling Howell what he was going to do he stopped the Hansom at a side street, got out and darted after the girl and at last overtook her. He reminded her of the promise she had given him and told her of his disap-

> pointment at her not coming and at last persuaded her to enter the cab and drive with him to Cheyne Walk.[64]

You have to wonder at the change of heart. Perhaps the encounter was less alarming the second time around; perhaps this had begun to feel like Fate; perhaps by now Alice knew who Rossetti was; perhaps there had also been a few hours in front of her looking-glass, wondering why she had been given this face and hair if to do nothing with it? In Rossetti's case, the explanation of his persistence is simpler—he was, in the words of the novelist Elizabeth Gaskell, "hair mad":

> If a particular kind of reddish-brown crepe wavy hair came in he was away in a moment, struggling for an introduction to the owner of said head of hair. . . .[65]

Thus Rossetti is forever immortalized as a classic example of Man with a Thing for Redheads. Aside from Rossetti and his obsession, however, why are there so many Pre-Raphaelite redheads? The term has become virtually synonymous, just as "Titian red" would be later in the century. They are there in works by Frederick Sandys, whose gloriously rufescent partner, the actress Mary Emma Jones, modeled for him first as the Magdalene in 1862, and then as Perdita, Proud Maisie, Helen of Troy, and countless other red-haired icons throughout his life. Arthur Hughes used his wife, Tryphena Foord, as his model for *April Love* (1856) and *The Long Engagement* (1854–9),

64 Jennifer J. Lee, "Venus Imaginaria: Reflections on Alexa Wilding, Her Life, and Her Role as Muse in the Works of Dante Gabriel Rossetti" (msaster's thesis, University of Maryland, 2006).

65 Quoted in Henrietta Garnett, *Wives and Stunners: The Pre-Raphaelites and Their Muses* (New York: Pan Macmillan, 2013).

both works depicting unhappy lovers, and both making much play of Tryphena's ghostly skin and shining red-gold hair. Henry Wallis depicts the eighteenth-century poet and suicide Thomas Chatterton on his deathbed, in his garret, neglected poetry torn to pieces on the floor beside his lifeless hand. Chatterton's bright red hair leaps out from the painting's chilly gray and green palette and is perhaps used by the artist to allude both to his subject's sensitivity and to his passionate poetic spirit. John Collier, a late Pre-Raphaelite, creates an irresistible red-haired *Lilith* in 1887, wearing her symbolic serpent like a feather boa, and paints Lady Godiva as a redhead in 1898 (although seeming to miss the idea that her hair should be thick and long enough to hide her completely). Rossetti surrounded himself with redheads, both male and female. Dunn's successor as his assistant was the redheaded Manxman and later novelist Hall Caine, a man of such unusual yet engaging appearance he might have been designed by Mother Nature to play Merlin. And for a year from 1862, Rossetti shared his house with the flame-haired poet and all-round oddity Algernon Charles Swinburne (Fig. 20), of whom more in the next chapter. Joanna Hiffernan and Whistler were guests at the house in 1863; one wonders if they were entertained by Swinburne, whose party piece consisted of sliding nude down the bannisters, and who reportedly infuriated Rossetti by dancing all over his studio "like a wild cat."[66] And famously, Rossetti's first muse and eventually his wife had been the red-haired beauty Elizabeth Siddal, the model for Millais's *Ophelia* of 1852, who was described by Rossetti's brother William as:

66 Quoted in Leonard Shengold, *If You Can't Trust Your Mother, Who Can You Trust?: Soul Murder, Psychoanalysis, and Creativity* (London: Karnac Books, 2013).

> A most beautiful creature with an air between dignity and sweetness with something that exceeded modest self-respect and partook of disdainful reserve; tall, finely-formed with a lofty neck and regular yet somewhat uncommon features, greenish-blue unsparkling eyes, large perfect eyelids, brilliant complexion and a lavish heavy wealth of coppery golden hair.[67]

Others were less kind. There seems to have been something unsparkling about Lizzie Siddal altogether:

> She was passive. . . . This passivity helped bring them together. She trailed slowly towards [Rossetti], a melancholy doll, set in sluggish motion by the virile, expansive gestures of the warm Latin. His roar of laughter elicited from her a wan smile, his jests provoked a faint answering shade of humour, his ardour the ghost of passion. In the same contrary fashion, he loved her because she was so little responsive. No one knew what she was thinking of or if she thought at all. She had . . . the habit of "keeping herself to herself" which deepened into an unfathomable reserve on being introduced into a clever and freakish group of artists. . . . In her mournful beauty, her natural silence, her frigid apathy, she was like a statue to be warmed into life. . . . [68]

That passivity probably had much to do with the fact that she suffered both from depression and an addiction to opium. Lizzie Siddal wrote poetry (not good), drew (not well), and died of an overdose of laudanum in 1862, leaving Rossetti, who had been neither constant as a lover nor compassionate as a husband, with a burden of lifelong guilt that one has to say he probably deserved. You have the feeling that he loved the passive face, the hair, the sad and empty eyes and all he could project onto them, but not so much the woman.

---

67 Quoted in Russell Ash, *Dante Gabriel Rossetti* (New York: Harry N. Abrams, 1995).

68 Quoted in Jennifer J. Lee, *op. cit.*

Perhaps his most affecting portrait of Lizzie Siddal is as *Beata Beatrix* (*c.* 1864–70; Fig. 21), created after her death and inspired by the Beatrice of that other Dante. It shows Lizzie as the loved one lost, eyes closed as if in death and much more otherworld than this, with her red hair gathered behind her like the tail of a pale comet, haloed with light. Now, in Alice, or Alexa Wilding, as his model was to rename herself, he had another Lizzie, just as beautiful, and apparently offering the same useful blankness too: "a lovely face," wrote Dunn, "beautifully moulded in every feature, full of quiescent soft mystical repose. . . . but without any variety of expression. She sat like the Sphynx. . . . "

Not all redheads are fiery. Despite her "repose," Rossetti used his new model, with her lovely face and unmissable hair, without stint. He made *The Blessed Damozel* a redhead, as if in her honor (his own poem, of 1850, had spoken of the Damozel as having "hair that lay along her back/. . . yellow like ripe corn"). He repainted her features over Fanny Cornforth's in his *Lilith* and *Venus Verticordia*, both of 1864–8, and had an agreement with her that she would sit to him exclusively. Perhaps, in fact, it was her blankness, along with her resemblance to Lizzie, that made her so inspiring a muse. Perhaps the only role she had to fulfill was to turn up on time and be decorative. Even *La Ghirlandata*, the painting in which she stars on the cover of this book, had no "underlying significance," according to Rossetti's brother William: "I suppose he [the artist] purposed to indicate, more or less, youth, beauty and the faculty for art worthy of a celestial audience. . . . " In other words, these paintings are to show off the painter, not the model. It is we, coming to them as their audience, who look for symbolic meaning in these yards and yards of red Pre-Raphaelite curls.

Certainly there is some. It's notable that it's almost always loose or loosened red hair that is depicted, hair so luxuriant that it's almost out of control. It may symbolize female sexuality in Eve or Lilith, female passion in Sandys's *Proud Maisie*, or be the inevitable attribute of fatal beauty in the same artist's *Helen of Troy*. It's suggestive of Bohemianism, the world in which these artists lived at least at the start of their careers. Many of their paintings are mythological in subject; perhaps playing on the notion of red hair as an attribute of the supernatural. They also undoubtedly give the artists an opportunity to show off their skill in depicting these twining tresses and shining locks. And they both draw and please the eye, in which case the artists are using red hair in exactly the same way as the advertising industry of the twenty-first century. In a bit of slang favored by the Pre-Raphaelite Brotherhood, which has survived unchanged from the 1840s to the tabloid press of today, it is the coloring of the "stunner." Used like this, of course, it starts to stigmatize the very characteristic that it thrusts at our attention. There's something a little prurient about the Pre-Raphaelites. Look at their treatment of Mary Magdalene. In their hands Mary Magdalene, the most important female saint of medieval Europe and, next to the Virgin herself, possibly the most empowered female figure in medieval art, becomes either merely an excuse for painting another "stunner," or in Rossetti's *Found* (Fig. 22), a prostitute for real. In his own words:

> The picture represents a London street at dawn, with the lamps still lighted along a bridge which forms the distant background. A drover has left his cart standing in the middle of the road . . . and has run a little way after a girl who has passed him, wandering in the streets. He has just come up with her and she, recognising him, has sunk under her shame upon her knees,

> against the wall of a raised churchyard in the foreground, while he stands holding her hands as he seized them, half in bewilderment and half guarding her from doing herself a hurt.

The "fallen woman" of Victorian cliché, in other words, here literally slumped to the ground.

"Magdalene," as a polite-ish bit of slang for a prostitute, had been in use since the late seventeenth century at least. What makes this painting particularly rich in allusion—so rich and complex, in fact, that for thirty years it defied the artist's attempts to finish it off, and was left as it is, uncompleted, on his death in 1882—is the fact that Rossetti's eventual model for the woman in *Found* was his mistress and housekeeper, Fanny Cornforth, she of the doubtful past. Fanny was no silent Sphynx, nor a woman who sounds likely to have had much time for disdainful reserve. As she and Rossetti grew fatter in their old age together, she called him "Rhino" and he called her "Elephant." In *Found*, however, the earthly Fanny is depicted as a pitiable being, in the last stages of consumption or perhaps of syphilis, with greenish pallor and face compressed in shame and agony. Her jaunty feathered hat has fallen back, revealing her coppery hair, by which perhaps her onetime sweetheart, the drover, is meant to have recognized her. (The sweet young thing come up from the country and ruined by the wicked city is another workhorse cliché of Victorian art and literature alike. Think of Little Em'ly in Dickens's *David Copperfield*.) The work was exhibited, even if unfinished, and found many an admirer, among them Lewis Carroll, who saw it at the Royal Academy in London in 1883 and called the face of the drover "one of the most marvelous things I have ever seen done in painting." A matter of opinion, perhaps; the painting's other great point of interest here is the way it reworks the iconography of the Colmar *Noli Me Tangere* of four centuries before. Here is the Magdalene as a Magdalene, disempowered and the lowest of the low; here it is the man who touches her against her will; here she who turns away. In Rossetti's nod to Colmar, the Garden of Gethsemane has become a single withered rose, lying in the gutter, the birds in the garden a pair of London sparrows.[69]

You get the sneaking impression that neither brains nor personality were much desired in the Pre-Raphaelite artist's model. Women with either tended to become bored and to move on, sometimes, in hope, like Lizzie Siddal, by marrying the artist. Some married and were bored still. A sixteen-year-old Ellen Terry sat for the artist G. F. Watts, a close associate of the Pre-Raphaelites in 1864, when he was forty-seven, and shortly after married him. The portrait he painted of her, *Choosing*, makes glorious play of her fair, schoolgirlish features and strawberry-blond hair, and is usually interpreted as showing Ellen's rejection of the artifice of the outside world for a life of married duty, but looking at it today, with its hit-you-over-the-head flower symbolism, it's possible to read the showy but scentless camellia that Ellen holds to her nose as life with Watts, and the tiny scented violets held close to her heart as her ambitions (like Alexa's, but much more successfully so) for a career on the stage. In 1889, at the height of Ellen's fame as an actress, John Singer Sargent would paint her in costume as Lady Macbeth, during the run of Henry Irving's production of *Macbeth* at the Lyceum Theatre in London.

69 Béatrice Laurent, "Hidden Iconography in *Found* by Dante Gabriel Rossetti." See http://www.victorianweb.org/painting/dgr/paintings/laurent.html.

Her extraordinary dress, in a peacock's-tail palette of green crochet and blue tinsel, sparkles with more than a thousand green iridescent beetle-wings. Willful, passionate, murderous, *and* Scottish, as she holds Duncan's crown over her own head, Lady Macbeth's ensemble is completed with two plaits of deep red hair, thick as hawsers, bound with gold, and reaching to her knees (Fig. 23).

And as an example of red hair equaling ethnicity, there is also Millais's *Martyr of the Solway* of 1871. Scotland had a particular importance for Millais. Some of his most significant works of the 1850s have Scottish settings or Scottish subjects; and it was here in 1853 that he fell in love with the auburn-haired Effie Gray, then Mrs. John Ruskin. Not that the story behind the *Martyr of the Solway* is itself an uplifting tale, nor does it have anything but the most tragic of endings. The martyr was one Margaret Wilson, an eighteen-year-old girl from Wigtown in Dumfries, who in 1685 had been sentenced to death for refusing to swear an oath accepting the Catholic James II of England, and VII of Scotland, as head of the church in Scotland. Despite the fact that a reprieve had been granted, Margaret Wilson nonetheless died as sentenced, by drowning, chained to a stake in the Solway Firth.

This too is a work with a unique backstory, which can only be told with reference to Millais's *Knight Errant*, of 1870. The redhead in this earlier work, hair blown across her naked, captive body as she is cut loose from a tree by the knight who has killed one of her attackers, has been interpreted as a victim of robbery, but rape seems just as likely (an early description speaks of the woman having been "despitefully used"). And then there is the knight's extremely prominent longsword, streaked with blood. The painting, which is otherwise rather Arthur Rackham–esque in its moonlit details and forest setting, trembles with sexual activity; and in it, in its original incarnation, the woman who now looks away from us had her face turned toward her rescuer. This was heady and subversive stuff. While a naked woman's body might be laid out for the viewers' delectation, her face was apparently better turned modestly away, in shame or private humiliation (as it would be in Lefebvre's *Mary Magdalene in the Cave*, for example). Millais's treatment was in fact held up for comparison with the "idealized" form the nude took in continental painting, and presumably the critics making this comparison must have had Salon painters such as Lefebvre in mind, rather than works such as Manet's *Déjeuner sur l'Herbe* of 1863, or his *Olympia*, of the same year. The fact that this, Millais's one full-length naked female figure is almost life-size, only added to the critic's discomfort when it was first exhibited. Her thighs are dimply, her waist a little thick; "too life-like," the critics called it, "too real."[70] It failed to sell until Millais cut away the central section and replaced it with new canvas on which the woman turns her face from us. What, then, to do with a half-length study of a nude redhead, arms bound behind her back? Create the *Martyr of the Solway* (Fig. 24) with it, of course, which, in a startling bit of artistic sleight-of-hand, is exactly what Millais did.[71] Margaret Wilson in the *Martyr* was originally the top half of the damsel being rescued by the knight. But having learned his

70 Anonymous review. *Art Journal*, June 1870, p. 164. Quoted in Linda Nead, "Representation, Sexuality and the Female Nude," *Art History* 6, no. 2 (June 1983) 233–6.

71 To see *The Knight Errant* as it would originally have appeared, see the cunning reconstruction created by Martin Beek on flickr and displayed by http://vadimage.wordpress.com/2010/11/08/too-life-like-the-knight-errant-1870-by-john-everett-millais.

lesson, perhaps, from the reception accorded to *The Knight Errant*, in her new incarnation Millais clothes her, in a very nineteenth-century-looking blouse and, tellingly, a plaid skirt. He also removes all possibility of rescue. There was to be no knight, no Perseus for this Andromeda. The dark waters of the Solway Firth are rushing in upon her; it is this, her fate, that she turns her eyes from, rather than her own nakedness. The viewer of the painting is left in the uncomfortable position of being forced to see what Margaret does not. But set against the dark waters as they roil in behind her and in contrast to the stormy and overcast sky is her red hair, used here like a flag, to suggest (with the tartan of her skirt) both her nationality and her defiance. This is one case where the subject's red hair and the meaning of the work are indivisible.

In fact the days of the smooth and perfect limbs of the Salon nudes were already numbered. Over the Channel the French Impressionists were also creating works in which redheads abound, in Renoir's marshmallowy nudes and in the pastels of Edgar Degas of the late 1880s in particular. There hadn't been so many redheads in art since the days of Elizabeth I. Degas's pastels of women sponging or toweling off after the bath, or combing through their long hair, have been criticized for their objectifying of their subjects, their equation of these women with so many cats washing and tending to themselves. They also have what one might view as at least a semi-exploitative subtext, in that such ablutions traditionally preceded or followed intercourse, and that the women are often naked, or almost so, and their faces are again often obscured. But Degas did produce one of the best depictions, indeed glorifications, of red hair ever, in his *La Coiffure* of 1896 (Fig. 25). Here one woman (older, a redhead, in apron and pinkish blouse) is combing through the hair of another (younger, wearing a red robe) who sits before her. The younger woman's long red hair is stretched between them, like washing going through a mangle. Everything in the image is red, from the curtain looped up in the top left-hand corner, to the color of the wall behind them. The beads of jewelry on the table are red. The young woman's cheeks are flushed. The painting shows what we should presumably take for a domestic space as being as red as a womb. It is as if the electricity one can almost hear crackling off that hair as it is combed has suffused the entire canvas with its color.

But the artist of red hair par excellence in this period must be Toulouse-Lautrec. All three of his most famous sitters from among the singers and dancers of the Folies-Bergère—Yvette Guilbert, Jane Avril, and La Goulue—were redheads. In his *Rue des Moulins* of 1894 he depicts one of his favorite models, a snub-nosed prostitute who was apparently named Rolande waiting insouciantly in line, shift held up above her navel, to be inspected for signs of syphilis. Her bright red hair is the hotspot of the painting. And of course this is another reason why artists place redheads in their works—for that vibrant dash of color, that ability of red to draw the eye; which is exactly the same reason why for a woman of the streets, or a tart in a brothel of the Belle Epoque, red hair works. It gets you noticed.[72]

72 The same principle has been used in the cinema, too. There is the little girl in her red coat in Steven Spielberg's *Schindler's List* (1993); and the heroine, Lola, in Tom Tykwer's *Run Lola Run* (1998), where her bright red head of hair becomes the pivot around which the film's alternative scenarios rotate.

Toulouse-Lautrec's paintings record a love affair between a particular kind of throwaway French chic and the blazingly artificial dyed red hair that can still be seen on the streets of Paris today. How did a characteristic once so linked to the lower end of the social scale become desirable and fashionable? One answer is the link forged by these artists between the image of the intriguing, independent, unconventional, Bohemian young women, and red hair either real or dyed; another can be found in a gradual change of attitudes toward red hair, evident in at least two specific examples of this period. The first is the courtesan Cora Pearl, a *grande horizontale* of the old school who was also known as "La Lune Rousse" (Fig. 26).[73] Her enormous wealth (in the 1860s she could command as much as 10,000 francs for an evening in her company) made everything she wore and every aspect of her show-stopping style worthy of emulation. Cora understood the importance of creating a spectacle in order to stay in the public eye, of reinventing her image; dyeing her hair on occasion not only red but also lemon yellow, to match the upholstery of a new carriage, and her dog blue, to match her gown. Rather more respectable was the opera singer Adelina Patti, who did much to popularize and make acceptable the use of henna as a hair dye and whose career was at its height in the 1870s and 1880s. But both women were still exploiting the ability of red hair to draw the eye and get the wearer noticed. As were Yvette Guilbert, Jane Avril, La Goulue, and Rolande her humble self. *Madame X*, the profes-

73 The name may also play on another meaning of the phrase in French, of an early spring moon before the last frosts had passed—deadly for young sprigs, just as Cora was to the young men she captivated, one of whom shot himself dead on her doorstep in despair. The character of Joséphine Karlsson in the French TV series *Spiral*—ruthless, manipulative, redhaired, and deadly—is another *Lune Rousse.*

sional beauty Virginie Amélie Avegno Gautreau, star of John Singer Sargent's portrait of 1884, was another much-emulated celebrity of fin-de-siècle Paris, also known to use henna to tint her hair. By 1881 Miss Maria R. Oakey could write of red hair in her *Beauty in Dress* not as something to be played down or even disguised but as a specific and desirable type, with its own palette of colors to show off the hair to best advantage: "White, of a creamy tone, black, invisible green [one assumes she means eau-de-nil], rich bottle-green, rich blue-green, plum color, amethyst," and so on. And by 1910, in *The History of Mr. Polly*, H. G. Wells can present an audacious red-headed schoolgirl as the object of the youthful Polly's chivalrous and most ardent admiration, and as an entity whose desirability would be wholly understood by his readers.

There is even a new vocabulary, to mark red hair's new status as socially acceptable. In 1890, on April 1 (All Fools Day) the Auburn Printing Works of Lightcliffe, Yorkshire, published *The Philosophy of Red Hair*. This work, of heavy Victorian humor, records the hapless state of mind of Rufus, a young man with red hair. Rufus reads in his sister's journal that "red hair if straight denotes ugliness," but if "given to curl" it denotes "deceit, treachery, and a willingness to sacrifice old friends for new or personal advancement." When traveling by train, Rufus is warned not to stick his head out of the window, lest he be mistaken for a danger signal. When invited to a costume party he is advised to wrap himself in brown paper and go as a lit cigar. He is presented with all the usual reasons for being an object of such ridicule: the dislike of red hair recalls the fear of the Danes; that Judas was a redhead; that it is a primitive characteristic (a favorite of the nineteenth-century anthropologist, this one). Yet at the same time Rufus notes that the typical female flirt is always presented with flaming red hair and green eyes as part of her charm. And most unfair of all: "A very curious trait with authors is that the red of the red-haired girl is transmuted to auburn, or golden, when she becomes an interesting young lady, whereas the red of the red-haired boy remains red to the end of the chapter." This is auburn in its modern meaning, an acceptable alternative to the pejorative carrots, or ginger, for a hair color that for women was well on its way to becoming positively desirable. "Auburn" has cachet.[74] "Titian," which also came into use in the late nineteenth century, according to the *Oxford English Dictionary*, has the same, plus the added advantage of suggesting familiarity with high art. Unsurprisingly, the term caught on. *The Dundee Advertiser* would note in 1904 that "twenty years ago hair with a reddish tinge was called 'carrots,' now Titian-colored locks are reckoned a definite beauty." By the time Elinor Glyn, the (green-eyed, red-haired) sensationally successful and knowingly risqué novelist of the pre–World War I period, was writing *Red Hair* in 1905, she could have her endearingly ditzy heroine Evangeline declare herself a social outlaw, "a penniless adventuress with green eyes and red hair" who is "bound to go to the devil" because of it, yet nonetheless also have her virtuous enough to not only frustrate the seductive wiles of her guardian but win over the hero's wealthy and aristocratic uncle and gain his blessing on their marriage. (And this despite the sight of her come-hither, pink silk nightdress scandalizing his killjoy relatives.)

74 Or, as the American humorist Mark Twain, a redhead himself, would put it in *A Connecticut Yankee in King Arthur's Court* (1889): "When red-headed people are above a certain social grade their hair is auburn." Until the 1980s, red-haired Barbie dolls were also sold as "Titian."

Fig. 12: Jan van Eyck, *The Crucifixion*, *c.* 1435–40. This painting is twinned with a right-hand panel showing the Last Judgment. It is the Crucifixion as narrative, and has been aptly called almost an eyewitness account. The Virgin is almost indistinguishable, shrouded in her blue robes as if these are her grief; Mary Magdalene, identifiable by her red hair, lifts up her hands in a gesture that combines horror, pity, and entreaty all at once.

Fig. 13: Jules Joseph Lefebvre, *Mary Magdalene in the Cave*, 1876. Now in the Hermitage, this work is typical of the highly finished Salon style of painting—so establishment on the surface, and at the same time so ripe for decoding and reinterpretation.

Fig. 16: Elizabeth I, painted by an unknown artist around 1575. Now in the National Portrait Gallery, London, this portrait epitomizes both Elizabeth's image and her elegance. The choice of colors in her dress and her accessories is perfect for a redhead, the masklike face somehow both ineffably sad and totally remote.

Fig. 17: Robert Peake's *Procession Picture of the Queen*, *c.* 1601. One of the last public images of Queen Elizabeth I, who was to die in 1603, showing her in one of the paler-colored wigs she favored at the end of her life. This is a snapshot of the Elizabethan court at the end of her reign, with the queen surrounded by just about all the men of influence and power of the day, bearing her aloft. At her death, an effigy, gowned and wearing one of her wigs, was carried on her coffin.

Fig. 18: James Abbott McNeill Whistler, *Symphony in White, No. 1: The White Girl*, 1862. Whistler came to object strongly to the notion that his paintings had any meaning beyond being art. He described this portrait of Joanna Hiffernan as ". . . a woman in a beautiful white cambric dress, standing against a window which filters the light through a transparent white muslin curtain—but the figure receives a strong light from the right and therefore the picture, barring the red hair, is one gorgeous mass of brilliant white."

Fig. 19: Gustave Courbet, *The Sleepers,* 1866. The naked curves, the tumbled, loosened hair… Joanna Hiffernan and an unknown dark-haired model, posed by Courbet for the delectation of Halil Bey, and for the rest of us ever since. The painting is now in the Petit Palais in Paris.

Fig. 20: Algernon Charles Swinburne, painted by G. F. Watts in 1867. At the height of his alcoholism, in 1879 Swinburne would be taken into to the Putney home of the poet and critic Theodore Watts-Dunton, who also provided a refuge for Henry Treffry Dunn after the latter quarreled with Rossetti.

Fig. 21: *Beata Beatrix* (1864–70). Lizzie Siddal as painted by Dante Gabriel Rossetti after her death. The red-robed figure might be taken as Dante Aligheri, waiting to escort her in the Underworld; the sundial is a centuries-old emblem of mortality. The poppy in the dove's beak alludes to her death by an overdose of laudanum.

Fig. 22: Dante Gabriel Rossetti, *Found*, *c.* 1869 (uncompleted). The calf is meant to typify the woman's plight. Rossetti published a highly charged poem, "Found," in his *Ballads and Sonnets* of 1881; the real Fanny Cornforth, who posed as his model, was photographed proudly wearing the gold earrings of the figure here.

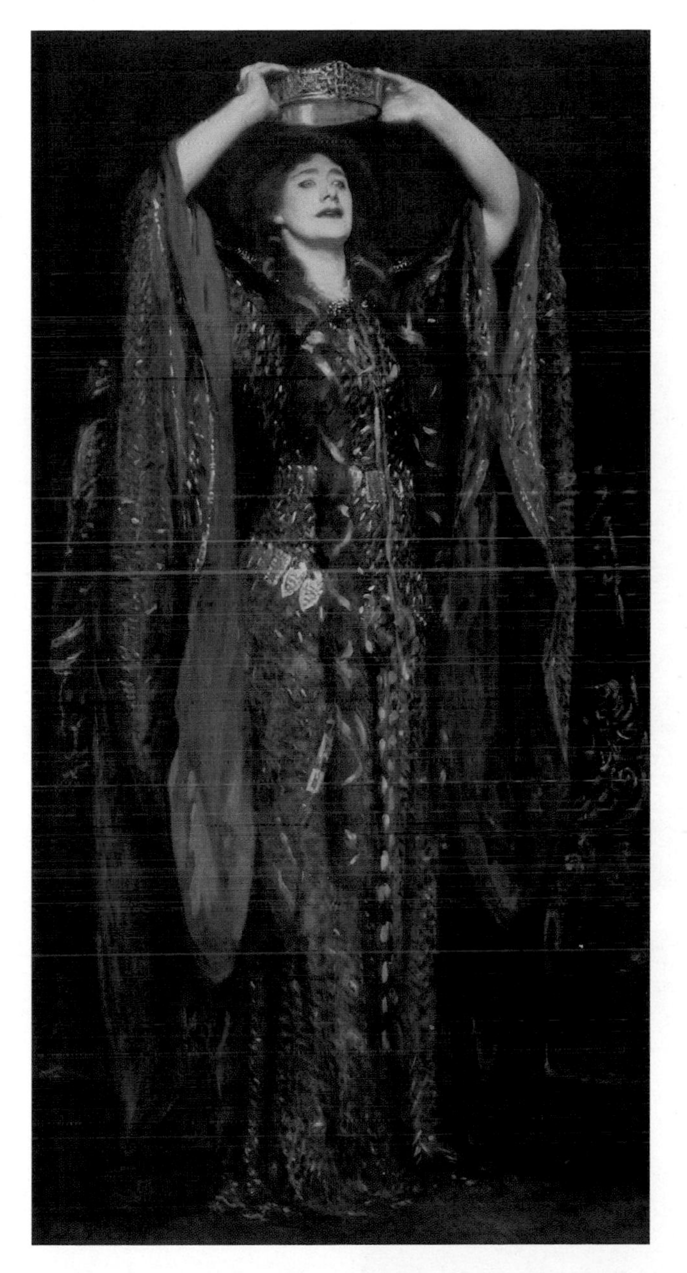

Fig. 23: John Singer Sargent, *Ellen Terry as Lady Macbeth*, 1889. Terry wrote of the painting, "It's talked of everywhere and quarreled about as much as my way of playing the part. . . . Sargent has suggested in this picture all that I should like to convey in my acting."

Fig. 24: Sir John Everett Millais, *The Knight Errant* (1870) and *The Martyr of the Solway* (1871). The critics had a field day with *The Knight Errant*, seeing evidence on the woman's body of "the ligature of draperies," meaning she had been undressed, and in her face a character not "over pure" or "refined." In other words, whatever is supposed to have happened to her, she asked for it.

The red-haired flirt as a toned-down, more socially acceptable version of the red-haired Pre-Raphaelite femme fatale was a new development for a new century, and owes much to Mrs. Glyn. But even more tellingly, by the time *Red Hair* was filmed, in 1928, complete with a very early Technicolor sequence (to do justice to the heroine's red hair, perhaps?) and starring Hollywood's first redheaded sexpot, Clara Bow, Clara's fellow starlet Mabel Normand could declare in an interview, "I'm shanty Irish"—and be proud of it.

A Scotsman, Tam Blake, was perhaps the first of the Celtic diaspora to make it to the New World, in 1540 (although there are legends that the aptly named St. Brendan the Navigator beat both the Vikings and Christopher Columbus to it, in the sixth century, thus opening up a whole new field of possibilities for those Native American stories of red-haired giants). Tam would be followed by more adventurers in both the sixteenth and seventeenth centuries. Red hair—as with the Normans in Sicily—has always been a convenient marker of human migration. So it was to be again. In the 1650s, under Oliver Cromwell, tens of thousands of Irish were shipped as slaves to the West Indies. In the eighteenth century many more were transported as convicts to Australia. In what has been called the final act in the great Celtic diaspora, in the nineteenth century there were immense migrations from both Scotland and Ireland to North and South America, to Canada, and to Australia and New Zealand. Once again, the gene for red hair went with them.

Clara Bow herself, the "It" girl of Hollywood in the Roaring Twenties (a sobriquet coined for her by none other than Elinor Glyn) owed her head of bubbling red curls to her Anglo-Irish and Scottish ancestry. Bow emerged from a childhood and adolescence of great hardship and tragedy with an underlying sense of herself as being set apart and thus extremely vulnerable, which she sought to disguise for all she was worth. She spoke of herself as being awkward and funny-faced and was teased for her coppery ringlets at school as many redheaded children are, yet in front of a camera she had "It" and to spare. Hollywood, and America, had discovered the redhead.

It is one thing to have your hair color as a badge of your underclass status in the Old World. It is quite another to carry it as a marker of your identity into the New. Transpose red hair into this environment and it comes to mean something completely different. It becomes a mark of authenticity as well as of identity. Rather than a stigma, it becomes something to celebrate, a bold visual claim to your heritage and history. And it must be remembered that in America there was already an underclass, marked out by the color not of its hair but of its skin. Immigrants from the Old World might indeed be regarded as of lowly social caste by those white-skinned Americans born in the New World, just as the Irish has been by the Anglo-Irish in the Old. They might still suffer what Noel Ignatiev in his *How the Irish Became White* defines as the "hallmark of racial aggression, the reduction of all members of the oppressed group to one undifferentiated social status." But whereas in the Old World there had been no buffer, no slave community beneath them, in America there was. In fact there were two: first the black community, and then as another wave of immigrants reached the New World from the Old, Russian Jews fleeing the pogroms of the 1880s.[75] The Irish in America could lift their social class simply by crossing the Atlantic. All the mighty difference created by their doing so is wrapped up in the one American concept of the despised "redheaded stepchild." When the Irish first began to arrive in numbers, the black community found itself referred to as "smoked Irish." The Irish in turn heard themselves stigmatized as "blacks turned inside out."[76] The Historical Society of Pennsylvania records the complaint of one black laborer: "My master is a great tyrant. He treats me as badly as if I was a common Irishman." The Irish had to decide, in effect, if they were going to identify with the slaves or with the oppressors. They chose the latter, and no difference, as Ignatiev points out, is fought for more fiercely than the thinnest. The phrase "beat you like a redheaded stepchild" in the States goes back very far and to somewhere very ugly, but I don't believe it has anything to do with the original lowly status of the Irish at all, nor as such with red hair. It indicates instead a child of mixed race, and originally very possibly the offspring of a black slave woman and a white slave owner. Hence the following exchange from Harper Lee's *To Kill A Mockingbird*, between Jem and his sister, Scout:

> "Jem," I asked, "what's a mixed child?"
>
> "Half white, half colored. You've seen 'em, Scout. You know that red kinky-headed one that delivers for the drugstore. He's half white. They're real sad."

75 Hollywood was to create something of a meme around this in the 1920s with films such as *The Cohens and the Kellys* (1926) and *Abie's Irish Rose* (1928).

76 See Ignatiev, *op cit.* Except of course that the word used wasn't "black."

> "Sad, how come?"
>
> "They don't belong anywhere. Colored folks won't have 'em because they're half white; white folks won't have 'em cause they're colored, so they're just in-betweens, don't belong anywhere."

In other words, unless they did something about it, the lowest and least-loved of the low. The magic of codominance means that the gene for red hair can certainly manifest itself with a black heritage—the red tint to Malcolm X's hair came from a Scottish grandfather. Nonetheless, and despite this unlovely phrase, there are, for me, very pertinent differences in the responses to red hair from one side of the Atlantic to the other. In the United States, where the legacy of black slavery has meant that social awareness of any sort of stereotyping as undesirable runs much higher than it does in Great Britain, I have often been told that there is no such thing as discrimination against redheads in the States (although many an American redhead may feel they reserve the right to disagree). Even more striking, however, are the differences in attitudes toward red hair between Canada and the States (where white immigration in the nineteenth century was elective), and Australia, where there was no urbanized underclass "other" already conveniently in existence, where immigration was often the result of judicial punishment, and where red hair carried with it all the connotations marking one out as very likely having been transported there not as an underclass but as a criminal.

Red hair in women might have become newly popular in the Old World in the nineteenth century, but there was no answering change in attitudes toward the redheaded man. In fact this period saw the creation of two of the nastiest redheaded wrong'uns ever to fall from between the pages of a book: Uriah Heep, in Charles Dickens's *David Copperfield*, of 1850, about as fell a villain as Dickens ever created; and in 1898 the character of Peter Quint, in Henry James's *The Turn of the Screw*.

Heep, as he is introduced to us, already has his ginger hair shorn convict (or we would say skinhead) short (Fig. 27). He has in his face, "in the grain of it . . . that tinge of red which is sometimes to be observed in the skins of red-haired people." In other words, he is another take on the ruddy-faced, red-haired impenitent thief of the Crucifixion. We are to understand that Uriah's true nature—grafting, scheming, leading every soul around him astray, if he can, like the devil in a medieval morality play—is announced by that telltale red coloring, seeping through the grain of his skin. Heep is sexual. He both fascinates and repels the hero, David Copperfield, in equal measure; he is Machiavellian, and he is very nearly triumphant. In 1898, Henry James would also draw upon this ancient connection between the color red and the whiff of brimstone for the character of Peter Quint.[77] Quint has a particularly sickly connection to the two young children in *The Turn of the Screw*, and he has returned in death to haunt both them and the novel's narrator, an unnamed governess. In the governess's description of him, Quint's red hair not only serves

77 The connection between the color red and the infernal is still being used today, in the *Hell Boy* comic book and film franchise, for example.

as a marker of his diabolical nature but identifies him, even if he should be in his grave, to her listener:

> . . . he has red hair, very red, close-curling, and a pale face, long in shape, with straight good features and little, rather queer whiskers that are as red as his hair.

Heep and Quint are nineteenth-century embodiments of far older fears, and in both cases their red hair marks them out as being apart from the normal, law-abiding, or even laws-of-nature-abiding society in which they wreak such havoc. These two come from somewhere very different indeed.

But there is a change of heart toward red hair in literature for children. First in France, in 1894, there is Jules Renard's semi-autobiographical *Poil de Carotte*, or *Carrot Head*, as it is usually known in translation; and then in the New World, *Anne of Green Gables*, in 1908.

Poil de Carotte is not a children's book, even though it is a stalwart of French literature classes in school. It is instead a book about a child, and what it is to be a child growing up in a household where your father hates your mother, and your mother takes out her hatred of her own life on her youngest, redheaded son. The fact that Poil de Carotte (he is hardly ever referred to by his given name, Francois) is a redhead is almost hidden; what the book does is present the interior workings of his life, his insecurities and fears, his attempts to make sense of the world as being no different to those of any other child. He is neither more nor less disobedient, neither more nor less hot-tempered. He is a heartbreakingly human and unhappy little boy who simply happens to have red hair. Meeting the actress

Sarah Bernhardt, once he too had become a celebrity, Jules Renard (fittingly, his surname means "fox") recorded in his *Journal* a faltering recognition of the shift to a more sympathetic awareness to which his book had contributed, when the divine Sarah tries to excuse the fact that he is a redhead: "Redheads are ill-natured. . . . You are rather on the blond side."

And then of course there is Anne. As envisaged by her creator, Lucy Maud Montgomery (a Canadian—would Anne have been the same had she been conceived in the Old World?), "she wore a faded brown sailor hat and beneath the hat, extending down her back, were two braids of very thick, decidedly red hair. Her face was small, white and thin, also much freckled."[78]

Anne, like David Copperfield, is an orphan, and as a character is related to those formulaic orphan children of so many children's stories (get the troublesome parents out of the way and have the hero or heroine stand in their own light), but Anne has a brain and a temper. Long before the ineffable Australian comic Tim Minchin, and his song, "Only a Ginger Can Call Another Ginger Ginger," Anne hits the nail exactly on the (red)head: "There's such a difference between saying a thing yourself and hearing other people say it." When Anne's coloring is commented upon in her hearing by the unpleasant Rachel Lynde ("Did anyone ever see such freckles! And hair as red as carrots!"), she fires up in her own defense at once.[79]

78 Anne was supposedly based on Evelyn Nesbit, a famous pin-up girl of the period, another onetime artists' model, and the center, in 1906, of a world-famous scandal (dubbed the "Trial of the Century") when her husband shot dead her lover, the renowned architect Stanford White, in Madison Square Garden. Notwithstanding, she was Montgomery's inspiration.

79 A perfect example of why prejudice against red hair is so pernicious. Thanks to this business of there

Yet, for all her spirit, Anne too longs for her hair to be "a handsome auburn" when she grows up.

Anne's quick wit and precocity are, if you like, an acceptably infantilized version of the red-haired flirt (Evangeline, Clara Bow), the girl with a twinkle in her eye and a smart riposte but with her virtue intact. Via Anne, the redhead female juvenile lead descends to Little Orphan Annie herself, in the 1920s, and to the redheaded tomboy's all-time heroine Pippi Longstocking, living adult-free in her Villa Villekulla with her monkey and her horse. Pippi was first published in 1945; to my mind Jessie the Cowgirl of *Toy Story 2* (1999) owes much to Pippi.[80] Thence and more recently we come to the independent-minded and very freckly Freckleface Strawberry of today. And let's not forget the Little Red-Haired Girl in *Peanuts*, object of as devoted an adoration on the part of Charlie Brown as ever Alfred Polly suffered for his red-haired sweetheart, sitting on her school wall, forever out of reach.

It's children's fiction that also finally creates a male redheaded character neither evil, ill-intentioned, nor a milksop. Tintin, first published in 1929, is resourceful, adventurous, and intrepid and has a signature cowlick of red hair (Fig. 28). And, he is the hero. William, of the *Just William* stories of the 1920s, has his trusty sidekick, Ginger, but he is a sidekick only. (William also has a redheaded and

---

being no "cultural barriers," no obvious visual difference between the victimizer and the victim, just as there are none between Rachel Lynde and Anne, it doesn't look like the prejudice it is.

80 Following criticism of Barbie's sexually mature face and figure, early versions of Midge were closer to the tomboy model, with freckles and a rounder face, in an attempt to make her (and Barbie) seem less adult. Midge today is rather more of a flame-haired siren than she was. The same freckles equals cute and wholesome theme is at work in the logo for the Wendy's fast-food chain.

at least by implication devastatingly soigneé and attractive older sister, Ethel, whose popularity with any number of eligible bachelors in the neighborhood is an eternal mystery to her younger brother.) *Biggles*, first published in 1932, also has a "Ginger" as his sidekick, thus continuing a tradition still used by writers today—think Ron Weasley, second-in-command to Harry Potter. But Tintin, despite being the youngest character in the books, is the protagonist, has a personality full of all the stalwart virtues of the Boy Scout, and is a first. Do these juvenile leads offset the centuries of prejudice against redheaded men in art and literature, their use as shorthand for villainy of every sort? I think that they at least start such a transformation, and they certainly record a shift in society's attitudes, away from "redheaded woman good/redheaded man bad" to something less unthinking and rather more nuanced. But it takes many, many Tin-Tins, Gingers, and Rons to expunge the centuries-long prejudice reflected in a single Uriah Heep. And what's Heep's come-uppance, when all his schemes and machinations are foiled? Transportation for life, of course—to Australia.

RAPUNZEL, RAPUNZEL

Well, listen up, stud,
Your life's been wasted
'til you've got down on your knees and tasted
A red-headed woman . . .

BRUCE SPRINGSTEEN, "REDHEADED WOMAN"

Augustin Galopin was a French man of letters, of philosophy, and of medicine. By 1886 he had published more than twenty works on subjects ranging from cremation to feminine hygiene. The health and well-being of the female sex was clearly a matter of great concern to Dr. Galopin, who from his writing it is very easy to imagine as a dapper Fernando Rey type, strolling along through the Luxembourg Gardens, savoring the smell of the horse-chestnut flowers (which, according to the Marquis de Sade, smell of spunk) and tipping his hat to any particularly *jolie dame* who happened to catch his eye. This was the Paris of Toulouse-Lautrec, after all. It was also the city in which Dr. Galopin let loose his *Le Parfum de la Femme* upon an unsuspecting world.

Le Parfum de la Femme is a winning mix of folk wisdom and high-blown science, spiced with anecdotes and Dr. Galopin's own observations and musings. He informs us for example as gospel truth that a plant, the Dutchman's pipe (*Aristolochia*), has often been observed to kill serpents—*Aristolochia* is remarkably toxic and possibly carcinogenic, but its anti-serpent properties are unproven, to say the least—and he then notes as an aside what sounds like a proper empirical scientific observation: that if you have been handling female toads, and place your hands in water, male toads will rush to them. Galopin claims that he is a materialist, hence all is to be proved by experiment and observation, which makes some of his assertions more than a little hard to take. While assuring us that exposure to tobacco provokes St. Vitus Dance, or chorea, and masturbation in children, he also claims to have broken the addiction to tobacco by substituting coffee and sugar (which for the anxious parent hardly sounds like an improvement). And then there are his views on female odor.

Galopin believed, rather charmingly, that each woman, classified by skin and hair color, had her terroir, like a wine, and gave forth a specific bouquet of scents. For example, women with chestnut hair, so he assures us (those chestnuts again!), give forth an odor of amber. Perfumiers do recognize a type of scent classified as amber. It's chiefly composed of vanillin and labdanum, a plant resin.[81] The good doctor may have meant this, as vanillin was a creation of the late nineteenth century, but as Galopin uses the term, and in the context of the passage paraphrased above, it seems more likely that he was talking about "amber" as shorthand for "amber*gris*."

Now ambergris, for those who have not sampled it, has an odor

81 Thanks to http://perfumeshrine.blogspot.com for this information, and many another fascinating insights into the history of scent.

so strong it's as if the human nose doesn't know what to do with it. It is salty and marinelike with base notes that are, frankly, fecal. To put it bluntly, if the sea emptied its bowels, the result would smell like fresh ambergris, and unsurprisingly, too, when you consider that this extraordinary substance is produced as some mysterious digestive aid in the stomach of sperm whales. Once it has aged and oxidized, however (which can take years—ambergris is soft and waxy and it floats, and ideally, once voided by a whale, would spend those years bobbing about in the ocean), its smell sweetens, although it never quite loses that animal tang of the midden, to something subtler and, for want of a better word, hormonal. It is impossible to smell ambergris and not think of sex. Its special value to the world is that it is an excellent fixative and carrier of other scents, making them last much longer, hence in an age when the science of artificial perfumes was in its infancy, it would have been much more commonly encountered than it is today. But according to Dr. Galopin, this was the natural scent of women with chestnut hair. "And some women with that hair color, who have very white skin, exhale a soft odor of violets from most of their sebaceous glands." Then "when they are hot . . . the coquettes pretend they don't know the ravages their perfume molecules make in the brains of those who breathe them in." The little minxes, shame on them. And that, you might imagine, would be the end of the matter; simply another pseudoscientific myth about red hair. You would be wrong. Redheads *do* smell different. Or rather, if you have red in your hair, anything applied to your skin is going to smell different from the way it will smell on anyone else.

The biochemistry of the human animal, as modern science is starting to unravel its secrets, is more complex and more fascinating than anything even Dr. Galopin might have imagined, and to those in the field it must sometimes feel as if every discovery yields up yet another mystery. Biochemists and geneticists are in something akin to the situation of explorers, attempting to understand the layout of a lost city in a jungle. Each starts in a different place. Each has a piece of a map. Gradually the main roads and byways and connections both expected and unimagined begin to emerge. And in the case of the boulevardier-biology of Dr. Galopin, the most surprising claims, the oldest of old wives' tales, can prove to be entirely correct.

Much of this, particularly for us layfolk, lies at a level of microscopically complex science that can best be dealt with by employing the principle of "Don't Worry About It."[82] We do not need to be capable of understanding, at a glance, the information that MC1R is a seven-pass G-coupled receptor located at chromosome 16q24.3 (although it is pleasing to have what you might call our red hair's postal address), or indeed that it is part of family of genes, from MC1R to MC5R. What is of interest here is the list of biological functions along with hair pigmentation in which these genes play a part. Among them are adrenal function; responses to stress; the fear/flight response; the pain and immune response; energy homeostasis (the body's chemical regulation of its use of energy); and sexual function and motivation. Briefly, all these fundamental functions of the human body are different for redheads from those of blonds or brunettes as a result of redheads' uniquely different biochemistry,

82 Professor Sydney Brenner, who won the Nobel Prize for medicine in 2002, created this hypothesis in 2011. In his words, "It forces you to keep going without losing your mind over mechanisms." *Metode*, interview with Sydney Brenner, http://metode.cat/en/Issues/Interview/Entrevista-a-Sydney-Brenner.

the consequences of which have all fed into the stereotyping and societal and cultural attitudes evinced toward the redhead for centuries. And while these differences are far more than skin-deep, that is where we start.

We all, on the surface of our skin, have a microscopically fine film known as the skin mantle. It acts as a barrier to bacteria and other contaminants, and on those possessing the gene for red hair it will very often be more acidic than on the skin of a blond or brunette. This is why any scent or cologne will smell different on a redhead from the way it smells on his or her non-redheaded girl- or boyfriend.[83] Nor does scent last as long on a redhead's skin as it does on that of a blonde or brunette, causing Dr. Galopin to bewail the fact that in the heady days of the first artificial perfumes, redheads used so much synthetic scent and products so concentrated "that they asphyxiate all those who approach them." The havoc redhead skin wreaks on the products of the perfumier's art is strange enough, but it also suggests another reason, one with a flawlessly scientific basis, for the perceived sensuality of the female redhead: pheromones. Those same sebaceous glands whose "perfume" molecules so disordered Dr. Galopin, and which secrete the skin's acid mantle, also produce pheromones, most generously from those parts of the body that still retain our original covering of hair, that is, the genitals and under the arms.

Pheromones—messages in a smell, basically—are an invisible

83 Susan Irvine, *The Perfume Guide* (London: Haldane Mason, Ltd., 2000), 9. Opium, on me, for example, the notorious spice-bomb scent of the 1980s, smells like a cat that has just come back from the vet's. If you want a scent created for redheads, and can find it, the Perfume Shrine blog recommends Patou's "Adieu Sagesse," created in 1925. It has a charming story and celebrates the throwing of caution to the winds in a love affair.

Morse code by which we share signals about our general state of health, and our receptiveness to a mate, unwittingly and with the whole of the world around us, twenty-four-seven.[84] If, on a redhead's skin, the scent from a bottle can be so transformed by the chemicals produced by their sebaceous glands, I suppose it's possible that redhead pheromones also differ, uniquely, from those of blondes or brunettes, and send forth some secret message of their own. There are endless articles on endless websites dedicated to promulgating the idea that redheads have more sex than blondes or brunettes, that sex with a redhead (inevitably a female redhead) is somehow "better" or more passionate, and that part of this derives from a redhead's particular bouquet. It has become a literary device. Jean-Baptiste Grenouille, the antihero of Patrick Süskind's novel *Perfume* (1985), commits murder to capture the scent of a redhead:

> Her sweat smelled as fresh as the sea-breeze, the tallow of her hair as sweet as nut oil, her genitals were as fragrant as the bouquet of water-lilies, her skin as apricot blossoms. . . . The harmony of all these components yielded a perfume so rich, so balanced, so magical, that every perfume Grenouille had smelled until now . . . seemed at once to be utterly meaningless.

Aristide Bruant, a cabaret singer and friend of Toulouse-Lautrec (he is the man in the red scarf in the artist's famous *Ambassadeurs* poster) contributes this, from his 1889 song "Nini Peau d'Chien":

> *She has soft skin*
> *And freckles*

84 Many years ago I was propositioned by a colleague who was married to one redhead, in the midst of a torrid affair with another, and declared he had reserved a room at his club for the two of us. Even for me, this was redhead overload. Refusing his offer as graciously as I could, but nonetheless intrigued, I asked him what was this thing with him and redheads. He answered, somewhat sheepishly, "You smell different."

And the scent of a redhead
That gives you the shivers.

Before either of them there was Charles Baudelaire and his beggar-girl, the "Blanche fille aux cheveux roux," whose body, dotted with freckles, "has its sweetness."[85] (The beggar-girl of the poem, with her dark chestnut locks, was painted by Baudelaire's friend the artist Émile Deroy around 1843–5. You can see her portrait in the Louvre.) And if we smell different, might we taste different as well? I have no idea, but The Boss seems to think so—and he would be a man to know.

We need to talk about sex.

The psychiatrist Charles Berg was a Freudian of the old school. To give you a flavor of his writing, from *The Unconscious Significance of Hair* (1951): "The Christmas tree has been associated with hair and with father's penis. We see Father Christmas (with his long beard) taking off the tree penises, which he benevolently gives to children. At the feast, the phylogenetic successor of the old totem feasts, his penis is eaten in the shape of an appropriate symbol—turkey or goose." (I promise I am not making this up.) Part of *The Unconscious Significance of Hair* deals with case studies, one of which,

85 From his collection of poems *Les Fleurs du Mal* (1857). Another poem from the same collection, "Delphine et Hippolyte," was a part of the inspiration for Courbet's *The Sleepers.*

as related by Dr. Berg, includes a dream told to him by a young male patient, in whose psyche Dr. Berg was rooting for any number of unacknowledged neuroses. In this dream, the young man was sitting on a bus, and putting out his hand to touch the red hair of the woman sitting in front of him, experienced what Dr. Berg describes as intense pleasure, and which you or I, if we wanted to be equally euphemistic, might term a visit from Lady Lilith. In other words, an erection, if not an ejaculation. And why? Because the young man had recently succeeded in convincing his girlfriend to let him remove her underwear (this was the 1950s, let's not forget) and on first seeing her pubic hair, had been delighted to discover that it had a reddish tint. Dr. Berg, I suspect, was not a fan of the redhead (according to him, redheads had "a supernormal capacity for rheumatism, chorea and TB," as well as "detumescence") and he notes nothing significant, unconscious or otherwise, in the young man's reaction. You or I might beg to differ. In fact I'll hazard a guess that the reason for the young man's delight was that, however respectable his girlfriend might be on the surface, this was proof-positive that in secret, and known only to him, she was hot stuff.

What *does* red hair mean? Not what does it mean for a redhead, but what does it mean for everyone else? Above all, what is this mysterious connection made with such constancy between redheaded women and sex? Or, to quote the writer Tom Robbins, in a favorite paean of redhead prose, "How are we to explain the power these daughters of ancient Henna have over us bemused sons of Eros?"[86]

86 The entirety can be found at http://www.angelfire.com/az/varuna/ode.html.

We've already speculated that red hair could have indicated to a mate that you would bear healthy children, and not die yourself in the process. Dr. John Cook, writing for the *Ladies' Magazine* in June 1775, offers this: "Red hair is not so agreeable, though this I can say, such women have the finest skins, with azure veins, and generally become the best breeders of the nation." Let's get down to the nitty-gritty here: what does it mean when you're naked?

To begin with, phaeomelanin, the chemical that colors red hair red, also colors those parts of the body chosen by Nature to stand out as pink. That is, the nipples, in women the labia, and in men the glans of the penis. Set against a redhead's normally pale skin, a naked redhead, male or female, is thus flashing a set of sexual super-stimuli at their partner, doubly so when aroused, when the coloring in these parts of the body deepens. (I am also told that again, due to the pale skin, when a redhead reaches orgasm, the skin flush is particularly noticeable and gratifying.) Could this be another reason why, to quote Grant McCracken, red hair in a woman is seen by (male) society as promising "sensual delights of extraordinary proportions"? Redheads are rule-breakers, rebels. Grant McCracken again: "We cannot rely on them [female redheads] to embrace stereotyped qualities of femaleness—sweetness, docility and politeness. . . . We imagine them ready to give vent to what we keep harnessed." This notion has been around for centuries. Jonathan Swift, in his *Gulliver's Travels* of 1726, has the red-haired members of his imaginary race the Yahoos being "more libidinous and mischievous than the rest." Red hair is a warning flag: here comes trouble. Sexually, this is a charged and potent mix. Redheads are different, so redheaded women, perhaps the thinking goes, *might* do things other women won't. . . .

There is of course a difference between the perceived and the actual experience of being a redhead, which one might liken to the difference between pheno- and genotype—except that where sex is concerned the two are not so easy to separate.[87] Does a partner's expectation affect sexual behavior? I would say yes, most definitely. If cultural attitudes toward your hair color give you license, as it were, to be articulate and confident in bed, will better sex result? Without a doubt. Will you be happier and more confident in your own sexuality, if you anticipate eliciting a positive response from your partners? What do you think?[88]

Then there is the part played by pheromones. There are good reasons to believe that we have retained body hair under our arms and around the genitals because the hair helps disperse pheromones into the air, and in the case of a redhead, one of the messages those pheromones is carrying is health. Again, this is a highly desirable quality in a mate, but, sadly, it has nothing to do with the red hair—red hair is simply the signifier; it's the pale skin that makes the difference, and the ability to synthesize vitamin D.

Most of the vitamin D we need is made in the skin, in response to exposure to UV radiation. The farther north you go, however, the fewer days there are in the year when vitamin D can be produced, and the more when vitamin D will be broken down by the body for

---

87 Briefly, your genotype is your genetic coding; your phenotype is a composite of that plus every other thing that acts on you to alter your observable characteristics.

88 A much-cited survey, conducted in 2006 by a Hamburg sex researcher, Dr. Werner Habermehl, appeared to suggest that redheaded women, at least in Germany, had a more active sex life than did those of other hair colors. Dr. Habermehl, however, confuses quantity with quality, which is a pretty startling error; for other issues with this survey see http://www.drpetra.co.uk/blog/do-redheads-really-have-more-sex.

use. And without enough vitamin D, everything eventually stops working. Hence the wretched fate of the Viking settlers at Herjolfsnes in Greenland.

Herjolfsnes (now Ikigait) was named for one Herjolf Baardsen, a follower of Erik the Red, and founded in 985. It was a tough and inhospitable place to live, but the Vikings were nothing if not tough themselves, and the point where the settlement was founded was the first landing place for ships that had trekked from Iceland or even Norway, so the community here should have been set to thrive. It did not. It simply disappeared. For many years the native population of Inuits were blamed for the settlement's demise. Then in 1921 the remains of Herjolfsnes were examined by archaeologists, who found there the ruins of a church and other buildings, and numerous burials, preserved by the cold.[89] Herjolfsnes is a Tarim of the snows.

The bodies in the graves told a sad story of decline. Shrouds were patched and reused; coffins, too. Those laid to rest were of noticeably short stature. The account of the excavation records with sadness how many of them were very young and notes too that "a conspicuously large number of the women were of slight and feeble build: they were narrow across the shoulders, narrow-chested, and in part narrow at the hips. Several showed symptoms of rachitis [rickets], deformity of the pelvis, scoliosis and great difference in the strength and size of the left and right lower extremities." Their teeth were worn from eating hard vegetable matter—these were people who had existed on a starvation diet for generations. It's been

89 William Hovgaard "The Norsemen in Greenland: Recent Discoveries at Herjolfsnes," *Geographical Review*, 15, no. 4 (October 1925): 605–16.

conjectured that by the time they needed to, the Viking settlers were too enfeebled to hunt seal and fish as did the Inuit. The climate had worsened. Fewer children were born. Those who were died young. The longboats from Norway had ceased. The community died of slow physical deterioration. It was killed, in fact, by the long Arctic winters and a lack of vitamin D, from which the Inuits' meat-rich diet protected them. The body of the last Viking was reportedly found by the Inuit in 1540, on the floor of his dwelling, where he had died, alone, his sheath-knife "much worn and wasted" by his side.

This is salutary enough. But vitamin D deficiency has also been implicated in an increased risk of certain cancers, of hypertension, cardiovascular disease, diabetes, autoimmune diseases such as multiple sclerosis, rheumatoid arthritis (*pace*, Dr. Berg), irritable bowel syndrome, and gum disease. Its role in preventing rickets has been known since the 1930s. And in the days before we all lived long enough to be so troubled by cancers of one sort or another, it was a weapon in the arsenal against the then all-time killer, tuberculosis. Plenty of vitamin D gives you a stronger immune system all around. So if your red hair comes with that pale skin, one of the messages your pheromones will certainly be carrying will be a message of health, and of an immune system boosted with resilience.[90]

Pale skin has also, in both East and West, been prized for centuries as an attribute of female beauty. It speaks of seclusion, of being kept apart, of Rapunzel's tower, and, to a degree, of ownership. It meant you did not have to earn your bread in manual labor.

---

90 A. W. C. Yuen and N. G. Jablonski, "Vitamin D in the Evolution of Human Skin Colour," *Medical Hypotheses*, 74, no. 1 (January 2010): 39-44.

It speaks, basically, of the harem, the seraglio. But in men, the message carried by pale skin is completely different, and it seems at least possible that one of the reasons why red hair is so gendered, why what is regarded as an acceptable if not indeed a highly desirable characteristic in one sex is seen as being so much less desirable in the other is because redheaded, pale-skinned men are presenting what many cultures have regarded for centuries as an attribute of female beauty.[91] Pale skin in man is a quality ascribed to the milksop (just look at the word)—someone too unhardy to go out into the world and make his way with his fellow men. In fact a 2006 study found that a significantly higher proportion than you would expect of CEOs had red hair, extrapolating from its rarity in the population at large, which may say much of interest concerning the effect on character of having to overcome being teased or bullied because of your hair color early on in life.[92] In our idiosyncratic human way, however, does the fact that something is nonsense stop it from being believed? No, it does not.

But then so much about hair is gendered, and is completely opposite from male to female. "If a man have long hair it is shame unto him." So wrote St. Paul in his first letter to the Corinthians. "But if a woman have long hair, it is a glory unto her." In his 1987 study *Shame and Glory*, the sociologist Anthony Synnott lists the following examples of contradictions between the sexes: hair on

91 Intriguing in this context that redheaded men are reportedly much more likely than women to describe themselves with the mildly pejorative term "ginger." Redheaded women are much more likely to describe themselves as "strawberry blonde." See Eleanor Anderson, *op. cit.*

92 Margaret B. Takeda, Marilyn B. Helmo, and Natasha Romanova, "Hair Colour Stereotyping and CEO Selection in the UK," *Journal of Human Behaviour in the Social Environment*, 13 (July 2006): 85-99.

the body is seen as good on men, but bad on women because it is regarded as a "male" characteristic, and is therefore with much labor and some pain removed (could this also be something to do with wanting to show off that pale, feminine skin?).[93] Men uncover their heads in a holy place; women cover theirs. Men rarely change their hairstyle—a man can have basically the same cut as he had as a boy in short trousers, which may also be pretty much the same hairstyle as his father had before him, and this is seen as entirely normal and acceptable; whereas women change their hairstyle with far greater frequency (Synnott gives a particularly telling example: when was the last time you saw a man change his hairstyle simply to go out to dinner?), and this is seen as entirely normal and acceptable for them. The norms of male hair tend toward unchanging uniformity—Synnott cites the example of a man with the same haircut JFK sported fifty years ago being wholly unremarkable today, whereas a woman styling her hair like Jackie O. would be regarded as making a statement that was very consciously retro. Women want their hair to be individual, to stand out. The conspicuousness of red hair, for a woman, its very rarity, thus makes it an advantage to her (which is no doubt why red hair dyes have such a large share of the market). It does exactly the opposite for a man, for exactly the same reason and by exactly the same means. Is this another factor behind red hair being regarded as less desirable both in and very often by the men who have it? Because it denies them that ability to

93 Another anomaly of being a redhead: laser hair removal rarely works, if at all, on red hair on pale skin—the laser needs the dark color of eumelanin at the root of the hair in order to heat it and destroy it. Redheads are doomed to the lengthy and unpleasant process of electrolysis if they want permanent hair removal.

blend in? Finally—and never mind numbers of redheaded CEOs—Synnott quotes another survey, this time from the United States and conducted in 1979. The finding here was that redheaded women were regarded as the executive type: brainy but no-nonsense, and slightly scary to the opposite sex (think Agent Scully: *The X-Files* played with this trope series after series); while redheaded men were regarded as "good but effeminate—timid and weak." It's as if the sexes had swapped their usual stereotypes entirely.

And there may be another reason why redheaded men have a reputation for a certain wimpishness. Once again it is a result of the redhead's unique biochemistry. Redheads feel more pain than do blonds or brunettes. Or rather, we feel the same amount of pain much more acutely, and thus require much more anesthesia to knock us out—20 percent more being the rule of thumb among anesthetists and surgeons I have spoken to. There is as you might imagine much discussion as to why this should be, how much it varies from redhead to redhead, whether some forms of pain (thermal, for example) are better or worse tolerated by redheads, and which and what anesthetic drugs are thus contraindicated. We do not bleed more than those of other hair colors. That is a myth (although I have also heard it stated as fact by one surgeon, at least, that we do). We do not bruise more easily than those of other hair colors. That is another myth, and very likely a result of the fact that bruises show up so much more noticeably on pale skin. But we do indeed require more anesthetic, and we all have horror stories of trips to the dentists as children when we weren't given enough. Unsurprisingly, redheads are notoriously bad at keeping dental appointments, having injections, and as children, having knots dragged out of our red hair (think of the poor woman in Degas's *La Coiffure*, wincingly holding on to her roots as her hair is combed out); but paradoxically, a normal level of pain for us would reduce many a blond or brunette to tears. The stoicism thus engendered nonetheless seems like a rather poor evolutionary trade-off to me. Why on earth should this pointless and painful adjunct of having red hair exist? Redheads also react badly to cold, reporting physical pain at temperatures perfectly bearable by non-redheads, although, paradoxically, we can eat highly spiced "hot" food with no discomfort at all. Madras curries and Scotch bonnet peppers hold no terrors for the redhead. There is also a belief, repeated on websites without number, that redheads are unusually susceptible to industrial deafness. Red hair is linked to brittle cornea syndrome, and there is an adrenal malfunction, one of the indicators of which is red hair, that is linked to early-onset obesity.[94]

With redheads, even the hair itself can be troublesome. Hair is made of keratin, the same substance as fingernails, and its character as well as its color derives from the shape of the follicle and the pigment-producing cells in the follicles. Beyond that, lively arguments are still under way as to why we have hair, why we have it where we do and not where we don't, and why there should be so many different types. Redheads have around 90,000 hairs on average, fewer than blonds and brunettes. Princess Merida's computer-rendered 1,500 separate strands, which would equate to about 112,000 actual hairs, are therefore well above average. Her

94 See http://www.ncbi.nlm.nih.gov/pubmed/9620771:

curls, however, would have to be natural; the keratin of red hair also contains more sulphur (up to twice as much) than hair of other colors, which makes it more difficult to perm. There are more disulphide linkages in red hair, which have to be broken down for the perm to take. This bolshy tendency of red hair to fight back against the hairdresser's art and its wearer's wishes has been known since the eighteenth century at least. In his *Art of the Wigmaker* of 1767, Monsieur François de Garsault explains, "Another kind of hair coming from Switzerland and England is also sold, this is red hair which has been bleached in the fields as cloth is bleached, and which for this reason is called 'field hair.' It will not frizz [that is, take a tight curl], and is only used for graduating the straight, smooth hair. It should *never*," he warns, "be mixed with a mass of frizzed hair." If you want red hair to behave, you need to show it a firm hand. Oh, and because of its color, redheads are also much more likely to be stung by bees.[95]

There is much in the strange and unusual connections afoot in a redhead's internal chemistry that in the words of Jonathan Rees still needs to be bottomed out.[96] To begin with, he notes an unexpected degree of diversity within that MC1R gene, seesawing between fully homozygous changes, where the recessive gene resulting in red hair is present on both chromosomes (a full-on redhead, as it were), and where it is only present on one and is only partially expressed (resulting in codominance and brown hair and freckles

95 Thanks to Tim Wentel for this and many another point of information in this chapter.

96 Jonathan Rees and Thomas Ha, "Red Hair—A Desirable Mutation?," *Journal of Cosmetic Dermatology*, 1, no. 2 (July 2002): 62–65.

or brown hair/red beard). Tim Wentel, a dermatologist at Erasmus University Medical Center in Rotterdam, suggests there may be as many as four hundred different genetic possibilities. Nor is MC1R on chromosome 16 any longer the only note in the chord. There is the part also played by HCL2, on chromosome 4.[97] One might anticipate that where there are two, there will be more, and as proof of the almost limitless discoveries waiting to be made as our DNA is finally uncoiled from its helix, labeled and laid out straight, you could hardly do better than point to the people of Melanesia.

The first definitions of what it was to be Melanesian were suggested by eighteenth-century European explorers of the region; even now there is no agreement on where the boundaries of Melanesia should be traced, or even whether the term is a geographic or a cultural entity. But to those people living on these islands, the word "Melanesian," which was once redolent of subjection and denigration, has become a term of affirmation and empowerment (redheads, take note). And on the Solomon Islands, 5 to 10 percent of the population, along with having very dark skin, have afros of the most striking shades of anything from a cinnamony-ginger to peroxide yellow (Fig. 29). Any number of explanations had been offered for this: natural bleaching by sun and salt water, diet, or the genetic legacy of early European explorers. But in 2012 it was found to be the result of another unique recessive gene, one totally separate from MC1R, and found nowhere else in the world. The geneticist Sean Myles, who finally identified the gene, has called it

97 Mutations on chromosomes 16 and 4 lead to brittle cornea syndrome 1 and 2, respectively. See http://www.omim.org/entry/229200.

"a great example of convergent evolution, where the same outcome is brought about by completely different means."[98] Sherlock Holmes and Jabez Wilson would have known all about that.

Unhappily, not all such recent genetic discoveries are quite so cheering. As with our prehistory, much of what follows is speculative and contentious. One only has to look at the cautious language of the science to realize that. Nonetheless, being a redhead can have very undesirable side effects indeed.

The biggest villain in the cast here is melanoma. Melanoma, as we all should know by now, is a particularly aggressive form of skin cancer, with a vindictive propensity to spread from the skin to other organs. One possible scientific explanation for this, for melanoma and a number of other conditions, suggests that despite the stronger immune system, redhead DNA is more fragile than that of other hair colors, less good at repairing itself, and therefore more prone to those disorders that, like melanoma, arise in damaged cells. As evidence of this, there is a link, although no one can state categorically where in the triangle is the cause and where the result, between red hair, melanoma and two serious medical conditions, the first of which is Parkinson's disease, and the second is endometriosis. Melanoma is a point in the triangle for both of them.

Parkinson's is a degenerative disease of the central nervous system. Its causes may be genetic; its development is associated with head injury and exposure to certain pesticides. In this particular

98 See http://news.sciencemag.org/2012/05/origin-blond-afros-melanesia.

Venn diagram of nastiness, Parkinson's is in one circle, melanoma in the second, and red hair in the third.[99] A history of melanoma is "associated" with an increased risk of developing Parkinson's. MC1R gene variants are associated with an increased risk of melanoma. But the lighter your hair color, equally the greater your risk of developing Parkinson's, with those with red hair being at the greatest risk of all—three times the risk of those whose hair is darkest. This is all deeply depressing stuff, no matter that Parkinson's is still a rare disease. It was thought at one time that the incidence of melanoma might be a side effect of one of the commonest treatment for Parkinson's, with the drug form of L-dopa, but the finger now points to the likely villain being the MC1R Arg151Cys allele. It probably makes this no easier to comprehend if I remind you here of those four hundred estimated possible variants of MC1R. Don't worry about it.

An equally unholy trinity exists between the incidence of red hair, melanoma, and endometriosis, another disorder of the immune system and a cripplingly painful condition where cells similar to those that line the womb (the endometrium) begin to grow on other organs in the abdomen, and just as if they were still within the womb, react to the menstrual cycle by swelling and bleeding. A study in 2000 of 3,940 college alumnae in the States found that among the group as a whole, 6.98 percent reported some degree of endometriosis. Of the 121 redheads in the group, however, this went

99 X. Gao, et al, "Genetic Determinants of Hair Color and Parkinson's Disease Risk," *Annals of Neurology*, 65, no. 1 (January 2009): 76–82; also C. Kennedy , et al., "Melanocortin 1 Receptor (MC1R) Gene Variants Are Associated with an Increased Risk for Cutaneous Melanoma Which Is Largely Independent of Skin Type and Hair Color," *Journal of Investigative Dermatology*, 117, no. 2 (August 2001): 294–300.

up to 12.4 percent, and again correlated with an increased incidence of melanoma. Or, as the researchers put it, "Among women with red hair there is an association between endometriosis and melanoma . . . which warrants further investigation." (I'll say.) They also speculate that part of this linkage may be bound up in the fact that the HCL2 gene is on chromosome 4 and thus close to the cluster of genes for fibrinogen, a protein necessary for the formation of blood clots.[100] As if redheads needed more reasons to stay out of the sun.

But the causal link, between red hair and blood and menstrual bleeding, harks back not only to the age-old slur that to have red hair means you were conceived while your mother was menstruating but to ancient Ayurvedic medicine as well, which also links red hair to disorders of the womb. Ayurvedic medicine has been around for roughly 3,000 years and is clearly akin to the Greek system of classifying humanity into types according to the four humors, which for most of recorded history was the basis of all Western medicine, too. The Greeks recognized four physiological types: you could be sanguine, choleric, melancholic, or phlegmatic, or any combination thereof, each of them being governed by one of the four "humors" or fluids within the body—respectively blood, yellow bile, black bile, or phlegm. (The origins of this system may go back even further, to Ancient Egypt or possibly to the Mesopotamians.) Ayurvedic medicine differs from the humors in that in has just three types, or *doshas*, with redheads being generally classed as the *pita* type, and being characterized as passionate, chivalrous, sensitive, and

---

100 G. Wyshak and R. E. Frisch, "Red Hair Color, Melanoma and Endometriosis: Suggestive Associations," *International Journal of Dermatology*, 39, no. 10 (October 2000): 798.

compassionate—all to the good. The *pita dosha* in women controls the health and well-being of the womb, and imbalances within it will manifest themselves there. Here is contemporary Western science confirming what traditional non-Western medicine has believed for centuries. Again according to Ayurvedic medicine, redheads are also prone to frustration, anger, arrogance, and impatience, and when out of sorts, it comes out in their skin, in eczema and dermatitis. Redheads sometimes have what is known as an "atopic" constitution (I certainly have), where the red hair goes along with a propensity for hay fever and any number of other annoying allergies as well.

One of the oldest stereotypes of the redhead is that they have a fiery temper. Again, how to distinguish between the geno- and the phenotype: if you are teased as a child you may very well react by losing your temper. If, like me, you work out that you have more leeway in displays of tantrum than your non-redheaded peers, that behavior is reinforced. But it is now thought that MC1R also plays a role in adrenaline production, and it has been suggested that redheads not only produce more adrenaline but that their systems can access it more speedily, too—in other words, they fire up more rapidly than others (that fear/flight response). Certainly Hans von Hentig, writing in the *Journal of Criminal Law and Criminology* in 1947, thought so. Tracing the history of the outlaw in the West from 1800, he lists examples ranging from "Big Harpe," who terrorized the Ohio Valley in the early nineteenth century and who had, apparently, "coarse hair of a fiery redness" to Wild Bill Hickok ("long auburn hair hanging down over his massive shoulders"). Von Hentig adds to the list of redheaded no-goods Sam Browne and Jesse James, speculating that the appearance of these men was remembered because with the color of their hair, they stood out, which is the sort of logic that sounds a little as if it might be getting into the all-too-familiar area of "they acted like redheads, so must have had red hair"; but he does also observe that "redheadedness also is often combined with acceleratedness of motor innervation," which, medically speaking, would seem to imply that redheads exist in a permanent state of heightened stimulation. I'm not so sure about that—but these were all gunmen, quick on the draw.

And suggestively, in light of this, science has very recently begun to explore a link between red hair and Tourette's syndrome.

Tourette's syndrome (Ts) is an astonishingly complicated, interwoven set of symptoms that can range from small tics and mannerisms to involuntary outbursts of obscene vocabulary. It's thought that Mozart may have been a sufferer. Technically it is termed a chronic, idiopathic, neuropsychiatric disorder, which should warn you that at present, almost nothing can be categorically ruled out as a cause. In 2009 an Australian pediatrician, Katy Sterling-Levis, whose thirteen-year-old son had just been diagnosed with the disorder, was attending a conference on the syndrome and was struck by the number of redheads in the room—or how they were "over-represented," to use the unemotive language of the scientific study. The pattern of inheritance for Ts is autosomnal recessive. In other words, two copies of the gene must be present for the condition to result, just the same as with red hair. A survey organized to test the strength of the connection revealed that 13 percent of those with Ts had red hair, compared to the normal average within the population of 2 to 6 percent.[101] More than half of the sufferers of Ts had one or more relatives with red hair. ADHD and hyperactivity are also associated with Ts, and a few pediatricians have drawn a much-contested connection between the pale skin/red hair phenotype and hyperactivity, to boot. In adults as in children this is characterized (among other symptoms) by impulsive and inappropriate behavior, mood swings, racing thoughts, and a craving for excitement. Spencer Tracy, one of Hollywood's leading men in its Golden Age, and one of the very few actors to make it seriously big despite an unmistakable tawniness in his natural coloring, was hyperactive as a child. It's also striking how many symptoms of hyperactivity seem to have been present in the poet Algernon Swinburne's behavior.

Swinburne was born in 1837. His mother was a daughter of the third Earl of Ashburnham, and Swinburne proudly traced his red hair back to Henry VIII. His well-to-do family sent him to Eton, where his cousin Lord Redesdale left this toe-curling description of him:

> What a fragile little creature he seemed as he stood there between his father and mother, with his wondering eyes fixed upon me!. . . His limbs were small and delicate; and his sloping shoulders looked far too weak to carry his great head, the size of which was exaggerated by the tousled mass of red hair standing almost at right angles to it. Hero-worshippers talk of his hair as having been a "golden aureole." At that time there was nothing golden about it. Red, violent, aggressive red it was, unmistakeable, unpoetical carrots. . . . His features were small and beautiful. . . . His skin was very white—not unhealthy but a transparent tinted white, such as one sees in the

101 Katy Sterling-Levis and Katrina Williams, "What Is the Connection between Red Hair and Tourette Syndrome?," *Medical Hypotheses*, 73, no. 5 (November 2009): 849–853.

petals of some roses. . . . Altogether my recollection of him in those school days is that of a fascinating, most loveable little fellow.[102]

The fascinating, loveable little fellow went on to scandalize the London literary scene with his poetry (technically some of the most accomplished in existence, still, but in its subject matter suggestive, erotic, and deviant, at the height, in the 1860s, of Victorian respectability), and outrage London society with his behavior as soon as he could.[103]

Swinburne's endlessly fluttering hands and feet had been diagnosed in childhood as "an excess of electrical vitality." As an adult he seems to have been willing to do anything to make himself the center of attention—that stunt of sliding nude down the bannisters at Rossetti's was one of many. He rapidly became both an alcoholic and a user of opium, and he posed (according to Oscar Wilde, whose gaydar, one would think, would be faultless) as a homosexual.[104] Medical opinion now is that he may well have suffered some degree of brain damage at birth and was possibly hydrocephalic. Such was his notoriety that when on July 10, 1868, he fainted in the Reading Room of the British Museum, the event made the papers. He filled perhaps something of the same role as the comedian Carrot Top does today—determined to be weird, and if possible, to make a good

102 Quoted in Christopher Hollis, *Eton: A History* (London: Hollis and Carter, 1960), 291–2.

103 Leonard Shengold, *op. cit.*, in his chapter on Swinburne makes a sympathetic and insightful case in his defense.

104 Wilde said of Swinburne that he was "a braggart in matters of vice, who had done everything he could to convince his fellow citizens of his homosexuality and bestiality without being in the slightest degree a homosexual or a bestialiser."

living out of it—but you do wonder with both what the image of the redheaded man on either side of the Atlantic might have been without them.

It seems almost a shame. Genetic science will one day solve so many of the mysteries that orbit the edge of our imaginations. The truth, or otherwise, of the "blond Eskimos" of Coronation Island in northern Canada, reported as recently as the early twentieth century, will be pinned down, one way or the other; the reconstructed faces of the Tarim mummies, so different from their Han Chinese neighbors, will be allied to their genome and end once and for all speculation as to whether they were European, Asian, or a coming-together of both. The capital of the Budini will be located; and the Udmurt people will be incontrovertibly assigned the same region genetically as they now inhabit geographically, poetically described as being between the Great Forest and the Great Steppe. Their language will give up its last Finno-Ugric secrets, and John Munro will rest easier for knowing he got one thing right: "A white freckled skin, greenish eyes and fiery red hair are characteristic of the Finns, Rühs, and other people of the Baltic highlands." The enormous graveyard of maybe a million mummies sitting on the edge of Fag el-Gamous, south of Cairo, will provide an answer as to why they seem to have been buried with blonds in one area and redheads in another, and who these blonds and redheads in first to seventh century AD Egypt were in the first place.[105] And (which would be particularly handy in this case) a simple means will be discovered of

105 See http://www.livescience.com/49147-egyptian-cemetery-million-mummies.html.

distinguishing true red hair in life from hair whose color in death has been altered by acids in the soil or fungi or bacterial decay. The reason for the red *scoria*-stone topknots on the statues, or *moai*, of Easter Island in the southeast Pacific will be found, and their links (or lack therof) to the cult of the redheaded Birdman that persisted right into the 1860s will be explained. An unguessable amount of the history and mythology of Easter Island has been lost to us, but it's hard not to see the use of red as connected in some way to those red-topped statues, gazing out over the Pacific for so many centuries before. The startling possibility that the first Maori settlers in New Zealand may have found a native people already living there, the Ngāti Hotu, who had fair skin, green eyes, and red hair, will be confirmed one way or the other.[106] And we may then be closer to understanding why it is that Edward Tregear (1846–31), one of the first to study the Maori, found so many terms among Polynesian speakers to describe red hair:

> Samoan, *'efu*, reddish-brown: Tahitian, *ehu*, red or sandy-colored, of the hair, *roureuhu*, reddish or sandy hair: Hawaiian, *ehu*, red or sandy hair, ruddy, florid; *ehuahiahi*, the red of the evening, or old age; *ehukakahiaka*, the red of the morning, or youth . . . Marquesan *hokehu*, red hair.[107]

And finally, the coppery topknot of the elongated Paracas skull and the skull itself, and its fellows, will be tested by a lab anybody has ever heard of, and be returned from the star-gazing world of

106 Kerry R. Bolton, "Enigma of the Ngati Hotu," *Antrocom Online Journal of Anthropology*, 6, no. 2 (2010): 221–26.

107 Quoted in Kerry R. Bolton, *op. cit.*

Indiana Jones and the Kingdom of the Crystal Skull to this small blue planet here.

When I was very young, but not too young to have noticed that I was the only redhead at my village school, I aligned myself, as many a redhead has done before, with a Celtic ancestry. I was convinced I must have much Irish and some Scottish blood to boot, and that was where the red hair came from. My mother had a vague idea that there was Irish blood in her family, and my father's family names were Ewart and Colliss, one of which was definitely Scottish, and the other of which I decided was also going to be Scottish, to keep things tidy. Besides, all redheads have Scottish or Irish blood, right? Show me a writer who won't take the nice smooth plotline over the untidy truth, and who doesn't start off by rewriting their own identity in some way. Anyway, I liked the idea of my Celticness; I thought it made me different from my friends and superior in some way to my enemies; and no one contradicted me, because obviously I had the red hair to back it up. But in one of those everyday marvels of our age, of course you can now test these assumptions out, with a DNA kit. So I now know that I appear to be 100 percent *un*-Celtic, that my haplogroup, centered in the North of England, is 46 percent Brit; 38 percent European (centered—who knew?—on Switzerland); 13 percent Scandinavian—Norwegian, by the looks of it; and 3 percent Eastern Middle East. I might as well be that Ur-redhead.

Just about everyone on the planet will have some DNA from the Middle East. We came if not from there then through there, and the haplotype will be there still. Norway and the North of England connect, via the Vikings. Switzerland, I have no idea. But according to the company who did the analysis, there is, on the database of this one company, twenty-seven pages' worth of people who have also had their DNA tested, and to whom I am genetically related in some way. This is not as revelatory as it may sound—you only have to go back six centuries or so and statistically there would be a European someone to whom everyone now alive in Europe would be related. And of course the analysis doesn't tell me how many of those twenty-seven pages' worth of folk have red hair. But there is a pleasing paradox here—that a science that begins by measuring difference ends by making brothers and sisters of us all.

So that is what I am in terms of cells and ancestors piled one atop the other. But to sort out what all that means in terms of being a redhead today, you need not the hard-and-fast of science but the loose-and-sloppy of everything else.

FREAKS OF FASHION

There is a deep-rooted and unaccountable
prejudice against this much-abused shade of colour,
which it is quite possible some unexpected
freak of fashion may one day change.

T. F. THISELTON-DYER,
FOLKLORE OF WOMEN, 1900

Every autumn it's the same—the days grow short, the skies are gray, the evenings dark, the nights too long. Leaves pile against the railings and crunch underfoot, and then are gone, and with them, color abandons the Northern hemisphere. The eye craves it, just as did the eyes of our ancestors before us, decorating their tombs with red-haired goddesses to give us life again, to bring back the sun. The glow of bonfires and the sparkle of fireworks across the UK in November doesn't only celebrate the last-minute discovery of Guy Fawkes ("a tall, powerfully built man, with thick reddish-brown hair, flowing moustache, and a bushy reddish-brown beard," according to the historian Antonia Fraser) and his barrels of gunpowder, and the foiling of the Gunpowder Plot, they also echo the festivals of All Hallows and of Halloween, of Diwali, the Day of the Dead, Hogmanay, and the world-turned-upside-down of the Roman

Saturnalia. We all feel it, that ancient winter hunger for warmth, for light; and every autumn, in answer, the red carpet—where else?—fills up with celebrities sporting newly dyed red hair. As *Vogue* puts it, "Mythologized, demonized, celebrated: every shade, from carrot to scarlet, conveys an inscrutable allure."[108] A wholly unscientifically assembled list includes, as I write, Jena Malone, Kirsten Dunst, Amy Childs, the models Suki Waterhouse and Karlie Kloss, Amy Adams (a great spokeswoman for the advantages of going red), Sofia Vergara, onetime Bond girl Olga Kurylenko ("It's been red for a couple of months . . . I'm definitely getting more attention from men"), Katy Perry, and Katie Yeager. Katie Holmes announces that she's always yearned for red hair, and the world stumbles to a halt. Meantime, in a pleasing twist on the notion of the undead redhead, a warmth of tone has been noted in the hair of the actor Robert Pattinson, Edward in the *Twilight* series. Nothing captures the eye like red hair.

We know we are drawn to the color red. Waitresses dressed in red supposedly make more in tips. Men are meant to find women more attractive if they wear red. Women are meant to find images of men more attractive when they are shown those images against a red background. And there is of course the age-old sexual allure of the flame-haired temptress, something neither Hollywood nor the rest of the celebrity industry needs to be told. But the media company Upstream Analysis suggests another reason for our fascination with red hair. As well as drawing the eye and grabbing an audience's

---

108 *Vogue*, "The Best Redheads," July 2014. See http://www.vogue.com/946492/best-redheads-jessica-chastain-amy-adams-julianne-moore-and-more.

attention, its rarity, Upstream suggests, sparks the reward-seeking instinct in us, firing up the center of our brain that is most highly sensitized to novelty.[109] When we see something unusual, we want to get close to it. We want to inspect it, or engage with it in some way, or even to touch it. In other words, we buy into red hair. This is the reason, Upstream concludes, why up to a third of all TV advertising features a redheaded character, when the actual percentage in the population hovers at a scarce (and thus noteworthy) 2 to 6 percent. If nothing draws the eye like a redhead, it would seem that nothing sells like one, either.

Every mother of every redhead the world over will know the experience of complete strangers coming up to comment upon or even to reach out and touch their children's hair. Growing up as a redhead, it sometimes felt as if the last person my red hair belonged to was me—the person from whose scalp it sprang. It was one of the many things that made growing up as a redhead so deeply confusing. Grant McCracken again: "Redheads become a handy plinth, a medium for the message, a carrier of the color." Once again, the hair overpowers everything else. It becomes all people see. The normal barrier, the invisible area around us that we all own, and into which others do not enter without our permission, apparently doesn't exist if your hair is red, and you're too small for your wishes to count. (And this we endure as children on top of our heightened flight response and hair-trigger adrenaline production. So next time you are tempted, all I'm saying is, *ask.*) People—like Rachel Lynde, in

---

109 See http://www.theatlantic.com/business/archive/2014/08/redheads-are-more-common-in-commercials-than-in-real-life/375868.

her encounter with Anne Shirley—talk about and comment upon your hair while you yourself are standing there beneath it, as if you were merely wearing it, like some kind of hat. And if, tiring of this as you grow older, as many a redhead does, and as your dominion over your body increases, you should cut your red hair, or dye it, there is outrage, as if the thing you had changed was everyone else's property, which you have damaged, willfully. These are such common behaviors in the non-redheaded world, they're convincing evidence that Upstream Analysis is on to something here.

And such primates are we that anything we admire we wish to emulate. The human race has been changing the color of its hair with henna, walnut juice, saffron, red wine, red ochre, lye, vitriol, indigo, and woad for millennia, but in the twentieth century the first commercially available hair dyes created an entire new industry and revolutionized society's response to dyed hair into the bargain. You can't emulate an image without there being one to emulate, so hand-in-hand, or rather chicken-and-egg with this, went Hollywood's remaking and shiny repackaging of the actresses who entered its studio system, none of which would or could have happened as it did without one Maksymilian Faktorowicz, or Max Factor, as he is better known.

Max Factor's life really should be made into a movie itself. One-time hairdresser to the Russian court—so prized that he conducted almost all his early professional life under guard—he fled Russia to hide the fact of a marriage for which he had failed to receive the tsar's permission, using his own makeup to fake the symptoms of jaundice in order to be allowed to leave Moscow. Arriving in America in 1904 with wife and very young family in tow, he was promptly fleeced of most of his savings by his partner in the Louisiana World's Fair, and then found his doorstep darkened by his ne'er-do-well half brother, John Jacob, or "Jake the Barber," to give him his mob name, a Prohibition gangster and con man who did once, literally, break the bank at Monte Carlo. Unsurprisingly, Max kept moving, heading west and arriving in Los Angeles in 1908 and setting up a business hiring out wigs to bedeck Hollywood's film extras. Then he did the extras' makeup, too. Then he did the makeup for the stars, as well. Then he invented the term "makeup." Within an impressively short span of time, the name Max Factor and the glamour of Hollywood had become just about synonymous. In 1935 Max opened the "Max Factor Make-Up Studio" (note the word "studio" in the title), with its color-coded makeup salons: peach for "brownettes" (a type that failed to catch on, and a rare example of a Max Factor marketing misstep); pink for brunettes, powder-blue for blondes, and a pale mint green for redheads (Fig. 30).[110] This Redheads Only room was officially opened by Ginger Rogers, and in painting it mint green Max created a color association that bedevils redheads to this day, and one that Maria Oakey would have taken issue with for sure: "Wherever there is red in the composition of the hair, green (*not* a pale green, which should only be worn by blondes). . . will be becoming." But no matter, Max's successes far outnumber the fails. He created Clara Bow's heart-shaped lips, and Jean Harlow's platinum blondness (if you thought those Renaissance hair dyes sounded undesirable, Jean Harlow's candle-flame white was achieved with a weekly wash of

110 Fred E. Basten, *Max Factor: The Man Who Changed the Faces of the World* (New York: Arcade Publishing, 2012).

ammonia, Clorox, and Lux soap flakes). And although the timing of the claim is hard to be exact about, by repute, Max also created Rita Hayworth's tumbling copper curls.

Hollywood has been graced by many a redheaded beauty, but nearly thirty years after her death, more than forty since her tragically early retirement, the first name that comes to anyone's lips when you put the words "Hollywood" and "redhead" together is Rita Hayworth. The second is *Gilda*, the part she plays in the film of the same name. The glorious, sensuous Gilda, undulating across the stage as she sings "Put the Blame on Mame" in elbow-length black satin gloves and a dress that required the real Rita Hayworth to wear both a corset to get into it and a hidden harness to keep it in place. The third name is of course Lucille Ball. Neither Rita nor Lucille was genetically a redhead, any more than Debra Messing or Christina Hendricks were born with the red hair they have worn so proudly, but no matter. All four, and many more, are a part of the image of the redhead as we all receive and respond to it today; what makes Rita Hayworth and Lucille Ball so extraordinary is that they managed to be so in monochrome. *Gilda* was filmed in 1945, when the film industry was still veering between the novelties of Technicolor and good old black and white. Lucille Ball's TV show (with its endnote sign-off "Make-up by Max Factor") went on the air in 1951. Color TV sets—or shows—did not exist in any numbers until the mid-1960s. What miles we have covered since then.

Here's a case history to ponder. In 1941, Rita Hayworth, coming up to the peak of her career, was in two movies, one—*Blood and Sand*, a cautionary tale of how badly bullfighting and hubris go together—in Technicolor; the other, *The Strawberry Blonde*, paradoxically, in black and white. In *Blood and Sand*, in color, and playing on her own "exotic" Mexican ancestry, Rita plays a socialite, Doña Sol, a sexual aggressor, the possessor of a kind of fantasy sexuality, as it is described by her biographer Barbara Leaming.[111] Darryl Zanuck, the film's producer, had originally wanted Carole Landis for the role, but Landis refused to dye her hair red for it, hence the part went to Rita. This gives some idea of the importance given to the aesthetic of Technicolor in the movie (wrong hair and you're out), with sets that were meant to evoke the works of the Spanish painters El Greco, Goya, and Velázquez. But you don't get the impression much thought went into the character of Doña Sol, who seems to be driven by nothing—albeit driven very beautifully by nothing—other than an urge to add notches to her bedpost. In *Strawberry Blonde*, Rita is a society girl again, but the all-American version thereof, and there's a neat if slightly queasy-making piece of story-telling/stereotyping in this film. As the cast list comes up on the screen, James Cagney is introduced as "Biff Grimes," Olivia de Havilland as "the girl he adores." When Rita comes up on the screen the line under her name is simply "M-m-m-m-m." She's not a character with a story, she's simply the audience's lip-smacking reaction; not a human being but a symbol, a vector for desire.

And ultimately, I would argue, one that is a dead end. The image of Gilda has been referenced in film after film, but in the end, what can you do with a sex symbol other than make it sexier and sexier, until it becomes a parody of itself? Once Jessica Rabbit, who

111 Barbara Leaming, *If This Was Happiness* (London: Sphere Books, 1990).

famously isn't bad—"just drawn that way"—had swung onto the screen in the film *Who Framed Roger Rabbit?* (1988), we'd gone as far as we could go. The association between redheaded women and sex has become so knee-jerk that advertising now doesn't even need a body for its message to be understood. In 2012 Sleepy's, the US mattress company, launched a range of gel-infused mattresses, guaranteed to keep you cool in the sweltering New York summer. The advertising image chosen was a sparkling white duvet, a pillow, and between the two a woman's head of rich red curls. All you saw of her was her hair and an ear. Sleepy's tag line was simply "HOT IN BED?" You start to understand Rita Hayworth's sad complaint that "men go to bed with Gilda, but they wake up with me." We are not all hot in bed. We do not all, always, want to be. Centuries and centuries of stereotyping, of social and sexual expectation, decades of advertising, can really make you feel the cold.

And did those same ads for *Strawberry Blonde* feature the redheadedness of Rita Hayworth's costar, James Cagney? No, they did not. In all this, where were the redheaded men?

Cagney had been a significant star since the 1930s, when Lincoln Kirstein described him as "a short redheaded Irishman, quick to wrath, humorous, articulate in anger . . . the semi-literate lower middle-class . . . mick Irish," and the *L.A. Record* as a "red-haired

Fig. 25: Edgar Degas, *La Coiffure*, *c.* 1896. Is this a servant preparing her prostitute mistress for the night's work, or an aristocratic young woman being tended to by her maid? Are the two mother and daughter? The one is impassive, absorbed in her task; the other holds on to the roots of her hair with one hand and raises her other in pain. Is the interior where this scene takes place a boudoir, or a brothel? Only one thing is certain: it's all about the hair.

Fig. 26: Cora Pearl (born Emma Elizabeth Crouch) as Cupid in Offenbach's *Orpheus in the Underworld*, of 1867. The operetta was notorious for its cancan; Cora by this time had also appeared in public as Eve, and at a dinner party at her château in the Loiret, had herself served naked on a silver platter.

Fig. 27: Dickens's Uriah Heep, as imagined by "Kyd" (Joseph Clayton Clarke: 1857–1937). Father of ten, Clarke worked as a designer of cigarette cards and as a book illustrator. His Dickens characters still influence how we visualize them today.

Fig. 28: Tintin, who first appeared in 1929. His creator, Hergé, may have been influenced by a freckle-faced red-haired Danish fifteen-year-old, Palle Huld, who in 1928 won a competition to travel around the world, Phineas Fogg–style, and who completed the trip in just forty-four days.

Fig. 29: Around 26 percent of the population of the Solomon Islands carry a unique gene found nowhere else on earth. This gives 5 to 10 percent of them fair hair, which ranges from bright blond to a light ginger.

Fig. 30: A detail of the decoration of the Max Factor mint-green "Redheads Only" room at the Hollywood History Museum, with a Max Factor advertisement endorsed by none other than Lucille Ball.

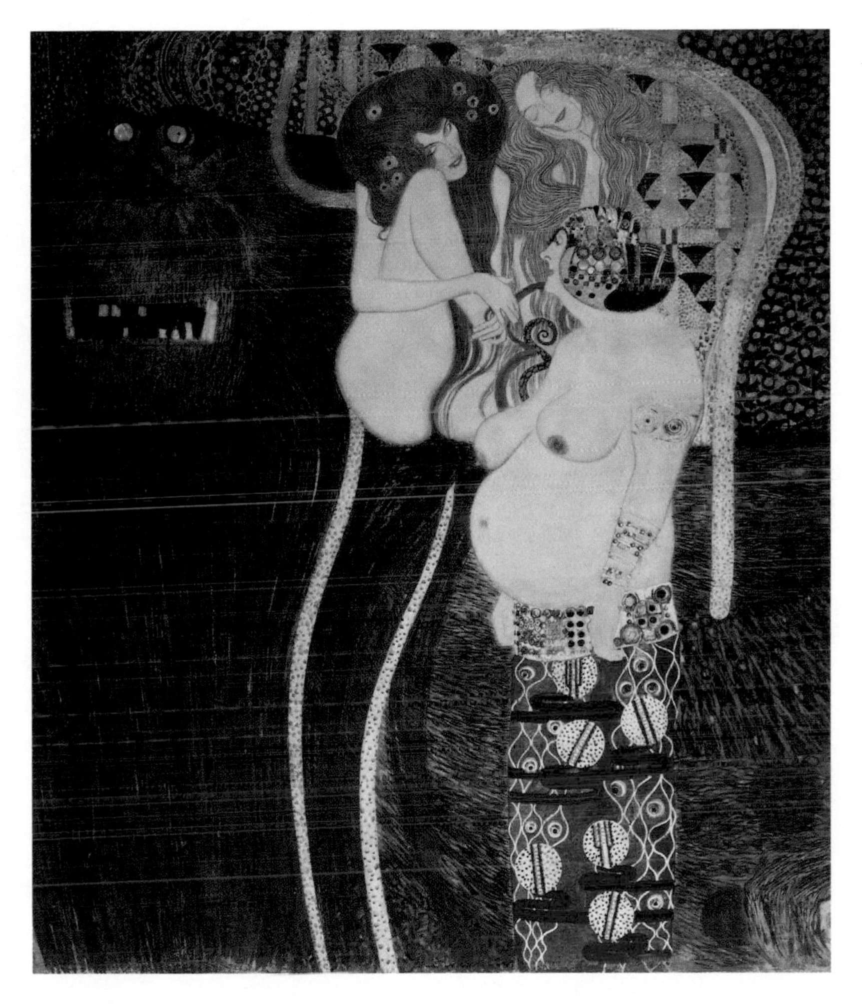

Fig. 31: Gustav Klimt, Lust, from the *Beethoven Frieze* in the Secession Building, an exhibition hall built in 1897 in Vienna. The building was created to show the work of artists who had seceded from the Austrian establishment, and was financed by Karl Wittgenstein, a steel tycoon and father of the philosopher, Ludwig Wittgenstein. The plump dark-haired figure is Intemperance; the sleeping blonde represents Debauchery.

Fig. 32: Redhead apocalypse. A detail of the group shot from Redhead Days in Breda in 2014. No fewer than six thousand redheads from all over the planet attended the festival.

Fig. 33: Sterra Vlamings, a redheaded model of Dutch and Senegalese descent.

Bowery Boy . . . fiery-tempered, but with a warm Celtic smile."[112] Note the "but," implied in one description, explicit in the other. We are being invited to like and admire Cagney despite the flaws in his presented image—his red hair being but one of them. We are to overlook that—in fact in the posters for *Strawberry Blonde,* his hair is brown. You can make an argument, however, that Cagney's career and the parts he played track the transformation of the pugnacious Paddy, the brawler, the laborer, the Celtic cliché of the nineteenth century into the twentieth century's sleek urban animal, albeit one who in Cagney's movies is very often on the wrong side of the law, ready and able to turn on that irresistible Celtic charm whenever needed. By the time Errol Flynn came along (chestnut-haired in life, since you ask) playing the wild Irish card could be the basis, almost, of an entire career. Again, we are to overlook that tint, or taint, of red. Another wholly unscientific list: Eric Stoltz, Ewan McGregor, David Caruso, Rupert Grint, Damian Lewis, and Benedict Cumberbatch. Simon Pegg says he is not; Michael Fassbender happily declares himself a ginger Viking, and who are we to argue? But unlike Rita, Lucille, Debra, or Christina, not one of these has dyed his hair red as a means to further his career; indeed Benedict Cumberbatch, deeply annoyingly, in his best-known role as Sherlock Holmes, has his dyed almost black—deeply annoying in that if ever there was a character who could and should be played as a redhead, it's the eccentric, intellectual, unconventional, alienated Sherlock. When the photographer Thomas Knights launched his *RED HOT*

112 Quoted in Ruth Barton, *op. cit.*

100 project in 2014, photographing redheaded men—staggeringly handsome redheaded men, fair enough—and encouraged them to talk about their experiences of going through life with red hair, one of his sitters confessed that he had found it easier to come out as gay than to come out as a redhead. What does that say about red hair as one of the last great social prejudices? It is, still, different for girls, but here is another paradox: we don't judge or value men by their appearance in the same way as we judge or value women. So whereas redheaded actors, with some clever manipulation by their image-makers, may escape the stereotyping that comes with red hair, and play against it, and have us overlook the color of their hair entirely in our approval of them, redheaded women, whether in the public eye or no, are still bound by stereotypes stiff as a straitjacket. As an adult it can feel as if you simply get to pick which one you'll don, out of a choice of three: Rita, Scully, or Lucille. Without Lucille Ball, that choice might have come down to just two.

Lucille Ball single-handedly created the third modern variant in the palette of redhead female types. Her character in *I Love Lucy* is a kook, a kind of back-formation from the winning, redheaded little girl characters of children's fiction at the start of the twentieth century, only this time in the body of a grown woman; a housewife who dreams of stardom, who talks fondly of her "henna rinse" on the air, who is unreliable and unpredictable, but human and believable and irresistible because of it. In real life Lucille Ball had her signature hair color created by her hairstylist, who declared "the hair may be brown but the soul is on fire!" Off set she was one of the canniest and most undauntable female studio executives there has ever been. She paid a price, of course, in chauvinism and name-calling, but if the soul was on fire, the business brain beneath the red hair was second to none, and she knew it. It's more than a little ironic therefore that the character she created—zany, unintellectual, and delightfully flawed—should be what she has been remembered for, referenced by innumerable actresses ever since: Debra Messing's Grace in *Will & Grace*, for one; Alyson Hannigan's character Lily Aldrin in *How I Met Your Mother*, for another—women whose comedy flows from the gap between what they try to be and what they are. And because she was flawed, and warm, and human, Lucy, the original, was within the reach of housewives, mothers, women everywhere. Hair dye would not work the miracle that might turn the average woman into Rita Hayworth. It might not give you the go-getting career of redheaded *Brenda Starr, Reporter*, who was at the height of her comic-strip fame in the 1950s. But it could bring you closer to the more achievable glamour of Lucille Ball. And we *knew* Lucy dyed her hair, and wore makeup, because in her show she told us so. No more the scandalized reaction of Aunt Marilla to Anne Shirley dyeing her hair ("A wicked thing to do!"). If it was okay for Lucy, it was acceptable for every woman. Her signature look even crossed the Atlantic into my parents' living room, with its shoe box–size television screen. And where demand leads, the advertising men are sure to follow. Except of course that by this time, some of those men were women. And one of those women was Shirley Polykoff.

In 1955 Polykoff was working at the ad agency Foote, Cone & Belding and took over the Clairol account, for whom she created the "Does she or doesn't she?" tag line for their range of do-it-at-home hair dyes.[113] What's particularly pleasing about this bit of advertising genius is the fact that Polykoff, who was born in Brooklyn in 1908, was Jewish and based her slogan on a perfect bit of Yiddisher-momma speak: "*Is* she or *isn't* she?" (complete with hand gestures). Thus one strong woman channeled generations of others, and changed women's relationship to their hair forever. If you wished to be a blonde, a brunette, or a redhead, now you could be, with a use-at-home hair dye that gave results so natural-looking, said the slogan, that your secret would be safe. And why would you do this? Because you're worth it—the second game-changing slogan in the field.

The original version of this, created by another woman advertising executive, Ilon Specht, for L'Oréal in 1973, was "Because *I'm* worth it"—an important difference to the warmer, fuzzier, we're-all-in-this-together variant used by L'Oréal today. "Because we're worth it" is several steps on from the hotly contested feminist debates of the 1970s. Of course we're worth it. We *know* we are.

And once that ball was rolling, once the products had been created that made dyeing your hair at home your choice, and so easy, and so commonplace, and so much something the stars in Hollywood were doing too, all the age-old censure was gone. An industry that had been worth $25 million became worth $200 million a year. The two shades of red offered by L'Oréal in 1970 were sixteen by 1989, with Redken offering twenty-nine and Clairol forty-three. In fact, it has been estimated that more red hair

113 Malcolm Gladwell, "True Colors," *The New Yorker*, March 22, 1999. See http://gladwell.com/true-colors.

dye is sold per annum off the shelves of supermarkets and pharmacies worldwide than any other shades.

Mrs. Roger Rabbit shares her ancestry between Rita Hayworth and another 'toon, the "little redheaded ball of fire, Red Hot Riding Hood," of 1943, with her floop of red hair over one eye, tiny scarlet skater's dress, and a Katharine Hepburn cut-glass accent. *Red Hot Riding Hood* is still considered one of the greatest cartoons of all time and has the rare distinction of having brought down the wrath of the censor on the director, Tex Avery, when the wolf, a night-club lothario (Grandma in the same cartoon becomes a frisky Park Avenue matron) became just a little *too* excited by Red Hot Riding Hood's performance. She was also supposedly influenced by the real-life Hollywood legend Lana Turner, a dazzling blonde. In the movies, if you want to show a blonde gone bad, you turn her into a redhead. It was done to Jean Harlow, of all women (of all blondes), in *Red-Headed Woman* in 1932. Harlow's character in the movie, wearing a flame-red wig throughout, was a home-wrecker, blackmailer, adulteress, and would-be murderess, thus ticking just about all the boxes for the sinful female redhead. The same transformation is still being used today in *The X-Men*—when Raven is Raven, she's a blonde. When she's the villainess Mystique, her hair turns copper.

This is a version of the redhead—the evil redhead—that began with Lilith. This is red hair as a marker for sorcery and the supernatural, nowadays with a pinch of existential angst thrown into the mix as well. Possibly one of the best and at the same time one of the most delectably evil redheads was created back in 1866 by Wilkie Collins, in the character of Lydia Gwilt in his Victorian mystery *Armadale.* Lydia is not only a proto-vamp, she seems to steal something of the vampire's eternal youth as well, seducing men much younger than her own advanced age of (God help us) thirty-five. Wilkie at this date was forty-two. The book kicks off with the deathbed confession of a murderer, and by the standards of the 1860s gets only more sensational from there on in. Here is Collins introducing his antiheroine:

> This woman's hair, superbly luxuriant in its growth, was the one unpardonable, remarkable shade of colour which the prejudice of the Northern nations never entirely forgives—it was red.

Never entirely forgives, eh? The description goes on:

> Her eyebrows at once strongly and delicately marked were a shade darker than her hair [every redhead knows the redhead eyebrow dilemma, of brows so fair they disappear; Lydia however is so flawless that she is spared it], her eyes large, bright and well-opened were of that purely blue colour. . . . Her complexion was the lovely complexion which accompanies such hair as hers—so delicately bright in its rosier tints, so warmly and softly white in its gentler gradations of colour on the forehead and neck.

In the complete description, we get her mouth and nose as well as her forehead, eyes, and neck, against which one can only imagine Collins resting, in his imagination, once his delineation of Lydia is complete, with a happy sigh. Lydia is poison, but she is the kind her victims imbibe willingly, and Collins's pen relishes every detail of her. She is a fortune-hunter, a seductress, possibly a bigamist, a murderess, and, ultimately, a suicide. This is the evil redhead, and they are never less than gorgeous, and always deadly in a particularly sexy way. Think of Edvard Munch's *Vampire*, of 1893, a painting the artist entitled *Love and Pain*, but his public, viewing the painting, took one look at the female figure nuzzling down into the back of her lover's neck, with her red hair dripping down around them both, and retitled it for him. Or there is Gustav Klimt's figure of redheaded Lust, in his *Beethoven Frieze* in Vienna of 1902, her long red hair curling between her thighs in a preposterously overt sexualizing of the much more modest pose, five hundred years before, of Botticelli's redheaded Venus, her head tilted, her predatory gaze measuring up the viewer (Fig. 31). There is Bevis Winters's 1948 hard-boiled gumshoe Al Rankin, encountering the redheaded Maisie Tewnham (also a bigamist) in *Redheads Are Poison*: "Her bright red hair ought to warn those foolish guys not to succumb . . . " The real Bevis Winter was an Australian, living in the genteel English seaside resort of Hove. In the same scarlet vein, there is Poison Ivy with her toxic kiss, who squares up even to Batman. Or even Bree Van de Kamp in *Desperate Housewives*, the perfect housewife and mother, with her spotless home and flawless hair and makeup—and her equally perfect willingness to kill to keep it that way. Her red hair is the viewer's clue that all with this woman is not as it seems.

Of course transformation works both ways. If you want to tame a redhead, turn her into a blonde. Orson Welles did this to Rita Hayworth when in 1947 he cast her as Elsa, the eponymous lady in *The Lady from Shanghai*. The film is heavy-handedly noir, and like many an Orson Welles movie, one of the most intriguing things about it is the way it pulls itself apart. Why Welles should have thought red hair unsuitable for a woman who manipulates every man around her is a mystery. Possibly in typical Wellesian fashion he was interested in defying one stereotype to remake another. Or, less pleasantly, it's a symptom of an attempt to control Hayworth as his marriage to her failed. But other actresses have done the same: Gillian Anderson has reverted to her natural blonde since playing Agent Scully—an excellent way to break the hold of that character on her career, perhaps. Christina Hendricks did the same to mark the end of *Mad Men.* Nicole Kidman, to the distress of many another female redhead out there, claims blondeness for herself these days, too. The red-to-blond and blond-to-red toggle switch: it comes in very handy. It was used by Disney in 1989, when their *Little Mermaid,* Ariel, was given red hair to differentiate her from Daryl Hannah's blonde Madison, the mermaid in 1984's *Splash.* A variant was employed in the 1960s sitcom *Gilligan's Island* in the opposing female types of Ginger and Mary Ann, one a redhead, one a brunette.

There is of course an argument, basically a chauvinistic argument, that changing your hair or hair color or lipstick or winter coat is proof that women are at the hopeless mercy of whatever the advertising industry last told them they must wear or be or do. There is in my view a much stronger argument that taking control of your appearance (your *body*), and having the freedom to make choices about it should be a part of the life of every human being on the planet, particularly every woman, and is a part of the ongoing emancipation of the female sex and, now, many a onetime minority group as well—redheads included.

Some years ago I spent a fortnight in the Ukraine, around the town of Sebastopol, famous as a locus of the Crimean War of 1853–6.

Much of Sebastopol looked as if the shelling from that conflict had only just ceased. This had clearly been a very tough place to live for a very long time. Almost every evening there would be a power cut, just to remind the Ukrainian people that Mother Russia had her finger on the switch. President Yushchenko was making his first public reappearances, still as heavyweight and charismatic a figure as ever, albeit pocked and scarred from someone's attempt to poison him with dioxin. The most popular fashion among the men was for black plastic peacoats, reaching to their knuckles and their knees, which gave them all the appearance of being low-ranking Myrmidons among the ranks of the KGB (for all I knew, from within my gaggle of Western tourists, they were). Far and away the most popular fashion among the young women of Sebastopol was for zinging, cranberry-colored hair, or orange of a beacon-like brightness, worn with red-carpet makeup applied with a professionalism that would have had Max Factor weeping tears of joy, and with vertiginous, stiletto-heeled boots and pencil skirts that had their wearers sashaying like Marilyn Monroe (who began life as a strawberry blonde, if her famous "Red Velvet" calendar shots of 1949 are to be believed). In this performance of hyper-femininity, these shining orbs of red or orange would go bobbing down the streets as their owners picked their way around potholes in the pavements and missing cobblestones in the roads. Talking to them, it was clear that this was the way they chose to counter the uncertainties and hardships of their lives, by expressing the same kind of intense, heightened femininity previously given shape in the "New Look" by Christian Dior, in the drab, gray, ration-regulated world of Europe post-1945. And if you are going to be as powerfully female, as ultra-feminine, as you possibly can, it seems the hair color you choose is red. Many Ukrainians have Tartar ancestry and dark coloring, and it is only human nature to want the opposite of what you've got, so you might imagine the hair color they would most desire would be blond, but not in this case. Not for the redheads of Sebastopol. For them, red was the only color that declared their pride in their gender, their defiance of all life might throw at them, and their solidarity with one another. Both the orange and the red were completely elective (there was no possibility of mistaking either color for anything found in nature) and thus said these woman not only had this choice but the freedom to exercise it, too. For the redheads of Sebastopol, red was the color of empowerment.

One of Shirley Polykoff's ads for Clairol showed a redheaded mother and daughter, the idea being that the mother's hair color would look as natural as the daughter's, while the red of their hair established for their audience the familial bond. In the movie *Atonement* (2007), for the wedding scene, the director filled one entire side of the church with redheads (that's me in the pale blue suit, dead animal over one shoulder, handsome RAF husband at my side), as a visual shorthand to establish that this was the family of the twin redheaded boys so central to the plot. Two redheads together find that people always assume they must be related to each other in some way. Evian mineral water made use of this idea in one of the recent ads in its "Live Young" campaign, with a pale-skinned, anxious-looking, red-haired young man twinned with pale-skinned, anxious-looking baby with a cockatoo quiff of orange curls. Elle Fanning and Naomi Watts, playing mother and daughter in the film *Three Generations*, are another example, with the truly red Susan Sarandon joining them in the role of grandmother. Even earlier: there is a story that Rossetti once took Lizzie Siddal and Algernon Swinburne to the theatre together. The boy selling programs was already unnerved by Lizzie's pallor and above all by her hair. Reaching the end of the row and encountering Swinburne, he is supposed to have dropped his armful of programs to the floor and let out a screech: "Here's another of 'em!" We crop up so unexpectedly that the non-redheaded world sometimes seems to find it easiest to assume we all belong together, that we must all be family, in some way. Which throws an interesting light on a recent change in attitudes toward red hair.

In 2011 the world's largest sperm bank, Cryos International (which, ironically enough, is based in the redhead-rich land of Denmark) announced that it would no longer take deposits (their word, not mine) from redheaded donors. This in turn inspired the Italian photographer Marina Russo to create a photographic matrix of forty-eight different types of redhead, published as *The Beautiful Gene*. Russo's starkly lit portraits, arranged according to the criteria most commonly used by clients of the sperm bank, stare back from page after page—the human race in all its non-symmetrical, never-ending variety, but presented here in test-tube-like ranks. This was also the period of the patronizingly infamous remark by the singer Taylor Swift, "I would do a ginger." (Try substituting "person whose skin, rather than hair, is a different color from mine" for "ginger" in that remark, and see how bad it tastes then.) But in 2014 the

Copenhagen Post reported that Cryos was now struggling to keep up with the demand for MC1R sperm. The turnaround was being driven by demand from couples both straight and gay in which one partner had red hair, so a redheaded child was desired to create the visual effect of consanguinity. And jolly good, too. But why, after centuries when society's response to red hair was mixed, to say the least, should it in our own day and age become desirable? Is it simply that appetite for rarity, as with those teenagers in Japan, where what was bad becomes good because it was bad? Or is red hair coming to stand for something other than the wimps, the barbarians, the kooks, the witches, the bluestockings, the sexpots?

There is this one question that our species has been asking itself ever since it stood upright: what are we? In our age, that too has undergone a subtle shift. We ask now: what am I? Our texts, our Facebook postings, our tweets to one another and to total strangers, people we have never met and never will, our blog entries, our websites, our Pinterests and Instagrams and uploads and likes and dislikes and all our multiple new means of communication are part of the attempt to answer that question. They're there, they're used, because we need them. In a way the network we create around ourselves duplicates our far-flung genetic connections to one another. And in all the words we now throw at one another, around and around the planet, day and night without ceasing, we are both all duplicating one another's behavior and all trying to define ourselves as apart, to give our likes and dislikes an individual value.

It isn't easy. We all see the same movies, are aware of the same celebrities and the details of their lives, have the same books, the same goods, the same fads and passing fancies thrown at us. Culture has become something that homogenizes us, rather than characterizes us, as it used to do. Against this background, perhaps red hair is starting to stand for something new and desirable. Perhaps it is starting to stand for individuality, for differentiation. And perhaps it is able to do this, even if twenty young actresses one after the other parade newly rubricated locks on the red carpet, because of the historic depth of its association with otherness and with the outsiders—the borderland, the liminal, the wildlings of society.

And perhaps the change in its status has something to do with redheads themselves.

After all, we know how to do this. If the decades since 1945 have taught us how to do anything, it is how you tackle prejudice head-on. You take hold of the stereotype used to define and to subjugate you, you confront and assess it. You reject its negative aspects. You go to war on them. The redhead's reputation for a fiery temper could come in very handy here. But you take the aspects of your image that you like, and you use them to fashion something positive and viable. For Conan O'Brien, one of the most successful, perspicacious, and quietly canny redheads of them all, that is his career. For many of us, it describes the arc of our life from childhood bullying to an adult sense of pride in our identity and our genetic inheritance.

Or at least it should. There have been some horrendous cases in recent years of redheaded children being bullied to the point of suicide. That has to stop. Would it be acceptable for a child to be bullied at all, let alone to death, because of their skin color? Their religion? Their own or their parents' race, their own or their parents' sexuality? Making it stop, making people more aware of the language they use and their attitudes toward red hair, was one of the impulses behind Thomas Knights's hugely successful *RED HOT 100* project. Redheads are also finding a voice for themselves. When in 2000 the UK energy company npower ran an ad featuring a redheaded "family" with the tag line "There are some things in life you can't choose," the flood of complaints against npower was dismissed. But things move on. The tag line was picked up and used by Eleanor Anderson, a redhead herself, as the title for her 2002 thesis on attitudes toward redheads. Two things to note here: that red hair should be the subject of a thesis, to begin with, and then the repurposing of this notorious ad to attack the very attitudes responsible for its creation. In Australia, where "ranga" is still supposedly an acceptable term for a redhead (hard to know who should be the most insulted by this, redheads or orangutangs), the Red And Nearly Ginger Association is doing the same thing. Picking up the pace, in 2014 the Australian company Buderim Ginger ran an "Australia's Hottest Ginger" competition as a marketing tie-in, and although some might say this was more a cynical reuse of a stereotype than a repurposing of it, it was not, refreshingly, for women only. There was a male hottest Australian ginger as well as a female. There are redhead websites, some of which, such as How to Be a Redhead, Everything for Redheads, and Ginger Parrot are models of how to e-market and thus by sleight of hand how to e-lobby, as well, as are Ginger with Attitude, Ingingerness, and Ginger Problems, all with a nice line in turning bias against red hair on its head. Bit by bit, attitudes start to waver, then to change. When in 2009 the Tesco supermarket chain in the UK offered a Christmas card showing a redheaded child sitting on Santa's knee with the legend "SANTA loves all kids. Even GINGER ones," the outcry and embarrassment to Tesco was such that the company not only apologized but withdrew the card from sale as well. So yes, attitudes are changing, but they never change fast enough. "Gingerism," so-called, is a truly ugly word for an ugly thing.

Part of the problem here is that gingerism doesn't *look* like it's racism, and in a way it's not, or at least not in the way we are used to thinking of it. Race may not be involved at all. And red hair does stand out, it can't help it, so for those of that befogged and bigoted understanding, in apparently calling attention to yourself it's as if you are asking to be picked on.[114] Then, redheads are not so other that we are going to turn out to be unexpectedly dangerous. We're a known quantity. Our skin is *white*. You see two people whose skins are different colors with one victimizing the other, you know what you're looking at. But with red hair there are likely to be none of those obvious clues between bully and victim; only the color of the hair. And who would pick on another human being just because of that?

But perhaps the most significant development is this: that redheads are no longer merely that 2 to 6 percent, those isolated individuals within the rest of society. Redheads are rapidly forming a community. There are redhead festivals in Russia, in Scotland, in Ireland, and one has even been started in Israel, on the aptly named Carrot Kibbutz. There are ginger pride events in the United States in Rome, Georgia; Portland, Oregon; in Chicago; in New York; and in Austin, Texas; in Milan, in Manchester in the UK, and in Montérégie,

114 See Druann Maria Heckert and Amy Best, *op. cit.*, for a more detailed discussion.

Quebec, in Canada. They are growing a new awareness of red-head identity and worth, of redheads' knowledge about themselves socially, scientifically, and culturally. There is naturally a sense of ganging together, of a rallying, of pride at these events. There is speculation of a redheaded moment not that far off, a redhead renaissance, indeed. And the biggest festival of them all is held in September in Breda, Holland.

REDHEAD DAYS

Hair . . . mediates between the individual and their culture. . . . It is a site of immense conflict—external authorities, parents, church, peer groups, schools gangs and fashion gurus all seek to impose their conventions on the individual.

JULIET McMASTER, "TAKING CONTROL," 2002

What began as the colour of children, comics and clowns is now a flag of determination.

GRANT McCRACKEN, *BIG HAIR*, 1995

The town of Breda stands where the River Aa broadens upon meeting the River Mark, about sixty-five miles south of Amsterdam. "Brede" means "wide," so the name of the town means "wide-Aa," basically, which is not only easy to remember but conjures up pleasing visions of an European Economic Community river-naming directive; one that might mean some other river, twenty-six water-courses distant, would be known only as the "Bb."

In its long history Breda has been bought, sold, taken in battle, inherited as a dowry, and (briefly, in 1795) French. It has been burned to the ground, immortalized by Velázquez, and laid siege to by the Spanish. In 1590 it was recaptured by the Dutch when a tiny force of sixty-eight men managed to get into the town by hiding under the turfs in a peat-boat, which is about as Dutch a version of a Trojan horse as you could ask for. Charles II lived in Breda during most of his exile and signed the Treaty of Breda in 1667, by which England gained the far-off territory of New York, but precious little else. Polish soldiers liberated the town in 1944. Such are the fortunes of war, however, that General Maczek, who led the liberating force, and who died at the age of 102 and is now buried in Breda, spent his old age working as a barman among the redheads of Edinburgh. The town is known for chocolate, lemonade, licorice, and beer. It has a park, the Valkenberg, and a splendid Grote Kerk, a building of such Gothic ribbiness, such piercings and knobbly crockets that it looks as if its bones are poking through its skin—a great gray Gothic pachyderm, quietly moldering away under those horizonless Dutch skies. But in all its long, eventful history, I doubt Breda ever anticipated any such happening as Redhead Days.

It is, after all, a very strange thing for a redhead to find him- or herself completely surrounded by other redheads. As the actress Julianne Moore has said, redheads notice one another, we become preternaturally alert to another redhead in the room—there's even a redhead look, a glance of complicity that passes between us. But put 6,000-odd redheads into the center of one small historic Dutch town, and complicity ain't in it. That's no longer a minority population of any sort, that's a tribe. Not bad for an event that started by accident, with the Dutch painter Bart Rouwenhorst placing an ad for models to act as inspiration. Rouwenhorst likes redheads. He paints redheads. In 2005 he wanted fifteen or so redheaded models for a sequence of paintings inspired by the works of Rossetti and Klimt; some 150 turned up. Faced with the choice between creating 150 paintings or a festival of redheads, Rouwenhorst went for the latter.

So here I am on Eurostar, heading for Brussels and ultimately to Breda and the largest gathering of redheads on the planet. I've been looking out for others of my rubified kin ever since manhandling my suitcase through security at St. Pancras, and thus far I haven't spotted a single one. I mean, I know we're rare, but. Six thousand redheads (the number the organizers of Redhead Days tell me they are expecting this year), at a very generous average of 6 percent of the population at large, would be the seasoning for 100,000 or so non-redheaded folks. Surely the number of people at St. Pancras should include one other, at least?

Apparently not. What my carriage can offer instead is a party of jolly sixty-somethings, on their way to Bruges. Their conversation is loud and joyous; it's one couple's anniversary, it's someone else's birthday, a fourth member of the group has just retired. They have crackers, pâté, wine. The men twit the women over the amount of chocolate being in Bruges will require them to eat. The women hoot with laughter, josh the men, and exchange sotto voce wisecracks of their own. The men are mostly dressed in affluent beige; the women are much more colorful—turquoise blues, purples, coral reds.

Red. I suddenly notice the queen of the group has hennaed hair. She has a voice so rich with damage (cigarettes, hard liquor, holding its own in God knows how many marital disagreements) that it almost carries its own static. The end of every sentence she utters, every tale she tells, is lost as the group dissolves with laughter, bending over their tabletops, lifting their plastic tumblers of wine up high as if to keep them safe. There's another woman seated beside her dressed in pale pinks and blues, round-faced, round-eyed, a little blonde duckling, tucked in beside this creaking, cackling scarlet parrot. *Red,* I find myself intoning inwardly, *is the color of dominance.*

The last time I was one among many redheads was as an extra, filming *Atonement.* A lot of titivating goes on in the longueurs between takes, and it passes the time, having your hat adjusted, lipstick renewed, 1940s pageboy recurled. "You have lovely hair," the woman with the curling tongs said, behind me, holding on to a hank of it with urethra-contracting tightness as all around us cables writhed across the floor like Laocoön's serpents, and the lighting guys cursed at the strength of the sun soaking into St. George's, Hanover Square. "Lovely color," she continued. "Is it natural?" She pulled the curling tongs free. "And really thick. Redheads always have lovely thick hair." I'm about to correct her—no, it's not thicker hair, it's thicker *hairs,* but seizing another hank, winding it around her tongs, she forestalled me. "It's a shame you're all going extinct. *Is* it your natural color?" The time before that was a family holiday near Balmoral, where everything was red—the deer, the squirrels, the grouse, the heads of hair on the people. We went to the Highland Games at nearby Ballater, and from a lifetime of being the odd one out in this family of blondies and brownies, suddenly I fitted in, and it was the other members of my family who were the anomalies, the Sassenachs, the ones who stood out from the crowd. Now here I am heading toward Redhead Days in Breda and there's not another redhead in sight.

I have to change at Brussels, wait for a train to Roosendaal.

I have to change again at Roosendaal for Breda. At Breda I am to speak, for an hour, to an audience of redheads about the history of the redhead, be interviewed by a documentary crew, and get myself photographed with a bottle of Gingerella ginger beer. These are not things I have ever done before, nor are they things I have ever envisaged myself doing. Even saving my receipts from this journey is a reminder that I am here in my capacity as a *professional* redhead. I am a tad nervous.

The men in beige, the scarlet parrot, and her blonde duckling girlfriend decant from the train still holding their plastic cups of wine on high. I locate the platform for Roosendaal, then go buy myself a beer.

And I'm standing there, on the concourse, drinking my beer and people-watching when I see a man, whitish hair in a long ponytail, denim shirt, South American striped and tagged and fringed and toggled waistcoat, fingers full of silver, and the reason I see him is because as he approaches, he is staring at me. He takes in my face, then his eyes go around my head as if checking on my aura, and as he passes, he pulls an imaginary hat from his head and tilts me a bow. "Enchanté."

Man with a Thing for Redheads.

On we go. I'm watching, on my iPhone, the blue dot that is me, traveling through these hinterlands of Belgium and on into the Netherlands, and through or past so many of the towns where the medieval artists, the medieval men with a thing for redheads lived. Brussels, where Rogier van der Weyden died. Bruges, which had Jan van Eyck. And now here we are at Antwerp, which had them all, including in the winter of 1885,Vincent van Gogh—cold, lonely, hungry, miserable, ill, ginger hair close-cropped as a prisoner's, sitting for hours in front of the paintings of Peter Paul Rubens (whose own surname in Latin means "colored" or "tinged with red"), worming his way into Rubens's paint swirls in his thoughts and bringing red into his own palette thereafter. The sky is boiling with clouds, the end of Hurricane Cristobal, flicking Europe with its tail. Strobes of sunlight pass across the fields, stride over the roof-scapes of the towns. Van Gogh was born near Breda. I feel as if I'm approaching redhead ground zero.

The train is much, much quieter than was the Eurostar. There's a discreet conversation in French bubbling like a water fountain to the right of me; I try to listen in as subtly as I can while taking notes. It seems one half of the conversation is recommending a hairdresser to her friend. I am reduced to the amateur subterfuge of examining both, as far as I can see them, in my handbag mirror. Writers are awful people, we respect nothing, and Sherlock Holmes would be proud of me. Sure enough, one purplish-reddish head, one orangey-brown. People have asked me all the way through the writing of this book, "Does dyed red hair count?" Of *course* it does. What more wholehearted acclamation could there be?

Roosendaal. Time to go grapple with the suitcase again.

Roosendaal is one of those stations where the platforms are as far apart as the coasts of a couple of continents, facing each other over a waste of rusty unused lines and wonky buddleia bushes, bright and lively in this Indian summer with bees. Traveling like this on my own to a place where I will be both new and everywhere, where I will know no one and everyone, is strangely nerve-inducing. My fellow redheads in Breda could include a little girl from the Kibbutz

Gezer (Carrot Kibbutz), a teenager from Afghanistan, and a Venus from Australia. It's ridiculous to think I am related to any of them in any way at all, or they to me, to presume that we will have anything in common, and yet we will. We do.

A train, grinding in. The sign hanging above the platform does a noisy electronic blink. BREDA.

There's something unshowy, self-effacing about Dutch towns, like the people—endlessly polite and, for a Brit, shamingly bilingual. They don't do look-at-me high-rise. Breda sits low to the landscape, the same height to the buildings as they have had for centuries. I do a little walkabout after checking in to my hotel (the Golden Tulip—how Dutch is that?) and register vague impressions of grayness and stone and mild hubbub from the bars. But the redheads, if they are here yet, are plainly all in hiding. I spotted two, in the park, from my taxi on the way to the hotel and that's it. Back at the hotel and the only thing orange is the Dutch football team on the television in the bar. I, like Breda, also have a Mark. I get into bed with my laptop and email mine assorted disconnected thoughts and a string of kisses.

According to my dog-eared program, I will be able this weekend to indulge in a weekend's worth of redheaded music, join a pub crawl, take a canoe trip or win one in a hot-air balloon, have my fortune told, my nails done, my colors done, get a makeover, speed-date, do a fire art workshop, a Lindy Hop workshop, a burlesque workshop, a cocktail workshop, buy more redheaded merchandise than

I ever knew existed, watch a catwalk fashion show, immortalize myself in the group photo, and learn how to keep my energy levels high by laughing at myself, which will no doubt come in very handy at two p.m. on Sunday afternoon. Before that, I have to meet with Papercut Films.

What the hell do you wear to be filmed? Solid colors, I am told, no black, no white, which leaves such an endless amount of space for picking the wrong thing you could fit my entire wardrobe into it. We start by walking me around the nave of the Grote Kerk, now bedecked with Thomas Knights's photographs, one of which apparently had to be removed when the church council took exception to the lowness of the model's low-slung jeans. The camera follows me, around and around the curve of the nave, and then behind me, up the winding stone stair to the room where the filming and interviewing proper is to take place. I manage to resist making the obvious and awful joke, as the cameraman labors up the stairs behind me, as to whether my bum looks big in this. I sit as directed and am powdered anew—apparently I'm shiny. The camera rolls. I keep such desperate eye contact with my interviewer that in the heat of the room my eyeballs start to dry out. Being interviewed seems to consist of being asked the same question three different ways until I give them an answer they can use, until Chris and Mark—another Mark?—are happy. I have a horrible sense that most of my answers are a disappointment to them, but I find an unexpected core of obstinacy within myself. Never mind the professional redhead, the redheaded activist has decided the time to assert herself is *now*. I have spent months surrounded by books and theses and journals and offprints, and in my opinion the true history of red hair is infinitely more fascinating than any of the myths. So no, redheaded young women were not hauled off by the hundreds to be burned at the stake as witches; no, we do not bleed more than other hair colors; no, we do not originate in Ireland; and no, we are not going to become extinct. The noise from the street outside is increasing. The crew is filming at the speed-dating session next, and there is some discreet checking of watches going on. The session wraps. There is lunch. I say that I hope I wasn't too useless. "You were fine," Chris tells me, very kindly. Music is bass-booming from a band in the town square. I head down the stairs, outside, and—

Whoa. *Whoa.* What the—what's the—whatever the collective noun is for six thousand redheads (a bonfire, a sunrise, a solar flare, a rubescence, an incarnadination, a conflagration, an incandescence, a frenzy, an *apocalypse* of redheads), let's hear it now. There are tall redheads, short redheads, plump redheads, thin; there are tiny little ones charging through the crowd at knee-height; there are redheads in baby carriers and redheads in wheelchairs. There are old redheads, gone sandy with age; there are redheads so new and young that to see any hint of color on their infant heads would be an act of faith only two redheaded parents could perform. There's an open-top bus with redheads on its top deck, an immense poster of the crowd from last year's festival completely covering one side, and redheads having their photographs taken in front of it, a sort of past, present, and future of red. There are redheads in costume—Rapunzels, mermaids, Magdalenes, vampires, Vikings. There's the redhead queen from the Irish Redhead Convention here somewhere; there are redheads in facepaint: foxes and squirrels. A man passes me with a dog, a red setter with a spotted handkerchief tied around its neck, padding along at heel, which makes me laugh—do redheads chose red pets? Ginger cats? Pomeranians? Dachshunds?

And there is hair, hair, everywhere. There is hair here that has plainly never been cut once in its life. There are *manes* of hair. There are braids down to hips; there are walking bushels of red, there is every shade of it describable. Tight deep red curls; curtains of pale, pale cinnamon; orange-hued skinheads; terracotta plaited like lawn edging; peppery cornrows; masses of ginger so prodigious they follow their owners a bounce or so behind with beards and whiskers to match. There are mothers and redheaded daughters, bonding as only mothers and daughters will. There are tribes overlapping—rockers and bikers and punks circulating around the merchandise tents side by side with neat little families in Peter Pan collars with babies in strollers. There are redhead dreads. You stop looking at the people—the first thing you measure up on everybody is the hair, and there is a certain air of rivalry in some of those sizings-up as well, a slight sense of "Who's the reddest?" It certainly isn't me. There are redheads here who put me to shame. One of the questions Papercut asked was where do redheads feel they belong? If we can come from so many different places, where's home? Here's the answer—right here (Fig. 33).

I wander the streets, sketching tableaux. The one redheaded sibling, for example, in a group of three, where it's the other two (bigger, older) who look awkward, who are now the odd ones out. A redheaded couple, doing that linked-together couple's walk, arms across each other's backs and fingers through each other's belt loops. She has her head on his shoulder and from behind their hair is meshing, too. A teenage girl in a costume made of a cloud of purple netting (this year's signature color), through which her face peers as if through purple stage smoke. A giant red plait, draped from the upstairs window of a bar. An older couple, must be in their seventies, in matching sweatshirts and sturdy hiking boots, who look both totally at home here and totally out of place—her hair faded with time, clipped sensibly short, his almost gone. They'd have been toddlers when the Nazis were in town, which reminds me of the one internet factoid that has eluded all my attempts so far to run it to ground—that the Nazis forbade redheads to marry. I can't believe red hair was any different a signifier for the Nazis than it was for the bureaucrats of the Spanish Inquisition. The way prejudice works is never more than predictable.

And standing here in the square outside the Grote Kirk of Breda is surreal. It hadn't occurred to me before how much a part of being a redhead is the business of being the only one on the train. It's so much a given of having the hair that no one else has the hair, but here everyone does. Breda makes the outsiders the tribe; the exception the rule. So if we are no longer the other, who are we?

Ruth is here from Bristol with her sister and her two sons. I find her in the Grote Kirk, admiring the men in Thomas Knights's photographs. It's very easy to fall into conversation on Redhead Day in Breda. Everyone seems keen to share the experience of being here and of being red in a manner that has something to it of catharsis. Her sister is a redhead too, but a toned-down version; Ruth is fire-engine red. Her two boys are darker, she tells me, pointing to one of the photographs, "like him." Teenagers, off on their own; she suspects they will be down at the redhead speed-dating tent. Do they like redheads? She laughs. "They like girls." It's kind of telling, that the graphic for the speed-dating venue has a gorgeous redheaded coquette, fluttering her lashes at a man whose hair is brown. She frowns when I point this out. Does she think it's different for girls?

It turns out Ruth is an army wife; when her boys were born, the family was in Germany. Her sons went all the way though kindergarten and junior school without a problem; then they moved back to the UK and everything changed. "They were big enough to stand up for themselves then," Ruth says, and from the sound of it, just as well they were. Do they get treated differently now? "There's two of them. They look out for each other." It was the same with her and her sister when they were young. Harder to pick on two. And attitudes are changing; she's noticed that. Things are different now.

Are they?

A long, long pause for thought. "Yes," she says at last, and then brightens. "I mean, there's Prince Harry, isn't there?"

Marius's family is half Hungarian, half Romanian. Despite his youth (when I offer to buy him a beer, he modestly requests a Coke), he's a Redhead Days veteran. Heard about it through the internet in 2006—could any of this be happening without social media? Helped out as a volunteer last year, brought along a gang of friends this. He and his mates have also been down at the speed-dating tent, which must be doing a roaring trade. He's also been to one of the gay bars in town, which was a good experience, as he calls it. He is trilingual at least; has moved with his family to Scandinavia and then with them to Germany. What was it like being a redhead in the land of the Vikings, I ask, anticipating a positive response, but Marius pulls a face. He was bullied, he says, "a little bit" in Norway, but that was because he wasn't Norwegian. His remark puts me in mind of a depressing suggestion made in their study by Feinman and Gill, that people have an inherent psychological need to dislike *something*.[115] All the same (Marius is thoughtful when he speaks, weighing his words), on balance he thinks red hair for guys is a bad thing, despite the fact that if you have red hair it makes you more tolerant, more aware of the feelings of others. His role model is the character played by Tim Roth in *Lie to Me* (to whom, it must be said, he bears more than a passing resemblance). Where, having lived in so many countries, does he feel at home? "Here."

Just to confirm my worst suspicions of the prevalence of being bullied in so many redhead lives: Kelly. Kelly comes from New Zealand, is a media student, and has deep red hair that truly does shine like polished copper and a tale of being picked on and marginalized all the way through her school days, with one of her worst persecutors being the only other redhead at the school. "Only," says Kelly, "she wasn't as red as me." Now, at college, her red hair is both envied and emulated. This is indeed the story for most redheads: you're teased as a child (and even the most well-meaning of teases, from those you know are fond of you, is wearing and unwelcome, please note—those teased always taste the vinegar more strongly than the honey). Then you grow up, and especially if you're a girl, your expe-

115 Saul Feinman and George W. Gill, *op. cit.*

rience of being red can transform with bewildering rapidity. Are you happy now with being a redhead, I ask her, thinking what a tragedy, with hair as beautiful as that, if she's not, and she nods her head, vigorous emphasis: yes. Now, it's her. Everything about it is her—even the interest and awareness it has given her, as a media student, in what makes other people tick. Happy ending. So should they all be.

Laura-May Keohane, crowned queen of the Redhead Convention at Crosshaven in Ireland in August. She's here in a red-and-gold cloak, like an old-time coachman, gowned and crowned, one of the stars of the show, and is so ridiculously pretty that when people stop to take photographs of us I want to get out of the picture. There is a huge amount of snapping going on. Two redheads need only stand together to be photographed by someone. You get the feeling this is not simply building up an archive of memories; there's some dedicated search for self-definition going on here as well, a crowd-sourcing exercise in personal classification. Laura-May is astonishingly gracious in her role; if I had been crowned queen of the Irish redheads at the age of twenty-one I doubt I would have been quite so mature about it. "Now where does the red hair come from?" she asks, smiling away as the camera-phones are flourished all about us. She has a lovely line in Irish lyricism, too. "There was a fella told me it was caused by the clouds and the rain."

Dr. Tim Wentel, my fellow speaker in the Grote Kirk at Redhead Days. He's the man to tell you where red hair comes from, if anyone can, but as well as being an expert on the hair, the skin, the freckles, he's something of an authority on beers as well. My notes from our meeting stray exuberantly up and down the page and at one point toddle off into the gutter of my notebook for a lie-down. We end up at an Indian restaurant where I preserve the redhead's reputation for dealing coolly with the hottest curries, and Dr. Tim gives me chapter and verse on the origination of the redhead extinction story—predictably, another internet myth, from years ago, but as he tells it, with a twist. Globalization may, eventually, many generations down the line, succeed where the makers of woozles have failed. We may all, as we mix, eventually revert to the phenotype of dark hair, eyes, and skin with which we once emerged from Africa. Given global warming, of course, we may all be thankful that we do, but if globalization should have this effect, *if*—does that mean no more red hair? Anywhere? Or might Mother Nature have more tricks up her sleeve?

I'm up at the business end of things now, tents selling t-shirts and wristbands, makeovers and do-overs, and it is beginning to feel a little odd, this relentless concentration on just the one body part. I have an urge to stand on the steps to Breda's noble town hall and shout, "I am not a redhead! I am a free woman!" I tell myself I at least will no longer gawp and stare at my fellow redheads. Instead, I step away and almost immediately find myself on the outskirts of a small circle of people gawping in amazement at a tall teenage girl who not only has the most wonderful skin, like illuminated bronze, but with it the most astonishing red, red, red, *red* afro, an aureole as bright as molten lava. This, so I learn, is Sterra. And Sterra is what Mother Nature can come up with if you only give her the right breaks (Fig. 33).

Sterra is here at Breda with her mother, Irmgard. They've been coming here since Sterra was a child. And while Irmgard's hair is now dark, when she was young, she tells me, it was red. Irmgard is Dutch, but Sterra's father is from Senegal, and somehow those two gene pools, as they mixed, created this thirteen-year-old standing here. I ask the obvious question: has Sterra ever experienced any racial prejudice (wondering, even as I ask the question, what on earth, given her coloring, the form of such prejudice could take)? No, Irmgard assures me, never. There's a line in Emily Cameron Walker's thesis to the effect that red hair is no more than a "genetic spandrel," that pale skin is the genetic driver, red hair the side effect, and technically yes, she is right. But in that case, looking at Sterra, all I can say is *quelle spandrel.*

Sunday. My talk is mere hours away. I'm brunching in Breda's main square, fortifying myself, and David and Anna are at the table next to mine. Anna has princess ringlets of the palest ginger, delicate as a cobweb, all the way down to her waist. She also has a wonderful story from her days at school. The rest of her class developed a game they called "carrot-busters" where every break time they would pile on top of or cannonade into Anna and the two or three other redheads in the playground. (Months later I'll find myself talking with a class of embryonic writers in a school in New York. The members of the class are about the age Anna was in her story, and we start talking about discrimination and prejudice and red hair. They all mention one girl in their year who has red hair, and thinking of Anna and her carrot-busters game, I ask if their classmate has ever been picked on in any way, and the whole of my audience recoils in horror. And that, I guess, is the difference, between growing up on the Upper West Side and growing up in the West Country.)

I turn to David and ask him if he noticed Anna because of her hair, if he would call himself an example of Man with a Thing for

Redheads. "Well," he says proudly, "Anna is the only girl I've ever asked to marry me." And Anna waggles at me on her left hand the most delicate diamond ring, so unmistakably new that even the hand has the look of still getting used to it.

The Grote Kirk in Breda can seat one thousand people, and I'm not saying it's full, I'm just saying that from the podium set up for us speakers it certainly looks that way. Row upon row of people fanning themselves with postcards of Alexa Wilding. I've been warned by the organizers that given the sensibilities of those who care for the Grote Kerk, I should mention that there will be nudity in the slides I'm going to be showing. So I'm now scrawling WARN THEM ABOUT THE COURBET in caps two inches high across the top of the first page of my talk when the hum and scrape of people settling into their seats is drowned out by a motorbike roar from outside. I look up. For the first time it registers quite how much studded leather as well as hair there is among my audience. Well, well. It seems I will shortly be explaining the historical significance of *L'Origine du Monde* to a gang of bikers. This is going to be interesting.

Joachim, who has gallantly volunteered to video-record my talk, gives me the thumbs-up. I take a deep breath.

"Okay. Ladies and gentlemen. Let me introduce you to King Rhesos of Thrace."

So after all that, where are we now in the history of the redhead?

First, I think Ruth is right. I think attitudes are changing, and I think it is redheads who are changing them. When, in response to the notorious *South Park* episode, "Kick a Ginger Day" was set up on Facebook by a particularly misguided fourteen-year-old in Vancouver, the Canadian comedian and activist (and redhead) Derek Forgie boldly co-opted the Canada Dry logo and pushed back by setting up, also on Facebook, "Kiss a Ginger Day." Nothing like turning your opponents' weapons against them. This year Kiss a Ginger Day became such a phenomenon it was trending on Twitter, although it certainly didn't hurt that it fell just before the announcement of the nominations for the 2015 Oscars. Lo and behold, redheads were so in the camera's eye that in the UK the *Mirror* newspaper ran the ebullient headline "It's Kiss a Ginger Day! Here are 13 red-haired celebs we definitely want to celebrate with," and went on to list them: Damian Lewis, Michael Fassbender, Emma Stone, Karen Gillan, Benedict Cumberbatch, Christina Hendricks, Prince Harry, Amy Adams, Eddie Redmayne, Ed Sheeran, Isla Fisher, Rupert Grint, and Lily Cole. Here is how you change a stereotype: you make it cool. You take what was marginalized and you make it desirable just by pointing out how unusual it is. You turn its downside upside. Simple. Comedians such as Catherine Tate get in on the act, with her "Ginger Refuge" sketch (270,000 views on YouTube and counting). Thomas Knights launches his *RED HOT 100* photographic show, to "rebrand the ginger male," as its creator puts it, and takes it around the world. Papercut Films creates their documentary, the catalyst being the experience of my interviewer, growing up as a boy with red hair. And in Breda, after his talk, I watched couple after couple come up to Tim Wentel, wanting a forecast of their chances of having a child with red hair, and asking this as something which they, like the clients of Cryos International, positively desired.

And this, I suspect, is the real Redhead Dilemma, and it doesn't have a thing to do with invisible eyebrows. We want to shake off the perjorative associations of being red, but we don't want to give up our so-called rare color advantage, the thing that makes us stand out, that sees us exchange the redhead look in public, that means we feel special, rare, unique. We want to have our ginger cake and eat it. We love Ed Sheeran, and that he calls his red hair his saving grace; we adore the fact the Michael Fassbender grows his ginger side-burns long; at the same time we wish to God that Putin weren't so carroty. We groaned in disbelief at *I Wanna Marry Harry* when that appeared on our television sets (to be fair, so did everyone else), yet when Ruth Wilson, Amy Adams, and Julianne Moore all triumphed at the Golden Globes and Julianne Moore went on the win the first redhead Best Actress Oscar, it felt like a tipping point. And I don't think this is something where you can pick and choose. If we were ever in a world where redheads *weren't* singled out for the color of their hair, where that wasn't the one thing about us that everyone remembers, would we really like it more? Looking at so many red-heads, all with such different stories, let alone with such different reds, brings it home all the more forcefully. Redhead Days is a cel-ebration of individuality just as much as it is of our one uniting factor. To see every redhead as being the same as every other is absurd.

And that goes for us all. I began working through the final draft of this book in New York, watching thousands of people march wearing t-shirts bearing the words *I Can't Breathe.* I ended it watching more silent marchers, this time wearing t-shirts reading *Je Suis Charlie.* And if all this seems to have become rather political all of a sudden, that's because when you drill down into it, dammit, it is. Sometimes it feels as if what will finish us off as a species is not climate change, is not running out of fossil fuels, is not some super-plague, is not even our deleterious habit of trashing the planet we live on. What will do for us in the end will be just two things: ignorance and intolerance. A world that can't deal with something as small and insignificant as people whose hair is a different color is one where there is little hope of dealing with any of the problems created by those far bigger issues, of different skins, different faiths, different loves, different lives. It's not simple at all.

But who wants to live in a world where we don't try?

READING FOR REDHEADS

Some of the works I have used in researching and writing this book really could have been cited on almost every page. I hope I have credited their insights wherever I have made use of them. For those readers interested in pursuing their own researches, these are the works, along with those footnoted in the text, that were of most value to me.

Eleanor Anderson: "There Are Some Things in Life You Can't Choose . . . : An Investigation into Discrimination Against People with Red Hair," *Sociology Working Papers*, (2002).

Ruth Barton (ed.): *Screening Irish-America: Representing Irish-America in Film and Television* (Dublin, Irish Academic Press, 2009).

Beth Cohen (ed.): *Not the Classical Ideal-Athens and the Construction of Other in Greek Art* (Leiden: Brill's Scholars' List, 2000).

Michelle A. Erhardt and Amy M. Morris (eds.): *Mary Magdalen, Iconographic Studies from the Middle Ages to the Baroque* (Brill, 2012).

Susan Haskins: *Mary Magdalen: Myth and Metaphor*, (New York: Riverhead, 1995).

Druann Maria Heckert and Amy Best: "Ugly Duckling to Swan: Labeling Theory and the Stigmatization of Red Hair," *Symbolic Interaction* 20, no. 4 (1997): 365–84.

Noel Ignatiev: *How the Irish Became White* (New York: Routledge, 2008).

Benjamin Isaac: *The Invention of Racism in Classical Antiquity* (Princeton, NJ: Princeton University Press, 2006).

Sandra R. Joshel: *Slavery in the Roman World* (Cambridge: Cambridge University Press, 2010).

Tara MacDonald: "Red-headed Animal: Race, Sexuality and Dickens's Uriah Heep," *Critical Survey* 17, no. 2 (2005): 48–62.

Catherine Maxwell: *Swinburne* (Plymouth: Northcote House, 2004).

Grant McCracken: *Big Hair: A Journey into the Transformation of Self* (New York: Overlook Press, 1996).

Juliet McMaster: "Taking Control: Hair Red, Black, Gold, and Nut-Brown" in *Making Avonlea*, ed. Irene Gammel (Toronto: University of Toronto Press, 2002).

Ruth Mellinkoff: *Outcasts: Signs of Otherness in Northern European Art of the Late Middle Ages* (Berkeley, Los Angeles, and Oxford: University of California Press, 1993).

Marion Roach: *The Roots of Desire: The Myth, Meaning, and Sexual Power of Red Hair* (New York: Bloomsbury, 2005).

Anthony Synnott: "Shame and Glory: A Sociology of Hair," *The British Journal of Sociology* 38, no. 3 (September 1987): 381–413.

Kelly L. Wrenhaven: *Reconstructing the Slave: The Image of the Slave in Ancient Greece* (London: Bristol Classical Press, 2012).

Kelly Wrenhaven, "A Comedy of Errors: The Comic Slave in Greek Art," in *Slaves and Slavery in Greek Comic Drama*, eds. Ben Akrigg and Rob Tordoff (Cambridge: Cambridge University Press, 2013).

ACKNOWLEDGMENTS

With some kind of organization, this list of names could stretch out to an extent that would be simply embarrassing, so . . .

At Black Dog & Leventhal, the wonderful J.P., and the just as wonderful Becky, Pam, Kara, Maureen, and Stephanie; also Becky Maines ("Red Becky") and Andrea Santoro, Rena Kornbluh, Mike Olivo, Christopher Lin, Ankur Ghosh, Nicole Caputo, Cindy Joy, and (for his splendid redhead map) Stefan Chabluk. At Fox and Howard: Chelsey and Charlotte. Jonathan Clements and Barbara Schwepcke—in at the birth.

The staff of the British Library and of the LSE Library, of the National Portrait Gallery and of the Royal Collection Trust; indeed all the friends and colleagues, past and present, who have been kind enough to interest themselves in this. It would have been a lesser book without you.

For advice and assistance: Nikolay Genov; Jeroen Hindriks; J.T. Leedson; Yvette Leur; Chris, Mark and Sara of Papercut Films; Professor Jonathan Rees; Joe Schick; Karin Schnell; Kirsty Stonnell Walker; Julia Valeva; Irmgard and Sterra Vlamings; and Dr. Tim Wentel. Like poor Ralph Holinshed, battling the exigencies of deadlines, I can only say that I have done what I could, not what I would, and any errors are mine. Heartfelt thanks also to Thomas Knights, to Bart and all the staff of Redhead Days in Breda, and all those many redheads who responded to the idea of this book with such enthusiasm and have offered details of their lives and experiences to me with such generosity.

For Millie, for listening so patiently. For being there: my family, especially Alice, Sam, Emma, Jack, and Ellie, and of course Nick.

And for all the support, the encouragement, the patience, and the wisdom any writer could ask for—Mark. Neither this, nor its writer, would be here without you.

ART AND PHOTOGRAPHY CREDITS

Frontispiece: Stefan Chabluk

Fig. 1: Copyright © REZA/Webistan

Fig. 2: Gusjer, via Flickr

Fig. 3: http://www.healthcare2point0.com

Fig. 4: Copyright © Nikolay Ivanov Genov

Fig. 5: Digital image courtesy of the Getty's Open Content Program

Fig. 6: © The Trustees of the British Museum

Fig. 7: © Bayerisches Nationalmuseum München. This object is a permanent loan from Bayerische Staatsgemäldesammlungen.

Fig. 8: *Calvary*, 1475 (oil on panel), Messina, Antonello da (1430–79)/Koninklijk Museum voor Schone Kunsten, Antwerp, Belgium/© Lukas—Art in Flanders VZW/Photo: Hugo Maertens/Bridgeman Images

Fig. 9: *The Coronation of the Virgin*, completed 1454 (oil on panel), Quarton, Enguerrand (*c.* 1410–66)/ Musee Pierre de Luxembourg, Villeneuve-les-Avignon, France/Bridgeman Images

Fig. 10: *The Last Judgement*, 1473 (oil on panel), Memling, Hans (*c.* 1433–94)/© Muzeum Narodowe, Gdansk, Poland/Bridgeman Images

Fig. 11: *Altarpiece of the Dominicans: Noli Me Tangere*, *c.* 1470–80 (oil on panel), Schongauer, Martin (*c.* 1440–91) (school of)/Musee d'Unterlinden, Colmar, France/Bridgeman Images

Fig. 12: *The Crucifixion* by Jan van Eyck (1390–1441)/De Agostini Picture Library/Bridgeman Images

Fig. 13: Copyright reserved

Fig. 14: Courtesy Southern Methodist University

Fig. 15: Courtesy of the British Library

Fig. 16: Queen Elizabeth I, *c.* 1575 (oil on panel), Netherlandish School (16th century)/National Portrait Gallery, London, UK/De Agostini Picture Library/Bridgeman Images

Fig. 17: Queen Elizabeth I (1533–1603) being carried in Procession (*Eliza Triumphans*) *c.* 1601 (oil on canvas), Peake, Robert (fl. 1580–1626) (attr. to)/Private Collection/Bridgeman Images

Fig. 18: *Symphony in White, No. 1: The White Girl*, 1862 (oil on canvas), Whistler, James Abbott McNeill (1834–1903)/National Gallery of Art, Washington DC, USA/Bridgeman Images

Fig. 19: *Le Sommeil*, 1866 (oil on canvas), Courbet, Gustave (1819–77)/Musee de la Ville de Paris, Musee du Petit-Palais, France/Bridgeman Images

Fig. 20: *Portrait of Algernon Charles Swinburne* (1837–1909) 1867 (oil on canvas), Watts, George Frederick (1817-1904)/National Portrait Gallery, London, UK/Bridgeman Images

Fig. 21: *Beata Beatrix* (oil on canvas), Rossetti, Dante Gabriel Charles (1828–82)/Birmingham Museums and Art Gallery/Bridgeman Images

Fig. 22: *Found*, *c.* 1869 (oil on canvas), Rossetti, Dante Gabriel Charles (1828–82) / Delaware Art Museum, Wilmington, USA/Samuel and Mary R. Bancroft Memorial/Bridgeman Images

Fig. 23: *Ellen Terry as Lady Macbeth* by John Singer Sargent/Tate Britain

Fig. 24: *The Knight Errant* by Sir John Everett Millais/Tate Britain and *The Martyr of the Solway*, 1871 (oil on canvas), Millais, Sir John Everett (1829–96)/© Walker Art Gallery, National Museums Liverpool/ Bridgeman Images

Fig. 25: *Combing the Hair (La Coiffure)*, *c.* 1896 (oil on canvas), Degas, Edgar (1834–1917)/National Gallery, London, UK/De Agostini Picture Library/Bridgeman Images

Fig. 26: *Cora Pearl* ©National Portrait Gallery, London

Fig. 27: *Uriah Heep*/Wikimedia Commons

Fig. 28: Tintin Press Club

Fig. 29: © Danita Delimont/Alamy

Fig. 30: Photograph by Pamela Tartaglio, courtesy of the Hollywood Museum in the Historic Max Factor Building. (PamelaTartaglio.com)

Fig. 31: *The Beethoven Frieze: The Longing for Happiness*, 1902 (mural), Klimt, Gustav (1862–1918)/ Osterreichische Galerie Belvedere, Vienna, Austria/De Agostini Picture Library/E. Lessing/ Bridgeman Images

Fig. 32: Picture by Colinda Boeren at the Redhead Days Festival in the Netherlands, www.redheaddays.nl.

Fig. 33: Photo made by Yvette Leur

INDEX

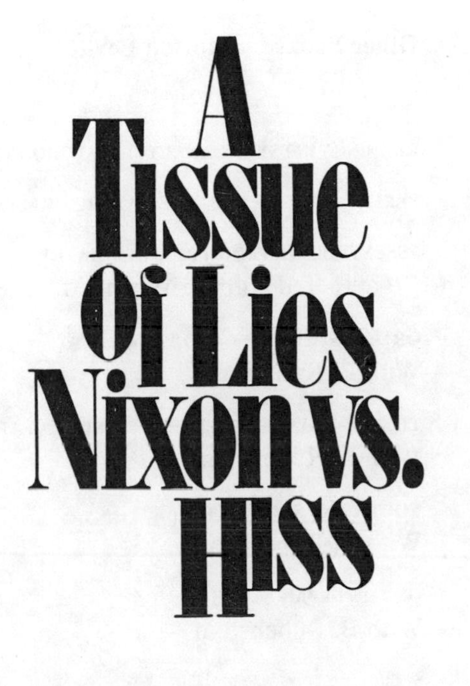
A
Tissue
of Lies
Nixon vs.
Hiss

Other Books by Morton Levitt

READINGS IN PSYCHOANALYTIC PSYCHOLOGY

FREUD AND DEWEY ON THE NATURE OF MAN

ESSENTIALS OF PEDIATRIC PSYCHIATRY
Co-authored with R. Meyer, M. L. Falick, and B. Rubenstein

ORTHOPSYCHIATRY AND THE LAW
With B. Rubenstein

THE MENTAL HEALTH FIELD: A CRITICAL APPRAISAL
With B. Rubenstein

YOUTH AND SOCIAL CHANGE
With B. Rubenstein

ON THE URBAN SCENE
With B. Rubenstein

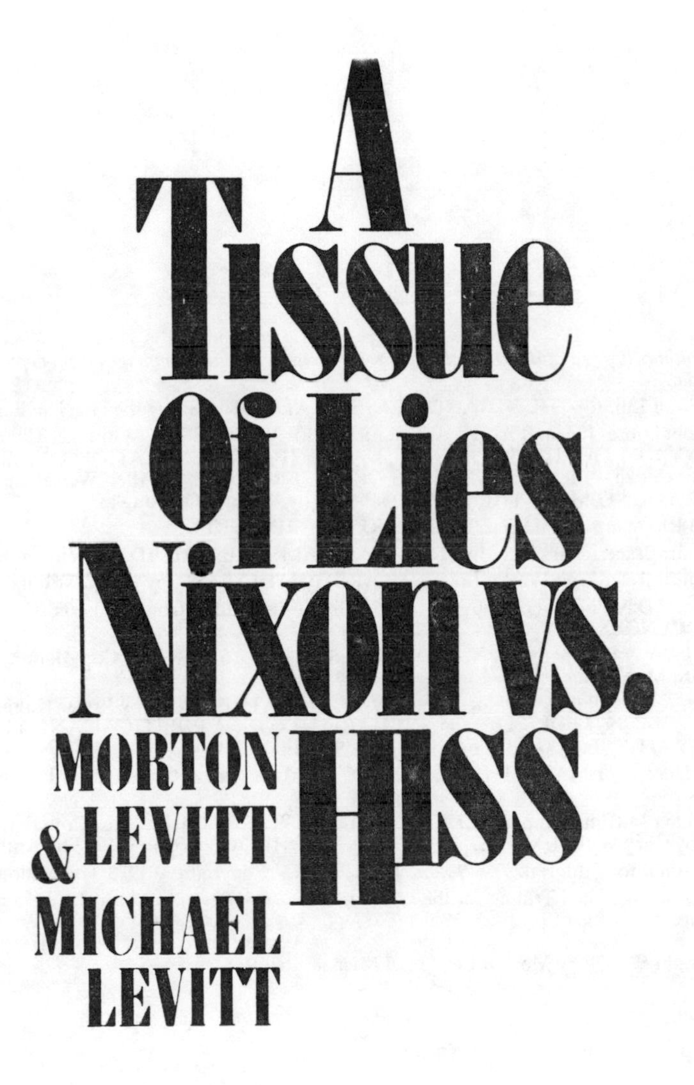

McGraw-Hill Book Company

New York St. Louis San Francisco
London Toronto Düsseldorf
Mexico Sydney

The authors are grateful to the following for permission to quote passages from copyrighted material:

A. P. Watt Ltd. for THE STRANGE CASE OF ALGER HISS by the late The Earl Jowitt

Random house, Inc. and Alfred A. Knopf, Inc. for WITNESS, by Whittaker Chambers, IN THE COURT OF PUBLIC OPINION, by Alger Hiss, A GENERATION ON TRIAL, by Alistair Cooke, PERJURY: THE HISS-CHAMBERS CASE, by Allen Weinstein, GO EAST, YOUNG MAN: THE EARLY YEARS, by William O. Douglas.

David Riesman for INDIVIDUALISM RECONSIDERED

Harcourt Brace Jovanovich, Inc. for A MAN CALLED INTREPID, by William Stevenson

Harcourt, Brace, and World, Inc. for CLAREMONT ESSAYS, by Diana Trilling, 1964

Stein and Day Publishers Copyright © 1955 by Leslie Fiedler from the book AN END TO INNOCENCE

Doubleday & Company, Inc. for SIX CRISES by Richard M. Nixon Copyright © 1962 by Richard M. Nixon.

Harper & Row, Publishers, Inc. for HOW THE WEATHER WAS by Roger Kahn, Copyright © 1973 by Roger Kahn, RICHARD NIXON: A POLITICAL AND PERSONAL PORTRAIT by Earl Mazo, Copyright © 1959 by Earl Mazo

Alyss Dorese for MARK THE GLOVE BOY by Mark Harris, Copyright © 1964 by Mark Harris

Houghton Mifflin Company for NIXON AGONISTES by Garry Wills, Copyright © 1969, 1970 by Gary Wills, LAUGHING LAST by Tony Hiss, Copyright © 1977 by Anthony Hiss

The Nation for Editorial, *The Nation,* May 28, 1973, Copyright © 1973 The Nation

The Estate of Lionel Trilling for the article in *New York Review* of April 17, 1975 by Lionel Trilling

Printed in the United States of America.

2 3 4 5 6 7 8 9 0FGRFGR7 9 8 3 2 1 0 9

LIBRARY OF CONGRESS CATALOGING IN PUBLICATION DATA
Levitt, Morton, 1920-
A tissue of lies
1. Communism—United States-1917- 2. Subversive activities—United States. 3. Nixon, Richard Milhous, 1913- 4. Hiss, Alger. 5. United States Congress. House. Committee on Un-American Activities
I. Levitt, Michael, joint author. II. Title.
E743.5.L47 322.4'2'0973 78-26542
ISBN 0-07-037397-3

Book design by Gloria Gentile.

For Lucille Keller Levitt

ACKNOWLEDGMENT

The authors wish to express their appreciation to Karen Long Schreiner, whose steadfast devotion, unlimited energy, high order of intelligence, and incredible sense of organization kept this project alive through some very rocky times.

Judge not, that ye be not judged.
For with what judgment ye judge,
Ye shall be judged; and with
What measure ye mete,
It shall be
Measured to you again.

MATTHEW 7:1–2

FOREWORD

In 1948, Alger Hiss was a former high-level State Department official who, among others, was publicly accused of being a Communist by a *Time* magazine editor named Whittaker Chambers. A reformed Communist himself, Chambers had tried unsuccessfully to interest the Democratic administration in his charges for several years. Finally the matter came to the attention of the House of Representatives' Un-American Activities Committee (HUAC) whose most junior member was Richard Nixon, elected to Congress just two years earlier on a tough-line-toward-Communism platform.

In unerring fashion, Nixon reviewed the national scene, recognized the political capital to be gained, identified a congenial issue, a ready victim, and struck for the jugular. Hiss became a stand-in for all that was weak and slothful in America during the New Deal, becoming thereby the stalking horse in Nixon's march to power.

By the end of the year, the original charges by Whittaker Chambers of simple party membership had escalated to espionage. Hiss was indicted by a New York grand jury. He was convicted after two trials and sent to prison for perjury, the statute of limitations having precluded a charge of espionage. The charges were upheld because he denied (1) turning State Department documents over to Chambers and (2) seeing Chambers after the beginning of the year 1937. The latter date was of central importance: presumedly the stolen documents in Chambers' possession all were dated after January 1, 1938.

Two years later, Richard Nixon's star was firmly fixed in the American political scene. His skill in prosecuting Hiss stood as the watershed event in his political life. Nixon would then go to the very top of the mountain, becoming one of the most powerful men in the world, before the Watergate imbroglio brought him plummeting down.

Although not from the same height, Hiss fell as far: forty-four months in prison, prolonged unemployment, underpaid positions, modest success as a stationery salesman, and, more recently in his seventy-second year, restoration of his license to practice law in the Commonwealth of Massachusetts. The justices of the Massachusetts Supreme Judicial Court unanimously concluded that Hiss was guilty as charged but that he was "reformed," as evidenced by his exemplary life since his conviction.

This is the first book to examine the strange vendetta that existed between Richard Nixon and Alger Hiss. Other books—a baker's dozen—have struggled over the years with complex evidentiary issues; the most recent victim of the literary need either to exonerate or convict Hiss is the flawed effort by Allen Weinstein, entitled *Perjury*.

The present authors have thoroughly examined the trial records and read the now released FBI papers relating to the Hiss case. They have interviewed Hiss on numerous occasions and examined original sources of Nixoniana available at the oral history project of the California State University at Fullerton. They are indebted to Alistair Cooke's fine political study, *A Generation on Trial,* as well as to The Earl Jowitt's scholarly legal treatise, *The Strange Case of Alger Hiss,* both of which came out very early in the game but still ring true. And although Nixon himself remained unavailable to the authors, his historical aide at San Clemente, Professor Franklin Gannon, was reported to have found a draft of this manuscript "too psychological" for historical comment, as well as too critical of Nixon.

The authors are the only ones to have examined the unique writ of *coram nobis* that Alger Hiss recently submitted to the federal court.

It was imperative for Nixon to convict Hiss so as to solidify his political career. Thus, Nixon personally revived the case against Hiss whenever it appeared to be in trouble. In doing so he surreptitiously consulted with Hiss's superior, John Foster Dulles, who was in turn advising Hiss how to defend himself. And with microfilmed evidence in hand, it was Nixon who made a dramatic appearance before the New York grand jury in its last hours of existence to ensure that the "right man [Hiss rather than Chambers] was indicted."

During the first trial, Nixon offered the prosecutor the benefit of his wisdom and experience in "breaking" Hiss down. After the first jury could not reach a decision, Nixon made an angry speech before Congress demanding that the fitness of the presiding judge be investigated and in particular criticized two of his crucial evidentiary decisions which (perhaps not by coincidence) were reversed by a different judge at the second trial. Nixon, apparently feeling Prosecutor Thomas Murphy was not up to the retrial task unaided, then proposed that a "special prosecutor" be appointed to bolster the government's case. This special office was to lose considerable appeal for Nixon two decades later.

It is ironic that the tactics Nixon employed to ensure the conviction and imprisonment of Hiss—lying, news leaks, midnight searches, attacks upon the press, invoking the separation-of-powers issue (although this time on the Congressional side of the fence), suppression and faking of evidence—were the same type of activities that toppled him from power a quarter of a century later.

Perhaps the supreme irony was that Nixon's flawed character could not be hidden away when his place in history was assured. Not even George McGovern's supporters could conceive of a Democratic victory in 1972. But the election was characterized by all the appurtenances of a Nixon campaign: breaking and entering, bugging, forged letters and political payoffs. By the time of the Watergate scandal, the President had pushed the public too far. People could identify with the hustling tactics of a poor boy on the political make—"hardball," as the President's men were wont to describe it—but they could not understand the need to continue this scary game when Nixon was entrenched at the top.

The authors argue that the evidence that convicted Hiss was, by today's standards, flimsy and controvertible. It is similarly contended, however, that Hiss's very demeanor, his persona, as it were, aroused suspicion. Oliver Wendell Holmes, for whom Hiss served as a law clerk, once wrote, "The life of the law has not been logic . . . it would seem therefore that the most important facts for a jury are the human ones." It was here that Hiss fared badly. He apparently concluded that purity and probity would win out over all other considerations—even sound legal preparation.

Hiss's biggest mistake was his refusal to admit that he knew a man named Whittaker Chambers, despite the fact that it was a name he had heard linked to his own several times before under most threatening circumstances. Alden Whitman reported that Hiss's wife, Priscilla, when handed a newspaper picture of Chambers by a neighbor while on vacation in Vermont, immediately responded, "I remember a dreadful man named Crosley or something like that." It took her husband eleven days longer specifically to link the photograph to Crosley, the name by which Chambers was known to Hiss. Thus, one denial led to other denials, and soon Hiss's credibility was at stake.

Was there something in the nature of Hiss's relationship with Chambers that had to be denied? Was it homosexuality? "No!" each said, although the FBI file revealed that Chambers admitted having such relationships with others. (While Hiss was denying the possibility to the authors, he was shortly thereafter telling an Associated Press reporter that Chambers was indeed a spurned homosexual who testified against Hiss out of jealousy and resentment.) But Hiss was certainly guilty about something in his relationship with Chambers—i.e., the two men were linked in some occult, even murky way. Otherwise what could have happened that made one man want to make the relationship explicit and the other want to conceal it? Hiss's initial refusal to admit knowing Chambers led to his prison term. The strait-jacketed life that followed, the absolute *noblesse oblige,* the pusillanimous public face, and the persistent lack of candor have done little since then to clear up the questions which persist.

Perhaps it is this last quality that is now and always has been the most distracting feature of Alger Hiss's personality. He came into the limelight straight and honest; three decades later, his appearance is markedly unchanged. Prison, dishonor, and poverty have hardly affected his outlook. Either Hiss was innocent of any wrongdoing in his entire life and had the strength associated only with the very pure, or he was the world's most accomplished artificer, the veritable Jack Dawkins of his era. It is this contrast between the man and the circumstance that divided both friend and foe from the very beginning. This division exists even today.

In their own time and in their own way, but in apparently contradictory fashion, both Hiss and Nixon argued for a high personal morality. Each behaved as if a crime had been committed, yet neither was willing to admit to any personal responsibility for its commission. Hiss portrayed himself as a beleaguered man being destroyed by an ungrateful acquaintance whose enmity was incomprehensible to him; Nixon, throughout his entire career, viewed himself as a man simply and completely dedicated to a single issue, that of national security.

Unfortunately, life does not allow for such simplistic constructions. There was about Hiss a curious, blind secretiveness, a purity that was so unreal as to seem opaque. His eternal optimism, his inability to believe the worst was occurring, his unending civility to an interrogator (Nixon) whose hatred for him was manifest are literally without precedent. Hiss's formidable rectitude over the past thirty years reflects the picture of a man determined not to look within himself. Clearly Hiss's "humaneness" had a self-serving quality. It projected a strong sense of innocence that was designed to deny culpability in matters large and small.

In comparison to Hiss, Mr. Nixon's demeanor was quite different. He was clearly Hiss's equal in piety. Sworn to defend the Constitution, he maintained a constantly visible high-mindedness. His explanations were reasoned and sober, and the occasional signs of irritation or flashes of anger were to be viewed as the signs of humanity in a sorely tried man.

Not unlike Hiss, Nixon cried, "Believe me, believe me!" but his voice was strident. Much taken by the trappings of high office, he sought to cloak himself in its protective colors. "When the President does something, it's not illegal" was one of his last public explanations.

This is a book about liars. Nixon's record over the years is dotted by lies. Hiss went to jail because he was convicted of lying to the New York grand jury.

Nixon lied early and late: When he entered public life, he misrepresented both his own naval record and the political policies of his opponent, Jerry Voorhis. And on the last day of his Presidency, he was still engaged in lying about his family background. He lied blatantly to Hiss about his prior knowledge of Chambers' testimony. He lied continuously about his personal sympathy for and support of Chambers and misrepresented evidence before the New York grand jury. In the end, Nixon was forced out of office when his own tapes revealed that he had lied in denying either knowledge of the Watergate break-in or culpability for the subsequent cover-up. Assessing the guilt of Richard Nixon is no particular problem.

Unfortunately, no such neat assay of guilt or innocence of Alger Hiss has been offered since his conviction. Twenty of twenty-four jurors who heard the case over two trials concluded that Hiss lied and was guilty as charged. Yet nearly three decades later people continue to argue about the case. Why should this be so?

Some, like Supreme Court Justice William O. Douglas, have concluded that Hiss was "framed." The FBI papers reviewed in this volume offer support for this contention. Since there was only a single witness, Whittaker Chambers—not without heavy guilt stains himself—who accused Hiss of espionage, the court was forced to rest its judgment on the nature of the corroborative evidence. In this instance, the evidence was the documents Chambers had in his possession, allegedly given him by Hiss, and the Hiss typewriter which presumably produced the copied materials.

Hiss admitted that he had read some of the documents, had indeed even initialed some, but there was no one who had witnessed an exchange of documents between the two men. Lacking that particular evidence, the weak case against Hiss was ably pointed out by journalist Alistair Cooke, who wrote, "If the question were whether some article of jewelry or the like had been stolen by A from B, or whether it had been handed by B to A as a gift, would the production of the article itself by A afford the slightest evidence one way or another?"

The fabled Hiss typewriter also had a checkered history, one which at this distance at least seems to render it highly suspect. At the trials, the most compelling piece of evidence was an old Woodstock typewriter, supposedly belonging to Hiss, which the government implied was used in the retyping of the State Department documents. This typewriter, which in Nixon's full-blown rhetoric was "the key witness in the case," has turned out to be a false witness.

The evidence surrounding this particular typewriter is so shocking that the authors have devoted an entire chapter to it. The FBI knew that the typewriter found by the defense and used most effectively by the prosecution at the trials could not have been the Hiss machine, because it was manufactured two years after the original Hiss machine was purchased.

Both the FBI and the prosecution went to great lengths to keep this exculpatory information secret, an unnecessary effort if the typewriter recovered by Hiss was the correct one. Finally, despite Nixon's certitude that a typewriter is like a fingerprint—"Every one is different and it is impossible to make an exact duplicate unless the same machine is used"—typewriters could indeed be forged. Intelligence operatives had been doing so since World War II. The FBI—and Richard Nixon, presumably, for that matter—was well aware of this.

Former Presidential Counsel John Dean related how Nixon instructed Charles Colson during a Watergate discussion, "Typewriters are always the key. . . . We built one in the Hiss case." It could be argued that this was the maunderings of a man under terrible pressure, but Nixon has said much the same thing during presumably more contemplative times.

There are the enigmatic references to the typewriter in Nixon's first book, in which he wrote that the FBI found the (real?) Hiss typewriter four months before the defense team located their Woodstock. No one has explained what happened to this first typewriter—if indeed there was such a machine—which, according to Nixon, "typed exact copies of the incriminating documents." When there was some inquiry about this serious discrepancy in *Six Crises,* Nixon discounted it as a "researcher's error" and had it deleted from the paperback edition.

Nixon's view of who found the typewriter is at odds with the evidence. It is even possible that neither Hiss nor the FBI found the Woodstock but rather Nixon and two HUAC investigators. Once again, Nixon, in speaking to John Dean on the Watergate tapes, said, "I conducted that [Hiss] investigation with two [characterizations omitted] committee investigators—that stupid—they were tenacious. We got it done. Then we worked that thing. We then got the evidence. We got the typewriter. We got the pumpkin papers."

The FBI, much like Nixon, apparently had a vested interest in Hiss's conviction. The Bureau had been aware of Chambers' accusations against Hiss for several years but took no real investigative action. The passage of time allowed Hiss the protection of the statute of limitations, which reduced the charges against him to perjury rather than espionage. There was much criticism of the agency on these grounds, and even a cursory reading of the recently released FBI papers reveals: (1) a prejudicial effort to convict Hiss, (2) an analogous effort to provide Chambers with evidence needed to testify successfully against Hiss—i.e., "coaching" Chambers and other witnesses, (3) the use of a double agent who infiltrated the Hiss camp and kept the FBI and United States Attorney Thomas F. Murphy abreast of defense plans and stratagems, and (4) the suppression of evidence which would have been exculpatory had the defense known of its existence.

Most significant in this regard was that, *before* the discovery of the Pumpkin Papers, Chambers, in secret testimony, told the FBI in March 1946 and the House committee in August 1948 that he broke from the Communist Party in 1937, and, "after 1937," read the FBI notes of this interview, "he lost all contact with Alger Hiss." As the contraband documents bore datelines of the first four months of 1938, Chambers could not have received them from Hiss. With the production of the Pumpkin Papers, Chambers recast his story: Now he last saw Hiss in mid-April 1938, when he broke from the Party and went into hiding in Florida with his family. This most exculpatory piece of evidence remained hidden in the FBI files until recently.

Justice Douglas wrote that the Hiss case spoke volumes on "the wisdom of having two witnesses on a perjury charge or if there is only one, as in the Hiss case, that the court ride herd on the nature of the corroborative evidence to make certain it has that 'trustworthy' character which will prevent an accused from being 'framed.''' It is here that Weinstein, author of *Perjury,* lost his way. For in his need to arrive at a summary judgment in this matter, he accepted the FBI's version of the evidence as "trustworthy." Another view of Weinstein's scholarship was offered by writer Philip Nobile, who described himself as "once Weinstein's friend" but who concluded in *The Nation* magazine that the former Smith College historian "fiddled with the evidence."

Witnesses whom Weinstein interviewed recanted, and prime among them was Czech historian Karel Kaplan, who provided the only other evidence linking Hiss to espionage through the confession of Noel Field. The latter was a friend of Hiss's from the State Department and an admitted Communist spy who was later imprisoned in Eastern Europe. The statement Weinstein attributed to Kaplan was baldly simple: "In all interrogations, Field named Alger Hiss as a fellow Communist underground agent in the State Department."

Since the publication of *Perjury,* Kaplan, whose testimony a New York *Times* reviewer found "most compelling," denied making such allegations. Kaplan, who read Field's testimony before and after he was incarcerated, wrote to Victor Navasky, "N. Field's testimony, as far as I can remember, did not contain any facts or explicit statements that A. Hiss was delivering U.S. documents to the Soviet Union."

Similarly, evidence in the voluminous FBI files which points to Hiss's guilt was accepted by Weinstein, but when it was discounted a few pages later in the self-same documents, the disclaimer was ignored. The famous typewriter is a case in point. Weinstein ignored the fact that there was no provable connection between the old Woodstock and the classified documents in Chambers' possession. Such tunnel vision was required by Weinstein—who had been saying for five years that he had unearthed the evidence that would at last erase any doubt about Hiss's guilt—for apart from the typewriter, it was, as Nixon concluded, "just one man's word against another." And who could believe Chambers?

The authors of this book conclude that Hiss lied about the nature of his relationship with Chambers. He also lied when he denied seeing Chambers after the year 1936, the deadline specified in the indictment. The need to maintain this position—for whatever disquieting reasons—accounted for Hiss's conviction for perjury involving espionage, lack of proof notwithstanding. Certainly lying is the worst thing a perjury defendant can do because it is *de rigueur* for a judge to instruct a jury, "If in a matter a witness is found to be telling an untruth on any question which is material and which is raised during the course of the court's proceedings, his credibility on other questions is also suspect." Still the government (like Weinstein) could not prove that Hiss told the crucial lie in denying his participation in espionage, and this was the very crime for which he went to jail.

Remarkably, Nixon arrived at the same conclusion, although he never informed the public of his posture. In a letter to John Foster Dulles (Hiss's former employer) that was written before the first trial, Nixon said, "Whether he [Hiss] was guilty of technical perjury or whether it has been established definitely that he was a member of the Communist Party are issues which may still be open to debate, but there is no longer any doubt in my mind that for reasons only he can give, he was trying to keep the [HUAC] committee from learning the truth in regard to his relationship with Chambers."

Nixon's career began with Hiss and espionage. It ended with Daniel Ellsberg and Watergate. The tactics that succeeded so well in 1948 were centrally related to his ultimate downfall in 1974. This is the Nixon tragedy. And Hiss, whose star burned so brightly, whose early career seemed to epitomize the best and the brightest, fell into disgrace because he either couldn't or wouldn't grasp the enormity of the crisis that faced him and respond to it in an effective manner. Hiss's failure to acknowledge even "knowing" Chambers in the simplest sense of that term came finally to be viewed as nothing save the act of an untruthful man.

The final irony in the Hiss-Nixon relationship is its absolute symmetry. A small lie escalates, and before the liar looks around, his fortune is imperiled. So it was with Alger Hiss and Richard Nixon. The former was an all-American boy laid low for an untruth. The scenario that brought about his downfall was masterminded by an ambitious young California Congressman whose own career was built on untruths and who finally ended public life in disgrace after being exposed in a monumental series of lies.

MORTON LEVITT
Davis, California

MICHAEL LEVITT
Norwalk, Connecticut

CHAPTER 1

I can understand those who say, "Gee, whiz, it just isn't fair, you know, for an individual to be, get off with a pardon simply because he happens to have been President."

I can only say that no one in the world and no one in our history could know how I felt. No one can know how it feels to resign the Presidency of the United States. Is that punishment enough? Oh, probably not. But whether it is or isn't, as I have said earlier in our interview, we have to live with not only the past but for the future and I don't know what the future brings. But whatever it brings, I'll still be fighting.

—RICHARD M. NIXON in a television interview with David Frost, May 25, 1977

Richard Milhous Nixon's rise to world prominence was strewn with the corpses of his political enemies. It began in 1946 in a struggle with a

Democratic Congressman from California named Jerry Voorhis, whom Nixon with no basis in fact labeled a Communist sympathizer. Nixon soon went to Washington in Voorhis' place as Representative from the 12th Congressional District. In 1950 Nixon described Helen Gahagan Douglas, the Democratic candidate for the U.S. Senate, as the "Pink Lady," an allusion to her liberal commitments, and shortly thereafter became the junior Senator from California. But he got the most mileage out of Alger Hiss, a State Department official during the New Deal whose most notable contribution was his work in organizing the United Nations. If there hadn't been an Alger Hiss to fit so neatly into Nixon's plans, the ambitious politician would have had to invent him.

For three years the Hiss name loomed large in the headlines across America. The case polarized the United States in the years 1948–1950 and became a worldwide symbol of American Cold War policies. Mr. Nixon dragged Alger Hiss along a path leading almost unerringly to this nation's political apogee, the Presidency.

For Alger Hiss, his experience with Nixon cost him more than his career. He lost his reputation and his liberty, spending four years of his life in prison. The ghost of Alger Hiss was soon forgotten. After being released from the Lewisburg Prison in Pennsylvania and having lost his license to practice law, Hiss went to live a nondescript life as a small-time stationery salesman. He had public visibility only when Nixon chose periodically to remove him from obscurity, and these rare appearances all seemed designed to reveal Nixon as a strong, moral, and forceful man. In the final two conflict-ridden years of his Administration, Nixon had much need to invoke the apparition of Alger Hiss.

Consider what happened when Nixon's troubles began in October and November of 1973 when there occurred a number of unprecedented events in the political history of this country. Vice President Spiro Agnew resigned from office after a plea-bargaining arrangement allowed him to plead guilty to a single charge of income-tax invasion. Then President Nixon's disenchantment with special Watergate prosecutor Archibald Cox's efforts to expand his investigation into the financial affairs of the President, his friends, and special conglomerates came to a head, and Cox was summarily fired.

The Attorney General of the United States, the peripatetic Elliot Richardson, who had hired Cox and promised him complete independence, resigned as a consequence, and his chief assistant, William D. Ruckelshaus, was then discharged for failing to execute a presidential order to dismiss Mr. Cox. Ruckelshaus remained adamant despite presidential aide Alexander Haig's strong statement that Ruckelshaus was disobeying a direct order from the Commander-in-Chief and that it must be carried out "on patriotic grounds."

These mid-fall happenings led to a hue and cry the country over, and the Administration, facing what it called a surprising "fire storm," moved hurriedly to ready a new Nixon team. Congressman Gerald Ford was to go in for Agnew, and the most egregious bad taste was seen in the pomp and circumstance which surrounded the announcement of the Ford nomination. The little-lamented Agnew, so recently "deep-sixed" that he was scarcely under the waves, was all but invisible; the Ford selection was made with all the aura of another major coup by the President.

As the President's travails ground on, Mr. Agnew was reportedly in Chicago organizing a new professional football league in cooperation with Frank Sinatra, while Roy Cohn, onetime counsel for Senator Joseph McCarthy, wrote an angry, open letter to the former Vice President:

> If you had stood your ground as you promised . . . your chances for legal and political survival were excellent. . . . In resigning and taking a plea, you surrendered the fiber that had brought you worldwide respect. Alger Hiss and Daniel Ellsberg can still argue their innocence. You no longer can.

It is difficult to assess the irrepressible Mr. Cohn's historical talents, but his linking of the well-known Ellsberg with the little-known Hiss has a certain persuasive quality to it. Both men ran afoul of Richard Nixon; one escaped with his life and the other did not. Nixon professed to be unable to understand either man and sent his minions to a psychiatrist's office in order to find out about Ellsberg, while Hiss went to jail because he could not make his behavior intelligible to an ambitious young Congressman. Both men were associated with the release of confidential documents, and each case became celebrated by its association with a popular name: Ellsberg with the Pentagon Papers and Hiss with the Pumpkin Papers (an allusion to a place where his accuser had hidden the evidence which presumably confirmed his allegations against Hiss).

Ellsberg, now a folk hero, could be seen and heard everywhere. But who was Hiss and how did he fall out with our former President? Alger Hiss was accused of being a Communist by a *Time* magazine editor, Whittaker Chambers. Chambers, who was a reformed Communist, had tried unsuccessfully to interest the Democratic Administration in his charges until the matter came to the attention of the then beleaguered House of Representatives Un-American Activities Committee (HUAC) and in particular to Richard M. Nixon, elected just two years before on the promise to be tough on Communists.

The life and times of the House Un-American Activities Committee can only be understood in relation to what was going on in this country in the immediate post-World War II era. The long-standing and deep suspicion of world Communism had been accelerated by Russia's invasion of Finland in 1939. Many in the United States saw this act as revelatory of the true Russian character, and this fairly large group was never fully able to rationalize the event in the detente that developed during World War II. This temporary cordiality between the Allies rapidly dissipated in the Cold War between the U.S. and Russia when there was a hardening of attitude with reference to the Soviets on the part of Prime Minister Churchill and President Truman.

The House committee during its lengthy history had continued to fight what it considered the persistent encroachment of Communism. By 1948, its contention that Franklin Roosevelt had been surrounded by Communists in the federal government found common bond with the pessimism generated by the events of the Cold War. Even liberals were disappointed that the country seemed once again to be on a near-war footing, this time with the Communist nations, and there was growing disillusion as the idealism associated with World War II was being dissipated. This was a period of public cynicism and uncertainty, and most of America was ready to believe that it had been misled by someone somewhere along the road. The resurrection of Chambers' earlier testimony provided the fuel for what followed.

Only Hiss and his younger brother, Donald, among those accused as being Communists in Roosevelt's Administration, insisted on an opportunity to defend themselves; only Hiss was subsequently prosecuted.* The Congressional hearings brought Hiss face to face with Nixon.

Whittaker Chambers' original charge of simple party membership had by year's end escalated to espionage, and Hiss was indicted by a New York grand jury. Hiss was finally convicted after two trials and sent to prison for perjury—a conviction for espionage was precluded by the

---

*Harry Dexter White, Assistant Secretary of the Treasury during the New Deal, also voluntarily appeared. His death three days after his HUAC hearing prevented him from becoming a central figure in this matter.

statute of limitations—because he denied (1) turning State Department documents over to Chambers and (2) seeing Chambers after the beginning of the year 1937. The very foundation of Nixon's political career can be traced to his skill in prosecuting Hiss.

The House Un-American Activities Committee's hearings about Chambers' accusation bore many similarities to the more recent evidence of misdeeds in high places. There were claims of telephone taps, missing documents that later appeared in the committee's hands, midnight searches, perpetual news leaks, attacks on the press, and finally a demand by Nixon for the censure of President Truman and for the dismissal of the United States Attorney General, the Secretary of State, and the federal judge in the first Hiss trial, which resulted in a hung jury. Nixon was surely involved in the separation-of-powers issue, but in this first instance he was on the legislative side of the line attacking the executive branch.

Central to both Watergate and the Hiss case was the matter of truth-telling. One observer wrote in 1948 that

> They [the jurors in the Hiss trial] were faced with a contradiction so gross that they had simply to decide whether Chambers' story was a wholesome fabrication or whether Hiss was making a blanket denial of an experience it would have been fatal to admit in part. By any definition of the truth, however naive, one or the other was on a lying spree.

By 1973, a similar issue of controversial testimony and shattered morality was so rampant that almost every public utterance by the Nixon forces raised the question of credulity. In testifying before the Watergate grand jury in late July 1973, the redoubtable John D. Ehrlichman answered, "I can't recall" or "I don't remember" or "I have no idea" 125 times. As for Nixon himself, his position changed almost every time he appeared in public. He knew nothing about Watergate, and it was an event of small import; he knew something about it, but only after it had become public knowledge, and the subject was still without much importance; he had asked that a complete investigation be conducted, and "no one presently employed in the White House" was anywhere involved. And so it went down to the summer of 1974; each dawn brought a revised story the public was expected to swallow whole. The final denouement came in July when the House Judiciary Committee was confronted with irrefutable taped evidence that Nixon had known of the Watergate affair almost from the very beginning and had actively led the planning for the cover-up. Despite largely partisan wrangling earlier, this discovery finally united the committee, which then voted unanimously to recommend impeachment proceedings against the President. A few days later, Mr. Nixon resigned and flew off to San Clemente.

While varying standards of "truth-saying" seemed both acceptable and normal to the Nixon Administration, those who fell out of presidential favor were judged most severely. John Dean was described as an unreliable person and his attachment to the truth was ridiculed, while Elliot Richardson—another former presidential favorite—was treated similarly. Mr. Nixon was reported to have said in a suggestive aside to Republican Senators who were questioning him about conflicting statements made by Nixon and Elliot Richardson on the events that led to the Cox dismissal: "Who's going to get him [Richardson] on perjury?"

Going back to the record of those days in 1948–50 makes it very clear that for Nixon absolute loyalty to the country was the sticking point. If you had it, you belonged to the American Club; if not, you were an outsider. And the worst man of all was the one who once belonged but had turned in his membership card. Perhaps this is why Hiss, a product of a fine family, the best schools, high governmental office, a man who had everything handed to him so easily, was so hated by Nixon. And perhaps this is why Nixon could so easily identify with Whittaker Chambers, who lived on the fringes and who, like Nixon, seemed determined to slay those who so easily threw away things that he himself had never had.

Nixon came to feel that his Hiss experience earned him the undying enmity of the press, saying, "Considering the amount of time I spent on it, it's as difficult an experience as I've ever had . . . from the standpoint of responsibility . . . the resourceful enemies I was up against . . . the battle day in and day out . . . the terrible attacks from the press, nasty cartoons, editorials, mail. . . ." Similarly, during the 1952 Vice Presidential campaign, Nixon made the following comment: "I remember in the dark days of the Hiss case some of the same columnists, some of the same radio commentators who are attacking me now and misrepresenting my position were violently opposing me at the time I was after Alger Hiss. . . ."

On another occasion, Nixon made a more encompassing charge. The Hiss case "left a residue of hatred and hostility toward me—not only among the Communists but also among substantial segments of the press and the intellectual community—a hostility that remains even today, ten years after Hiss's conviction was upheld by the United States Supreme Court." The residue Nixon referred to included "an utterly unprincipled and vicious smear campaign. Bigamy, forgery, drunkenness, insanity, thievery, anti-Semitism, perjury, the whole gamut of misconduct in public office, ranging from unethical to downright criminal activities—all these were among the charges that were hurled against me, some publicly and others through whispering campaigns which were even more difficult to counteract." In his darkest moment before Watergate, the defeat by Pat Brown in the California gubernatorial race in 1962, Nixon made another assay of the event (in his infamous you-won't-have-Nixon-to-kick-around-anymore speech), saying to reporters, "For sixteen years, ever since the Hiss case, you've had a lot of . . . a lot of fun . . . you've had an opportunity to attack me." This inspired his victorious opponent to offer an optimistic assay: "Nixon will never live that speech down!"

Still, Mr. Nixon did well for himself with the conviction of Alger Hiss. It became a political plum for both Nixon and the Republican Party. When General Eisenhower introduced his Vice Presidential nominee to the Republican National Convention in 1952, he spoke of Nixon as "a man who has shown statesmanlike qualities in many ways, but has a special talent and an ability to ferret out any kind of subversive influence wherever it may be found, and the strength and persistence to get rid of it."

Observers of that campaign noted that a casual visitor to this country might well have assumed that Alger Hiss was the Democratic candidate for the presidential office. Nixon was not loath to state that a Democratic victory would assure America "more Alger Hisses, more atomic spies, more crises." In a national television address in October of that year, Nixon used the first public confrontation between Chambers and Hiss as the lengthy central theme in his presentation. As the nation watched, Nixon held up some documents he identified as those Chambers had obtained from Hiss and declared that the Russians had been given hundreds like them "from Hiss and other members of the ring." This, he concluded, "meant that the lives of American boys were endangered and probably lost because of the activities of a spy ring."

Nixon then offered this forceful evaluation: "That shows what just one man can do to injure the security of his country when he owes his loyalty not to his own government but to a foreign power . . . and again this case is a lesson, because we see the action of the administration in covering it up rather than in bringing Hiss to book many years sooner as they should have."

It was only at the very end of the speech that the actual Democratic candidate's name appeared:

> Mr. Stevenson was a character witness, or should I say a witness for the reputation, and the good reputation, of Alger Hiss. He testified that the reputation of Alger Hiss for veracity, for integrity, and for loyalty was good. It was voluntary on Mr. Stevenson's part . . . it was given at a time when he was Governor of Illinois and the prestige of a great state and the Governor of the state were thrown in behalf of the defendant in this case. . . . It is significant that Mr. Stevenson has never expressed any indignation over what Mr. Hiss has done and the treachery that he engaged in against his country.

At one point in the campaign, Nixon, showing a bent for alliteration which his Vice President Agnew would press to its limits years later, referred to the Democratic candidate as "Adlai, the Appeaser . . . who got a Ph.D. from Dean Acheson's College of Cowardly Communist Containment." In a major television address, Nixon contrasted the Roosevelt–Truman Administration with Eisenhower's military staff, which, according to the Vice Presidential candidate, had "never had an instance of infiltration." He continued, "We can assume because of the cover-up of this administration in the Hiss case that the Communists, the fellow travelers, have not been cleared out of the executive branch of the government."

Thus, over the years, Mr. Nixon's public utterances were still dotted with references to this once famous case.* But most Americans remember very little about the issues involved. Hiss remained important to Nixon notwithstanding the fact that the old Communist conspiracy mattered far less to a President who met at the summit in friendly concourse with leaders of Communist nations.

From all of the above, it is understandable that Richard Milhous Nixon regarded his relationship with the Hiss–Chambers case "as the

---

*One most curious mention came on March 15, 1973, when Nixon contrasted what he considered his judicious stance in refusing to allow members of the Administration to testify before Congressional committees investigating the Watergate affair with Harry Truman's executive order which kept a FBI report on its interrogation of Hiss from the House Un-American Activities Committee. Nixon commented that the Truman order "cannot stand from a constitutional standpoint on the merits of the case." The implication was clear that Nixon's refusal was in his opinion much more solidly based.

first major crisis of my life," saying, "My name, my reputation, and my career were ever to be linked with the decisions I made and the actions I took in that case, as a thirty-five-year-old freshman Congressman in 1948." Speaking in a somewhat different vein, Richard Nixon wrote in his book *My Six Crises:*

> "If it hadn't been for the Hiss case, you would have been elected President of the United States." This was the conclusion of one of my best friends after the election of 1960. But another good friend told me just as directly, "If it hadn't been for the Hiss case, you never would have been Vice President of the United States or candidate for President."

To expand upon Nixon's historical view: Without the Hiss case there probably would not have been a President Nixon nor the host of abuses during his Administration which came to be known collectively as the Watergate scandal. And without the Hiss case and a victorious Nixon who came to national prominence because of it, this country might have been spared its terrible McCarthy era as well.

From 1948 to 1950, the Hiss case exposed an American fear of Communism that Senator Joe McCarthy was to exploit with horrifying results. One month after Alger Hiss was convicted, the junior Senator from Wisconsin announced to the Women's Republican Club of Wheeling, West Virginia, "I have in my hand a list of 205, a list of names that were known to the Secretary of State as being members of the Communist Party and who, nevertheless, are still working and shaping policy in the State Department." Senator McCarthy's speech, given in February 1950, precipitated a national state of internal agony, and he often used the Hiss case to justify his devastating campaign.

Whittaker Chambers, the *Time* editor who set the Hiss case in motion, said of McCarthy, "I hand my mantle to him," and Nixon did what he could to see that it was an orderly transition. He supported McCarthy's re-election in 1952 and lent him his files on Communism when the Senate was investigating McCarthy's charges.

The editors of *The Nation* concluded on May 28, 1973, that

> The conviction of Alger Hiss was a key event—perhaps the single most important event—in setting in motion the witch hunt which provided much of the domestic political support for cold-war policies. To place the case in its proper perspective, one need merely note that the national reputation of President Nixon was largely based on the role he played in initiating the prosecution and in securing the conviction of Hiss.

Now nearly thirty years after the Hiss trials, the scales have tipped again. The deposed President is relatively silent in his Elba by the Pacific, while the seventy-two-year-old Alger Hiss is enjoying a kind of renaissance. His license to practice law has been restored by the Commonwealth of Massachusetts. Hiss is working vigorously with the National Emergency Civil Liberties Foundation (NECLF) for another hearing on his case, where he expects to be vindicated. His work with NECLF revolves around a little-used legal issue called *coram nobis*. If Hiss can show that the government failed to introduce key pieces of evidence in its possession, a federal judge can wipe the Hiss slate clean.* Despite serving nearly four years in prison, Hiss has never wavered in maintaining his innocence.

Now there is an inescapably tragic quality about Alger Hiss. He presents a hollowed-out appearance: the gauntness of his face and the sunken quality of his eyes create the distinct impression of a man who has suffered much. Alistair Cooke described the forty-four-year-old Alger Hiss in his excellent book on the case, *A Generation on Trial,* as he first appeared before the District Court in New York City in 1949:

> He had what anyone must envy who has come to know that youth is a bloom that sags and vanishes if there is no good bone to let it rest on. He had a fine articulation of chin and mouth and brow and nose that would defy softening tissue and leave him handsome at eighty. Only the eyes failed to match the serenity of his bone structure. They were deep-set and agile.

Today as Hiss talks about his experiences, he becomes progressively more animated, as if all his energy has been caught up in this case. Personal vindication appears to be his *raison d'être*. Only at times like this is one reminded of Cooke's perception, the lean good looks so often seen in the newspapers during that period. It was this physical quality that men like Nixon and Whittaker Chambers found so disquieting. "The parts were miscast," said the corpulent *Time* editor. "The world's instinctive feeling was against the little fat man who had stood up to

---

*The Petition for a Writ of Error *coram nobis* was filed on July 27, 1978. Hiss's attorney of record is Victor Rabinowitz. The document asks that the conviction (January 25, 1950) be set aside because "the prosecution withheld and concealed evidence which was exculpatory; that it employed informers who infiltrated the counsels of Hiss's attorney; that it misrepresented material facts to Hiss and to the court and jury, and that it suffered testimony it knew to be perjurous to be presented to the court and jury." This is, says the petition, a violation of Hiss's rights as guaranteed by the Fifth and Sixth Amendments to the Constitution.

testify for it, unasked. The world's instinctive sympathy was for the engaging man who meant to destroy it . . . Alger Hiss."

Although in his seventh decade, Hiss is on the lecture stump to raise $50,000 for what has to be his legal—if not his life's—final moment of truth. Hiss said proudly that he has reached the third category of lecturers, $2,500 a performance, which puts him in the same league as 1976 Olympic decathlon medalist Bruce Jenner.

Part of the reason for the Hiss renaissance is the release of the previously confidential FBI papers about the case under the Freedom of Information Act of 1966. To date, 40,000 pages of information have been declassified following a civil action by the ACLU for Professor Allen Weinstein of Smith College. The total number of pages available is still far from complete. According to Professor Athan Theoharis of Marquette University, who has studied the matter, it is now known that the FBI devised separate filing procedures as early as 1942 where documents that pertained to "illegal, sensitive or embarrassing activities, which involved the Bureau," were isolated from the central file.

Not surprisingly, much of the information released is peripheral at best. The papers are peppered with deletions like V-mail letters during the war, many of the photographic copies are illegible, and there is no smoking pistol to be found. Nevertheless, some of the questions about this case which have puzzled America for almost thirty years are at last answered.

Another reason for the new interest in Alger Hiss is the need of America to come to some understanding of the fall of Richard Nixon. The Hiss case certainly accelerated Nixon's rise. Could the case similarly have contributed to Nixon's fall? The President's Watergate tapes are full of references to Alger Hiss.* They show that when the President's world was crumbling, he seemed to find solace in reliving it, as if this event were his port in the fire storm of Watergate. As Hiss later said in this very regard, "I make him [Nixon] feel strong when he is afraid of being weak. I am really that man's King Charles's head."†

---

*Garry Wills, a professor at Johns Hopkins—Hiss's alma mater—and now a political columnist, counted seven references to Hiss in the first 120 pages of the transcripts, an accounting that encouraged him to say to Hiss, who was lecturing to his class, "You are a famous man."

† "A phrase applied to an obsession, a fixed fancy. It comes from Mr. Dick, the harmless half-wit in Dickens' *David Copperfield,* who, whatever he wrote or said, always got round to the subject of King Charles' head, about which he was composing a memorial—he could not keep it out of his thoughts." (*Brewer's Dictionary of Phrase and Fable,* 1970).

There are, of course, the dramatic elements of the case itself which continue to confound observers. It had all the elements of an international thriller—espionage, stolen documents, Russian and American spies, microfilm hidden in a hollowed-out pumpkin, documents placed in an abandoned dumbwaiter for ten years, a typewriter which seemed to ooze guilt on display at the trials, and an ambitious young United States Representative spearheading the prosecution. On a dramatic level alone the strange case of Alger Hiss refuses to be put down.

What follows is an effort to understand the relationship between present events and the past in the perplexing and fragmented world of Richard Nixon as seen through the prism of his relations with Alger Hiss. Incidents such as the Hiss affair and Watergate can well provide insight into structures and activities that are normally protected from public scrutiny. There is a necessity to do so at this time, for there exists the real danger that anything short of a comprehensive view will encourage a public need to read the Watergate events in isolation—that is, to assume that a minor aberration has been uncovered and that some small tinkering with election laws will prevent a reoccurrence in the future.

As an extension of these theses, some have argued that the country and the Presidency have been troubled enough, that Watergate and its associated activities should be allowed to fade from memory. This was Mr. Nixon's expectation. His public-relations campaign in 1973 to tell his story to the American people (he said soberly, "I'm not a crook" to the Associated Press managing editors' meeting) expressed this point of view very well indeed. Watergate and the furor it occasioned interfered with government operation and with efficiency, and therefore we were enjoined to be rapidly done with it.

What if Watergate had succeeded? The final evidence that Mr. Nixon learned of the events almost immediately and attempted to obscure them bears directly on the President's notions of moral values and legal norms. What was at stake was not so much the prestige and efficiency of the government, as was so often proclaimed, but rather the question of so-called "moral behavior" in our society as a whole.

CHAPTER 2

When Nixon—a first-term Congressman, as lowly a creature as exists in Washington—pushed the Hiss case, he seemed to be taking a great risk. It was less than it looked. He had cards all up and down his sleeves and inside his vest.
—GARRY WILLS

To summarize briefly the pertinent facts in the case from almost thirty years ago, Alger Hiss, a former high-level official in the State Department, was found guilty of perjury on January 21, 1950, after two trials for *denying* that he had handed over secret documents to a Communist agent, Whittaker Chambers, more recently a senior editor of *Time*. Hiss also denied that he had seen and conversed with Chambers after January 1, 1937. Hiss could not be tried for espionage because of the statute of limitations; he could only be tried for perjury in refusing to admit to committing espionage. Moreover, the more serious implications emerged *only* after Hiss filed a libel suit against Chambers and there were indications that the government might drop the charges against Hiss for lack of evidence.

At that time Chambers, despite earlier protestations on his part that espionage was not involved, produced the famous Pumpkin Papers which forever altered the thrust of the case. The earlier charge of membership in a Communist study group was significantly changed. Hiss was now accused of taking documents which were the property of the government and to which he had access by reason of his employment and handing them over to Chambers, who was, by his own confession, a Communist agent.

The information that Hiss was alleged to have handed over consisted of documents, some of which admittedly crossed his desk and others which presumably had been typed on Hiss's own typewriter. Some of the documents reportedly bore handwritten notations by him, and others were initialed by Hiss; and the case revolved around efforts by the prosecution to establish the probability that Hiss had personally conveyed the documents to Chambers, the defense contending that Chambers and others had illicit access to them during periods of admitted lax security in the State Department.

The matter was mightily complicated by the fact that Julian Wadleigh, a State Department employee during that time, confessed that he had indeed passed documents to Chambers (whom he knew as Carl Carlson) for transmission to the Communist Party. C. S. Darlington, Wadleigh's superior in the department, described him as having a "well-developed curiosity" and said that he often found Wadleigh reading papers on his desk on his return from lunch. Darlington also confirmed the fact that Hiss's door in the State Department was always open. Adding to the confusion was Wadleigh's statement that Chambers had never mentioned Hiss's name to him although he had named the other sources of Communist support that existed in the State Department.

Chambers had, over a period of several years, attempted to interest the government in his story of a Communist ring with members in high official places. Despite unofficial discussions among government officials and even FBI questioning of Hiss on two occasions in 1947 and 1948, it was apparently decided that the basis for Chambers' allegations was not substantial enough to launch a full-scale inquiry. Chambers was mentioned in both Hiss interrogations, but the name reportedly meant nothing to the State Department official. The House Un-American Activities Committee, learning of Chambers' charges, subpoenaed him on August 3, 1948, in order to corroborate the testimony of another witness, Elizabeth Bentley, a woman of some mystery, whom the press quickly characterized as "Mata Hari." If the label were correct, this

Connecticut-born, Vassar-educated former member of the Communist Party was certainly a plainer-looking version of the legendary *femme fatale*. Elizabeth Bentley said she collected "secret information" from twenty or thirty people who worked for the United States government. She also said the Party had spies in all government offices except the FBI and the Navy.

When asked to name names, Elizabeth Bentley mentioned, among others, Harry Dexter White, former Assistant Secretary of the Treasury during the New Deal (whom Chambers also would name), and Laughlin Currie, one of Franklin Roosevelt's closest advisers. Besides securing confidential information, she alleged, these two men—neither of whom she had ever met—helped place other Communists in strategic jobs in the government. Both White and Currie denied her charges. Bentley did not mention Alger Hiss. During Chambers' appearance before the committee, he named Alger Hiss and his brother Donald as Communists, along with six other individuals occupying relatively high positions in the federal hierarchy.

Over the years of testifying formally and informally, Chambers identified at least a dozen prominent government officials as Communist Party supporters. During the committee hearing on the charges, five of the group (Lee Pressman, Nathan Witt, John Abt, Henry Collins, and Charles Kramer) were interrogated about their relationship to Hiss. All refused to answer on Fifth Amendment grounds. Two years later, all but Collins were summoned once again. This time Pressman admitted that he was a Communist and identified Witt, Kramer, and Abt as Communists but stated clearly that Hiss was not a member of the group.

There can be no question that Alger Hiss was one of two senior men in the group named by Chambers (the other being Harry Dexter White, whose death shortly after testifying before the House committee profoundly disturbed Hiss). Quite naturally Hiss became central to the interests of the House Un-American Activities Committee. Whether or not Alger Hiss knew Whittaker Chambers and under what circumstances became the issue around which the entire case turned.

Hiss had originally been shown a newspaper picture of a man named Whittaker Chambers. He denied any knowledge of the man. Prior to the third meeting of the committee which was arranged for the purpose of confrontation, Hiss was privately able to reconstruct a resemblance to someone he had known a dozen years before as George Crosley. According to Hiss's recollection, Crosley was an itinerant journalist with a passing interest in the work of the Nye Committee, a senatorial committee concerned with munitions traffic, to which Hiss had been a legal consultant. Hiss described Crosley as something of a deadbeat whose impoverished status impelled Hiss both to lend him money and to sublet an apartment to him. Somewhat later Hiss had also offered him a "worthless" car and had received a gift of a rug from Crosley in lieu of repayment of rent and other debts.

At that dramatic confrontation of the two men, a minute examination by Hiss of Chambers centering largely on his dental status—the original Crosley had extremely bad teeth, and the reconstituted Chambers had excellent teeth—led Hiss finally to admit knowing Chambers in the past. The following dramatic colloquy finished that session:

MR. HISS: *Did you ever go under the name of George Crosley?*

MR. CHAMBERS: *Not to my knowledge.*

MR. HISS: *Did you ever sublet an apartment on 28th Street from me?*

MR. CHAMBERS: *No, I did not.*

MR. HISS: *You did not?*

MR. CHAMBERS: *No.*

MR. HISS: *Did you ever spend any time with your wife and child in an apartment on 28th Street in Washington when I was not there because I and my family were living on P Street?*

MR. CHAMBERS: *I most certainly did.*

MR. HISS: *You did or did not?*

MR. CHAMBERS: *I did.*

MR. HISS: *Would you tell me how you reconcile your negative answers with this affirmative answer?*

MR. CHAMBERS: *Very easily, Alger. I was a Communist and you were a Communist.*

MR. HISS: *Would you be responsive and continue your answer?*

MR. CHAMBERS: *I do not think it is needed.*

MR. HISS: *That is the answer.*

MR. NIXON: *I will help you with the answer, Mr. Hiss. The question, Mr. Chambers, is, as I understand it, that Mr. Hiss cannot understand how you would deny that you were George Crosley and yet admit that you spent time in his apartment. Now would you explain the circumstances? I don't want to put that until Mr. Hiss agrees that is one of his questions.*

MR. HISS: *You have the privilege of asking any questions you want. I think that is an accurate phrasing.*

MR. NIXON: *Go ahead.*

MR. CHAMBERS: *As I have testified before, I came to Washington as a Communist functionary, a functionary of the American Communist Party. I was connected with the underground group of which Mr. Hiss was a member. Mr. Hiss and I became friends. To the best of my knowledge, Mr. Hiss himself suggested that I go there, and I accepted gratefully.*

MR. HISS: *Mr. Chairman.*

MR. NIXON: *Just a moment. How long did you stay there?*

MR. CHAMBERS: *My recollection was about three weeks. It may have been longer. I brought no furniture, I might add.*

MR. HISS: *Mr. Chairman, I don't need to ask Mr. Whittaker Chambers any more questions. I am now perfectly prepared to identify this man as George Crosley.*

During the session Hiss asked Chambers to make the above statements public so that he might sue for damages. Shortly thereafter on the radio program "Meet the Press" Chambers did indeed state his charges, which brought a suit against him by Hiss some three weeks later.

Although Nixon was not chairman of the committee and was in fact its most junior member, it is instructive to note that he took the lead in questioning the witness, a procedure that was to become standard practice throughout the remainder of the hearings. Nixon and committee investigator Robert Stripling formed the prosecution team and revived the flagging case against Hiss each time it seemed ready to dissolve.

Nixon concluded that since the fact of Hiss's Communist Party membership could—and probably would—be concealed, the committee must at least prove that a close relationship existed between Hiss and Chambers. If they were perceived to be close personal friends, Chambers' accusations were all the more persuasive. If, as Hiss maintained, there had been very few meetings between them, the basis for the charges was imperiled. The bulk of the early proceedings before the House committee revolved around Chambers' offering proof of close and enduring friendship with Hiss. Chambers was asked to document his personal knowledge of Hiss, and he did this abundantly, citing home addresses, pet names for each other used by Mr. and Mrs. Hiss, his assessment of the quality of Hiss's library, Hiss's hobbies, and the fact that the Hiss house was decorated with pictures of rare birds.

Central to these issues was the exotic matter of Chambers' name. Surely if the two men knew each other well, how could Hiss initially deny any knowledge of Chambers? That this failure was crucial to Hiss's credibility in Nixon's eyes soon became clear. The reader will recall that in the dialogue presented above, Chambers replied to Hiss's question "Did you ever go under the name of George Crosley?" with the oblique answer "Not to my knowledge." It later developed that Chambers stated that he was known as "Carl," and in response to further questioning about a last name he answered, "Just plain Carl." He maintained this position throughout both the Congressional hearings and the trials.

Chambers denied that the Hisses knew his last name but acknowledged that they had never heard the name Whittaker Chambers. Moreover, Chambers admitted that it was "possible" that he had used the name George Crosley, saying that he "might have done so" or that he "couldn't remember"; but as The Earl Jowitt in his book *The Strange Case of Alger Hiss* has pointed out, Chambers never denied that he used this name. Even more strange is the fact that in Chambers' own book *Witness,* Chambers wrote that "it is possible" that he used the name of George Crosley, though he had no recollection of it. He added that he had recalled all of his other aliases without effort.

During trial cross-examination when asked about the name he had used in staying in the Hiss apartment, Chambers answered in inexplicable fashion, "I have never been able to remember; it may have been Crosley." Mrs. Chambers during the trial testified that the Hisses called the Chambers Carl and Liza, saying, "We never had a last name to them." When Hiss's attorney persisted in asking how the Hisses would locate the Chambers when visiting them since there was a common doorbell in the apartment lobby, Mrs. Chambers answered that she would have known approximately when the Hisses were coming and would have intervened before they rang the bell.

As continued evidence of a close relationship between the two men, Chambers cited a lunch in a Georgetown restaurant when a woman known to the Hisses as "Plum Fountain"* was introduced to him. Hiss's attorney, Claude Cross, was quick to ask what name Hiss had used for Chambers. Chambers could not remember, and the following interrogation ensued.

MR. CROSS: *You were introduced to her?*
MR. CHAMBERS: *That is right.*
MR. CROSS: *And you can't tell what name you used?*
MR. CHAMBERS: *That is right.*
MR. CROSS: *Well, you know it wasn't Carl, don't you?*
MR. CHAMBERS: *It certainly was not Carl.*
MR. CROSS: *Was it Mr. Crosley?*
MR. CHAMBERS: *I don't believe so.*
MR. CROSS: *Well, what name was it?*
MR. CHAMBERS: *I have no idea, as I have told you.*
MR. CROSS: *You wouldn't say it was not Mr. Crosley, would you?*

---

*When the case went to trial, Mrs. Olivia Fountain Tesome, whose nickname was "Plum," testified that she remembered neither the introduction nor the man, now known as Whittaker Chambers.

MR. CHAMBERS: *I think it very unlikely.*

MR. CROSS: *Well, you have testified here that it is possible that you used the name and they knew you as George Crosley, haven't you?*

MR. CHAMBERS: *Yes. To hasten matters, let us say that it is possible that they could have introduced me that way, but it is improbable.*

MR. CROSS: *Do you know how they did introduce you?*

MR. CHAMBERS: *I do not.*

MR. CROSS: *You know they did use the name?*

MR. CHAMBERS: *I believe they did.*

MR. CROSS: *And you know they did not use the name Mr. Chambers?*

MR. CHAMBERS: *That is definite.*

MR. CROSS: *That they did not use the name Mr. Chambers?*

MR. CHAMBERS: *That is also definitely right.*

MR. CROSS: *And you know they did not use the name of Dwyer, don't you?*

MR. CHAMBERS: *That they did not know.*

MR. CROSS: *Did they use the name Lloyd Cantwell?*

MR. CHAMBERS: *I don't think so.*

MR. CROSS: *You can't help at all beyond what you have said, that you can't remember?*

MR. CHAMBERS: *I am sorry, I cannot help.*

It slowly emerged that Chambers was known by a variety of last names over the course of his adult life. He finally admitted openly or by inference to the use of at least eight—Adams, Crosley, David, Cantwell, Dwyer, Breen, Kelly, and Land—but frankly stated that Alger Hiss would never have heard his real name, Jay Vivian Chambers.*

The matter assumes considerable importance, for much of Hiss's early relationship with the House Un-American Activities Committee was clouded by the name problem. Chambers was not present when Hiss first appeared before the committee on August 5, 1948, and the transcript reveals the following testimony:

MUNDT: *. . . it is extremely puzzling that a man who is senior editor of* Time *magazine, by the name of Whittaker Chambers . . . should come before*

---

*Perhaps Chambers was destined to be a man of many names. His own father, long estranged from the Chambers' family, employed Whittaker after he graduated from high school. Apparently afraid of being accused of nepotism by others at his firm, the father identified the new worker as Charles Whittaker. And even this did not mark the first time young Chambers used a pseudonym, for he had used the name Charles Adams just prior to this employment while working as a laborer in Washington, D.C. No explanation has ever been offered for this strange choice; Chambers' only comment was that he was beginning his "proletariat period." He was known as Vivian while growing up, a name he detested. In college he began calling himself Whittaker, after his maternal grandfather, Charles Whittaker.

this committee and discuss the Communist apparatus working in Washington, which he says is transmitting secrets to the Russian government, and he lists a group of seven people—Nathan Witt, Lee Pressman, Victor Perlo, Charles Kramer, John Abt, Harold Ware, Alger Hiss, and Donald Hiss.

HISS: *That is eight.*

MUNDT: *There seems to be no question about the subversive connections of the six other than the Hiss brothers, and I wonder what possible motive a man who edits* Time *magazine would have for mentioning Donald Hiss and Alger Hiss in connection with those other six.*

HISS: *So do I, Mr. Chairman. I have no possible understanding of what could have motivated him. There are many possible motives, I assume, but I am unable to understand it.*

MUNDT: *All we are trying to do is find the facts.*

HISS: *I wish I could have seen Mr. Chambers before he testified. . . .*

STRIPLING: *You say you have never seen Mr. Chambers?*

HISS: *The name means absolutely nothing to me, Mr. Stripling.*

STRIPLING: *I have here, Mr. Chairman, a picture which was made last Monday by the Associated Press. I understand from people who knew Mr. Chambers during 1934 and 1935 that he is much heavier today than he was at that time, but I show you this picture, Mr. Hiss, and ask you if you have ever known an individual who resembles this picture.*

HISS: *I would much rather see the individual. I have looked at all the pictures I was able to get hold of in, I think it was, yesterday's paper which had the pictures. If this is a picture of Mr. Chambers, he is not particularly unusual looking. He looks like a lot of people. I might even mistake him for the chairman of this committee.*

Nixon was destined to make a great deal out of Hiss's careful statement made during that same hearing that "I have never known a man by the name of Whittaker Chambers." According to Nixon, " . . . looking over my notes on his [Hiss's] testimony, I saw that he had never once said flatly, 'I don't know Whittaker Chambers.' He always qualified it carefully. . . ." Still, in his previous testimony before the committee, Chambers did not mention that he was known to Hiss only as Carl, and Nixon was to find this out after the first Hiss hearing through a phone call to Chambers.

At the conclusion of Hiss's first appearance before the committee, there was great concern that the Congressman had made a serious error. Representative John Rankin rushed over and shook Hiss's hand, and Mr. Karl Mundt expressed the chair's appreciation of Hiss's "very cooperative attitude." Nixon, however, was strangely unmoved. He derided Hiss's testimony, calling it dishonest, a "virtuoso performance." To Nixon, Hiss left the impression that his accuser's testimony was either a "terrible case of mistaken identity" or a "fantastic vendetta" launched for reasons he was unable to fathom. Nixon wrote: "From considerable experience in observing witnesses on the stand, I had learned that those who are lying or trying to cover up something . . . tend to overact, to overstate their case. . . . Hiss was much too smooth. . . . Much too careful a witness for one who purported to be telling the whole truth without qualification."

The press, however, found Hiss's testimony that day more convincing. Nixon's determination to push on was apparently not shaken by their reaction. Immediately following the hearing, a reporter approached him and asked, "How is the committee going to dig itself out of this hole?" Nixon relates how another reporter whom he respected said, "The Committee on Un-American Activities stands convicted, guilty of calumny in putting Chambers on the stand without first checking the truth of his testimony."

Following such reactions, the committee—primarily Richard Nixon—was forced to go out to prove Chambers' story. The committee met that evening in executive session, and Nixon noted that "virtual consternation reigned among the members." All except Nixon and Robert Stripling, the committee's chief investigator, felt that Hiss was telling the truth; F. Edward Hebert, a Democrat from Louisiana, suggested that the committee wash its hands of the case at once, while acting chairman Mundt counseled finding a collateral issue to take the committee off the spot it found itself in. Nixon remained steadfastly determined to continue the hearings, and his position was clearly outlined.

> Although we could not determine who was lying on the issue of whether or not Hiss was a Communist, we should at least go into the matter of whether or not Chambers knew Hiss. . . . Hiss was a particularly convincing witness that day . . . the committee had no real facts to use as a basis for cross-examining him and consequently, he was able to dominate the situation throughout.*

Nixon's request that a subcommittee meet with Chambers in executive session to seek more proof of his relationship with Hiss was

---

*There is a difference of opinion between Nixon and Stripling about who was the prime mover in persuading the committee to continue the investigation. Nixon wrote, "I was the only member of the committee who expressed a contrary view, and Bob Stripling backed me up strongly and effectively." In Stripling's words, it went like this: "But as I pointed out, with the aid of Representative Nixon, a *prima facie* perjury was somewhere involved."

reluctantly agreed to, and with Mr. Nixon in the chair and pressing hard, Chambers was intensively interrogated two days later, on August 7.

Had Chambers seen Hiss's picture in the newspaper, and was he the accused man? Chambers had seen the picture, and, yes, it was Hiss. How could he be sure that Hiss was a Communist? Chambers was told this by his superior, J. Peters, who was the head of the entire Communist underground of the United States. Chambers knew Hiss as a dedicated and disciplined Communist. Did Hiss carry a party card? For obvious reasons, there were no cards for undercover agents. What did Chambers know personally about Hiss?

At this point a wealth of incriminating detail was produced by Chambers. The Hisses called each other "Hilly" and "Dilly"; they had moved several times while living in Washington; a stepson lived with them and attended a private school; Hiss was a bird watcher and once mentioned seeing "a prothonotary warbler at Glen Echo Park"; the Hisses were both Quakers* and spoke to each other in "plain" speech. Nixon was very impressed by this last revelation, for he was a Quaker himself.

> I knew from personal experience that my mother never used the plain speech in public but did use it in talking with her sisters and her mother in the privacy of our home. Again I recognized that someone else who knew Priscilla Hiss could have informed Chambers of this habit of hers. But the way he told me about it, rather than what he said, again gave me an intuitive feeling that he was speaking from first-hand rather than second-hand knowledge.

In response to Nixon's question on the "life-style" of the Hisses, Chambers described it in detail. Hiss was "a man of great simplicity and a great gentleness and sweetness of character." Their life-style was in keeping with this temperament. Chambers said the furniture in their apartment on 28th Street in Washington "was kind of pulled together from here or there. . . . Nothing lavish about it." Also the Hisses cared little about food. "It was not a primary interest in their lives."

Alistair Cooke was impressed by how much Chambers knew about the Hisses. Such intimate knowledge, he felt, could only come "from the closest and most observant friendships *or* from a tireless detective job."

---

*Hiss has categorically denied that this was true. He described himself as a lifelong Episcopalian, but added that his wife, Priscilla, considered herself a Quaker.

Nixon was similarly troubled by the fanatic detail; after all, "overstating" the case was what he found suspicious in Hiss's testimony. Still, Nixon overcame his scruples, explaining that

> Chambers' memory of minute details was one of the very things, incidentally, that raised doubts in the minds of some Committee members as to his credibility. How could he possibly recall names, places, and events with which he had last been associated over ten years before? In retrospect, I believe that two factors contributed to his ability to do so. First, even his most bitter enemies had to agree that Chambers was a man of extraordinary intelligence. In addition, as a Communist underground agent, he had to train himself to carry vast quantities of information in his head so that he could reduce to the minimum the risk of ever being apprehended with documents on his person. As a result, his mind's retentive capacities were developed to an astonishing degree.

While he rode back from New York that night on the train to Washington, Nixon's doubts once again began to surface. "Could Chambers, by making a careful study of Hiss's life, have concocted the whole story for purposes of destroying Hiss—for some motive we did not know?" Nixon was troubled by the contrast between the two men—one a paid functionary of the Communist Party who slept with a gun under his pillow in fear of assassination, and the other, coming from "a fine family, had made an outstanding record at Johns Hopkins and Harvard Law, had been honored by being selected for the staff of a great justice of the Supreme Court, Oliver Wendell Holmes, had served as Executive Secretary to the big international monetary conference at Dumbarton Oaks in 1944, had accompanied President Roosevelt to Yalta, and had held a key post at the conference establishing the United Nations at San Francisco."

Unable to trust his own judgment, Nixon decided that the wealth of biographical data would have to be checked out before the committee next met. The staff was assigned the task, and Nixon himself met privately with a Pulitzer Prize-winning journalist, Bert Andrews* of the New York *Herald Tribune,* whose opinion he valued, and also with Allen Dulles, then head of the CIA, and his brother, John Foster Dulles, who as chairman of the Board of Trustees of the Carnegie Endowment

---

*Andrews' Pulitzer Prize was awarded for a series of articles he wrote critical of HUAC's handling of the "Hollywood Ten."

for International Peace was Hiss's superior. Hiss was at that time president of the Endowment. All agreed that Chambers *must* have known Hiss intimately. Nixon was nonetheless still uncertain and decided impulsively to drive to Chambers' farm accompanied by Andrews, who was to "grill him [Chambers] as only a Washington newspaperman can." According to Nixon, "Chambers met the test to Andrews' complete satisfaction."

Andrews did, however, know Hiss prior to the trip with Nixon, and there was some bad blood between the two men. Andrews, although a "Washington newspaperman" in Nixon's eyes, was not an entirely objective reporter in Alger Hiss matters. Their altercation was over a series of stories Andrews had written on the Yalta Conference, which Hiss attended as a minor functionary—not as one who "helped to write the political agreements at Yalta," as Representative Karl Mundt later overstated it.* The issue centered around the makeup of the General Assembly of the United Nations. The Russians were demanding three votes for the three largest Soviet republics: Ukraine, Lithuania, and Byelorussia. They reasoned that the Commonwealth nations would vote with the United Kingdom, the Latin American countries would vote with the United States, and they wanted extra votes to counter these blocs. The dispute was left to President Roosevelt, Prime Minister Churchill, and Marshal Stalin to settle.

Roosevelt agreed to this (without having been fully briefed),† and when he learned what he had agreed to he was not particularly troubled by the Soviet's demands, although a compromise was worked out giving them two rather than three votes. In turn, the United States would get two extra votes—if we wanted them—in the names of Hawaii and Alaska, not yet states. Roosevelt, recognizing that such an agreement might easily be misconstrued, felt it should remain a secret until after the charter was drafted in San Francisco. He did, however, privately brief the U.S. delegation on the compromise at a meeting at the White House.

---

*Actually Hiss wrote only one paragraph at the Yalta Conference which dealt with the trusteeship of the Axis nations' colonies.

†At Yalta, Roosevelt was being briefed by Under Secretary of State Edward Stettinius about the foreign ministers' meeting he had just attended. He told the President that "We are in total agreement with the Russians." Stettinius was about to say "except on the matter of the extra votes in the General Assembly that the Soviets were demanding" when Stalin entered the room. Roosevelt greeted Russia's leader by commenting that he understood we "have reached an agreement on everything." Stalin, somewhat surprised, inquired about the extra votes. Here Stettinius tried to interrupt, but Roosevelt said, "Sure."

Less than a week later, the entire story appeared under Andrews' by-line, and it created suspicion about the surreptitious goings-on at the conference.

Hiss was not pleased with this particular story. He charged Andrews with "improper journalistic practice." Recently Hiss commented to the authors, " . . . He [Andrews] should have known to check with me or somebody else. It was minor, but the story could only damage the United Nations. It could be played up sensationally, and I called him on it because I had good relations with him before that, as I think I had with the rest of the press."

Nixon's strange vigil with Chambers was repeated once again two days later in the company of Robert Stripling (described by Nixon as "having a sixth sense in being able to distinguish the professional 'Red-baiters' from those who were honestly trying to help the Committee in its work in exposing the Communist conspiracy"). Stripling also came back convinced that Chambers and Hiss had been close friends.

By the time Hiss was next interrogated on August 16, newspaper stories had appeared that alluded to Chambers' anecdotal testimony. Hiss, unaware that John Foster Dulles had been in contact with Richard Nixon, discussed with Dulles the possibility of walking to the *Time* office and confronting Chambers but was dissuaded from this course by the possibility of legal complications.

Only later was Hiss to discover that Dulles had met with Nixon, "who," said Hiss, "showed him Chambers' HUAC testimony [and] presumably his then-secret testimony about me." Many years after the event, Hiss was to say with understandable bitterness, "John Foster Dulles was not only advising me on how to proceed in my defense but was simultaneously advising Nixon and the House Committee on how to prosecute me as well." Asked what he would have said to Chambers had he confronted him at his *Time* office, Hiss's reply was surprisingly straightforward and at the same time remarkably naïve: "I would have asked Chambers, 'Who are you and what have I done to you?'"

During the month of August 1948, Hiss thought considerably about his past life and talked with old friends about the possible identity of Chambers. Although he was still uncertain, one possibility kept recurring. It was on the train trip to Washington for his second committee appearance on August 16 that Hiss wrote the name "George Crosley" on a pad.

Although Congressman J. Parnell Thomas was in the chair, Nixon once again asked most of the questions. The committee was confused,

Nixon reported, about the testimony of two witnesses who said so many contradictory things. As a consequence, it was necessary to go over old ground, and Nixon politely asked Hiss to "bear with me" in this tedious effort. Nixon then interrogated Hiss on the matter of names. When was it that he first heard the name "Whittaker Chambers"? Hiss recalled that it was a name which came up when he was first questioned by the FBI in 1947, adding defensively that it was only one among many names he had heard on that occasion. Nixon, still courteous, continued: Had Hiss known somebody named Carl? Hiss allowed he had known a Carl or two in his life but not this particular Carl. He also didn't know anyone named J. Peters but had known Henry Collins since they were boys in camp together. Yes, he had visited Collins and only for social reasons in his Washington apartment but didn't know for sure who else might have been present at these times. (It was evident that the central issue here was Chambers' testimony that Collins had collected Communist Party dues at such times, and Hiss was clear in his own recollection that he had indeed not turned any money over to Collins or to anyone else in the latter's apartment.)

Then came the photographs of his accuser. There were two pictures, and Hiss began to waffle. They really didn't bring anyone to mind, but then again the likeness was "not completely unfamiliar." Hiss repeated his frequent statement that he wished for an opportunity personally to confront this man Chambers, and Nixon agreed that this would surely be arranged but made clear his inability to accept the idea that Hiss could forget someone who stayed overnight several times at his home.

Nixon felt that the opaque statement "not completely unfamiliar" was the "first tiny crack in his [Hiss's] statement that he did not know Chambers" and began to prod Hiss. According to Nixon, Hiss's answers became "increasingly lengthy and evasive." After some fruitless questions about old Hiss addresses and domestic help, Nixon finally asked whether Mrs. Hiss would come to Washington to testify. Hiss, who, in Nixon's view, had begun to argue with the committee, requested permission to make a statement. Looking directly at Nixon, he spoke of his worry that he was not acquainted with what Chambers had said to the committee at its previous private meeting.

> By the attitude you have been taking today that you have a conflict of testimony between two witnesses—one of whom is a confessed former Communist and the other is me—and that you simply have two witnesses saying contradictory things as between whom you find it most difficult to

> decide on credibility. I do not wish to make it easier for anyone who, for whatever motive I cannot understand, is apparently endeavoring to destroy me. I should not be asked to give details which somehow he may hear and then may be able to use as if he knew them before.

Nixon testily replied that Hiss was implying that "the committee's purpose today is to get information with which we can coach Mr. Chambers so that he can more or less build a web around you." Hiss, who had felt that there existed from the beginning of the second session "a proprietary attitude toward Chambers, as if he were the committee's witness and I an outsider," now spoke accusingly: "I have seen newspaper accounts, Mr. Nixon, that you spent the weekend—whether correct or not, I do not know—at Mr. Chambers' farm in New Jersey." Nixon answered shortly, "That is quite incorrect."

What is fascinating is that Nixon chose to deny a patently true accusation dealing with events which he freely spoke of in his own book and which were confirmed by Bert Andrews in a series of newspaper articles as well. Nowhere, however, in Nixon's account of the trial is mention made of this strange scene, and one is left with the feeling that perhaps Nixon used Hiss's confusion about the locale of Chambers' farm—actually Maryland, not New Jersey—or the length of time he spent there—an evening, not a "weekend"—to elude Hiss's assertion that the committee had already made its decision about his guilt prior to the second hearing. Recently Hiss commented on Nixon's play-acting. "I thought that meant he'd never seen him. Later I decided that this was Tricky Dick."

Hiss tried again to make his point: "I would like him [Chambers] to say all he knows about me now . . . let him tell you all he knows, let that be made public, and then let my record be checked against the facts." Mr. Thomas, the committee chairman, replied sternly, "Questions will be asked and the committee will expect to get very detailed answers."

Alistair Cooke wrote the following summary of what then transpired:

> They started again to check old Hiss addresses, and Hiss once again put forward his plea. "If this Committee feels"—he began. Mr. Thomas broke in with "Never mind feelings. You let Mr. Nixon ask the questions and you go ahead and answer it." At least two members of the Committee were losing their patience. Mr. Stripling said: "Let the record show, Mr. Hiss, you brought up this ex post facto business. Your testimony comes as ex post facto testimony to the testimony of Mr. Chambers. He is already on record." Mr. Stripling didn't want to infer, though, that Hiss knew what Chambers had

said. Mr. Nixon made the counter-plea to Hiss to accept the good faith of the Committee. They were there only to "test the credibility of Mr. Chambers," and you are the man who can do it. "Frankly," he said, "I must insist." Hiss said if they insisted he would "of course" answer. But then he told what was troubling him. Had not the Committee heard secret testimony from Chambers before his first appearance in public? "No, sir," said Mr. Stripling. "But," said Hiss, "there was a press report to that effect: didn't he meet with you in executive session?" Mr. McDowell put in that on the morning of Chambers' accusation they had had about two minutes with him to get his name and job. Still incredulous, Hiss asked: "Didn't you know his testimony—that he was going to testify about me?" "No," said Mr. Stripling.

Nixon said nothing at this point, and this fact is worthy of notice. Garry Wills in *Nixon Agonistes* contends that Nixon knew of Hiss and his alleged involvement with Chambers well in advance of August 1948. Wills quotes an interview with Father John Cronin, a priest who was actively involved with the labor movement in Baltimore in the Forties. He discovered that "Communist cadres" were responsible for rigging union elections, and Cronin was soon in communication with the FBI. As a consequence, Cronin became known as a Communist expert in Catholic circles, and he was visited one day by Congressman Charles Kersten of Wisconsin, of whom Nixon once said:

Charlie Kersten is a deeply religious man, whose anti-communism is of a philosophical sort. It's too bad he came from that terrible district (the old fifth), where he couldn't get reelected after his third term. He taught me most of what I know about communism.

Kersten brought Nixon to meet Father Cronin in 1947, and a close personal relationship developed between the two men. Cronin told Nixon about "certain Communists in the State Department." Wills interviewed Cronin at some length:

I asked him how he got involved with Nixon and Hiss. "Early in the Forties, when I was working with the dockside unions in Baltimore, some of my friends came to me with complaints that they were being voted out of union offices by suspiciously packed meetings. I did a little investigation and found these were Communist cadres at work. About that time the FBI approached me to find out what I knew about this. Soon I was in touch with Bill Sullivan [later an assistant director of the FBI]. I kept track of what was going on from them. And I got to know many agents intimately. . . . Cardinal—then Archbishop—Mooney heard of my knowledge in this area: so he asked me to prepare a secret report on Communism for the American bishops, and I was able to use classified material that had come my way. . . . By this time, I was known, in Catholic circles, as something of an expert on Communism. Charlie Kersten heard this and came to see me. Later, he brought Nixon, and I told them about certain Communists in atomic espionage rings and in the State Department." Did you name names? "Yes." Was one of the names Alger Hiss? "Yes. This was a year and a half before Whittaker Chambers was called by the House Committee to confirm testimony given by Elizabeth Bentley—when Hiss was first named publicly as a Communist."

Father Cronin was even more helpful to Nixon as the hearings progressed, for he had retained connections with the FBI. In the post-Watergate vernacular, he might be considered Nixon's "Deep Throat." Of this period he says:

> Ed Hummer was one of the FBI agents I had worked with. He could have got in serious trouble for what he did, since the Justice Department was sitting on the results of the Bureau's investigation into Hiss—the car, the typewriter, etc. But Ed would call me every day and tell me what they had turned up; and I told Dick, who then knew just where to look for things and what he would find.

Wills draws an interesting and cynical conclusion from the above material, one that clearly stands at some distance from the picture which has routinely been offered by Nixon.

> When Nixon—a first-term Congressman, as lowly a creature as exists in Washington—pushed the Hiss case, he seemed to be taking a great risk. It was less than it looked. He had cards all up and down his sleeves and inside his vest.

As the August 16 HUAC hearing continued, additional efforts were made to connect Hiss and Chambers personally, and in Nixon's view Hiss "attacked on another front" by trying to shift the focus of the inquiry away from the question of whether an intimacy existed between the two men to "whether I am a member of the Communist Party or ever was," an affiliation which Hiss had publicly denied, and "whether he [Chambers] had a particular conversation that he said he had with me," which Hiss also denied.

(This discussion, according to Chambers, occurred when he decided to break from the Communist Party. The only man he told about his decision was his friend, Alger Hiss, whom he tried to persuade to do the same because "I was very fond of Mr. Hiss," said Chambers. Hiss was reported to have wept at this request, but he "absolutely refused to break," claimed Chambers.)

Nixon, however, was unwilling to change his thrust. He stated firmly:

> When Mr. Chambers appeared, he was instructed that every answer he gave to every question would be material and that answers to a material question would subject him to perjury. Membership in the Communist Party is one thing, because that is a matter which might be and probably would be concealed. But items concerning his alleged relationship with you can be confirmed by third parties, and that is the purpose of these questions.

Then, Hiss "came to the end of the road," as reported in Nixon's book, and surfaced his suspicion that Chambers was Crosley. As in the Japanese movie *Rashomon*—where the observers had very different reactions to the same occurrence—Hiss had his own view of that particular crisis:

> To the Chairman's question as to whether I would recall someone who had been a week at a time at my house, I said that I would, but insisted:
>
> . . . I am not prepared to testify on the basis of a photograph.
>
> At this point Nixon stated that Chambers had told the Committee that he had spent a week in my house. The vague image of Crosley and his broken tooth returned. He must have some connection with the story Chambers was telling. But I could not be sure what that connection was and I was therefore reluctant to cause Crosley the embarrassment that might come from mentioning his name. I replied: "I have written a name on this pad in front of me of a person whom I knew in 1933 and 1934 who not only spent some time in my house but sublet my apartment. That man certainly spent more than a week, not while I was in the same apartment. I do not recognize the photographs as possibly being this man: If I hadn't seen the morning papers with an account of statements that he knew the inside of my house, I don't think I would even have thought of his name. I want to see Chambers face to face and see if he can be this individual. I do not want and I don't think I ought to be asked to testify now to that man's name and everything I can remember about him. . . .

At this point the committee sent Hiss out of the room and held an executive session. What went on there is unknown. During the intermission Hiss conferred with two attorneys, actually friends from his old law firm. Suddenly it dawned on him that "I had left the pad there with the name [George Crosley]. I had said before, 'I've written down a name.' When I came back in, I thought, gee, the way that everyone is behaving they'd probably seen it. I'm not going to protect him anymore so I gave the name."

Professor Allen Weinstein, author of *Perjury*, the recent book which concludes Hiss was guilty, relates how Hiss's wife Priscilla had no difficulty recognizing a photograph of Chambers and linking it to Crosley. Edmund F. Soule, a neighbor from Peacham, Vermont, where the Hisses summered, brought Priscilla the New York newspapers which first linked Hiss to Chambers and which featured a picture of the latter on August 5, 1948. Soule remembers that after seeing the photograph Priscilla responded either to his mother or to his mother and him, "Wait a minute. . . . I remember a dreadful man named Crosley or something like that. . . ."

When questioned about this, Hiss speculated that Priscilla's instant recall was an example of "feminine intuition," and he said, "That would not have satisfied me." He is not sure "when Priscilla got the word of Chambers' testimony," but he is sure he talked to her by telephone between August 3 and 5. "Now if she told me then that it might have been Crosley, this would not have affected my testimony on the fifth [of August] because I said I wished to confront the man." Hiss now remembers discussing the matter with Priscilla prior to his second appearance before HUAC on August 16, for he had returned to Vermont the previous weekend.

It took eleven days for Hiss publicly to identify the photograph and link Chambers to Crosley, a process that took Priscilla but moments. Also rather than remembering the name "Crosley" on the train while traveling to the hearing on August 16, as Hiss recounts in his book, Hiss now states he made the connection at least the weekend before.

And so it was finally out. Hiss, although reluctant, had linked Chambers to Crosley and went on to describe what he had been able to recall about Crosley, and then almost out of nowhere came the prothonotary warbler, a rare songbird that in the committee's hands turned positively predatory.

Chambers had testified in the preceding hearing on August 7 that the Hisses had the same hobby; they were ornithologists: "They used to get up early in the morning and go to Glen Echo, out the canal, to observe birds. I recall they once saw, to their great excitement, a prothonotary warbler. . . ." Representative John McDowell had then asked, "A very rare specimen?" And Chambers replied in enigmatic fashion, "I never saw one. I am also fond of birds."

The transcripts of the committee hearing on August 16 contain the prosaic notation that a short intermission followed Hiss's recollection of George Crosley. This time Hiss remained in the room. The record then reveals the following dialogue:

NIXON: *What hobby, if any, do you have, Mr. Hiss?*
HISS: *Tennis and amateur ornithology.*
NIXON: *Is your wife interested in ornithology?*
HISS: *I also like to swim and also like to sail. My wife is interested in ornithology, as I am, through my interest. Maybe I am using too big a word to say an ornithologist, because I am pretty amateur, but I have been interested in it since I was in Boston. I think anybody who knows me would know that.*
MCDOWELL: *Did you ever see a prothonotary warbler?*
HISS: *I have right here on the Potomac. Do you know that place?*
THE CHAIRMAN: *What is that?*
NIXON: *Have you ever seen one?*
HISS: *Did you see it in the same place?*
MCDOWELL: *I saw one in Arlington.*
HISS: *They come back and nest in those swamps. Beautiful yellow head, a gorgeous bird.*

Meyer Zeligs, in his book *Friendship and Fratricide: An Analysis of Whittaker Chambers and Alger Hiss,* has some interesting things to say about this event. He interviewed Donald Appell, an investigator for the House committee, who described how the "prothonotary trap was set for Hiss." The committee had by this time come to regard Hiss as a master spy, one who would try to outwit them at every turn, and they were worried that Hiss had already recalled telling "Carl" of the sighting of the prothonotary warbler. This might lead to a denial by the witness of the testimony, and Chambers' case would be weakened. The decision was made to pose the question in a decidedly informal fashion. At the planned moment, Nixon and Appell excused themselves, and McDowell and Hiss were left alone.

As Nixon and Appell were returning to the hearing room, McDowell, himself an ornithologist, was in the process of speaking informally to Hiss about birds and casually asked him about the bird. The official record makes clear that a reprise took place after Hiss responded, yes, he had.

For many, this was the clinching testimony. *Time* featured a picture of the now well-known bird, and while many were prepared to doubt stories of person-to-person relations, man-to-bird congress seemed irrefutable. Despite Hiss's statement that

> We have never been to Glen Echo, an amusement park, to observe birds [as Chambers had claimed]. I doubt if anyone else has. . . . It was not a special haunt for bird watchers. Prothonotary warblers . . . can only be seen in the vicinity of Washington in a swampy woods in Virginia south of Alexandria. . . . I might certainly have mentioned to anyone with whom I talked about birds that I had seen a prothonotary warbler.

the bird-sighting anecdote proved very hard to wash out. In Hiss's view, "it was [in the eyes of the committee], the prime alleged proof of a close relationship between Chambers and me—as if an enthusiast boasts of his finds only to intimate friends."

At a later hearing on August 25 when Hiss was openly regarded as a hostile witness, Congressman Hebert charged that Chambers knew "all about Hiss's private life" and that Chambers had "unhesitatingly answered every question in the minutest of details which, as Mr. Mundt has indicated, comes back and checks. . . ." The testimony continued:

> HISS: *Who would remember—how would any man remember all those details about any other man after 14 years?*
> MR. HEBERT: *Unless he knew him extremely well.*
> MR. HISS: *Unless he was studying up on it.*

The record then indicates that Hebert returned to the fabled bird, citing it as evidence that Chambers knew that Hiss had seen a prothonotary warbler:

> MR. HEBERT: *Now, that is not from* Who's Who.*
> HISS: *I have told many, many people that I have seen a prothonotary warbler, and I am very, very proud* [*of having done so*]. *If Mr. McDowell has seen it, he has told very, very many people about it.*

Nixon was no less impressed by the prothonotary warbler story, and he used it to illustrate the extent of Chambers' personal knowledge of Hiss. In a speech before the House of Representatives, Nixon recalled

---

*What Hebert was here referring to was the fact that although *Who's Who* listed Hiss's hobby as ornithology, it obviously failed to provide a potential enemy with the name of one of Hiss's major findings.

Chambers' bird-watching recollection. He said, "We knew from that answer alone that one person could not know such intimate details about another unless he had known him some time in his life."

The committee's doubt about its course of action vanished forever with the August 16 hearing. After further sparring on that fateful day about the matter of confidentiality of testimony (Hiss's lingering concern about the *ex post facto* use of his testimony), Representative Hebert broke in angrily:

> Mr. Hiss, let me say this to you now—and this is removed from all technicalities, it's just a man-to-man impression of the whole situation. . . . I will tell you exactly what I told Mr. Chambers so that it will be a matter of record, too: either you or Mr. Chambers is lying . . . and whichever one of you is lying is the greatest actor that America has ever produced. Now, I have not come to the conclusion yet which one of you is lying and I am trying to find the facts. Up to a few moments ago you have been very open, very cooperative. Now, you have hedged. . . . Now if we get the help from you and, as I say, if I were in your position, I certainly would give all the help I could, because it is a most fantastic story. What motive would Chambers have? You say you are in a bad position, but don't you think that Chambers destroys himself if he is proven a liar? What motive would he have to pitch a $25,000 position as a respected Senior Editor of *Time* magazine out the window?

Hebert's attack signaled the beginning of the end for Alger Hiss. It was a significant defection because up to this point, sentiment for or against Hiss had been divided pretty much along party lines. Prior to Hebert's outburst, it was easy to write off the committee's investigation in this particular matter as party politics, with the Republicans, denied the Presidency for sixteen years, looking to discredit a Democrat of Hiss's stature in order to regain the national spotlight. As HUAC chairman J. Parnell Thomas would admit years later, the Republican National Committee chairman "was urging me in the Dewey campaign to set up the spy hearings . . . in order to keep the heat on Truman."

Of the four Democrats on the committee, two, John S. Wood and J. Hardin Peterson, didn't even attend the meetings; however, John E. Rankin, Democrat from Mississippi and cut from the more reactionary stripe of the party, used the hearings as a sounding board for his favorite theme—anti-Semitism.* Hebert was the only Democrat likely to

---

*Rankin spent a considerable amount of time at the hearing thumbing through *Who's Who in American Jewry* in the apparent belief that Whittaker Chambers was a Jewish "turncoat."

befriend Hiss, and this possibility blew up in the fire of Hebert's anger.

Today (and with the hindsight of better than two dozen years), Hiss feels that the committee's verdict was a *fait accompli*. "My impression now is that nothing I did could have changed the attitude of the committee," said Hiss. "They were seeking . . . which I didn't know at the time, namely something [with which] to hurt Truman."*

Nixon's report of what followed at the August 16 hearing is reproduced below. It is interesting both as an anecdote as well as for its language and style:

> Hiss was shaken to his toes by this blast. Up to this time he had, not without considerable support from the press and from President Truman himself, tried to imply that the entire hearing was a "Republican plot" to smear the New Deal. Now for the first time, a Democrat had begun to question his story. Hiss reacted by counter-attacking Hebert as hard as he could
>
> "It is difficult for me to control myself!" he exclaimed. "That you can sit there, Mr. Hebert, and say to me casually that you have heard that man and you have heard me and you just have no basis for judging which one is telling the truth. I don't think a judge determines the credibility of witnesses on that basis."
>
> But Hebert, not to be cowed, fired back: "I absolutely have an open mind and am trying to give you as fair a hearing as I could possibly give Chambers or yourself. The fact that Mr. Chambers is a self-confessed traitor . . . and a self-confessed former member of the Communist Party has no bearing at all on the alleged facts that he has told. . . ."
>
> "Has no bearing on his credibility?" interrupted Hiss.
>
> "No, because, Mr. Hiss, I recognize the fact that maybe my background is a little different from yours," replied Hebert, who had been a New Orleans newspaper editor for many years. "But I do know police methods, and you show me a good police force, and I will show you the stool pigeon who turned them in. We have to have people like Chambers to come in and tell us. I am not giving Mr. Chambers any great credit for his previous life. I am trying to

---

* As former President Harry Truman told author Merle Miller in the latter's book *Plain Speaking*, "I think if they'd had the goods on Hiss, they'd have produced it, but they didn't. All it was was his word against the word of that other fellow [Whittaker Chambers]. They didn't come up with any proof. . . . What they [HUAC and the Republicans] were trying to do, all those birds, they were trying to get the Democrats. They were trying to get me out of the White House, and they were willing to go to any lengths to do it. They'd been out of office a long time, and they'd have done anything to get back in. They did do just about everything they could think of, all that witch-hunting that year, that was the worst in the history of this country, like I said. The Constitution has never been in such a danger, and I hope it never will be again."

find out if he is reformed. Some of the greatest saints in history were pretty bad before they were saints. Are you going to take away their sainthood because of their previous lives? Are you not going to believe them after they have reformed? I don't care who gives the facts to me, whether a confessed liar, thief, or murderer—if it is facts. That is all I'm interested in."

Hiss had a bear by the tail. He tried to change the subject. "I would like to raise a separate point," he said. "The real issue," he again insisted, "was not whether Chambers knew him or he knew Chambers; it was whether he and Chambers had had the one particular conversation to which Chambers had testified."

I [Nixon] answered by saying, "If Chambers' credibility on the question of whether he knew you or not is destroyed, obviously you can see that this statement that he had a conversation with you and that you were a member of the Communist Party, which was made on the basis of this knowledge, would also be destroyed. And that is exactly the basis upon which this questioning is being conducted. If we prove that he is a perjurer on the basis of his testimony now, the necessity of going into the rest of the matter will be obviated."

There was an intermission to allow Hiss to phone his wife. After he returned, Hiss began to answer Nixon's questions which came from Chambers' earlier testimony. Nixon asked about family pets. Hiss had a brown cocker spaniel which he boarded at a certain kennel, a fact Chambers had mentioned to the committee. He was interrogated about his youth; had he, for example, as Chambers had related, as a boy of twelve hauled water from Druid Hills spring to sell in Baltimore? Yes. There were other questions about Hiss's mother, brothers, sisters, and his stepson and the schools the latter attended. Nixon concluded, "In virtually every detail, Hiss's answers matched those of Chambers." Nixon then told Hiss that Chambers was willing to take a lie-detector test and asked if he was also willing. Hiss declined, saying that according to the experts he had consulted, there was no such thing as a true lie detector. Hiss had been warned by two reporters that the committee was considering asking the principal witnesses to take such a test, so he had investigated and concluded, "It is an emotion recording test; that it is not scientific, and that nobody scientifically competent, including the [Federal] Bureau [of Investigation], regards it as a scientific test."

Nixon replied that "The Committee has contacted Mr. Leonardo Keeler." Since it was Keeler's polygraph machine Hiss had looked into, he commented:

Would it seem to you inappropriate for me to say that I would rather have a chance for further consultation before I gave you the answer? Actually, the people I have conferred with so far say that it all depends on who reads, that it shows emotion, not truth, and I am perfectly willing and prepared to say that I am not lacking in emotion about this business.

I have talked to people who have seen, I think, Dr. Keeler's own test and that the importance of a question registers more emotion than anything else. I certainly don't want to duck anything that has scientific or sound basis. I would like to consult further.

Writing later of this incident, Hiss expressed the fear that Nixon's request for the polygraph examination would confuse a naïve public who might not know of the "lack of scientific validity" of such tests.* These fears proved to be well grounded when the prosecution in the second trial interrogated Hiss on the matter and felt it was important enough to ask the question twice.

MURPHY: *Well, in any event, you didn't submit to a lie detector test?*
HISS: *No.*
MURPHY: *Well, in any event, you did not take the test? Well, you didn't come forward and insist upon it?*
HISS: *No.*

As the August 16 meeting ended, it was apparent to Nixon that the time had come to confront Hiss with Chambers. Hiss was indifferent to whether it was to be a public or private meeting. Nixon preferred a private session, saying, "If you have a public session, it will be a show." Stripling added, "It will be ballyhooed into a circus." Hiss remarked that if they were worrying about him "after what has been done to my feelings and my reputation, I think it would be like sinking the Swiss Navy." Since Hiss had some experience with committee secrecy, he pushed for an open meeting.

It was finally decided that someone ought to find out what Chambers' wishes were. It was also arranged that Mrs. Hiss would have her day with the committee sometime before the date of the confrontation, set at that time for August 25. The committee report contains the homey assurance "with absolutely no publicity" and the last entry on the record was by the chairman and read: "We will see you August 25."

---

*According to Allen Weinstein, Dr. Carl Binger, a psychoanalyst who testified for the defense at the second trial, offered to inject Hiss with a truth serum (scopalomine) and interview him about the case. This offer was likewise declined.

CHAPTER 3

I would like to say that to come here and discover that the ass under the lion's skin is Crosley . . . I don't know why your committee didn't pursue this careful method of interrogation . . . before all the publicity. —ALGER HISS

*T*hat evening, Nixon met with Stripling and they spent several hours going over the testimony. They reached two conclusions: (1) that Crosley and Chambers were the same man and (2) that Chambers and Hiss surely knew each other. Still unresolved, however, was the key question: Which man was telling the truth? Even after Stripling left Nixon's office at midnight, the young Congressman continued to look through his notes:

> I knew that we had reached the critical breaking point in the case. Timing now became especially important. If Hiss' story about Crosley were true, why had he not disclosed it to the Committee when he first appeared in public session? Why had he first tried so desperately to divert the Committee from questioning him on the facts Chambers had previously testified to? The longer I

> thought about the evidence, the more I became convinced that if Hiss had concocted the Crosley story, we would be playing into his hands by delaying the public confrontation until August 25, thus giving him nine more days to make his story fit the facts. With his great influences within the Administration, and among some of his friends in the press, he might be able to develop an enormous weight of public opinion to back up his story and to obscure the true facts in the case. The more I thought about it, the more I became convinced that we should not delay the confrontation. Only the man who was not telling the truth would gain by having additional time to build up his case.

Having made that decision, Nixon called Stripling at 2:00 A.M. and told him to summon Hiss and Chambers to appear before the committee in New York at 5:30 P.M. that very afternoon (August 17). Nixon was subsequently accused by Hiss and others of arranging the sudden meeting to cover the possibility that the committee would be accused of contributing to the death of another witness, Harry Dexter White. White had voluntarily testified before them with a heart condition, and his subsequent death was announced earlier on this day.

White, a former Assistant Secretary of the Treasury, had been a rather eloquent witness before the committee on August 13. He denied the charges of both Whittaker Chambers and Elizabeth Bentley. Chambers had said of him, "He was certainly a fellow traveler so far within the fold that his not being a Communist would be a mistake on both sides."

Before testifying, White had passed a note to the committee chairman which read, "I am recovering from a severe heart attack. I would appreciate it if the chairman would give me five or ten minutes' rest after each hour." Chairman J. Parnell Thomas was rather uncharitable to this request; in fact, he read it aloud (a procedure which offended White) and indicated sarcastically that he doubted its veracity, saying, "For a person who had a severe heart condition, you certainly play a lot of sports." White, angered now, said he played "a lot of sports" *before* this attack, and the chamber erupted in applause. Three days after his appearance, White was dead, and sympathetic reporters described him as a victim of the committee's tyranny.

Hiss had this to say about the committee's treatment of Harry Dexter White:

> Flying back from Washington after my testimony of the afternoon before, I had had time to read the fully reported accounts of his courage in voluntarily facing, despite his illness, the ordeal of a public grilling in the circus-arena atmosphere of klieg lights and flash bulbs. I had found unpalatable the

Committee's badgering of a sick man and its implication that he was malingering in privately asking for an occasional intermission.

Nixon, responding to Hiss's charge that the confrontation was moved up to take the heat off the committee for White's death, said, "This accusation—like so many others made against the committee—while plausible is completely untrue. I myself had made the decision on the confrontation well before I learned of White's death."

Feeling lonely on August 17, Chambers came to Washington and went to look up the committee, whose members "were the only people left in the world with whom I could communicate." There is a mystical quality to Chambers' writing as he described his dramatic passage to the confrontation. Like Macbeth's "foul and fair a day," Chambers sensed that something of historical importance was about to occur, as if on this day the confrontation between the two antagonists was preordained. "I felt a curious need to go see the Committee," he writes in *Witness*. So Chambers canceled a planned trip to New York and went to Washington instead in search of his only friends, the committee.

It was high noon when Chambers arrived at the Old House Office Building. Starting up the steps, he noticed a crowd of reporters and quickly made for another entrance—one he had never used before. Out of that very door at that very moment emerged Robert Stripling; Ben Mandell,* the committee's chief researcher; Donald Appell, an investigator; and other HUAC staff members. It was apparently a joyful reunion, for the committee had been frantically trying to contact Chambers that day to set up the rescheduled confrontation. Stripling peered at Chambers, almost transfixed, and then announced to his associates, "I believe he must be psychic."

They bundled Chambers into an automobile without telling him, he claimed, where they were taking him or even the nature of the mission. On the way to the railroad station, Appell took a newspaper from his pocket and showed Chambers that day's hearings about the death of Harry Dexter White. Chambers was strangely—if somewhat gratuitously—upset by this. He later wrote of White's appearance. "He had

---

*In Chambers' post-trial book *Witness*, he belatedly identified Mandell "as the man who issued to me my Communist Party card." When Chambers appeared before the House committee for the first time, the two were, in Chambers' words, "resuming a close acquaintance."

skirmished brilliantly with the Committee, turning their questions with ridicule and high indignation."

From Washington's Union Station they rode with Congressmen Nixon and McDowell to Pennsylvania Station in New York City, where they all waited for the train to empty before disembarking. "They did not want the press to discover prematurely that they were in New York," Chambers wrote. Still not knowing the precise reason for this trip, Chambers, left in Appell's hands, loitered about the streets before catching a cab for the Hotel Commodore in midtown Manhattan.

Appell led him to Room 1400, where they entered one bedroom of a suite. Chambers soon realized a subcommittee of HUAC was meeting in the other bedroom. Suddenly one of the committee's investigators opened the door between the rooms of the suite. He pointed at Chambers and motioned for him to follow. With some trepidation, Chambers entered the room where the subcommittee had assembled. He saw Congressman McDowell in the chair and Representative Nixon seated nearby flanked by Robert Stripling. They watched silently as Whittaker Chambers came in.

Other staff members milled about the left side of the hotel room. Chambers noticed one other man sitting on a couch with his back turned toward him. He concluded that this was Alger Hiss. As Chambers entered the room, Hiss did not look at him. As Chambers sat down heavily, Hiss stared straight ahead.

Hiss's arrival at this historic confrontation was marked by none of the melodrama that characterized Chambers' appearance. That morning while at his office at the Carnegie Endowment, Hiss was called by Donald Appell, who said that Congressman McDowell would be in New York late that afternoon and wanted to see Hiss for ten or fifteen minutes. When Hiss inquired whether this was in relation to committee business, Appell said he did not know. Later that day Hiss received a wire from McDowell stating that he would call Hiss about 5:30 that afternoon.

Shortly before 5:30, according to Hiss, McDowell called and surprised Hiss by inviting him to come to the Commodore Hotel and meet with him, Nixon, and *one* other person. Already angry about the White incident, Hiss was further annoyed by the reports in the morning newspapers of his testimony on the previous day which he felt "were a plain and prompt repudiation of protestations of secrecy by the committee." He asked Charles Dollard, a colleague at the Carnegie Endowment, to accompany him to the Commodore so that an outsider "would be able to give his version of any further relations I might have with the committee."

When the two men arrived at Room 1400, they found that it was hastily being converted into a hearing room. Nixon and McDowell were there along with Appell and a stenographer. J. Parnell Thomas, the committee chairman, arrived much later. Nixon opened the meeting by announcing that its purpose was to bring Hiss and Chambers together, saying that "the case was dependent upon the question of identity," but did not mention why the meeting had been suddenly moved up. Hiss, resenting "the hypocrisy of the elaborate assurances of secrecy which had been given me the day before only to be cynically violated by immediate leaks slanted to injure me, as well as the disingenuous manner" by which he had been summoned to the hearing, responded by saying that the death of Harry Dexter White was very disturbing to him, and although he was not "in the best mood for testifying," he certainly would not miss this opportunity to face his accuser.

Then Hiss turned to the matter of the hollow oaths of secrecy taken by all present at the hearing on the previous day. Much of what was said in that supposedly secret hearing had appeared that morning in the New York *Herald Tribune*. Hiss said with obvious disgust and sarcasm, "I read in the papers that it was understood that in the course of my testimony yesterday the committee asked me—the subcommittee asked me—if I could arrange to have Mrs. Hiss be examined privately. You will recall, and I hope the record will show, that Mr. Nixon assured me with great consideration that you desired to talk to Mrs. Hiss without any publicity. This [the story in the newspaper] was less than twenty-four hours after you had been so considerate." Then Hiss mentioned other news leaks, such as the fact that the committee had asked him to take a lie-detector test. "They [the leaks] could only have come from the committee," said Hiss. "They did not come from me."

McDowell responded for the committee, saying:

> Obviously, there was a leak, because the story that appeared in the various papers I read was part of the activities of yesterday afternoon. As a member of Congress, there is nothing I can do about that. It is a regrettable thing, and I join you in feeling rather rotten about the whole thing.

Nixon was somewhat more defensive about committee responsibility for the "leak" and stated that he was "pretty sure that the *Tribune* had used sources outside the Committee and outside the Committee staff." What other possibilities this left were unclear, since Hiss was the aggrieved party, and the meeting was off to an unhappy start.

The confrontation was stark drama: Hiss's almost pathological caution, Chambers' curious sense of righteousness, Nixon's increasing impatience and virtuous anger at Hiss, and the irrevocable hardening of the committee's attitude toward the former State Department official.

It began with Richard Nixon directing Whittaker Chambers to a couch. Across the room was Hiss, looking nearly transfixed. With his back turned to his accuser, he did not resemble the man who had been demanding such a confrontation but more like a trapped pigeon well aware that the moment of truth with the hawk had arrived. With the final guest seated, Nixon asked Hiss and Chambers to rise and said, "Mr. Hiss? Mr. Hiss, the man standing here is Whittaker Chambers. I ask you now if you have ever known that man before." Hiss, wearing his caution like armor, inquired if Nixon would ask Chambers to speak. "Yes. Mr. Chambers, will you tell us your name and your business?" said Nixon. As Chambers responded, Hiss stalked menacingly toward his accuser, saying, "Would you mind opening your mouth wider?"

Chambers complied, saying, "My name is Whittaker Chambers." Hiss, wanting a better look at Chambers' dentures, repeated his request for Chambers to open his mouth wider. Turning to Nixon, he said, "You know what I am referring to, Mr. Nixon." To Chambers he said, "Will you go on talking?" Chambers, revealing a bit more of his orthodontia, told those assembled in the hotel room about his position with *Time*. "May I ask whether his voice, when he testified before, was comparable to this," asked Hiss, "or did he talk a little more in a lower key?" Representative McDowell said that Chambers sounded much the same.

Still Chambers' voice was troublesome to Hiss—who today still characterizes himself as a "man not given to snap judgments"—and so he asked Chambers to continue speaking. Mr. Nixon, with no irony intended, handed the *Time* editor a copy of *Newsweek* and asked him to read an article, which turned out to be about President Truman—certainly no favorite of Nixon's—and his search for a replacement for a cabinet member. "I think he is George Crosley," Hiss said. "But I would like to hear him talk a little longer." He asked if Chambers were Crosley. "Not to my knowledge. You are Alger Hiss, I believe" was the response.

"I certainly am!"

"That was my recollection," said Chambers, who turned to his magazine and read, "Since June—"

Mr. Nixon interrupted, "Just one moment. Since some repartee goes on between these two people, I think Mr. Chambers should be sworn."

"That is a good idea," agreed Hiss. Mr. McDowell administered the oath to Chambers, who was asked if he promised to tell the truth, the whole truth, and nothing but the truth. Immediately on the heels of Chambers' "I do," Nixon jumped back in and said with anger, "Mr. Hiss, may I say something? I suggested that he be sworn, and when I say something like that I want no interruptions from you."

"Mr. Nixon, in view of what happened yesterday," Hiss said, referring to the news leaks, "I think there is no occasion for you to use that tone of voice in speaking to me, and I hope the record will show what I have just said." Nixon responded portentously, "The record shows everything that is being said here today."

Committee investigator Stripling asked politely for Chambers to continue reading, and like a Greek chorus of one, he continued. "Since June, Harry S. Truman had been peddling the labor secretaryship left vacant by Lewis B. Schwellenbach's death in hope of gaining the maximum political advantage from the appointment." Hiss interrupted the recital by inquiring if he might interrupt. Permission was granted, and Hiss commented, "The voice sounds a little less resonant than the voice that I recall of the man I knew as George Crosley. The teeth look to me as though either they have been improved upon or that there has been considerable dental work done since I knew George Crosley. . . . I believe I am not prepared without further checking to take an absolute oath that he must be George Crosley."

Nixon asked for permission to ask a question of Chambers, but Hiss broke in, saying, "I would like to ask Mr. Chambers, if I may—"

"I will ask the questions at this time," retorted Nixon with considerable pique. Nixon inquired if Chambers "had any dental work since 1934 of a substantial nature." Chambers said he had some teeth pulled and wore a plate. Hiss wanted someone to ask the name of the dentist who performed the work and in a more subdued fashion asked if such a request were "appropriate." Nixon agreed that it was appropriate and Chambers answered, "Dr. Hitchcock, Westminster, Maryland."

Hearing this, Hiss concluded, "That testimony of Mr. Chambers, if it can be believed, would tend to substantiate my feeling that he represented himself to me in 1934 or 1935 or thereabouts as George Crosley, a free-lance writer of articles for magazines. I would like to find out from Dr. Hitchcock if what he has just said is true, because I am relying partly—one of my main recollections of Crosley was the poor condition of his teeth."

Nixon, losing patience with Hiss's caution, said, "Before we leave the teeth, do you feel that you would have to have the dentist tell you just what he did to the teeth before you could tell anything about this man?"

"I would like a few more questions asked," said Hiss. "I didn't intend to say anything about this, because I feel very strongly that he is Crosley, but he looks very different in girth and in other appearances—hair, forehead, and so on, particularly the jowls."

The emphasis by Hiss on oral hygiene may seem incongruous for a House subcommittee hearing—even one held in a hotel room—but the ruin in Chambers' mouth was his dominant physical characteristic as a youth. The late novelist and literary critic Lionel Trilling, who was acquainted with Chambers when they were both students at Columbia University in the early 1920s, based a character (Gifford Maxim) on him in his novel *The Middle of the Journey*. Dental decay was in fact such an unmistakable physical trait of Chambers that Trilling chose to ignore it because " . . . to do so would have been to go too far in explicitness of personal reference," the author wrote in *The New York Review of Books* (April 17, 1975). Trilling describes the physical Chambers in this article:

> The moral force that Chambers asserted began with his physical appearance. This seemed calculated to negate youth and all its graces, to deny that they could be of any worth in our world of pain and injustice. He was short of stature and very broad, with heavy arms, and massive thighs; his sport was wrestling. In his middle age there was a sizable crop of belly and I think this was already in evidence. His eyes were narrow and they preferred to consult the floor rather than an interlocutor's face. His mouth was small and like his eyes, tended downward, one might think in sullenness, though this was not so. When the mouth opened, it never failed to shock by reason of the dental ruin it disclosed, a devastation of empty sockets and blackened stumps . . . that desolated mouth was the perfect insigne of Chambers' moral authority. It annihilated the hygienic American present—only a serf could have such a mouth, or some student in a visored cap who sat in his Moscow garret and thought of nothing else save the moment when he would toss the fatal canister into the barouche of the Grand Duke.

Hiss was then asked for his version of the apartment rental to Crosley and the gift of the car. Hiss offered gratuitously the recollection of receiving a rug from Crosley in payment. Stripling, possibly feeling that

Hiss might carry the day by producing a series of voluntary recollections, brought him up short:

> I certainly gathered the impression when Mr. Chambers walked in this room and you walked over and examined him and asked him to open his mouth, that you were basing your identification purely on what his upper teeth might have looked like. Now here is a person that you knew for several months at least, a man you had had in your home, to whom you had given a car and leased an apartment.

Hiss was not abashed; in his governmental capacity, he had seen hundreds of persons daily. This man denied he had ever used the name Crosley. Hiss reminded Stripling that he had wanted to question Chambers but Mr.Nixon had denied him that opportunity. The committee agreed that Hiss could ask Chambers whatever he wished. A few questions later, Hiss made the dramatic statement "I don't need to ask Mr. Whittaker Chambers any more questions. I am now perfectly prepared to identify him as George Crosley."

Even after that, there was litigious byplay:

> MR. NIXON: *Would you spell that name?*
> MR. HISS: *C-r-o-s-l-e-y.*
> MR. NIXON: *You are sure of one "s"?*
> MR. HISS: *That is my recollection. I have a rather good visual memory, and my recollection of his spelling his name is C-r-o-s-l-e-y. I don't think that would change as much as his appearance.*
> MR. STRIPLING: *You will identify him positively now?*
> MR. HISS: *I will on the basis of what he has just said positively identify him without further questioning as George Crosley.*

After some discussion about the identities of those who might be able to support Hiss's assertion that Chambers was known generally as Crosley in Washington, Hiss made the following statement for the record, and then Chambers was asked to identify Hiss. This seemingly simple procedure was not without complications:

> MR. HISS: *I would like to say that to come here and discover that the ass under the lion's skin is Crosley, I don't know why your committee didn't pursue this careful method of interrogation at an earlier date before all the publicity. You told me yesterday you didn't know he was going to mention my name, although a lot of people now tell me that the press did know it in advance. They were apparently more effective in getting information than the committee itself. That is all I have to say now.*

MR. MCDOWELL: *Well, now, Mr. Hiss, you positively identify . . .*

MR. HISS: *Positively on the basis of his own statement that he was in my apartment at the time when I say he was there. I have no further question at all. If he had lost both eyes and taken his nose off, I would be sure.*

MR. MCDOWELL: *Then, your identification of George Crosley is complete?*

MR. HISS: *Yes, as far as I am concerned, on his own testimony.*

MR. MCDOWELL: *Mr. Chambers, is this the man, Alger Hiss, who was also a member of the Communist Party at whose home you stayed?*

MR. NIXON: *According to your testimony.*

MR. MCDOWELL: *You make the identification positive?*

MR. CHAMBERS: *Positive identification.*

MR. HISS [who began to walk toward Chambers]: *May I say for the record at this point, that I would like to invite Mr. Whittaker Chambers to make these same statements out of the presence of this committee without their being privileged for suit for libel. I challenge you to do it, and I hope you will do it damned quickly. I am not going to touch him [addressing Mr. Russell].*

MR. RUSSELL: *Please sit down, Mr. Hiss.*

MR. HISS: *I will sit down when the chairman asks me, Mr. Russell, when the chairman asks me to sit down.*

MR. RUSSELL: *I want no disturbance.*

MR. HISS: *I don't . . .*

MR. MCDOWELL: *Sit down, please.*

MR. HISS: *You know who started this.*

MR. MCDOWELL: *We will suspend testimony here for a minute or two, until I return.*

As the hearing began again, Stripling then asked Hiss whether he was fully aware that "the public was led to believe that you had never seen, heard, or laid eyes upon an individual who is this individual . . . and now you know him?" Hiss carefully stated that he had never laid eyes on "Mr. Whittaker Chambers" and accused Stripling of "stating your impression of public impression." The committee was now angry, and there were a number of small quarrels with the witness. When Hiss protested the sudden change in date of the hearing, Chairman Thomas, who had just arrived, asked Hiss if he really didn't expect the date to be advanced when "you built up this Mr. Crosley." Hiss hotly denied this expectation and then, questioned by Stripling about his association with Chambers and other alleged conspirators, responded with a series of sharp "I did not"s.

There was a short intermission, and the committee sequestered itself in the bedroom for a private meeting. Upon their return, there was confused discussion about a public confrontation planned for August 25 between Chambers and Hiss which apparently had been decided on by the committee in private. The question about the necessity of serving a subpoena on both witnesses brought an angry outburst from Hiss, who wanted the record to indicate that he had appeared voluntarily and would continue to do so. As the meeting was ending without clear determination about Mrs. Hiss, who had been asked to come down from Vermont to testify, Hiss wanted clarification about committee plans. This final exchange reflected the full flavor of the deteriorated relations between Hiss and the committee:

MR. HISS: *Does the committee still desire to hear Mrs. Hiss in executive session or have you changed your mind?*

THE CHAIRMAN: *There is no decision on that.*

MR. HISS: *Yes, there was a decision. I have asked her to start down from Vermont.*

THE CHAIRMAN: *Well, you asked her to start down from Vermont.*

MR. HISS: *At your request.*

THE CHAIRMAN: *Believing that she would appear on what date?*

MR. HISS: *As early as possible was the request you made of me, considering her own convenience and whether she could get somebody to stay with our child.*

MR. CHAIRMAN: *Is she on the way from Vermont?*

MR. HISS: *I hope she is on her way by now.*

MR. CHAIRMAN: *If she is on her way now, I think the subcommittee would be glad to hear her.* [*They agreed to see her in New York, in the same hotel room.*]

MR. HISS: *What would be the most convenient hour for you?*

MR. NIXON: *Ten o'clock in the morning.*

MR. MCDOWELL: *If she is on her way.*

MR. HISS: *I cannot be sure she is on her way.*

MR. NIXON: *If you could tell us she is going to be here, we would be willing to stay over.*

MR. HISS: *I cannot guarantee it.*

THE CHAIRMAN: *Can she be in Washington on Monday night?*

MR. HISS: *God, she just made arrangements, if she succeeded at all, to get somebody to stay with the kid two or three nights.*

THE CHAIRMAN: *You don't know whether she has made arrangements or not.*

MR. HISS: *I believe so.*

THE CHAIRMAN: *You don't know; you just believe so.*

MR. NIXON: *I will stay over tonight. There is no objection to this. Just let us know. I don't want to stay a week.*

MR. HISS: *I don't want her to stay a week. Where can I reach you tonight?*

MR. NIXON: *You can reach me at this hotel; and if you will simply let me know if she will be here any time tomorrow, I am perfectly willing to be here.*

MR. HISS: *Vermont trains are unpredictable. May I ask if she is privileged to have anybody with her.*

MR. NIXON: *Absolutely.*

MR. HISS: *May I come with her?*

MR. MCDOWELL: *Yes.*

MR. HISS: *Thank you. Am I dismissed? Is the proceeding over?*

THE CHAIRMAN: *Any more questions to ask of Mr. Hiss?*

MR. NIXON: *I have nothing.*

THE CHAIRMAN: *That is all. Thank you very much.*

MR. HISS: *I don't reciprocate.*

THE CHAIRMAN: *Italicize that in the record.*

MR. HISS: *I wish you would.*

And so the sad tableau ended. Hiss was furious, and Alistair Cooke portrays him as "stewing with impatience" and at the end "with all his defenses down."

Eyes were of much importance in the confrontation. Nixon faulted Hiss's credibility because the latter "did not once turn around and look at his accuser—the man he was so anxious to see 'in the flesh'" as Chambers walked behind Hiss's chair on entering the room. Chambers noticed the same thing. "When I entered the room, Alger Hiss did not turn and look at me. When I sat down, he did not glance at me." Hiss criticized Chambers for not meeting his eyes when his dentures were examined, saying, "Chambers . . . stared fixed before him or up to the ceiling."

Chambers recalls being shocked at the sight of Hiss, feeling a strong sense of pity for the "trapped man." He saw himself as "a killer only by necessity" and prayed that Hiss would here admit the truth so that Chambers would not be compelled to testify "to worse about him and others." He saw that Hiss was in a bad mood, really a "conspicuously bad humor," and Chambers felt that Hiss's colleague, Dollard (there at Hiss's request to verify the former's version of what went on), played the role of the "friend to the Duke in a Shakespeare play" as if the two of them were "native to another atmosphere" and were condescending to the summons of the "earthlings" in the room.

In *Witness,* Chambers has described the confrontation as "great theater," saying:

Not its least horrifying aspect was that it was great theater, too; not only because of its inherent drama, but in part because, I am convinced, Alger

> Hiss was acting from start to finish, never more so than when he pretended to be about to attack me physically. His performance was all but flawless, but what made it shocking, even in its moments of unintended comedy, was the fact that the terrible spur of Hiss' acting was fear.

When Hiss examined his teeth, Chambers writes:

> By then, I felt somewhat like a broken-mouthed sheep whose jaws have been pried open and are being inspected by wary buyers at an auction.

All in all, Chambers felt that Hiss's recital of the biographical facts of their relationship was a fairly accurate representation of the truth, although he resented the transformation of Carl, the Communist agent, into Crosley, the deadbeat, and ascribed this to Hiss's cleverness, contending that "very few people knew anything about Communists, but everyone knew about deadbeats."

Chambers felt that Hiss was finished when the latter advanced on him with fists clenched. When the hearing was over and Hiss and Dollard had left, Chambers wrote that Mr. Stripling turned to him and "completely dead-pan, but in his broadest Texas brogue he drawled, 'Hi-ya, Mistah Crawz-li?'"

Nixon also felt the struggle was all but over at that point. Once again, his account is laced with strangely simplistic idioms for such a portentous occasion. Hiss "was on the defensive"; he was "fighting every inch of the way"; Nixon could "hardly keep a straight face" when Hiss examined Chambers' dentures; Hiss "fought like a caged animal"; there was "a look of cold hatred" in Hiss's eyes. For Nixon, the crucial hearing brought an ambivalent reaction. He wrote:

> I should have been elated. The case was broken. The Committee would be vindicated and I personally would receive credit for the part I had played. We had succeeded in preventing injustice being done to a truthful man and were now on the way to bringing an untruthful man to justice. Politically, we would now be able to give the lie to Truman's contemptuous dismissal of our hearings as a "red herring."
>
> However, I experienced a sense of letdown which is difficult to describe or even to understand. I had carried great responsibility in the two weeks since August 3, and the battle had been a hard one. Now I began to feel the fatigue of which I had not been aware while the crisis was at its peak. There was also a sense of shock and sadness that a man like Hiss could have fallen so low. I imagined myself in his place and wondered how he would feel when his family

> and friends learned the true story of his involvement with Chambers and the Communist conspiracy. It is not a pleasant picture to see a whole brilliant career destroyed before your eyes. I realized that Hiss stood before us completely unmasked—our hearing had saved one life, but had ruined another.

Still, the Nixon forces had come off well. Nixon, who was moved by the fact that at one of the previous hearings "Chambers' voice broke and there was a pause of at least 15 to 20 seconds during which he attempted to get hold of his emotions before he could proceed," now understandably regarded Hiss as "much too smooth . . . much too careful . . . I felt he put on a show." Nixon felt it necessary to correct the impression that he had lost his temper with Hiss about the swearing in of witnesses, saying, "Hiss actually interrupted me as I made the suggestion and of course his manner and tone were insulting in the extreme." If he were Chambers, wrote Nixon, he would have bitten Hiss's finger when the latter pointed it at Chambers, and, like Chambers, Nixon felt that Hiss spoke in a loud and dramatic voice as if "he were acting in a Shakespeare play."

When asked recently to comment on the strange proceedings in the Hotel Commodore, Hiss was quite defensive about his handling of the situation. He pointed out that Chambers had gained forty pounds since he was last seen, that his teeth were entirely reconstructed, and that, moreover, there had been a change in the sound of Chambers' voice which was very deceptive. The old Chambers with his bad teeth spoke in a kind of "paw, paw, paw," walruslike fashion, while the new version had a higher, somewhat nasal quality.

Hiss was also asked why he was so terribly cautious about identifying Chambers prior to the face-to-face confrontation at the Hotel Commodore. It was an act that cost Hiss heavily, at least in the court of public opinion. "After all," Hiss replied, "I am a lawyer and lawyers are by temperament given to caution. I was not willing to be stampeded in the face of my continued requests that I have an opportunity to meet this man face to face."

When the authors asked Hiss what he was thinking about when he rushed toward Chambers with clenched fists, he said, "Literally I wanted to open his mouth to get a look at his teeth." When asked, "Were you going to hit Chambers?" Hiss said, "No, not at all. I'm not a violent man."

As Hiss stalked toward Chambers, he was stopped by an aide, Louis

Russell. Hiss said that this was merely "a stage show." "He soon took his hands off me as I demanded. So if I wanted to hit Chambers, there was nothing to prevent it."*

Chambers spent the evening after the confrontation having dinner with his former employer, Henry Luce, editor-in-chief of *Time, Life,* and *Fortune,* who spoke to Chambers of the parable of St. John, the young man born blind. Upon restoration of his sight, the Lord said, "Neither this man sinned nor his parents, but that the work of God should be manifest in him." Chambers suddenly saw that Luce was saying to him: "You are the young man born blind. All you had to offer God was your blindness that through the action of your recovered sight, His work might be made manifest." Chambers was much buoyed by this connection and later wrote that "in the depths of the Hiss case, in grief, weakness and despair, the words that Luce had repeated to me come back to strengthen me."

Even at the end of the long day, Nixon and Hiss were not yet finished with each other. Hiss had been instructed to call Nixon on Mrs. Hiss's arrival in New York. He phoned Nixon at his room at the Commodore to settle the time for the morning's hearing, but Nixon's phone was continuously busy.

Dispensing with dinner that night in order to be certain of making the morning edition of the newspapers, Nixon spent hours on the telephone leaking stories to the press. There for the world to see the next morning would be Nixon's version of the confrontation, as well as Chambers' anecdotal (and supposedly confidential) testimony about Hiss from the previous session.

Not wishing to spend the night of his triumph alone in a strange hotel room—according to Weinstein—Nixon asked Donald Appell, a committee investigator, to stay with him. As Nixon placed a call to his friend Bert Andrews of the *Herald Tribune,* Appell fell asleep. He awoke much later to find Nixon still talking on the phone, looked at his watch, and "realized that the call [to Andrews] had already lasted for over three hours."

---

*Professor Weinstein discovered the following memorandum in the Hiss defense files, dated January 21, 1949. To the Hiss attorneys, Charles Dollard described the confrontation at the Commodore Hotel: " . . . Alger behaved very badly, was very irritable. He [Dollard] could not tell whether Alger really recognized Chambers before he admitted it or not. Dollard thought that both McDowell and Nixon were trying to be fair. He also thought that they would not have called Priscilla if Alger had not practically insisted on it."

Although Hiss did not get to speak to Nixon until much later, he was called by the Washington press corps, who told him that Nixon had "been on the phone all evening giving the press the committee's story of the hearings." Feeling outraged, Hiss summoned a press conference of his own which was held at his apartment on East Eighth Street in Manhattan. There, Hiss said that despite his brief acquaintanceship with Crosley/Chambers, it in no way changed his complete denial of the charges. "I do not believe in Communism," he said. "I believe it is a menace to the United States." Nixon, however, had captured the headlines, and Hiss came to recognize that his efforts to tell his version were not very effective.

It was well after midnight when Hiss and Nixon finally connected, and the hearing was set for 11:00 o'clock that morning. Nixon would be alone, and Hiss could accompany his wife and could also bring his friend Dollard. As the hearing began on August 18, Mrs. Hiss preferred to "affirm" rather than "swear" to the oath. Did she at any time, Mr. Nixon asked, between the years 1934 and 1937 know a man going by the name of George Crosley? She did. How did she first become acquainted with him? Well, it had been a business relationship with her husband; she didn't think she could pretend to an acquaintance with him. She didn't have the "vaguest" recollection where and when she met him. She faintly remembered "this man and his wife looking at the apartment which we sublet to them" and distinctly remembered their spending two or three days in the Hiss house before taking over the apartment lease. And what sort of a man was this Crosley? She had "a very dim impression of a small man, very smiling person—a little too smiley, perhaps." She didn't recall ever taking a trip with him or when she last saw him. Her only impression was of having been "a little put out." "About what?" Mr. Nixon asked. "Well, I think the polite word for it is probably I think he was a sponger."

That was all Mr. Nixon wanted. Hiss thanked him for his courteous treatment of his wife. Mrs. Hiss, at any rate, had had an easy time—no rankling, no badgering of her past. "It all," she remarked at the conclusion, "seems very long ago and vague."

Nixon came to see the morning as a failure, a missed opportunity to sew up the case. His reaction to some of their exchanges provides an interesting insight to his suspicious character:

> She played her part with superb skill. When I asked her to take the oath to tell the truth, she inquired demurely if she could "affirm" rather than "swear."

Subtly, she was reminding me of our common Quaker background. . . .

Offering her vague impression of Crosley, she said, "I think the polite word for it is probably I think he was a sponger. I don't know whether you have ever had guests, unwelcome guests, guests that weren't guests, you know."

She succeeded completely in convincing me that she was nervous and frightened, and I did not press her further. I should have remembered that Chambers had described her as, if anything, a more fanatical Communist than Hiss. I could have made a devastating record had I also remembered that even a woman who happens to be a Quaker and then turns to Communism must be a Communist first and a Quaker second. But I dropped the ball and was responsible for not exploiting what could have been a second break-through in the case.

Hiss's view was somewhat less sanguine: "The hearing lasted but ten minutes—yet it had taken my wife a full day on a local train to comply with the committee's request that she come to New York immediately."

CHAPTER 4

One personal word. My action in being kind to Crosley years ago was one of humaneness, with results which surely some members of the committee have experienced. You do a favor for a man, he comes for another, he gets a third favor from you. When you finally realize he is an inveterate repeater, you get rid of him. If your loss is only a loss of time and money, you are lucky. You may find yourself calumniated in a degree depending on whether the man is unbalanced or worse. —ALGER HISS

The next week proved to be a very busy one for Congressman Nixon. Between the private confrontation on August 17 and the public session on August 25, 1948, Nixon put in what he described as the longest hours of his life. He tried to anticipate how Hiss would "explain the mass of contradictions in his story" and sought ways "to plug up each and every loophole with documentary proof." Nixon was a man possessed, a prizefighter peaking for a title bout. As the day of the public session approached, he stepped up his workload even further, putting in "eighteen to twenty hours" at the office, for "the more I worked the sharper and quicker my mental reactions became," he wrote.

The attitude taken by his legislative colleagues had changed as a consequence of the August 17 confrontation, and according to Nixon, from "being way behind, now we were about even." During the week the committee staff was assigned two major jobs: (1) to track down anyone who might have known Chambers as George Crosley and (2) to find critical evidence bearing on the story of the transfer of a 1929 car from Hiss to Chambers.

While neither the committee investigators nor even Hiss himself (it later became clear) could find confirmatory evidence of the Crosley identity, there were indeed exciting developments relating to the Ford roadster. A motor vehicle transfer certificate dated July 23, 1936, and signed by Hiss was found which conveyed the car to a William Rosen rather than to Chambers. The FBI identified Rosen as a member of the Communist Party. A check of the apartment leases signed by the Hisses also brought out the fact that the apartment which reportedly provided the locale for the "loan" or "gift" of the car to Chambers had been vacated almost a year prior to when the Hiss car was conveyed to Rosen.

In Nixon's words, he had finally "hit pay dirt." He noted these anomalies in Hiss's story: Why, if Hiss had "given" or "sold" an automobile to "Crosley," did that name not appear on the title? Why, if Chambers had reneged on the rent agreement and they had parted company with "strong words," as Hiss described it, did Hiss give the car to him a year *after* the friendship supposedly ended?

Despite the above evidence, which confirmed his faith in Chambers, Nixon underwent a strange physiological reaction following its appearance. By his own testimony, he was "mean to live with at home and quick-tempered with members of my staff." He ate and slept poorly and friends warned him that he "looked like hell." Nixon himself was more philosophical:

> I suppose some might say that I was "nervous," because I knew these were simply the evidences of preparing for battle. There is, of course, a fine line to be observed. One must always be keyed up for battle but he must not be jittery. He is jittery only when he worries about the natural symptoms of stress. He is keyed up when he recognizes those symptoms for what they

are—the physical evidences that the mind, emotions, and body are ready for action.

The night before the hearing began, Nixon took a sleeping pill for the first time in his life, slept twelve hours, and awoke refreshed, "ready for the most important test that I had had up until that time."

Hiss was having a different experience. He had become convinced that the August 25 confrontation would find him in an adversary relationship with the committee and had sought help from John Davis, a Washington lawyer, whose office staff attempted to follow much the same path as that traced by committee investigators. They, too, went to real-estate agencies, the motor vehicle bureau, and in search of anyone who might have known George Crosley.

The latter effort was notably unsuccessful. The secretary of the Nye Committee was away on vacation, his receptionist had died, the resident manager of the apartment subleased by Hiss to Chambers couldn't recall the events which had transpired over a decade ago; only one person, a man named Samuel Roth, volunteered that he had published an erotic poem written by Chambers, who had used the pen name of George Crosley. For reasons that are not entirely clear, Hiss chose not to mention this incident, and Roth's name appears nowhere in the testimony despite Hiss's comment in his own book:

> Roth had been convicted under the obscenity laws—a fact which supports the credibility of his information, for he would hardly have dared to invite further trouble for himself by making a false statement in a highly publicized case.

The rest of the preparatory effort was equally unsuccessful. Davis's people found that the House committee was everywhere it turned, and most importantly, the soon-to-be-crucial registration certificate was no longer available for scrutiny. Still, there were some clues even in these facts, and Hiss concluded that the impounding of the public records signaled that the committee had found his memory was faulty regarding events that transpired more than a dozen years before.

Hiss considered the possibility of insisting on viewing the official records before testifying again but rejected this tack because " . . . the sequence of minor personal events in my life was in itself unimportant." This decision was to prove extremely important, perhaps pivotal, and it can be said the die of the Hiss case was thereby cast.

In Hiss's words:

This impounding of public records suggested to me that the Committee had found my memory to be inaccurate in some of the details about the Ford and would probably use this material to my disadvantage in the forthcoming hearing. I now consider that before testifying again, I should perhaps have insisted on seeing the official records, which were, as a matter of course, available to me by law. Instead, knowing that the sequence of minor personal events in my life was in itself unimportant, I decided that I would continue to rely on my unaided memory, but would not permit the Committee to lead me into committing myself to specific dates or to a particular sequence of these long-past occurrences.

Although Davis was of much help to Hiss in preparing for the August 25 hearing, the counsel became a target for committee acrimony early in the hearing itself. Davis tried to get chairman J. Parnell Thomas to clarify which pictures he was referring to in an interrogation of Hiss and was admonished sharply by the chairman, "Never mind, you keep quiet!" From that point on, he largely did. Davis and Hiss had agreed, however, to the strategy of an opening statement in which both placed an inordinate amount of faith. It clearly represents Hiss's perception of his past difficulties with the House committee as well as his correct anticipation of the future. Hiss read:

This charge [that I was or had been a Communist] goes beyond the personal. Attempts will be made to use it, and the resulting publicity, to discredit recent great achievements of this country in which I was privileged to participate.

Certain members of your committee have already demonstrated that this use of your hearings and the ensuing publicity is not a mere possibility, it is a reality. Your acting chairman, Mr. Mundt, himself, was trigger quick to cast such discredit. Before I had a chance to testify, even before the press had a chance to reach me for comment, before you had sought one single fact to support the charge made by a self-confessed liar, spy, and traitor, your acting chairman pronounced judgment that I am guilty as charged by stating that the country should beware of the peace work with which I have been connected. . . .

One personal word. My action in being kind to Crosley years ago was one of humaneness, with results which surely some members of the committee have experienced. You do a favor for a man, he comes for another, he gets a third favor from you. When you finally realize he is an inveterate repeater, you get rid of him. If your loss is only a loss of time and money, you are lucky. You may find yourself calumniated in a degree depending on whether the man is unbalanced or worse.

The hearing unfortunately began in a carnivallike atmosphere. It took place in the caucus room of the old House Office Building on a very hot day, and television cameras, a new phenomenon on the American political scene, were to be found in profusion. Their lights intensified the already sticky atmosphere, and crowds had collected several hours in advance of the hearing time. About 1,000 spectators were admitted, and another 300 were held outside by the Capitol police. Chambers came alone, and his description of his arrival is illuminating:

> My approach to the hearing was distressing. The plaza in front of the Capitol was alive with people. A newsboy was holding up a paper with a banner headline in mock *Time* style: C (FOR CONFRONTATION) DAY! I repeated to myself the words that Luce had repeated to me a few days before: "Neither this man's nor his parents' is the sin, but that the works of God might be made manifest."
>
> As I took my place, a cold wave of aversion and hostility swelled out toward me from the spectators. When I stood up, a few minutes later, to identify Alger Hiss, there were titters among them. "Good people," I thought, "it is yourselves you are laughing at. For I stand here not for myself, but for you."
>
> A little girl about seven years old was sitting next to me. She had come with her father, mother and a woman whom I took to be her grandmother. Evidently, they had come to spend the day, for they had brought their lunch in a pasteboard box. When I sat down, the father got up and changed places with his daughter so that the child would not have to sit beside me.

The hearing began with a dismaying byplay that was predictive of the lines that were rapidly being drawn. Hiss, feeling that he had entered into a steel trap that was likely to close shut that day, held to his previous determination to avoid categorical statements. The committee, largely convinced by this time that Hiss had been lying to them, was anxious to make him provide direct answers to its questions. The committee counsel began by asking Hiss if he were there in response to a subpoena. Yes, he had received a subpoena. But he was here in response to it, was he not? Hiss's reply to this opening gambit conveys his wariness. "To the extent that my coming here was quite voluntary after receiving the subpoena 'is in response to it,' I would accept that statement."

Hiss and Chambers were then asked to stand up. Did Hiss know this man? Hiss replied, "I identify him as George Crosley." When did he know him as Crosley? "According to my best recollection, and I have not had an opportunity to consult records, I first knew him some time in the winter of 1934 or 1935." When did he last see him? "Prefacing my answer with the same remarks, I would think some time in 1935."

Then it was Chambers' turn. Unbothered by the ambiguities which so sorely troubled Hiss both before and during the harrowing day, he answered the questions in a markedly unconcerned fashion. Did he know this man? He did. Who was he? Alger Hiss. When did he first meet him? He thought about 1934. When did he last see him? About 1938. Chambers was asked to return to his seat.

There followed a lengthy and often vitriolic exchange which developed over questions of Hiss's earlier failure to identify Chambers from newspaper pictures and moved rapidly to committee irritation over Hiss's repeated use of the cautious phrase "according to the best of my recollection, and without being able to check the record." After much circumlocution, Mr. Mundt finally exploded:

> I would just like to register a protest at this continuous evasion on the part of these witnesses. I am getting tired of flying halfway across the country to get evasive answers. If the gentleman doesn't know who told him, let him say, "I don't know." If he knows, let him say, "I do know." Let's not say, "I believe" or "I think."

Matters for Hiss were not much improved when he reported that he had been unable to locate anyone who knew Chambers as Crosley, and then the committee turned finally to the extraordinarily important issues surrounding the relationship between the 1929 Ford "thrown in" with the sublease of the apartment. Mr. Nixon began to take Mr. Hiss over the variety of places and dates which were relevant to the housing history of the Hiss family.

In keeping with his plan, Hiss's responses often seemed evasive and occasionally patronizing. Again he struck that airy pose, as Chambers so accurately framed it, "as one native to another atmosphere." Hiss had not been able to locate all of his leases. During the critical years of the case, the Hisses had lived at five different addresses. He had been able to contact the real-estate people by telephone, and information about the leases, "which the committee considers of importance, is being compiled but it is not yet ready."

Stripling read from Hiss's testimony from the August 16 meeting about subletting of the 28th Street apartment to Chambers and asked Hiss for confirmation. Hiss said testily, "That was the best recollection I had on the day I testified and that is why I so testified." Since checking the records, however, he would amend this. Apparently the lease began and ended earlier than he had previously described. "The overlap, which I remembered, and which was the main thing in my memory, was, according to the best records I have so far been able to check, accurate,'' said Hiss.

Stripling pursued the question of the other Hiss residences in Washington and in reply to Hiss's repeated statement that ''considering the length of time [that had elapsed], I could not be sure of the exact day or even month,'' Stripling then read into the record statements the committee had earlier obtained from landlords and rental agents giving a month-by-month summary. Hiss, feeling that it was ''plain that in asking me about facts he [Stripling] had before him he was not seeking information from me but errors of recollection, as would a prosecutor bent on discrediting a defendant,'' made the comment: ''May I say it is apparent that the committee has been better staffed with people to inquire into records than I have been. . . .''

Mr. Nixon was now annoyed. He said angrily that the committee was merely following Hiss's lead. ''I understood you to say,'' Nixon retorted, ''that you thought the committee should check the leases. . . . I wanted to be sure I heard you correctly.'' Hiss countered that the move from Washington to New York ''meant a certain contraction of possessions,'' and apparently the leases were a victim of this purge. Hiss tried to turn the focus of the hearing away from what he called ''trivial housekeeping details of fourteen years ago'' to the question of whether he was or had ever been a Communist.

Nixon was not to be moved. The issue was, he said, perjury. Either Chambers or Hiss had lied before the committee. Again Nixon sounded the leitmotif of the hearings. ''It isn't the intention of the committee to hold to exact dates on matters that happened years ago, but it certainly is the intention of the committee to question both Mr. Hiss and Mr. Chambers very closely on the matter of their acquaintanceship, because it is on that issue that the truth or falsity of the statements . . . will stand or fall.''

So saying, Nixon dragged the reluctant Hiss back to the confusing matter of leases. Citing the standard instructions that a judge gives to the jury in determining the credibility of a witness, Nixon explained, '' . . . If in any matter a witness is found to be telling an untruth on any question which is material and which is raised during the course of the court's proceedings, his credibility on other questions is also suspect.''

Hiss, by now thoroughly discouraged, was asked by Stripling to turn his attention to the Ford car. He recalled telling Crosley that the car had no practical value, that he (Hiss) had two cars, and he didn't know what disposition Crosley had made of the car. Mr. Stripling then produced the title certificate showing that Hiss had purchased the second car, a Plymouth, on September 7, 1935. The significance was unmistakable: Since the records produced that day proved that Chambers had been out of the 28th Street apartment by the end of June 1935, the new revelation seemed to suggest the possibility that Hiss had not given Chambers a car as part of the apartment arrangement.

A committee investigator then came to the stand and produced evidence that a year after Hiss reportedly gave the car to Chambers, Hiss had signed the car over to the Cherner Motor Company and that the car was transferred to William Rosen, whose given address proved to be false. Hiss was now recalled to the stand, and in Alistair Cooke's view, "This was a gun barrel of explosive evidence and Mr. Nixon's finger was on the trigger for an hour."

Hiss was once again an elusive target. He did not accept the photocopy as evidence of his signature. Chairman Thomas, in one of his few moments of levity, inquired, "Well, if that were the original, would it look any more like your signature?" The committee then produced an official in the State Department who had notarized Hiss's signature. Hiss knew the man and acknowledged that he must have signed the bill of sale. Mr. Hebert asked if Hiss's memory, now refreshed, allowed him to recall the transaction. "No," said Hiss, "I have no present recollection of the disposition of the Ford." Hebert replied, "You are a remarkable and agile young man, Mr. Hiss."

As the sad hearing went on, Hebert enlarged on this judgment:

> I repeat you are a very agile young man and a very clever young man and your conduct on all appearances before this committee has shown that you are very self-possessed and you know what you are doing and you know yourself why you are answering and how you are answering. Now, that is the reason why I am trying to find out exactly where the truth lies. I can't understand and I can't reconcile and resolve the situation that an individual of your intellect and your ability who gives to casual people his apartment, who tosses in an automobile, who doesn't know the laws of liability, who lends money to an individual just casually, is so cautious another time.

The official who notarized the sale of Hiss's Ford to the Cherner Motor Company was attorney W. Marvin Smith. Less than two months after his testimony before HUAC, he took his life by plunging down a five-story circular stairwell at the Justice Department, where he worked.

Hiss feels Smith's suicide had nothing to do with his case. "I'm sure it was for personal reasons," he said. "At the time the papers liked sensationalism, [and] a lot was made of it. He had notarized the sale. He worked a couple doors down from me. I think it had nothing to do with that. Marvin Smith had personal problems."*

The hearing rapidly took on the character of a court trial. Hiss's effort to aboid entrapment inadvertently contributed to this ambience. He used the expression "according to the best of my recollection" almost two hundred times, and spectators found it hard to believe that this was only a Congressional investigation. By the conclusion, Hiss, after six hours of testimony, was angrily demanding that the committee subject Chambers to the same inquisition, that they ask him

> . . . where he now lived, where he had lived from 1930 on, and how long at each place; what was his given name and what names he had used since; what was his complete employment record with the Communist Party, for a bibliography of all his writings, ask him if he had been convicted for any crime. Ask him if he had ever been treated for mental illness, ask him about his marriage and how many children he had and where his wife lived now. Ask him the circumstances under which he came in contact with this committee and to make public all written memoranda which he may have handed to any representative of the committee. I would like to know whether he is willing . . . to make the statements so that I may test his veracity in a suit for slander or libel.

The committee happily did Hiss's bidding, although it did omit the last two questions. Chambers' responses, which consumed another three hours, were summarized in a comparative fashion by Alistair Cooke:

> Chambers was a very different witness. Placidly, directly, he ran through names and places, nodded assent, recited the whole charge with the air of a man sportingly reiterating a list of vital statistics before an insurance company that was sorry it had misplaced them.

---

*There may have been—as Hiss maintained—no connection between Smith's appearance before the committee and his subsequent suicide, but there were strange developments related to the old car which came out later. On the very day (July 23, 1936) Hiss traded the car in at the Cherner Motor Company, it was sold to William Rosen, whom the FBI identified as a Communist. Although there is a state title notarized by Smith, the invoices at the dealership do not list the transaction. There are at least two conclusions which can be drawn: (1) Someone who worked at the dealership was friendly to the Communist Party and, to protect either the seller (Hiss) or the buyer (Rosen) or both, did not enter the transaction in the records, or (2) the title itself is a forgery. These matters are discussed more extensively in Chapter XI.

Hiss, in describing that unhappy day, concluded the committee had "exploited three mistakes" in his recollection involving the dates of a lease and when he acquired one car and disposed of another. Describing himself as having an "average" memory, Hiss wrote, "As these dates were all linked in point of time, one misdating led to another in my memory." His first mistake involved the sublease to Chambers of his apartment on 28th Street. Hiss testified that the Chamberses had stayed there from June to September of 1935, while actually they were there from April to July of that year.

The second mistake involved the purchase of the new Plymouth. He testified it was bought during Chambers' tenancy; actually it occurred about two months later. "The new car," Hiss wrote, "the first my wife and I had jointly acquired, had been the subject of much discussion and shopping tours over a number of months. What had still been only a family plan when Crosley was my tenant became in memory an accomplished event."

His third error was the date of the gift of the old Ford to Crosley/Chambers. Hiss testified he had given it to him during his tenancy (April–July 1935); actually it was not until the end of 1935 or the early spring of 1936. Hiss said, "Anticipating the new car, I had told him at that time that he could have the old one when I should no longer need it. He also borrowed the Ford once or twice that spring. Thus, my association of his tenancy, the acquisition of the new car, and giving him the old one was basically correct although I confused intentions with their fulfillment."

For Richard Nixon, the August 25 hearing came close to being a complete vindication. Writing with an obvious sense of relief, he commented:

> The confrontation was over. The tide of public opinion which had run so high in favor of Hiss just three weeks before had now turned against him. Critics who had condemned the Committee for putting Chambers on the stand now congratulated us for our perseverance in digging out the truth.

Nixon resented Hiss's efforts to make the hearing a political one and felt that linking himself in his opening statement to Presidents Roosevelt and Truman, to Senators, Congressmen, Secretaries of State, and to federal judges was "in the highest degree innocence by association." Nixon was very moved by the contrast between the two witnesses and described Chambers' answers to his question at the end of the long day as "one of the most eloquent statements made in the history of the Committee on Un-American Activities." It went like this:

MR. NIXON: *You were very fond of Mr. Hiss?*
MR. CHAMBERS: *Indeed I was; perhaps my closest friend.*
MR. NIXON: *Mr. Hiss was your closest friend?*
MR. CHAMBERS: *Mr. Hiss was certainly the closest friend I ever had in the Communist Party.*
MR. NIXON: *Mr. Chambers, can you search your memory now to see what motive you can have for accusing Mr. Hiss of being a Communist at the present time?*
MR. CHAMBERS: *What motive I can have?*
MR. NIXON: *Yes, do you, I mean, is there any grudge you have against Mr. Hiss over anything he has done to you?*
MR. CHAMBERS: *The story has spread that in testifying against Mr. Hiss, I am working out some old grudge or motives of revenge of hatred. I do not hate Mr. Hiss. We were close friends, but we are caught in a tragedy of history. Mr. Hiss represents the concealed enemy against which we are all fighting, and I am fighting. I have testified against him with remorse and pity, but in a moment of history in which this nation now stands, so help me God, I could not do otherwise.*

Chambers was exhausted by this concluding remark and reported that he was at the point of an emotional breakdown when Nixon asked that last fateful question. When Chambers finally left the hearing room with a friend, it was well after eight o'clock in the evening, and he heard footsteps behind him. A young man ran up. Chambers recognized him as being Jewish and likely not over seventeen years of age. He stopped the *Time* editor and blurted, "I want to thank you. That part about the tragedy of history—you don't know what it means to young people like me."

Hiss was visibly angered by Congressman Mundt's summary statement which concluded the August 25 hearing. The Congressman, undertaking what was certainly a difficult task, chose to present a political indictment of Hiss, accusing him of malfeasance in creating U.S. policy toward Nationalist China, the Yalta agreement, and the Morgenthau plan and concluded with the spiteful statement, "[All three policies] . . . are hopelessly bad and I shall consider them hopelessly bad even though you prove yourself to be president of the American Daughters of the Revolution." Mundt then delivered a final broadside by alluding to the fact that the State Department, which had admitted to having "134 Communists in its employ" [no such admission was ever made], now had one more, saying, "So it is not inconceivable that the number could just as well have been 135 as 134." In a masterpiece of understatement, Hiss writes, "I did not feel that the committee's attitude toward me [at that time] was either fair or impartial."

Chambers' reaction the next day was a very complicated one. He had been reluctant to testify prior to the hearing:

> Another public hearing, especially another confrontation with Hiss in front of hundreds of strange people, was more than I could endure. . . . Another great circus under eager, avid eyes, under batteries of news cameras and the hard stares of a prevailing hostile press was too much.

When he told Nixon of his feeling, Nixon replied:

> It is for your own sake that the Committee is holding a public hearing. The Department of Justice is all set to move in on you in order to save Hiss. They are planning to indict you at once. The only way to head them off is to let the public judge for itself which one of you is telling the truth.

Surprisingly, Chambers came later to doubt Nixon's assertions about the Department of Justice ("Regardless of whether or not Richard Nixon was mistaken—and there are many who claim that he was—I believed him, and his words weighed heavily upon me."), but he did testify, and the sympathetic attitude of the House committee toward him in contrast to their antipathy to Hiss was commented on by James Reston of the New York *Times* on the day following the hearing.

The hearing did mark a change in Chambers' self-perception:

> I began to grasp the degree to which I was not merely a man testifying against something. I was first of all a witness for something. The turn the struggle had taken made it clear that what most of the world supposed it to be—a struggle between the force of two irreconcilable faiths—Communism and Christianity—embodied in two men, who by a common experience in the past, knew as few others could know what the struggle was about, and who shared a common force of character, the force which had made each of them a Communist in the first place, and which I had not changed when I changed my faith.

Chambers finally found an identity with which he was comfortable; the man with the multiple names and many careers became for one and all "a witness" and the conversion had, as do most conversions, a mystical character.

CHAPTER 5

Any SOB that gives Congressman Mundt any information gets his ass kicked out of this building. . . . I want you to get the word around that anyone giving information to the Committee is out, O-U-T!

—ATTORNEY GENERAL TOM C. CLARK'S instructions to the FBI

The next 100 days were described by Chambers as the most important of his life. He saw himself as "one single man, seeking to do what was right to the limit of his understanding and strength, upon whom beat a surf of pressures, rumors, warnings, so that to keep my footing in the inevitable backwash of doubts and fears was a daily feat." His feelings for Hiss changed promptly to hatred. Hiss was not alone: he was clearly being supported by the Communist Party, which was using him to pass its threatening message to Chambers: "This is the showdown. Stop testifying. Stop it or we will destroy you." It was the fight against what

Chambers termed "that rally of force" (Hiss and the Communist Party) that led him to make use of his secret weapon.

On August 28, the House committee released an interim report on the Hiss–Chambers matter. The balance had now clearly shifted. Hiss was described as one who had "changed his position on the car and testified in a manner which to the Committee seemed vague and evasive"; that this "evasive testimony . . . raises a doubt as to other portions of his testimony . . . while Chambers, on the other hand, was for the most part forthright and emphatic in his answers"; that "the confrontation of the two men and the attendant testimony from both witnesses has definitely shifted the burden of proof from Chambers to Hiss, in the opinion of this Committee. Up to now, the verifiable portions of Chambers' testimony have stood up strongly; the verifiable portions of the Hisses' testimony have been badly shaken and are primarily refuted by the testimony of Hiss versus Hiss."

The next event in the tangled skein was Chambers' public statement on Lawrence Spivak's radio program "Meet the Press": "Alger Hiss was a Communist and may still be one." Despite previous proclamations, Hiss, however, was not so quick to file a libel suit against Chambers. A week after Chambers' "Meet the Press" appearance, the Washington *Post,* which up to this point had been very critical of the committee, editorially asked Hiss about the delay. Said the *Post,* "Mr. Hiss himself has created a situation in which he is obliged to put up or shut up. . . . Each day of delay in making it known that he will avail himself of the opportunity Mr. Chambers has afforded him does incalculable damage to his reputation." The New York *Daily News,* never very civilized, inquired, "Well, Alger, where's that suit?"

Shortly thereafter, Hiss filed for damages of $50,000. When an Associated Press reporter asked Chambers for comment, Chambers said, "I welcome Alger Hiss' daring suit. I do not minimize the ferocity or the ingenuity of the forces [the Communist Party] that are working through him. But I do not believe that, ultimately, Alger Hiss, or anybody else, can use the means of justice to defeat the ends of justice." Because of this statement, the Hiss attorneys increased the damage claim to $75,000.

Nixon now felt the committee's work was at an end and turned to the political campaign of 1948. Although his own re-election was assured when he won both the Californian Democratic and Republican parties' nomination in the primaries after having destroyed the liberal opposition in 1946 in his heavily controverted campaign against Jerry Voorhis,

Nixon devoted himself to helping less fortunate colleagues, and the Hiss case was a favorite topic when he spoke.*

Nixon found himself disappointed by the failure of the Dewey–Warren ticket to make more of the Communist issue and later referred to the Dewey team as a bunch of amateurs when they were managing the Eisenhower campaign in 1952. It is striking to discover that in this bleak moment of Republican defeat occasioned by Truman's re-election, Nixon chose to visit "the witness":

> The result on Election Day, 1948, was an unpleasant surprise for me and all Republicans. It really jolted Whittaker Chambers. A few days after the election I stopped to see him and his wife in Westminster on my way to York County, Pennsylvania, for a visit with my parents. He was in a mood of deep depression. He was not concerned with the election results from a partisan standpoint—he had never mentioned his partisan affiliation with me in our many discussions. What worried him was that the whole investigation of Communist infiltration in the United States and particularly in our government might be allowed to die.

Chambers was deeply troubled by Hiss's libel suit, and after the first of two depositions given to Hiss's attorney he began to wonder whether he might lose. He wrote: "The sum of $75,000 was fantastic as compared with the ability I had to pay it." Chambers had been pressed at the second deposition to produce letters or papers from the Hiss family which would support the theory of close association. On the next day, William Marbury, serving as Hiss's attorney, said, "Yesterday, at the close of the hearing, I asked you if you would produce any papers or notes or correspondence from any member of the Hiss family this morning. Have you got any such papers with you, Mr. Chambers?"

Chambers answered shortly, "No, I do not," and his attorney offered the following information: " . . . Mr. Chambers has advised us that he has not explored all of the sources where some conceivable data might be." As a consequence of this pressure, Chambers wrote that his attorney warned him "that if I did have anything of Hiss, I'd better get it."

While being interrogated by Marbury, Chambers said he was overcome by a deep lethargy. He had indeed hidden away some memoranda from Hiss but felt they were of little consequence. Nonetheless, he was

---

*Under California's cross-filing system, then in effect, primary candidates were usually listed in *both* the Democratic and Republican columns of the ballot, despite their affiliation, and voters could vote for either candidate.

aware from the two-day deposition that "the Hiss forces had turned the tables with the libel suit. The issue had ceased almost completely to be whether Alger Hiss had been a Communist. The whole strategy of the Hiss defense consisted in making Chambers a defendant in a trial of his past, real or imaginary, which was already being conducted as a public trial in the press and on the radio."

While the libel suit was pending, another Hiss attorney, Edward McLean, held a pretrial interview with Chambers at his farm in Westminster. McLean wanted to see certain items which Chambers claimed Hiss had given him as gifts. Responding to McLean's request, Chambers showed him a disheveled love seat, which, Hiss later said with some humor, "may or may not have some Freudian significance." It was part of the furniture Hiss had left in the 28th Street apartment he sublet to Chambers. Another "gift" was a book, which, commented Hiss, "I think he had swiped. I do not remember giving him that book." Abruptly, Chambers left the room. He returned with a small folded piece of cloth. With some ceremony, he unfolded it and said to the attorney that this used to be part of a slipcover belonging to Alger Hiss.

"My hair stood up," McLean told Hiss. He knew it was strange behavior but was not sure precisely what it signified. Hiss, more sophisticated in the world of psychiatry, said, "Ed [McLean] had never heard of fetishes and didn't have any idea what it meant." Hiss said McLean was "absolutely antipathetic to all psychiatry." Incidentally, it was McLean who later counseled wisely but unsuccessfully against calling a psychiatrist as a witness for the defense.

The specific nature of the relationship between Hiss and Chambers occupied a certain amount of public attention at that time. Chambers posed it this way at the HUAC confrontation: "The story has spread that in testifying against Mr. Hiss, I am working out some old grudge or motives of revenge or hatred." A story soon emerged that explained the "grudge" or need for "revenge." It was that Chambers was a homosexual and was aggrieved because his advances toward Hiss were rejected, or were not, but in any event the relationship was broken off by Hiss.

Hiss said that during the first trial his attorney, Lloyd Paul Stryker, asked him, "Didn't Chambers make a pass at you because it is so clear that this man is a homosexual?" Stryker felt proof of Chambers' homosexuality would be useful because rejection of a homosexual's advance was a motive a jury could understand. Despite persistent prodding to that effect, Hiss remembered no such advances, and the defense was unable to get any concrete proof of Chambers' homosexuality.

It is now known that Chambers, fearing that the Hiss forces might use presumed homosexuality to discredit his testimony, concluded an interview with the FBI by handing an agent a sealed envelope and requesting that it not be read until after he left. The envelope contained a lengthy confession of his homosexual life. He described this chapter with much embarrassment as "of that character that it should be discussed with a priest." Consequently, he requested that the FBI have one agent rather than two when conducting the expected follow-up interview on the subject. The FBI files say of his confession, "He indicates this so-called 'confession' is made because he expects to tell the truth despite any instinct he has to self-preservation. . . ."

The FBI interviewed Chambers about this matter on February 17, 1949. Chambers said, "In 1933 or 4, a young fellow stopped me on the street in New York and asked me if I could give him a meal and lodgings for the night. I fed him and he told me about his life as a miner's son. I was footloose, so I took him to a hotel to spend the night. During the course of our stay at the hotel that night, I had my first homosexual experience. . . . It was a revelation to me. As a matter of fact, it set off a chain reaction in me which was almost impossible to control. Because it had been repressed so long, it was all the more violent when once set free. I engaged in numerous homosexual activities, both in New York and in Washington, D.C. At first I would engage in these activities whenever by accident the opportunity presented itself. However, after a while, the desire became greater and I actively sought out the opportunities for homosexual relationships. . . . I generally went to parks or other parts of town where these people were likely to be found. I am positive that no man with whom I had these relations . . . ever knew my true identity. . . . The possibility exists, however, that since my photograph has appeared in the public press . . . someone with whom I had previously had homosexual relations might remember me."

Chambers told the Bureau that he kept his "secret" from his associates in the Communist Party, and, he said, "I emphatically state that at no time did I ever so much as hint about this to Alger Hiss." Chambers stated that he ceased his homosexual activities at the time of his break from the Communist Party. "I do not believe," he said, "that the cessation of my homosexual activities and my break from the Communist Party are in any way connected. . . . However, both of these activities on my part were more or less simultaneous with the advent of religion and God in my life."

Still, Hiss had failed to recall any sexual advances on the part of

Chambers. Recently when Hiss was asked if he would have been sensitive to such advances in those less sexually liberated times, Hiss answered affirmatively. He felt that he would because he had such an experience when he was eleven or twelve during the First World War. It was with an army officer, a friend of one of his older sister's suitors. Hiss's mother, as part of her patriotic duty, was always happy to entertain soldiers in the family home in Baltimore. This captain stayed overnight after taking the young Hiss to see the Washington Senators play baseball. They slept in the same room and the captain climbed into the boy's bed. "I knew—while I was asleep at the time—I just knew [what he wanted]," said Hiss of the experience. Hiss said he understood what was happening to him. He described himself as a savvy twelve-year-old, "a tough street kid . . . so for me it was something bestial and silly." But he was also frightened.

There was a second meeting with this captain. After another baseball game, Hiss was obliged to spend the night in his quarters at Camp Meade. This time "I decided to be very Machiavellian," Hiss related. "In the middle of the night, he made a pass. I got up and pretended that I had to go to the john. Got dressed and went down the stairs and told the guard on duty. He said, 'That guy always got kids around here. I can't do anything. I'm only a private.'" Hiss concluded, "I think if Chambers had insinuated—I would have known about it. . . ."

During this period Chambers was reluctant to leave his farm for any purpose. However, pressured by his attorney about the Baltimore libel deposition, he finally decided to go to New York and reclaim an envelope from his wife's nephew who had agreed to keep it for him over ten years earlier. Chambers had some vague recollection about what he might have hidden in the envelope, but he was unsure of whether the material would aid in establishing evidence of the "required" close relationship with Hiss. When the envelope was handed over to him, Chambers was puzzled that it was so bulky, and a closer examination revealed many surprises. In addition to some State Department documents, there were three cylinders of undeveloped microfilm and one spool with two strips of developed film. Chambers described himself as stunned:

> I knew that the documents and the film meant much more than any part they might play in the libel suit. They challenged my life itself. They meant that there had been given into my hands the power to prove the existence of the Communist conspiracy. They meant that I must decide once and for all

> whether to destroy that documentary proof and continue to spare those whom I had so far shielded, or to destroy the conspiracy with the means which seemed to have been put into my hands for that reason by the action of a purpose that reached far back into the past to the moment and the impulse that had first led me to secrete the film and papers.

Chambers then made a strange decision. He turned all the documents over to his lawyer, Richard Cleveland, but kept the microfilm for himself. By Chambers' own testimony, Nixon's later subpoena for evidence was unnecessary, for Chambers had withheld them from his own lawyer in order to turn them over to the House committee:

> No act of mine was more effective in forcing into the open the long-smothered Hiss Case than my act in dividing the documentary evidence against Hiss, introducing the copied State Department documents into the pre-trial examination (which, in effect, meant turning them over to the Justice Department), and placing the microfilm, separately, in the pumpkin. It was my decisive act in the Case. For when the second part of the divided evidence, the microfilm, fell into the hands of the Committee, it became impossible ever again to suppress the Hiss Case.

Just prior to the above-mentioned subpoena being served by Stripling on Chambers, Nixon felt the need for a rest. He decided to take a long-delayed vacation and set sail on a ten-day cruise through the Panama Canal Zone on December 2, 1948. As he prepared to leave, he read a dispatch from United Press International in a Washington newspaper which reported that the Justice Department was likely to drop its investigation of the Hiss–Chambers controversy "unless additional evidence is forthcoming."

Nixon immediately called Stripling, and "on a long hunch" Nixon persuaded the committee counsel to drive with him once again to Chambers' home in Maryland. They showed Chambers the newspaper, and he looked out the window a long time, shaking his head. Finally, he said, "This is what I have been afraid of."

After another brief pause, Chambers made a surprising comment:

> In a deposition hearing two weeks ago, I produced some new evidence in the case—documentary evidence. It was so important that Hiss' attorneys and mine called the Justice Department. Alex Campbell, Chief of the Criminal Division, came to Baltimore and took the documents back to Washington. Before he left he warned everybody present to say nothing whatever about tnese documents and that if we did divulge any information, we would be

guilty of contempt of court. So, I can't tell you what was in the documents. I will only say that they were a real bombshell.

Despite the importunities of the two men, Chambers refused to reveal the contents of the documents. Nixon was distressed to think that a Democratic Administration official had the only copies of the material to which Chambers had referred. He was only partially reassured when Chambers indicated that his attorney had photostatic copies but was considerably buoyed by Chambers' next statement that he had deliberately withheld from Campbell some evidence which he called "another bombshell." Nixon cautioned Chambers to keep that material away from the Chief of the Criminal Division of the United States government—"Don't give it to anybody but the Committee"—and on the way back to Washington decided to stop at his office and signed the subpoena requiring Chambers to produce "any and all documents in his possession relating to the Committee hearings. . . ." And so as Nixon was boarding the *Panama* the next morning, Stripling was serving the subpoena on Chambers.

How different the story is in Chambers' book! Not only is Richard Nixon something less than the prime mover in this pivotal scene, he is, in fact, nowhere to be found.* From Whittaker Chambers' vantage point, it happened this way: One night in early December while he was milking his cows, Chambers' work was interrupted by Stripling. On the way back to the house, Stripling was struck by the size of the pumpkins growing in the yard and commented on them. This pleased Chambers, who now divided his time between farming and testifying. But Stripling, who didn't come to Westminster from Washington to discuss autumnal fruits, handed him a clipping from the Washington *Post,* which described "startling developments" in the case but did not reveal what they were.

This was apparently a reference to the documents Chambers had turned over to the Justice Department, of which Stripling was unaware. "I hear that there has been a bombshell in the suit," Stripling said, hoping to find out what had happened. Chambers, legally silenced by the judge in the libel suit, refused to comment and also neglected to tell Stripling he had divided the evidence and still had the microfilm in his

*In his own book, Stripling said Nixon was with him at the farm, and it was the Congressman who asked the bombshell question. This contradiction in recollection will unfortunately never be resolved.

possession. Pumpkins were apparently the only fruitful topic of discussion that day, and the committee's counsel left Westminster no wiser.

As Chambers was to drive to Washington the next day to testify in a loyalty case, he decided the microfilm warranted a better hiding place. He had kept it in his bedroom for two weeks, but "I felt," he wrote dramatically, "[in my absence] the investigators might force my wife to let them ransack the house." Where to hide it? For Chambers, whose life was lived—and later judged—on the torturous road between fact and fiction, the answer to the conundrum lay in the arts.* "Hurriedly, I tried to think of any classic examples of concealment," he wrote. Then he remembered a Russian film where underground Communists hid weapons in pumpkin-shaped figures. This rumination was broken by a phone call from Stripling, who, knowing Chambers was coming to Washington, requested that he stop by his office. Never one to take coincidence lightly, he juxtaposed the "pumpkin-shaped idol" from the movie, Stripling's comments about his pumpkins and his unexpected phone call, and concluded the microfilm must be hidden in a hollowed-out pumpkin.

In Stripling's office in Washington, Chambers was served a subpoena *duces tecum*, ordering him to turn over to the committee any evidence in his possession in any way connected with the case. That evening he led Donald Appell and another HUAC staff member to his pumpkin patch, rummaged about for a moment, located the right pumpkin, removed the top, and handed over the rolls of microfilm to the incredulous investigators, flippantly saying, "I think this is what you are looking for."

There exists a third version of this pivotal scene as well. This is the one offered by the other principal mover, Robert Stripling. Nixon and his chief aid, HUAC investigator Stripling, were to have a long but not always amicable relationship. Over the years their memories began to differ significantly on many important matters dealing with the Hiss case. As can be predicted, the controversy revolved around who was the major figure in keeping the case alive before HUAC. In *Six Crises*, Nixon clearly assigned himself the lion's share of responsibility; he led

---

*Chambers' life was a strange mixture of fact and fiction. Psychiatrist Meyer A. Zeligs wrote in *Friendship and Fratricide*, "Chambers believed his own fiction to be the truth. . . . What he did when he wrote fiction or autobiography was very similar to what he did as a child when he read fiction, especially Dostoevski. . . . These forms of literary mastery and manipulation were among Chambers' ways of preserving his own identity by hiding it inside someone else's." Zeligs saw Chambers as an impostor; and Hiss when interviewed was obviously quite taken with this characterization.

the chase, "smelled" out inconsistencies, rallied the HUAC staff and committee, and led a largely compliant Stripling hither and yon over the Maryland countryside in search of new evidence. Stripling pictures a somewhat different scene in his own book and in interviews with Howard K. Smith, then of CBS News, and Allen Weinstein.

Stripling's version of the major events is recounted below. A visit on December 1, 1948, by Nicholas Vazzara, a lawyer associated with Chambers' defense in the libel suit, brought Nixon and Stripling information that Chambers had turned over a number of confidential documents to the Justice Department. The previously dying case was now about to come alive again, but Nixon was scheduled to leave on his long-delayed ocean vacation. The Congressman asked Vazzara's opinion on whether to delay his trip.

Vazzara had no advice to offer on the matter and soon left. Stripling found Nixon to be irritable and nervous and proposed that they drive out to Maryland that very day and confront Chambers about this withholding of evidence. Nixon exploded, saying angrily, "I'm so goddamned sick and tired of this case. . . . I don't want to hear any more about it and I'm going to Panama. And to hell with it and with you and with the whole business."

Nixon resisted all Stripling's entreaties, even the one which identified the fact that unless they acted immediately, the new Congress about to be seated would likely dismantle the committee. They parted with Nixon still angry and dispirited and Stripling still perplexed at the Congressman's behavior. The investigator decided to risk a final phone call and after enduring another furious explosion was able at last to persuade Nixon to make the fateful drive. After being confronted at his Maryland farm, Chambers volunteered the information that he had kept "some" other documents from his attorney and that these were equally as important as those already surrendered to the Justice Department.

On the way back, Stripling once again found Nixon dispirited and still ruminating on his vacation plans. Nixon said listlessly, "I don't think he's got a damned thing. I'm going right ahead with my [vacation] plans." There was no mention of a subpoena. But when Nixon arrived home that evening, he called Bert Andrews, who in his view, as expressed in his own book, *A Tragedy of History,* then took over the case. Andrews wrote that he was openly critical of Nixon for his lackadaisical handling of the evening's events and literally pressured Nixon into signing a subpoena requiring Chambers to produce all documents relating to the case.

The next morning the subpoena was signed by Nixon and served by Stripling, and then Nixon left on his soon-to-be-interrupted vacation. The evening before, apparently still unable to make up his mind, Nixon encountered William "Fishbait" Miller, the Congressional doorkeeper, in the hall of the Capitol. According to Miller, Nixon said enigmatically, "I'm going to get on a steamship and you will be reading about it. I am going out to sea and they are going to send for me. You will understand when I get back, Fishbait."

Once the case moved into court, Stripling and Nixon had understandably little to do with each other. Meeting during one of the Eisenhower campaign trips in New York City, Nixon recalled the old days: "Strip, those sons of bitches are out to get me, and they'll try to get anybody that had anything to do with the Hiss case."

But despite their early collaboration, or perhaps because of it, they drifted apart actually as well as ideologically. When Howard K. Smith reviewed Nixon's career after his defeat by Pat Brown in the California gubernatorial campaign in 1962, the appearance of Alger Hiss on the program was bitterly protested by Nixon. Stripling assumed a different stance: "As far as I'm concerned, I want everybody who is an American citizen always to have the opportunity to be heard."

While Stripling had little sympathy for Hiss, describing him as a perjurer, he later displayed little sympathy for Nixon as well. Apart from understandable pique at Nixon for exaggerating his own role in the Hiss case, Stripling expressed the belief that Nixon was motivated by a personal antipathy in his pursuit of Hiss. Stripling said in 1975: "Nixon had his hat set for Hiss. It was a personal thing. He was no more concerned about whether or not Hiss was [a Communist] than a billy goat!"

So the picture of Nixon as the cool and collected prime mover in the Hiss affair is open to serious doubt. Stripling's and Andrews' versions present a very different view of the Congressman, one which shows him as cautious, mercurial, and, most importantly, indecisive in the most crucial moments of the HUAC investigation. This is not too far removed from the later Watergate impression of Nixon under fire.

The documents that Chambers had "put by" when he left the Communist Party were in his terms "a life preserver" which "should the Party move against my life, I might have an outside chance of using as a dissuader." Most surprising was the fact that Chambers had forgotten the microfilm since the rest was characterized by him as "two or three scraps of Alger Hiss's handwriting, and perhaps something of Harry

Dexter White.'' The copied State Department documents and the spools of microfilm ''had sunk from my memory as completely as the Russian regiments in World War I sank into the Masurian swamps.''

Hiss's evaluation of the hidden material is instructive:

> These documents are of three categories: four small slips of paper with penciled notes of a kind I made regularly in my office routine; sixty-five typewritten pages which I have consistently denounced as ingenious forgeries; and two strips (totalling fifty-eight frames) of developed microfilm containing photographs of eight State Department documents.

While there was no public notice of the November 17 depositions which concerned the documents Chambers had initially given to his attorney, Richard F. Cleveland, except for the tantalizing reference in the *Post,* the events that ensued as a consequence of Nixon's subpoena were of a different order of magnitude. On December 4, the newspapers published a report that Chambers had passed onto the House committee what he swore were films of secret State Department documents given to him by the spy ring. The microfilm galvanized the sinking House committee, which was developing a peculiar tilt in view of the surprising Republican defeat in the November elections.

The Democrats, in reorganizing the House, were trying finally to lay the controversial committee to rest; Congressman McDowell and Vail, both committee members, were defeated for re-election, and committee chairman J. Parnell Thomas was about to be indicted for systematically padding his Congressional payroll, an offense that led to a prison term just a year later. It is thus understandable that the committee immediately announced that it was posting a twenty-four-hour guard on the ''pumpkin papers,'' and Mr. Mundt cut short his visit back home to repair political fences and flew immediately back to Washington to take charge.

Nixon was away but not out of touch. On his very first day at sea, he had received a wire from his friend, Bert Andrews of the New York *Herald Tribune*, alerting him to the possibility that Chambers was offering new evidence, which came as no surprise to either of them since, after returning from Chambers' farm, Nixon had discussed the day's strange occurrences with Andrews until deep into the night. It was the reporter who had counseled Nixon to serve Chambers with a subpoena. Nixon's response had been, ''I'll think about it.'' That same evening brought another wire, this time from Stripling, which read:

Second bombshell obtained by subpoena 1 A.M. Friday. Case clinched. Information amazing. Heat is on from press and other places. Immediate action appears necessary. Can you possibly get back?

When Nixon read the second message to those seated at the captain's table (a strange move under the circumstances), he reported that "Pat threw her hands up and said, 'Here we go again!'" The denouement was foreordained when Andrews wired again in jocular fashion:

Documents incredibly hot. Stop. Link to Hiss seems certain. Stop. Link to others inevitable. Stop. Results should restore faith in need for Committee if not in some members. Stop. New York Jury meets Wednesday. Stop. Could you arrive Tuesday and get a day's jump on Grand Jury? Stop. If not, holding hearing early Wednesday. Stop. My liberal friends don't love me no more. Stop. Nor you. Stop. But facts are facts and these facts are dynamite. Stop. Hiss' writing identified on three documents. Stop. Not proof he gave to Chambers but highly significant. Stop. Love to Pat. Stop. (signed) Vacation-Wrecker Andrews.

In a mind-boggling bit of political drama, Stripling persuaded Democratic Secretary of Defense James Forrestal to send a PBY to the Caribbean to bring the young Congressman back to Washington. The *Panama*'s captain agreed to pilot the ship to a stretch of water calm enough for the pontoon plane to land, and Nixon was lowered to a lifeboat and rowed over to the Coast Guard plane. Nixon writes tersely this time: "I climbed aboard and was on my way to Miami."

He was not surprised when he was met at Miami by reporters and photographers but began to have doubts about Chambers when the reporters told him of the Pumpkin Papers: "Now I wondered if we really might have a crazy man on our hands." When Stripling met the plane in Washington and told Nixon the full story, Nixon no longer doubted Chambers. "I soon learned that Chambers was crazy—like a fox."

William Reuben described Nixon's return in more ironic fashion in his unflattering portrait, *The Honorable Mr. Nixon:*

On December 5, Nixon, with photographers somehow miraculously able to record every phase of the drama, was picked up at sea from a cruise ship by a Navy crash boat and then rushed to a waiting sea plane and flown back to Washington. His first word on the "pumpkin papers" was issued as he sped back to Washington. With no need to even look at the documents, or to ask Chambers a single question for an explanation as to how their existence had never previously been mentioned or even hinted at, Nixon held his first press

> conference aboard a speeding Navy crash boat. The AP quoted him as saying that, "it is no longer just one man's word against another's"; and that: "The hearing is by far the most important the Committee on Un-American Activities has conducted because of the nature of the evidence and the importance of the people involved. It will prove to the American people once and for all that where you have a Communist, you have an espionage agent."

When Nixon returned from his vacation, he and Stripling worked feverishly one night examining and cataloguing the microfilm and the photostats of the typed documents and handwritten notes that made up the Chambers cache. It was decided that Stripling would check with Eastman Kodak to verify that the microfilm was indeed as old as Chambers had claimed. The return phone call from Rochester brought disquieting news: The emulsion numbers on the film indicated a recent manufacturing date.

Nixon, according to Stripling and others, became abusive and hysterical, shouting, "Oh, my God, I'm ruined." Weinstein, who interviewed Stripling extensively, provides a very moving picture of the California Congressman on the brink of political oblivion. Nixon finally phoned Chambers and asked him carefully about the dates of the microfilm: "Am I correct in understanding that those papers were put on microfilm in 1938?"

"Yes," came the quiet answer. According to Nixon's account, this was the time he lost control. He wrote:

> Then I took out on him all of the fury and frustration that had built up within me. "You'd better have a better answer than that. The subcommittee's coming to New York tonight and we want to see you at the Commodore Hotel at 9:00 and you'd better be there!" I slammed the receiver down without giving him a chance to reply.

The subsequent news that an error had been made by Eastman Kodak and that the emulsion figure was correct brought profound relief to all involved, except Chambers. He was aware of the importance of the Eastman Kodak incident as it related to questions of veracity, and it served further to compound his sense of futility and emotional exhaustion. He purchased a can of cyanide poison and sat down at his mother's home to write farewell notes to his family. Finishing with a statement that his testimony against Hiss was truthful, Chambers attempted to take his own life by inhaling the gas. The attempt was unsuccessful, and the next morning found him still alive and now determined to push on.

While the House committee was assembling to review those late developments, the Justice Department called the New York grand jury back into session and presented them with the typewritten documents Chambers had earlier surrendered to Mr. Campbell. Not to be outdone, the House committee summoned Sumner Welles, former Under Secretary of State, who, in response to a direct question, stated that the material Chambers had turned over to the committee would have proven perilous to the United States if it had been released to a foreign power. Before the committee could exploit this news, the grand jury subpoenaed both Hiss and Chambers, and the committee was forced to cancel its own hearing.

Perhaps in retaliation, the House group began to release the story of the first group of documents Chambers had turned in at the libel deposition and thereby drew a rebuke from the Justice Department denouncing this practice as "premature and ill-advised." In an unfortunate piece of timing, President Truman refused at a press conference to comment on the issues and declared that the House committee was "a dead agency."

With their star witnesses otherwise engaged, the House committee had to contend with some decidedly second-stringers to corroborate parts of Chambers' story. A closed session was held on December 8 where the committee heard Isaac Don Levine, an anti-Communist editor of the magazine *Plain Talk*. Chambers had approached Levine in 1939 with the idea of writing some articles for him about the Communist underground. Nothing came of this project, but Levine, having heard some of Chambers' fascinating tale, convinced him that he should tell his story to the proper authorities. It was Levine who brought Chambers to the attention of Assistant Secretary of State Adolf Berle—a meeting in 1939 that eventually set the Hiss–Chambers case in motion.

Levine testified before HUAC on that day that Lawrence Duggan, a former chief of the State Department's Latin American Division, was a member of a six-man apparatus that was passing confidential documents to the Soviets. In fact it was Hiss, Chambers would later write in his book, who tried to recruit Duggan and his friend, Noel Field, for such work.

Duggan and Field were best friends from 1930, when Duggan joined the State Department, until 1936, when Field left the government to work in Switzerland. They had much in common: Both were handsome, came from fine families, were well educated in Ivy League schools, were committed to helping the less fortunate, and were gentlemen, yet each had a bohemian strain as well. Their informality and mutual regard caused Noel Field and his wife Herta to share a house with Larry

Duggan and his wife Helen—a custom that was rare in the nation's capital in those days.

A popular social activity in Washington then was the study group, where domestic and world affairs were argued about late into the night. Called "cross-fertilization," such study groups brought together bright young New Dealers in different jobs to exchange ideas on the better world they were sure they were creating. At one such informal gathering—"informal" because wives were included—the Fields met the Hisses. The couples got on well. So close would they become that Priscilla took to calling Field in the Quaker way of "thee" and "thou." Soon the Fields introduced the Hisses to the rest of the "family"—the Duggans. Enjoying one another's company, the three couples were often together on social occasions.

Not so, wrote Chambers, who claimed that Hiss's interest in Duggan and Field was political rather than social. It was one of Hiss's duties, said Chambers, to attract people of culture to the apparatus. Under this charge, Hiss went after Noel Field and, failing there, after Duggan. (This allegation first sounded before HUAC by Isaac Don Levine was to have tragic consequences for Duggan.) The FBI papers say that Hiss was not alone in this search for talent for the Communist Party. A woman, Hede Massing, who served as an agent of the Soviet Intelligence Service in the 1930s, also tried to "cultivate" Field and Duggan. Mrs. Massing, a former Viennese actress, and her third husband, Paul, became, what the FBI files characterize as "valuable informants" of the Bureau.*

In interviews with the FBI and later in the courtroom at the second trial, Mrs. Massing testified that she made several trips to Washington from New York to importune Field to furnish her State Department information. Eventually he told her that someone else had approached him for the same type of work. The other person was wrong, she said, but she would like to meet him in any event.

She allegedly met the competition at a supper hosted by the Fields in 1935. The man was, she claimed, Alger Hiss. The FBI files describe Massing's impressions and recollections of this meeting: Mrs. Massing was much impressed with Hiss's good looks, charm, and intelligence,

---

*Hede Massing's second husband was Gerhart Eisler, once described in the New York *Times* as "the nation's foremost Communist." Eisler fled the country, forfeiting federal bail, for a life in Communist Poland. Incidentally, the first speech Nixon ever made in Congress (February 18, 1947) was a request for a contempt citation against Eisler for his refusal to testify before HUAC.

and she said they got along famously. At the first opportunity after the meal, she drew him aside and, according to her, they had this conversation:

> HISS: *Well, you are the famous girl who's meddling in my affairs.*
> MASSING: *And you are the man who's meddling in my affairs.*
> HISS: *What is your apparatus?*
> MASSING: *I wouldn't ask that . . .*
> HISS: *Well, we'll fight it out and see who gets Noel.*
> MASSING: *I'll beat you in this game because I am a woman.*

Finally it was decided it made no difference who recruited Field because, as one of them said, "We're both working for the same boss." Mrs. Massing wasn't sure just who said that, but she was certain the term *boss* meant the same thing to both of them: the Communist international movement. The oft-married Mrs. Massing, apparently suiting feminine wiles to words, ultimately won Field.

Hiss heard Mrs. Massing's story in December 1948 just before the first trial when the FBI staged a confrontation between them. As a result of this meeting, the defense hunted for Noel Field to "contradict," as Hiss said, this story. By this time, unfortunately, Field had mysteriously disappeared in Soviet-controlled Eastern Europe.

Having recruited Field, Mrs. Massing was unable to induce him to steal any State Department documents. It was not until Field went to Geneva to work for the League of Nations—in essence becoming a citizen as well as an employee of the world—did he rid himself of his scruples and attempt espionage. Even then Field was not particularly successful at it. Two of his Party "contacts" gave him assignments and then promptly defected. One of these men, Ignatz Reiss, who upon his break in 1936 denounced Stalin, was assassinated by the GPU (Soviet Secret Police) in a plot in which Field boasted to Paul Massing (Hede Massing's husband) and another ranking Communist that he had played a key role. There was probably more bravado than truth to that boast.

This was a time when Stalin's purge trials were having a chilling—or far worse—effect on Soviet agents the world over. Field, like most foreign agents, was under suspicion then, and no new assignments were forthcoming. In 1937 Field and his wife traveled to Russia to find out why. There they met the Massings, in Russia for questioning, who were having difficulty getting exit visas. From the Fields's hotel room, Hede Massing called the GPU to demand their visas. On securing them, the Massings returned to America and defected from the Soviet apparatus.

They cooperated with the FBI in the Hiss case. Later, Hede Massing wrote a book about her experiences in the underground, entitled *This Deception*.

Four years later (1941) Field was again contacted by a Soviet operative, who instructed him to write down everything he knew about his first two contacts—Reiss and Walter Krivitsky—and also the Massings, all of whom defected. Field dutifully completed this assignment and hoped it would return him to a position of trust within the Party. However, the contact never returned for it.

Field found work with the Unitarian Service Committee in Europe as their director, until his dispersing of aid to Communists became an embarrassment to the Boston-based church which sponsored the project. As a result, the office in Geneva was closed in October 1947. Then Field tried to find work as a free-lance journalist. He felt his contacts in Eastern Europe would be invaluable for such work. He wrote in 1948 to Hiss at the Carnegie Endowment and to Lawrence Duggan, now at the International Institute for Education, about using their influence with editors to help him line up some assignments. Neither could help however, because both were aware that rumors about their politics were making their own positions rather untenable. Hiss also wrote to Field that there was little hope of his finding work with the Carnegie Endowment. To add to his troubles, Field's passport was about to expire and his health was poor.

While traveling alone to Eastern Europe, Field was arrested in May 1949. The order for his arrest came from Moscow, perhaps—it was later discovered—even from Stalin himself, who, after Yugoslavia's break from Russian Communism, was demanding total obedience to the Moscow line. At this time even the most loyal Communists from Eastern Europe were under grave suspicion, and Field, as a Westerner with many contacts in these satellite countries, was an easy target. After his arrest in Prague, Field was sent to a Budapest prison, where he was held in solitary confinement for four years.

Close friends and even those who knew Field under the most innocuous circumstances or those who knew people who knew him were incarcerated, even executed. As Flora Lewis, who wrote a biography of Field called *Red Pawn* and is now Paris bureau chief and European diplomatic correspondent for the *Times,* described it:

> From people directly involved with the Fields in one way or another, the purges spread to the contacts of these people, and on to their contacts and

friends. It was insane but there was a strong inner logic, for one person did lead to another, though it was sheer haphazard chance that had allotted the center to the Fields. . . . Hundreds and then thousands and then tens of thousands of people were carried off to prison or death in Poland and in neighboring countries on this wave of closed logic set off by the purposeful pebble splash of arresting the Fields.

Although the Hiss attorneys were unable to locate Noel Field, who was by then missing, his wife Herta was contacted in Geneva in July 1949 by a Swiss attorney representing Hiss. Herta Field acknowledged that she entertained both the Massings and the Hisses in her house but never at the same time. She added, even if this recollection were wrong, "The small size of it [the apartment] would have excluded a private talk," such as that described by Mrs. Massing. Herta Field was arrested in Prague less than two months after this interview (August 26, 1949); her brother-in-law, Hermann H. Field, had been arrested two days before in Warsaw.

After Noel Field's release from prison in 1957, he and Herta remained by choice behind the Iron Curtain. There he read Hiss's book, *In the Court of Public Opinion,* and wrote in a two-page letter to Hiss:

It was, of course, not until I came out of jail that I learned of the part played in your second trial by false testimony of a perjured witness [Mrs. Massing] with regard to a purported meeting and conversation, neither of which ever took place, either within or without the confines of our Washington apartment. My definite and absolute personal knowledge of the complete untruth of this particular bit of "evidence" is the clearest proof to me—aside from my experience of your personality and outlook—of the falsehood of the rest of the "evidence" on which you were convicted.

Allen Weinstein was to learn from a Czech historian, Karel Kaplan, who in 1968 studied the Stalin purge trials under the auspices of the Dubček government, that Field told a different version of the story when interrogated by Czechoslovakian and Hungarian officials. According to Weinstein, Kaplan said that even months before Field's arrest, he named Hiss as a fellow Soviet agent. During this same interview conducted in 1948 when Field was applying for an extension of his Czech visa, the reason he gave for not returning to the United States was to avoid testifying in the Hiss case. Kaplan purportedly said that in every subsequent interview, after his arrest, "Field said he had been involved [in espionage] and that Hiss was the other involved." This information was described as "most striking" in the New York *Times* review of the Weinstein book.

Assuming that Weinstein quoted his source accurately, assuming that the source accurately reported what he saw, and assuming the documents Kaplan saw were authentic, how much attention should be paid to the testimony of a man under arrest in Eastern Europe, or a man anticipating or fearing such an arrest? Having turned his back on America for underground work with the Soviets and finding no great reward in the pursuit, Field must have felt like the proverbial "man without a country," an unwelcome visitor nearly everywhere. Before he was jailed, he was out of work and his health was failing. Flora Lewis reports that while in prison he was tortured. Under these circumstances, Field might say (or do) whatever he thought his interrogators wanted him to. Finally, if what Field said under those circumstances was true, why should a man who decided to remain behind the Iron Curtain take the time, trouble, and risk, too, to write Hiss and characterize Mrs. Massing's testimony as "an outrageous lie," a "perjury," and a "complete untruth"?

Professor Kaplan has since disavowed the notion that Field implicated Hiss. To Victor Navasky, editor of *The Nation,* Kaplan, now employed by Radio Free Europe, wrote, "N. Field's testimony, as far as I can remember, did not contain explicit statements which would indicate that A. Hiss was delivering U.S. documents to the Soviet Union." Weinstein responded that three others had heard Kaplan's original remarks: an Italian journalist, a translator, and Stanford University professor David Kennedy. Weinstein also noted that Kaplan wrote about the Hiss–Field connection in an article in *Panorama,* an Italian magazine.

Navasky contacted Professor Kennedy, who described the interview with Kaplan as "troubling." He told Navasky what bothered him about the interview and Weinstein's use of it was that Kaplan offered no documentation. Even if he had, there would be no way of judging the authenticity of such documentation. Other things that disturbed Kennedy were that the Italian journalist who was present had ghosted sections of Kaplan's article and also that the article made ambiguous shifts from Kaplan's research to Mrs. Massing's book, *This Deception.*

After Isaac Don Levine's testimony linking Larry Duggan to a six-man Communist apparatus, the FBI, apparently alerted by the House committee, interviewed Duggan and his wife, Helen. The FBI papers summarized the interviews:

> Lawrence Duggan did not know Hiss as a Communist. . . . Did not know or suspect that Alger Hiss ever engaged in Soviet espionage, was never approached by Alger Hiss to work in the interest of the Soviets. . . . Duggan stated that he knew Noel Field very well, both as a co-worker and as a neighbor since they lived in the same apartment house during 1930–35; never approached by Noel Field to work for the Soviets. . . . He denies that Hede Massing ever propositioned him.

Ten days after his interview—a session the FBI characterized as "congenial"—Lawrence Duggan, president of the Institute of International Education, "jumped or fell" (the coroner's conclusion) sixteen floors to his death from his office building in New York City. He was wearing his overcoat, a scarf, and one boot. The other boot was found in front of an open window on the sixteenth floor. Immediately upon hearing this news, Mr. Nixon and Mr. Mundt decided to release Isaac Don Levine's testimony about Duggan to the press. The death, which so closely followed Duggan's FBI interview, and the hastily released testimony by the Congressmen precipitated another internal battle between HUAC and the Justice Department.

Asked by the press when the committee might release the other five names of the alleged "six-man apparatus" described by Levine, Congressman Mundt said, "We'll release them as they jump out of the window." Whittaker Chambers said by way of correcting Levine's testimony that he (Chambers) merely "believed" Duggan was cooperating with the Communists but had no documentary proof of this.

Finally with the Christmas holidays nearing, Nixon and Mundt, perhaps in a more charitable mood, conceded that their actions in this matter warranted "some honest criticism." They decided Duggan had been "cleared" of the charges, and Mundt said that, as far as he was concerned, the case was a "closed book."

Just after the suicide, Attorney General Tom C. Clark, who concluded that Duggan was "a loyal employee of the United States," encountered Representative Mundt at a radio station. The Congressman told him he was receiving confidential FBI information "under the table." Clark was furious. The next day he made an angry phone call to the FBI and talked to J. Edgar Hoover's aide, assistant director of the Bureau D. M. Ladd. Clark shouted at him, "Any SOB that gives Congressman Mundt any information gets his ass kicked out of this building. . . . I want you to get the word around that anyone giving information to the Committee is out, O-U-T!"

Isaac Don Levine's appearance in December 1948 was not his first hearing before HUAC, or at least before Richard Nixon. On August 18, 1948—the day after the historic Hiss–Chambers confrontation in Room 1400 of the Commodore Hotel—Nixon, a subcommittee of one, interviewed Levine in that same room. Their meeting lasted from 11:45 A.M. to 12:40. It is curious that fewer than five pages of transcript emerged from this fifty-five-minute session. Normally a meeting of such length would have produced perhaps five times as many pages, so all is not known—or presumably even much—of what was said. The transcript shows, however, that Levine told Nixon that at the Berle–Chambers meeting in 1939, which he also attended, "Mr. Chambers was opening up the inside of the State Department and various other departments in Washington where he had underground contacts who supplied him documentary and more confidential information for transmission to the Soviet Government."

Amazingly Nixon ignored this significant statement, at least as far as the printed record shows. This was at some distance from Chambers' allegations about Hiss and others as he first described it on August 3 and maintained this position for months thereafter:

> These people (Ware, Pressman, Witt, Perlo, Kramer and Donald and Alger Hiss) were specifically not wanted to act as sources of information. . . . These people were an elite group, an outstanding group, which it was believed would rise to positions—as, indeed, some of them did—notably Mr. White and Mr. Hiss—in the government, and their position in the government would be of very much more service to the Communist Party.

Although it is not known what Nixon and Levine discussed during the unrecorded interval of forty-one minutes, the public record gives ample evidence of Levine recalling Chambers linking Hiss to espionage and specifically to microfilmed documents. The December 1947 issue of *Plain Talk,* the anti-Communist magazine edited by Levine, features an article entitled "Stalin's Spy Ring in the U.S.A." The article, written by Levine at least eight months before Chambers' appearance before HUAC and a year before the fateful scene at the pumpkin patch, reads, ". . . Certain high and trusted officials in the State Department, including one who had played a leading role at Yalta and in organizing the United Nations, *delivered confidential papers to Communist agents who microfilmed them for dispatch to Moscow.*" (Authors' italics.)

Should Nixon and his committee have missed that reference there was also the one in the Washington *Times-Herald* on August 3, 1948—early on the day that Chambers would appear before HUAC—by James Walter, who wrote, "The most astounding story yet revealed about Communist intrigue and infiltration of spies into important government agencies is scheduled to be unfolded today before the House Committee. . . . The details will come from Whittaker Chambers." Although Hiss is not mentioned by name in this article under the subhead "Red Spy Goes to Yalta," he is described in unambiguous detail. Most fascinating is this reference:

> Chambers has related how he finally broke from the Communist Party and returned to the Catholic Church. He tried to persuade many other top Communists to recognize the error of their ways, he declared, but had little success. On one occasion, Chambers declared, he risked his life because of his insistence in trying to induce one man, *supplying the party with vital information about U.S. government* [authors' emphasis], but the man was adamant. . . .

The man whom Chambers had that conversation with would be named by him later that day as "Alger Hiss."

It is inconceivable that the House committee did not know of Hiss's alleged link to espionage and microfilmed documents months before the most dramatic scene at Chambers' pumpkin patch in December 1948 and the even more dramatic scene of Nixon's triumphant return from the *Panama* when he announced without even looking at the microfilm, "It will prove to the American people once and for all that where you have a Communist, you have an espionage agent." It is apparent that the House committee withheld Hiss's link to espionage waiting for the right moment: the eleventh hour of the grand-jury proceedings and the right prop (the Pumpkin Papers) to be produced. Thus it is not so surprising that Nixon told "Fishbait" Miller before he left on his vacation that he expected to be called back.

Perhaps this is what Mr. Mundt was referring to when he commented to Hiss at his first appearance before HUAC:

> I want to say for the committee that it is extremely puzzling that a man who is senior editor of *Time* magazine, by the name of Whittaker Chambers, whom I had never seen until a day or two ago, and you say you have never seen . . . should come before this committee and discuss the Communist apparatus working in Washington, which he says *is transmitting secrets to the Russian Government* [authors' italics].

The significance of that remark was hardly noticed at the time.

As intriguing were Nixon's comments to Bert Andrews on the latter's radio program on January 21, 1950, the day Hiss was convicted for perjury. Nixon and Andrews had this self-congratulatory conversation:

> ANDREWS: *I am going to begin this broadcast with highly immodest words. They are immodest as far as Mr. Nixon is concerned—and as far as I am concerned. But they are true. For the words are these—if it had not been for Mr. Nixon—and for me—I doubt that Alger Hiss would ever had to face trial.*
>
> NIXON: *I remember it very well, Mr. Andrews. It was back in August 1948. We had heard from Mr. Chambers and Mr. Hiss, at two separate sessions. Mr. Chambers said that Mr. Hiss supplied him with confidential papers that Mr. Hiss was a dedicated Communist, that Mr. Hiss knew the information was going to Soviet Russia. . . .*

Whether Levine told Nixon about Hiss and his connection to illicit microfilmed documents is open to debate, but we do know that on November 10, 1953, Levine wrote in the New York *Telegram and Sun* that Chambers showed Berle in 1939 the documents that would later be known as the Pumpkin Papers. Levine wrote:

> I took Chambers to Berle on September 2, the day that Hitler attacked Poland. Mr. Chambers had in his possession the documents which a decade later became known as the Pumpkin Papers. These included, in addition to the many documents involving Alger Hiss, four notes in Mr. White's own handwriting, which had been delivered by him to Mr. Chambers.

On December 8, the same day that Isaac Don Levine was testifying before the committee, Hiss appeared before the grand jury. While in the waiting room with his brother Donald, Hiss encountered Henry Julian Wadleigh, a former State Department economist. Wadleigh had confessed to turning over documents to Chambers, whom he knew as "Carl," for transmission to the Russians. Hiss knew Wadleigh from their government work, and although their relationship was, at best, casual, he remembered him as a "nervous eccentric."

To Hiss, Wadleigh seemed, if anything, more nervous that day. He was chattering uncontrollably. He initiated the conversation by telling Hiss he couldn't speak to him. "My lawyers tell me I am not supposed to recognize you because I am going to have to refuse to answer on grounds that it might incriminate me whether I knew you or not," said Wadleigh.

As far as Hiss was concerned, their conversation was finished, but Wadleigh went on. As Hiss described it in an interview:

> He was distraught. He said to me, "I have admitted that I have given Chambers some papers. But my lawyers said they weren't confidential. So if I keep quiet about it, I'll be all right. I don't know what to do."

Hiss, about to appear before the grand jury and recognizing that he might be questioned about Wadleigh, cut him off. He said, "Julian, don't talk to me. . . . Don't tell me anything in confidence because I can't accept it in confidence."

Chambers and Hiss resigned their positions in the fall of 1948. Chambers thanked his colleagues for their loyalty to him but added that "no one can share with me this indispensable ordeal." Much to his surprise, he felt that his resignation from a $30,000 position created a "sudden general surge of sympathy for me." Hiss's resignation from his $20,000 position was tabled by the Carnegie Endowment, which gave him three months' leave of absence with pay.

Nixon was most concerned during this period that the grand jury might indict Chambers and "thereby probably destroy the only opportunity to indict other individuals because the star witness will be an indicted and convicted person." Unbeknownst to all, the grand jury was struggling with an immense problem that had to do with a typewriter. From the time of the receipt of the original documents on November 17, the FBI had been diligently searching for the machine that had been used to produce the copies that Chambers had alleged were typed by the Hisses. The FBI was unable to locate the typewriter itself nor any typed letters for comparison purposes, although they were later supplied four such documents by Alger Hiss himself. After what later seemed to have been a cursory review, the FBI announced that the Hiss standards and the Chambers documents were typed on the same machine.

The above finding was so dismaying to Hiss's attorneys that they felt his indictment was inevitable, but a friend of Hiss's, William Field, who was soon to run the Hiss Defense Fund, found him to be "cool, collected, and unruffled." Field was talking to Hiss at the Harvard Club and discovered that Hiss, too, expected to be indicted but still insisted that he had great faith in the judicial process and didn't see how a jury could believe Chambers. Field concluded that such an attitude was consistent with either an innocent man or a foolhardy one.

On December 10 the House committee was hearing from Nathan

Levine, Chambers' nephew, who had kept the hidden documents in his mother's home in Brooklyn for ten years. After his testimony the committee made a series of statements lauding Chambers. Nixon's comments are particularly ingenuous in placing the blame for Chambers' espionage on Hiss's willingness to "cooperate" with him:

> I for one wish to say that Mr. Chambers apart from the disservice that was rendered to the country and a disservice which was rendered *only* [authors' emphasis] because there were people in this Government who gave him the information that he was able to turn over to the Communists and which he couldn't have rendered without the cooperation of those people, that apart from that, that Mr. Chambers has willingly and voluntarily, with no necessity at all upon his part to do so, rendered a great service to the country by bringing these facts before the American people at this time.

Throughout the crucial next several days Mr. Nixon was very unhappy that "the witness" had been removed from the committee's jurisdiction by the grand jury. In an effort to stay even, the committee decided to meet with Chambers at night after he completed his testimony before the New York grand jury. Upon arrival in New York, Nixon and others were met by representatives of the Justice Department who demanded that the microfilm be turned over to them for continued investigation. Nixon was adamant in his refusal, and his statement that evening bears an interesting comparison to his attitude toward events that transpired some twenty-five years later when he was in the President's office and under attack during the Watergate affair. He wrote in 1948 that

> I made it clear that we had the greatest respect for lower echelon Justice Department officials who were just as interested in getting at the truth in this case as we were. But I also made it clear that I had no confidence in some of their superiors who were under great political pressures and who so far had made a record which, to put it politely, raised grave doubts. Did they intend to bring out any facts that might be embarrassing to the national Administration?
>
> In short, we did not trust the Justice Department to prosecute the case with the vigor we thought it deserved. The five rolls of microfilm in our possession, plus the threat of a congressional public hearing, were our only weapons to assure such a prosecution. In retrospect, I imagine that some Justice Department officials suspected our motives were primarily political and that we were impeding the regular law-enforcement agencies by withholding evidence.

A compromise was arrived at which resulted in the Justice Department being furnished with copies of the microfilm, while the committee was allowed to question Chambers despite the fact that he was no longer their witness. It was this arrangement that made it possible for both inquiries to continue. That night Chambers, according to Nixon, finally told the "full story to the committee." In directly contradicting his previous testimony to the grand jury and HUAC that espionage was not involved, Chambers accused Hiss and others of being members of an espionage ring.

Chambers said that Hiss took home confidential documents each night. His wife Priscilla then retyped them, summarizing where possible, and Alger returned them to his office in the State Department the next morning before they were missed. Chambers said that this division of family labor was not only functional, but it allowed Priscilla to get involved. "Mrs. Hiss usually typed them," Chambers said. "I am not sure she typed them all. Alger may have typed some of them himself. [Hiss testified he couldn't type at all.] But it became a function for her and helped to solve the problem of her longing for Communist activity."

Then, still according to Chambers, "every week or ten days, on what I called the 'transmission' date, I would go to the Hiss home in Washington and take back to Baltimore the previously typed documents and any originals Hiss had pilfered that day." There, two photographers known to him by their Party names of "Felix" and "Keith"* would photograph them on microfilm. Priscilla's typed copies would be destroyed

---

*The FBI discovered that the photographer known to Chambers as "Felix" was Felix August Inslerman. They made the identification by matching an old Leica camera, still in his possession, to the microfilm from the pumpkin. When Inslerman was identified, he was working for General Electric in Schenectady, New York, on an "extremely confidential" project under the U.S. Army's control. Inslerman's work involved the hydraulic tail assembly of missiles and rockets. Inslerman did not cooperate with the Bureau.

The photographer known to Chambers as "Keith" turned out to be William Edward Crane, who joined the Communist Party in 1932. After Crane was linked with "Keith," he became a most cooperative FBI subject. Shortly after his identification, an article written by Gerry Greene appeared in the New York *Daily News* with this lead: "A red spy suspect long sought as a vital link in the famous case of the Pumpkin Papers has been found in California by the FBI and is singing his head off." When the story broke, Crane was so upset that he consulted his attorney. An FBI office memorandum concluded this about the leak: "My frank opinion is that Chambers was the original source. That he probably went to Mundt or Nixon or even some of his other friends and in that manner Gerry Greene picked it up."

after that, and Chambers would return to the Hiss home late that night with the originals, which were returned to Hiss's office the next morning. Chambers handed the microfilm over to his superior in the Party, Colonel Bykov, who then sent the film to Russia.

Just before Christmas of 1936, Chambers said that his superior, Colonel Bykov, was impressed by the quality of the material turned over by Hiss—whom Bykov called "*Der Advokat*" ("lawyer" in German)—and the rest of Chambers' apparatus that he decided to give them all Christmas gifts. "Big money" was the Russian's first suggestion, but Chambers told him that that would be improper. "They would be outraged," he said with some passion. "You do not understand. They are Communists. . . ."

Colonel Bykov said in German, the language they often used, "No, you, Chambers, didn't understand; who pays is boss. Who takes money must give something." According to Chambers, he was able to persuade his superior that this would be politically as well as socially improper. Bykov yielded, saying, "You will buy four rugs, big expensive rugs. You will give [Harry Dexter] White, [Abraham George] Silverman,* Wadleigh, and *Der Advokat* each one a rug. You will tell them that it is a gift from the Soviet people in gratitude for their help." Bykov's presumptions clearly offended Chambers. He said, "You know these are good people, but they are not stupid people." Bykov, who seemed to enjoy this type of exchange with his subordinate, gave Chambers about $1,000 for the purchase.

Knowing nothing about rugs and having little stomach for the project, Chambers turned to an old school chum, Meyer Schapiro, then a professor at Columbia, and asked him to be the purchasing agent. After consulting with a friend who was more knowledgeable in these matters, Dr. Schapiro bought four Oriental rugs from the Massachusetts Importing Company in New York City. Years later as the Hiss case was breaking, the FBI was able to find two tickets on the sale of the rugs; one

---

* Abraham George Silverman, then research director of the Railroad Retirement Board, according to Chambers, got him a WPA research job with the government. There, Chambers claimed, all his superiors were Communists. Unknown to Silverman, Chambers took this job in preparation for his break with the Party. He worked there under his real name—a name he rarely used during his years as a Communist—to establish "an identity, an official record of the fact that a man named Chambers had worked in Washington in the years of 1937 and 1938." And he also needed the salary to finance his break. Chambers wrote of Silverman's function in the Party, " . . . his chief business, and a very exacting and unthankful one too, was to keep Harry Dexter White in a buoyant and cooperative frame of mind."

was dated December 23 and the other December 29, 1936. The latter ticket from the importing company indicated that Dr. Meyer Schapiro was the purchaser and the person to whom the rugs should be shipped. The rugs and their date of purchase and delivery were to prove central to the government's case.

In his book Chambers described White and Silverman as being "clearly impressed" by both Bykov's gift and accolade. He was not sure how Hiss felt. But the rug was never prominently displayed in the Hiss home. This could have been a factor of taste; it was "a little vivid," Chambers recalled, or perhaps it was evidence of dangerous association. In any event, the red rug was rolled up in a Hiss closet for a time, was apparently stored commercially for a time, and ultimately landed on the floor in the room of Hiss's stepson.

Bykov's use of the capitalist's *quid pro quo* greatly offended Chambers. "For the first time," he wrote of the experience, "I felt a riffle of disgust at my comrades, that is to say, at myself. I was one of them."

This, then, was the alleged story of the rug which Hiss said was originally "thrown in" in lieu of repayment of the overdue debt. Even though Nixon found the story "almost too fantastic to believe, . . . we had learned that where Communist espionage was concerned, we had become accustomed to actions which stretched our credulity."

Still, the Justice Department was not entirely convinced, and Nixon learned on December 8 that the department was about to ask the grand jury to indict Chambers, not Hiss, for lying during earlier testimony when he had denied under oath that he had committed espionage against the United States. It is important to note that this information about grand-jury intent came to Nixon from "some of the Justice Department employees in lower echelons who were so infuriated by their superiors' handling of the case that they apprised the Committee staff of every action that was being taken."

Nixon decided to make an open fight of the issue, and in a direct statement to the press, made after the late-night meeting with Chambers, Nixon said:

> We have learned from unimpeachable sources that the Justice Department now plans to indict Chambers for perjury before any of the other people named by Chambers in this conspiracy are indicted. It is clear that the Justice Department does not want this Committee to hear any witnesses scheduled to go before the Federal Grand Jury and is bringing pressure on the Committee to drop its investigation. Chambers has confessed. He is in the open. He is no longer a danger to our security. If Chambers is indicted first, Hiss and the

> others will go free because the witness against them will have been discredited as a perjurer. The Committee will not entrust to the Justice Department and to the Administration the sole responsibility for protecting the national security in this case. We intend to do everything we can to see that the Department does not use the device of indicting Chambers as an excuse for not proceeding against Hiss who has continued to decline to tell any of the truth up to this time.

Still struggling to make sure that the "proper witness" was indicted and now operating in the "interest of national security," a preoccupation that seemed to become an *idée fixe* in his career, Nixon appeared before the grand jury with microfilm in hand. He said he was there "to assist them in bringing to justice those who fed this information to Chambers." By now apparently convinced that Chambers, the confessed spy, had been the passive dupe of those he had admittedly persuaded to spy for him, Nixon was markedly undeterred by the fact that he was openly supporting a man who had just confessed committing both perjury and espionage. He was now firmly on the side of the Congress in the separation-of-power issue:

> Justice Department officials demanded that I leave the microfilm with them. By this time, I believed that they would prosecute the case with diligence, but the full Committee had given me instructions that under no circumstances was I to surrender the microfilm without Committee approval. Alex Campbell threatened to ask the Judge to cite me for contempt. I, in turn, warned him of the constitutional question that would be raised if a member of Congress, appearing voluntarily before a grand jury, were so cited while carrying out a mandate of the Committee which he represented. After a few anxious moments, I was allowed to return to Washington with the microfilm still in my possession but with the understanding that, in the event Hiss was indicted, I would take the responsibility of seeing that the Committee would make the microfilm available as evidence in the trial.

When Hiss came to the grand jury to testify on December 15, he talked with Alexander Campbell, the Assistant U.S. Attorney General, out in the hall. Campbell told him that the typewriter specimens secured from Hiss-family correspondence had been definitely linked with the documents produced by Chambers. Mr. Campbell made what Hiss called "prejudicial declarations," and Hiss told him that he intended to relate the incident to the jurors when testifying. Hiss's notes for the day contain the following summary:

I then told the Grand Jury that at noon today Mr. Campbell had asked to speak to me and had told me that I would be indicted. I said . . . that naturally I did not know all of the evidence which the jury had but that I did know that much damaging evidence had been produced by my own direct efforts and had been turned over to the government officials at my direction. . . . I said that whatever the evidence, I knew that I had done nothing that was a breach of trust or a dereliction of my duty, that I was proud of my years of government service.

At this point, Mr. Whearty said, hadn't Campbell said I would *probably* be indicted? I said no. I was confident that he had not. . . . Mr. Whearty asked if Mr. McLean, my lawyer, had not been present. I said that he had. . . . I then testified from the notes I had made as to the conversation as follows:

Mr. Campbell said in practically these words, "The FBI has cracked the case. You are in it up to your eyes. Your wife's in it. Why don't you go in there and tell the jury the truth?" I said that I had replied that I had continuously told the truth and that I will continue to do so. Mr. Campbell had then said, "You are going to be indicted. I am not fooling. There are five witnesses against you." . . . Mr. Campbell then said, "This is your government speaking." That concluded the very brief interview but later Mr. Campbell had called me back in again and had said, "I want to make it plain, your wife will be included."

I said I thought Chambers was a person of unsound mind and not normal, that he had for some reason which I did not pretend fully to understand, acquired a grudge against me and was trying to destroy me. I said that I considered that he had deliberately tried to frame me.

Thomas J. Donegan, special assistant to the Attorney General who was responsible for the grand-jury proceedings that day, planned to surprise Hiss with the fact of the typewriter identification. Prior to Hiss's appearance, the grand jury had heard the testimony of an expert linking the documents to Hiss correspondence. As Donegan was building to the critical point, Hiss interrupted him and said he wished to make a statement. Donegan hoped Hiss was about to make a full confession, but Hiss said he knew the typewriter had been identified, as Mr. Campbell had told him so out in the hall. He also said that since this was the case, Chambers must have borrowed his typewriter to produce the documents.

The grand jury wanted to know how Mr. Campbell could be so certain they would return an indictment since it was not within his province to know. They called for Mr. Campbell, who admitted that Hiss was telling the truth about their conversation.

FBI agent Belmont, who was the Bureau reporter for the secret grand-jury proceedings, concluded that the jurors were not pleased with the opportunistic Mr. Campbell. The following quote is from the FBI's summary of Belmont's intelligence:

> Mr. Belmont stated that apparently Campbell was trying for headlines, and in this instance was trying to get in the position of saying that he was the man who made Hiss tell the truth. Further the action of Campbell obviously gave Hiss the opportunity to think and come up with the answer he did—that he must have loaned the typewriter to Chambers. . . .

Mr. Campbell was not easy to categorize, for Hiss's attorney, Edward McLean, said in a sworn affidavit that Campbell had in speaking to him that very same day characterized Whittaker Chambers as an "unstable and abnormal" individual and further that Campbell had daily expected "to pick up the paper in the morning and find that Mr. Chambers had jumped out of the window."

Since Hiss allegedly disposed of the unwanted Ford to William Rosen, an organizer of the Communist Party in Baltimore, the FBI reasoned that he might have disposed of the unwanted typewriter in a similar fashion. An office memorandum to J. Edgar Hoover from assistant D. M. Ladd reads, "It is believed that the local Communist Party possibly has Hiss' typewriter, or has had it in the past." For this reason, the Bureau secured typed Party correspondence from "confidential sources" and tried to link it with Chambers' documents or the Hiss correspondence, which came to be known as the "Hiss standards." No such correlation could be made.

HUAC must have heard this rumor, for their witness on this same December 15 was a woman named Marion Bachrach, who was employed as a writer for the Communist Party. She refused to answer questions relating to any personal knowledge of the Hisses and denied ever lending or delivering a typewriter to Mrs. Hiss. With very little real news to release to the press, the committee's new chairman, Mr. Mundt, delivered himself of the startling judgment, "I would like to state that the crime involved is very definitely a capital crime. It is either treason in wartime or treason in peacetime."

Events in early December seemed to assure the indictment of Chambers for perjury in at first denying and then later admitting to the jury acts of espionage. This would, of course, invalidate his status as a "witness," for his claims of reformation would appear cynical in the light of recent evidence of deceit. There is certainly the likelihood that the committee's persistence in continuing their own hearings during the grand-jury proceedings was self-serving for this purpose. It also follows that Nixon's appearance before the jury was similarly motivated, and his statement at that time clearly documented the need to "save" Chambers.

Related to the above problem was another central issue. If Chambers were indicted, the government could not expect any further informants to come forward with reports of Communist infiltration. Perhaps the entire McCarthy era, one of the nation's saddest chapters, might never have occurred if Chambers had been indicted. Suffice it to say that the government, stung by disclosures that secret documents had indeed vanished from high offices, was now belatedly putting on a new face.

It is not surprising, therefore, to discover that the same Mr. Campbell was to make a last-gasp appearance before the grand jury and ask for Hiss's indictment rather than Chambers'. According to the jurors, Campbell's logic was simple: An indictment of Hiss would only guarantee that a trial be held, which was, after all, a salutary thing and was not in and of itself evidence of existing guilt; but the indictment of Chambers would be seen as nullifying the first indictment by invalidating the witness to the espionage.

The next day saw Hiss's final visit to the jury. He was once again asked for an explanation of how the documents produced by Chambers in Baltimore could have been typed on Hiss's machine. Hiss expressed the same confusion as before: Chambers had either gotten access to the typewriter while in Hiss's house or he had obtained possession of it later after it had been disposed of. Hiss wrote:

> Four or five of the jurors appeared very frankly to be completely skeptical of my testimony with respect to the possibility of Chambers' having had access to the typewriter. . . . I said that I felt his use of the typewriter, if the identity of the typed exhibits were established, was simply a form of forgery less easily detected than an attempted forgery of handwriting and was part of the pattern of attempting to frame me.

This seemed to end Hiss's affairs with the grand jury, but as he made ready to leave he was suddenly summoned back to the room. A juror wanted to ask him what he meant by his statement that Chambers was trying to frame him. Hiss replied, "I said that for some psychological reason that I did not understand, he was trying to destroy me." Then followed a momentous exchange as Hiss was asked two final questions: Had he ever turned over any State Department documents or any other government material to Chambers? Other than the title to the Ford, Hiss answered, he had not. Hiss was also asked if he could definitely say that he never saw Chambers again after January 1, 1937. Hiss replied that he could definitely testify to that fact.

On the last day of its statutory life, the grand jury delivered itself of the following judgment:

COUNT ONE *read that:*

The Grand Jury charges that, at the time and place aforesaid, the defendant Alger Hiss testified falsely before said Grand Jurors with respect to the aforesaid material matter, as follows:

Q. *Mr. Hiss, you have probably been asked this question before, but I should like to ask the question again. At any time did you, or Mrs. Hiss in your presence, turn any documents of the State Department or any other Government organization, or copies of any other Government organization, over to Whittaker Chambers?*

A. *Never. Excepting, I assume, the title certificate to the Ford.*

Q. *In order to clarify it, would that be the only exception?*

A. *The only exception.*

JUROR: *To nobody else did you turn over any documents, to any other person?*

THE WITNESS: *And to no other unauthorized person. I certainly could have to other officials.*

That the aforesaid testimony of the defendant, as he then and there well knew and believed, was untrue in that the defendant, being then and there employed in the Department of State, in or about the months of February and March, 1938, furnished, delivered and transmitted to one Jay David Whittaker Chambers, who was not then and there a person authorized to receive the same, copies of numerous notes and other papers the originals of which had theretofore been removed and abstracted from the possession and custody of the Department of State, in violation of Title 18, U.S. Code, Section 1621.

COUNT TWO read that Hiss, while under oath, also perjured himself in this exchange with the Grand Jury:

Q. *Now, Mr. Hiss, Mr. Chambers says that he obtained typewritten copies of official State documents from you.*

A. *I know he has.*

Q. *Did you ever see Mr. Chambers after you entered into the State Department?*

A. *I do not believe I did. I cannot swear that I did not see him some time, say, in the fall of '36. And I entered the State Department September 1, 1936.*

Q. *Now, you say possibly in the fall of '36?*

A. *That would be possible.*

Q. *Can you say definitely with reference to the winter of '36, I mean, say, December '36?*

A. *Yes, I think I can say definitely I did not see him.*

Q. *Can you say definitely that you did not see him after January 1, 1937?*

A. *Yes, I think I can definitely say that.*

MR. WHEARTY: *Understanding, of course, exclusive of the House hearings and exclusive of the Grand Jury.*

THE WITNESS: *Oh, yes.*

That the aforesaid testimony of the defendant, as he then and there well knew and believed, was untrue in that the defendant did, in fact, see and converse with the said Mr. Chambers in or about the months of February and March, 1938, in violation of Title 18, U.S. Code, Section 1621.

It was later discovered that the vote for indictment was one more than that needed for a majority. There is extant a letter to Zeligs from a grand juror who wrote:

Hiss' indictment was a close vote, not a unanimous one. I was never convinced that Hiss was guilty of the crime we indicted him for. Chambers perjured himself many times, but the final decision of the jury was, "He's our witness, we are not going to indict him." It was a politically inspired matter.

How had these strange events come to pass? How had the *wunderkind* from Johns Hopkins, Harvard, and the State Department found himself the subject of a criminal indictment? To answer these questions it is necessary to examine the mood in the government and in the country when this matter was being argued before the grand jury.

It was no accident that the Hiss case surfaced in the year 1948. This was the year of the Russian's blockade of Berlin and the end of democracy in Czechoslovakia. It would see the Cold War battle lines drawn. At home, there were loyalty oaths, political harassment, and the deportation of Communists. The Republicans controlled Congress, and their political interests were being defended by the Republican-dominated House Committee on Un-American Activities. The party, denied the oval office for sixteen years, sensed victory in what was also a presidential election year and were availing themselves of every opportunity to embarrass the Democrats.

The Republicans asserted that Roosevelt gave away half of our security at the Yalta Conference and that Truman had given away the rest at the United Nations Charter session in San Francisco. Because Hiss had been at both conferences, the Republicans were able to snipe at both Democratic Presidents. On the day of Hiss's first and most successful appearance before the committee (August 5, 1948), Truman, never one to avoid a fight, concurred with a reporter's description that the House committee's investigation was a "red herring" and wondered why the Republican Congress didn't forget the Communists for a moment and do something important—such as passing his legislative proposal to control inflation. Much later he would describe the House committee as "more un-American than the activities it is investigating."

The year 1948 was a critical period in the life and times of HUAC as well. It began well enough with ten screenwriters, whom the committee had investigated in hearings which resembled nothing save a shouting match, being brought to trial. The group known as the "Hollywood 10" was eventually convicted in 1948.

There were some embarrassing failures as well, most notably that involving Dr. Edward Condon, who had worked in the atomic bomb project. Chairman Thomas, without holding a hearing, in fact from a hospital bed where he was recovering from gastrointestinal hemorrhages, labeled the scientist "one of the weakest links in our atomic security." Such charges against Dr. Condon, who was then director of the National Bureau of Standards, were supported only by the rather dubious facts that Condon's wife was of Czechoslovakian descent and that he was recommended for his present position in 1945 by the Secretary of Commerce, Henry Wallace, who Thomas calculatingly described as the Vice Presidential candidate of the Communist Party. Condon fought tenaciously to have himself cleared of Chairman Thomas' charges. With his acquiescence—in fact his blessing—Condon became one of the most investigated men in Washington. Eventually he was cleared of all charges by the Atomic Energy Commission.

After the Condon fiasco, the New York *Times* and other liberal newspapers strongly objected to the committee's tactic of presuming guilt (often before even hearing the "guilty" party) and then requiring the accused to prove his innocence. In an editorial the *Times* wrote, "The Thomas Committee may now proceed, as it threatens to do after more than a year of baseless rumor-mongering. . . . If good and faithful servants so judged by those who best know them and their work are to be persecuted in this fashion, our government research is likely to fall into the hands of drudges and time servers."

The year 1948, which began so brightly for HUAC, was looking rather tarnished by the summer. The committee was floundering after the Condon embarrassment and had a genuine need for a new issue. It was fortunate that Chief Investigator Stripling remembered hearing about Whittaker Chambers and the story he had been telling for nine years. In a last-gasp effort, the committee hitched its wagon to Whittaker Chambers. History demonstrates that they never had to look back. For the committee, 1948 was the "vintage year," so much so that it gave the controversial committee momentum for the next twenty-seven years. In one form or another the Committee on Un-American Activities existed until 1975, when the 94th Congress, the first to serve after President Nixon's resignation, finally laid it to rest.

The Hiss case made tidal waves throughout the government. Certainly HUAC, the FBI, the Justice Department, the State Department, and the President's office had a vested interest in this matter. The House committee in general and Richard Nixon in particular, who was certainly its most astute and agile member, saw the political capital to be gained by keeping Chambers' name in the headlines. The controversy kept HUAC alive at a time when the committee was in danger of losing its mandate from Congress. If Hiss were indicted and convicted, the House committee could claim the credit as the country's major bastion against subversion.

August 1948 was a particularly trying time for both the executive and legislative branches of government. As it often is in summertime Washington, the weather was unbearably hot and humid. It was also the summer before the fall presidential and congressional elections, and those legislators who were up for re-election were using the recess to campaign back home. President Truman fueled congressional discontent by calling it back into session that August for an emergency session.

Inflation was a national obsession at this time. Automobile prices had gone up for the third time in a year, and meat prices were so high that housewives from across the nation took to picketing their butcher shops demanding that something be done. President Truman, up for re-election and looking for all the world like a one-term President, presented Congress with eleven measures to control inflation. Only their passage, he implied, would save the 80th Congress from being branded "the do-nothing Congress."

Things were bad at home, but they were no better internationally as the victorious Allies, back from the latest edition of "the war to end all wars," were again on a near-war footing. In this climate President Truman's program stood little chance of passage by a Republican Congress who expected the election of one of their own, Governor Thomas E. Dewey of New York, in November.

Consequently, Congress, not caring much for what Harry Truman was doing, was seized by a midsummer's malaise. Only two rather innocuous measures were passed before the session became terminally mired in a filibuster. Then, as Alistair Cooke wrote, "Back to the steaming grass roots went the Democrats to blame the Republicans for inflation. And back went the Republicans trumpeting some alarming testimony . . . which seemed to show, or could be made to show that the administration was criminally 'soft' toward Communists, if it was not actually riddled with them."

With campaign rhetoric heating up, President Truman delivered his "red herring" remark and described the investigation as a smokescreen to make people forget the ineffectual Republican Congress. Truman's charge followed Hiss's first appearance before the committee. As pointed out earlier, Hiss so deftly handled the committee's accusations that virtual consternation reigned on his departure. All of the members of HUAC were up for re-election that fall, and with Hiss's virtuoso performance no one save Richard Nixon (assured of re-election since he was unopposed) and Robert Stripling (who was a staff member and thus not subject to elections) wanted to pursue the Communist hunt any further. Only Nixon and Stripling argued for additional hearings and fought tenaciously and successfully to see the committee vindicated.

The President, elected despite the polls, found new license to disparage the committee's work. With Truman's victory the Congress had become Democratic, and HUAC was about to disappear. The Democrats were able and, more importantly, willing to put the controversial committee to rest by not renewing its mandate. At this nadir Chambers produced the Pumpkin Papers. Undaunted, Truman stuck fast to his counteroffensive even after Chambers produced his "life preserver," which significantly buoyed up the enfeebled committee.

The Justice Department was another target of the House committee's wrath. Unable to conquer the Department of Justice, Nixon and his colleagues divided the lower-echelon officials from their superiors and in this way were able to obtain confidential information. HUAC maintained a pipeline of informants reaching the halls of Justice, which provided, among other things, information on secret grand-jury proceedings. Such inside information placed Nixon, microfilm in hand, in front of the grand jury in its eleventh hour.

War was waged over the microfilm Chambers had saved for the committee, with both the FBI and the Justice Department clamoring for its possession. But Nixon held on firmly, as if Chambers' "life preserver" had become the sole support of the committee as well. In a classic example of putting the cart before the horse, Nixon promised to release the films *after* the right man—i.e., Alger Hiss—was indicted.

The committee was interested in a trade with the FBI—the microfilm for Bureau cooperation—but J. Edgar Hoover was wary of such an exchange, citing HUAC's inability to keep secrets. From FBI papers comes Hoover's caveat:

> As you very well know one of the basic problems that would always be present in an exchange of information with the Committee would be the security of that information. The Committee, in the past, has not been too scrupulous in selection of its staff members with the consequent result that there have been premature leaks of information to embarrass both the Department [Justice] and the FBI.

However, Nixon saw it differently. While on a campaign stop four years later in Bellefontaine, Ohio, the Vice Presidential candidate said to 4,000 people that Truman was responsible for the stand-off between the committee and the FBI. Nixon claimed that the President threatened to fire any agent who cooperated with HUAC. In this speech, which preceded by one day an appearance by President Truman, who was accompanying Adlai Stevenson, Nixon said, "If Mr. Truman had his way, Alger Hiss would still be free and active today."

The Justice Department and its investigative arm, the FBI, nursed a wound of their own in the Hiss–Chambers affair. Chambers had first told the story to Assistant Secretary of State Adolf A. Berle in 1939. After that, the FBI had interviewed Chambers numerous times, but no action was ever taken. Questions were asked by the press as well as by the others: Why had it taken almost a decade for something to come of it, and why was the prosecution being spearheaded by HUAC rather than the Justice Department? This ten-year hiatus precluded Hiss from being tried for espionage under the statute of limitations.

After the case broke, Chambers was again interviewed by the FBI when he discussed their lack of interest in his story. The interview was summarized by the agency:

> . . . [Chambers] was very much impressed with the interest that Adolf Berle manifested in his story. On September 2, 1939, Berle took approximately two pages of notes,* and then he [Chambers] went up to his office in *Time* magazine and expected the FBI to show up the following day. When the FBI did not show, he said he was the most disappointed man on earth. Some years later, when the FBI did come to see him, two agents talked with him in general terms, primarily about the late General Kravitsky. [General Walter Kravitsky had been head of the Fourth Section Communist Party in Western Europe.] It was suggested by intimation that possibly Chambers had heard of Kravitsky, and when the agents left, one of them made the observation, "You [Chambers] interest me greatly."

Years later when the Hiss–Chambers case exploded on the American scene, the FBI was acutely sensitive to the time lag. A reporter, Victor Lasky, who would eventually cover the trials for the New York *World Telegram,* called the FBI on this matter. After checking their files on Lasky ("He has written anti-Communist articles"), the FBI decided he might be the right reporter to use as a sounding board for their version of the story. From the Bureau's files comes this public-relations strategy:

> I do think that after seeing Lasky, if it appears he has the right viewpoints, you should go so far as to protect the Bureau on the matter. I think what we should do is to point out . . . our jurisdiction, the time-honored practices and policies of the Bureau, the fact that we first investigated this in 1942; that Berle did not tell us of the Chambers interview until we secured his notes in 1943, that we interviewed Chambers in 1941 (that's been corrected to 1942), neither at that time nor until December, 1948 did he even as much as hint of the existence of microfilm . . . and in interviews he [presumably Chambers] did not point out the vigorousness of our investigation, which led to the development of the handwriting and typewriting examinations. I think we can

---

*Adolf Berle's notes on this meeting were introduced at the second trial. This is what he wrote in regard to Hiss: "Ass't to Sayre—CP—1937; Member of the Underground Com.—Active Baltimore boys—Wife—Priscilla Hiss—Socialist; Early Days of New Deal."

hint to him that we'll identify photographers and the like, but cannot go beyond this, as all of this will come out in the trial of the case.

The FBI was also sensitive about an interview conducted with Alger Hiss. In 1946 Hiss, still in the State Department, had come back from the United Nations General Assembly meeting in London, where he had been the principal adviser to the United States delegation. Secretary of State James Byrnes had told him he was in danger of being called a Communist by Senator James Eastland of Mississippi and Congressman Edward Cox of Georgia. The legislators got their information about Hiss from a former FBI agent who was working as an investigator for HUAC.

Byrnes said this was a very serious situation and advised Hiss to see J. Edgar Hoover immediately. When Hiss called the FBI the director was out of town, and he saw his assistant, D. M. Ladd, in March of that year. According to Hiss, he was "courteously received" by the FBI, where, among other things, he recited a list of organizations he belonged to. Hiss had written for one organization on the FBI's subversive list, the International Juridical Association (IJA), which he described as an "editorial group" that published notes on labor cases. Hiss was also asked if he knew various people from a list of about forty or fifty names.

Some he knew from his school days at Harvard or his government work; he did not remember being asked about Whittaker Chambers. Presumably Hiss didn't know then that Chambers and Crosley were one and the same, and the name would have meant nothing to him. To Byrnes, Hiss described the session as "rather perfunctory." He told his superior that if this situation would cause any embarrassment to the State Department, he would resign. This was a time when many people in government were being called Communists, and Byrnes did not ask for Hiss's resignation.

Then in 1949 newspaper stories appeared which said that, as a result of this interview, Hiss had been "cleared" by the FBI of all charges. This, too, proved awkward for the Bureau, and they checked with Byrnes. From the FBI papers comes this summary of the interview:

> Mr. Byrnes stated that he has some recollection that he either received a copy of the results of the interview with Hiss or was advised by telephone of the results by either the Director or one of the Director's assistants. . . . The FBI searched their voluminous files and concluded, "Nothing was found as a result . . . which would even suggest that Alger Hiss had been given clearance by the Bureau."

Today, with the hindsight of nearly thirty years, it seems entirely possible that the Bureau's early inactivity in this matter—which caused the delay that saw Hiss tried only for perjury rather than espionage—might well have provoked the FBI to take extraordinary and illegal actions to secure the conviction of Hiss. The FBI papers related to this case leave the inescapable conclusion that the Bureau (much like HUAC) badly wanted a conviction in order to redeem themselves.

CHAPTER 6

"A prince must imitate the fox and the lion, for the lion cannot protect himself from traps, and the fox cannot defend himself from wolves. One must therefore be a fox to recognize traps, and a lion to frighten wolves. Those that wish to be only lions do not understand this. . . ."

—MACHIAVELLI, *The Prince*

When Hiss was finally indicted by the grand jury on the charges, Nixon once again felt relief and thought the fight was over. He stated that the indictment was a "vindication of the Committee," but there was abundant evidence of the group's continued interest in the case. For a number of days after the grand jury announced its indictment, the committee released daily installments of the "Chambers documents," which were reported to be very detailed, "in order to make possible a comprehensive understanding of their contents."

In a philosophical mood, Nixon wrote:

> We succeeded in the Hiss case for three basic reasons. First, we were on the right side. Second, we prepared our case thoroughly. Third, we followed methods with which few objective critics could find serious fault. This is not

> the easy way to conduct a congressional investigation and certainly not the best way to make sensational headlines. But it is the way which produces results. In dealing with Communists, any other procedure can play into their hands and usually does.

Chambers read the grand-jury portents another way. He was convinced that he had lost the case and would be indicted. On the last day of the hearings he telephoned his lawyer and said, "I believe I've just been indicted. Will you be on hand with a bondsman at eleven o'clock tomorrow morning? I'll need your help in going through this business." When Chambers visited his mother that afternoon, he was much surprised to hear her comment that she was so thankful.

> "About what?" I asked. Her words seemed to make it more difficult to break the news I had for her. She was halfway down the stairs now and could see my face. "You mean," she asked incredulously, "you don't know that Alger Hiss has been indicted?" She threw her arms around me. I said something about Alger's mother. "Oh, I do think of her," said my mother. "Poor woman! Oh, poor woman!"

Hiss, who understandably had a bitter reaction, nonetheless found some reason to be of good cheer:

> The last act of the grand jury was to indict me on the charge of perjury. They acted by a divided vote, only one more than a bare majority. They wished to indict Chambers, but were persuaded not to. But all the various influences on them did not obtain the indictment of my wife. There was at least this one victory for sanity at this crucial stage of the case.

Hiss spent the weeks before the trial going over the evidence with Lloyd Paul Stryker, an experienced criminal lawyer who had been recommended to him by his friend, Justice Felix Frankfurter. Hiss continued on his own to search for the details that would corroborate his recollections of his past life. He wrote to friends and acquaintances, to family and strangers, and even tried to understand a character presumably based on Whittaker Chambers in the then popular book *The Middle of the Journey* by Lionel Trilling. He submitted his analysis of the fictionalized Chambers to his lawyers.

Since he felt that the key to Chambers' behavior was psychological, Hiss consulted with a psychiatrist, Dr. Carl Binger, for an expert opinion. It is apparent that Hiss remained preoccupied with the problem of explaining to himself and to others the etiology of Chambers' attack on him. Hiss said in a recent interview that he got word from the late columnist Joseph Alsop, "who doesn't think politically the way I do, and he knows it," that the key to Chambers' actions could be found in Franz Werfel's novel *Class Reunion,* which Chambers had translated into English. About the reading, Hiss commented that it was "helpful to nourish my understanding of Chambers' motivation. I'm someone, I suppose, who is more of a positivist than people like to be nowadays. Something that is a void is disturbing to me. I puzzled and puzzled why would this man do this to me? What did I ever do to him that would rationally justify what he was doing to me?"

Dr. Carl Binger was as taken with the Werfel translation as Hiss was and spent some time testifying about it at the second trial. He found what he called "extraordinary analogies" between the book translated in 1929 and the Hiss case. The psychiatrist said, "There are two characters, principal characters, in the book. One is Sebastian and the other is Adler. The name Adler is very close to the name Alger. . . . Adler is described in the book as . . . the closest friend of the other character Sebastian, but later became the enemy whom he is trying to destroy. There is a forgery, and Sebastian signs a paper that Adler, not he, is responsible for it."*

When interviewed by the FBI about his translation, Chambers said that he hadn't read it in twenty years and described the novel with disdain as "unnecessary." Chambers, who had translated a dozen books from German and French (his best-known translation being the popular children's book *Bambi*), belittled the suggestion that this story gave him the idea for destroying Hiss. He wondered why so many found this theory so compelling. "I never quite understood [this]," he wrote, "since it always seemed to me that if I had been bent on ruining Alger Hiss from base motives, the idea might well have occurred to me without the benefit of Franz Werfel."

Six weeks before Hiss's trial began, Edward McLean, in collaboration with Hiss's brother Donald, finally located what was assumed to be the long-missing Hiss typewriter and purchased it for $15 from a night watchman who had it in his possession after it had passed through the

---

*The Franz Werfel analogy created considerable trouble for Binger. Clever questioning by Prosecutor Murphy during the second trial brought out the fact that Binger himself had likely not read the book, relying instead on a plot memo prepared by one of Hiss's attorneys. The entire effort became largely unsupportable at this time, and Binger's credibility as a witness was imperiled.

hands of several other persons. This find—*seemingly* accomplished by relatively sparse efforts of the two men when compared to the thirty-six FBI agents who had searched for it and were manifestly unsuccessful—was to play a significant role in the final trial determination. In point of fact, Meyer Zeligs concluded in this regard:

> It was an irony, indeed, hardly mentioned at the time, and forgotten by most through the years, that it was the Hisses and their investigators who discovered and introduced into court the material evidence that was eventually to damn him.

Zeligs here was referring to the fact that the typewriter provided by the Hiss defense team was regarded by many as becoming the most potent weapon in the prosecution arsenal. However, in the learned opinion of The Earl Jowitt, the real case against Hiss "rests upon the documents, and only upon the documents that the necessary corroboration of the story told by Chambers is to be found." Still, for any number of unpersuaded people, the presence of the "guilty" machine itself loomed as immutable evidence of Hiss's complicity. It was only after his conviction that even Hiss himself was able to offer any kind of meaningful explanation, and this will be recounted later.

Nixon's recollections of the circumstances surrounding the finding of the typewriter were spotty. In his book *Six Crises,* he wrote that "on December 14th [1948], FBI agents found the typewriter." The final report on the Hiss matter published by the House committee on December 30, 1951, contained a statement congratulating the FBI for "the location of the typewriter." When such matters were brought to Nixon's attention, he placed the blame on faulty staff work, and the statement disappeared from the paperback edition of the book. Nonetheless, there is no attribution to the Hiss forces in the substituted statement ("Actually, the typewriter was found several months later and produced during Hiss's first trial for perjury") and certainly no evidence that Nixon was either aware of or moved by the ironical circumstance that Hiss contributed to his own destruction. This is a matter of no small interest in the overall scheme of things, since this is precisely the judgment commonly offered about Nixon's own later difficulties.

The trial under the grand-jury charges was originally set for January 24, 1949, but was postponed a number of times. Events had to be reconstructed from a dozen years back; many witnesses were no longer alive or available; records had disappeared or had been thrown away in the daily course of events. It officially began on May 31, 1949, in the United States District Court, Southern District of New York, with the Honorable Samuel H. Kaufman as presiding judge.

Thomas F. Murphy, Assistant United States Attorney,* was the major prosecution figure and a major figure of a man as well, standing some six feet four inches tall, with a walrus mustache that almost completely covered his mouth. His bulk alone would have made him the center of attention in any arena. Thus, the splendid mustache seemed almost gratuitous—an explanation point on a man who physically lacked little. Only his appearance was out of the ordinary, however, for when Murphy spoke it was common speech without pretension and often without regard for syntax or grammar. There were no airs about Mr. Thomas Murphy. He was the kind of man any jury could trust.

As the prosecutor was a man of the people, Hiss's attorney Stryker was a man of the elite—an American Brahmin. His manners and speech were elegant. While Murphy was methodical and often plodding, Stryker was dramatic and flashy. He had been known to be moved to tears in the courtroom when painting in vibrant colors an injustice supposedly being inflicted on his client.

It took just over two hours for the prosecution and defense to agree on a jury. The ten-man, two-woman jury professed the requisite attitudes: They swore they had no connections with, opinions of, bias or prejudice against counsel and/or the publisher of *Time* magazine and no preconceptions about or antipathy to or sympathy for any of the public figures who might appear, any law-enforcement agencies, the House Un-American Activities Committee, or the principal witnesses. They could also weigh the testimony of a Communist impartially.

This last point—the believability of a Communist—kept more prospective jurors out of the box than any other. Interestingly enough, this concern ceased to be an issue at the second trial, following the blitzkrieg of publicity pumping up Chambers—much of it created by Richard Nixon. Alistair Cooke wrote, "By the autumn of 1949, the reformed Communist was in some places the most trustworthy of American patriots." Only one prospective juror was excused at the second trial for being unable to believe a sworn Communist.

---

*Presently Murphy is a United States District Judge sitting in the Southern District of New York. Ironically it is this court which will rule on Hiss's *coram nobis* petition, submitted July 27, 1978.

Mr. Murphy's opening was, in keeping with the man, matter of fact. He briefly identified the issues mulled over by the grand jury and by the House committee before them, the documents, the typewriter, and the credibility of the witnesses. It was on this last point, Murphy said, that the case would turn. "Now when you've heard all of this testimony, I want you to go back . . . and see on which side the truth lies. I want you to examine Chambers. I want you to listen attentively; watch his conduct on the stand; watch the color of his face; watch the way his features move; because if you don't believe Chambers, then we have no case under the Federal perjury rule."

This was a point that Hiss's flamboyant attorney, Lloyd Paul Stryker, was happy to endorse. It was also a point Murphy chose not to make at the second trial. As Stryker said in his summation, echoing Murphy's speech in his opening argument, "You need one witness plus corroboration, and if one of the props goes, out goes the case." Chambers was to be that prop.

Murphy said that he would prove Hiss lied before the grand jury. "How are we going to prove that?" he asked rhetorically. "Lying, as we know, is a sort of mental process. It goes on underneath bone and hair. . . . It is not like taking a photograph. We are not going to give you a photograph of this man lying. . . . Let us suppose a child of yours came home and said he or she was in school on Tuesday, and you had some slight doubt about it, but you decide to investigate. So you went to school and you found that the child was marked absent the whole day in question. That would make you begin to doubt, at least, that the child was telling the truth. But then if you pressed it a little further, and you found your brother saw the child in a movie that day, I daresay you would come to the conclusion . . . that the child was lying. . . ."

Like the prosecutor had before him, Mr. Stryker started slowly, welcoming the "calm and quiet" of this court as opposed to the "mob scene" as he characterized the committee's hearings. He bid adieu to the days of "klieg lights, the television, and all the paraphernalia, the propaganda which surrounded the beginning of this story. . . ." Those who had seen Stryker perform before recognized these words as posturing. If ever there was a man of the "klieg lights," it was Lloyd Paul Stryker, as dramatic a personage as ever stalked the halls of justice.

Stryker compared the protagonists. Hiss was a paragon of virtue, a golden boy, one of the country's best and brightest. After a brilliant academic career in two of America's foremost institutions of higher learning, he became clerk and confidant of a Supreme Court Justice.

"Alger Hiss was good enough for Oliver Wendell Holmes . . ." Stryker said. Next he discussed Hiss's secretaryship of the Dumbarton Oaks Conference, where the United Nations Charter was hammered out. "He was weighed in that crucible and not found wanting." He concluded Hiss's impressive résumé with this line: "I will take Alger Hiss by the hand, and I will lead him before you from the date of his birth down to this hour, even though I would go through the valley of the shadow of death, I will fear no evil, because there is no blot or blemish on him." Only Stryker could deliver as unctuous a line as that.

Recalling the prosecutor's opening soliloquy about a child's lying, Stryker said, "You would want to know . . . who was the person that made this charge against your child? What kind of person is he? Is he an honest man? Had he been a God-fearing, truth-loving man? How did he live? What had he done?"

Then he turned to Chambers. "Now who is the accuser?" A man of a thousand names. Someone who launched into this alias gambit even before becoming a Communist. "Alias Adams, alias Cantwell, alias a great many other things." Then he joined the Communist Party, this "nefarious filthy conspiracy . . . against the land that I love and you love." This conspiracy was pledged to overthrow the government. For Chambers, this was not a spur-of-the-moment thing; it was a "considered choice." This same man was thrown out of college for writing a "filthy despicable play" about Jesus Christ. And then he lied his way back into Columbia University.

For Stryker the comparison was plain: It was Alger Hiss the good versus Whittaker Chambers the evil. The antagonists clearly looked the part. (Even Prosecutor Murphy felt the need to explain away this physical typecasting.* He described Hiss as "a clean-cut, handsome, intelligent . . . male," while Chambers "is short and fat and he had bad teeth.") "It was this evil man," continued Stryker, who cast the first stone *ever* at Alger Hiss. "The first man in the world!" he cried to the jury. He described Chambers as a moral leper, "unclean, unclean!"

The foreman of the jury was Hubert Edgar James, a General Motors Acceptance corporate executive. The other eleven jurors included an accountant, a superintendent, an office manager, an unemployed hotel

---

*Hiss, too, thought his appearance and demeanor would be an asset in the courtroom. He said to the authors, "I am an arrogant man, and I don't suffer fools gladly. I just couldn't believe that anyone seeing us [Chambers and Hiss] face to face would believe that creep."

manager, a credit analyst, a real-estate broker, a clerk, a dressmaker, a secretary, an advertising man, and a production worker. Not precisely a jury of Hiss's peers, but his attorney felt he could work with them.

As the trial began, the wife of Hubert James was a patient at the St. Francis Health Resort in Denville, New Jersey. In a conversation with other guests, Mrs. James commented that her husband was the foreman of the jury in the Hiss case. "If they knew his sympathies," she is alleged to have said, "they wouldn't have picked him. If it's up to him, Hiss would get away with it."

This conversation was reported to the FBI on the second day of the trial. Mr. James was the subject of a private conversation between Judge Kaufman, Prosecutor Murphy, and Defense Attorney Stryker. Murphy argued that Mr. James was obviously biased and an alternate juror should be substituted for him. An article that appeared in the now defunct Hearst paper the New York *Journal American* described the conference: "Mr. Murphy, in protesting, contended a juror must be like Caesar's wife: above suspicion, and therefore this juror should be excused." Judge Kaufman, however, was not persuaded, and James remained on the jury.

When the first trial ended with the jury unable to reach a decision, Nixon was angered by Judge Kaufman's ruling on James. He wanted to see Mr. James prosecuted for sitting on the jury with an *a priori* bias, perhaps to serve as a warning to the next dozen jurors. Nixon called the FBI to advise them about him. An interview was arranged. Quoting from the FBI files:

> Congressman Nixon was interviewed by Special Agent Courtland J. Jones on July 11, 1949, at which time the Congressman stated that he had received information from a reliable source concerning Hubert E. James, foreman of the Hiss jury. According to the Congressman's informant, James' superior, James Schick, was interrogated . . . concerning an allegation that Hubert James had played an active part in the Madison Square Rally for Henry A. Wallace a few months ago. . . . Mr. Nixon said that his source also indicated that Mrs. James is a member of the "left-wing group" of the League of Women Voters. . . . He feels that if proof could be obtained, reflecting that James became a juror with the preconceived idea of Hiss' innocence, that would be a violation and James should be prosecuted.

The prosecution began its case against Alger Hiss. After some laborious preliminaries where they checked Hiss's leases and read the indictment—but not before it was witnessed by the grand-jury secretary and two court stenographers who received it—Murphy called for Whittaker Chambers. A corpulent man in a rumpled suit took the witness stand and began telling the story of his life, beginning with his birth in 1901 in Philadelphia. Stryker, for one, was not much interested in this story and exploded with a series of objections during the recitation, which set the tone for the rest of the autobiography. It became a sour litany of "immaterial(s)" and "irrelevant(s)," which soon wore very thin.

They limped through Chambers' years at Columbia; his joining the Communist Party, where he had the lowly task of collecting unsold copies of the *Daily Worker* from newsstands. Soon he became a writer for this paper and rewrote stories from other papers, giving them "a Communist slant." Next he was promoted to foreign-news editor, the same post he was to hold with *Time*. Stryker objected tirelessly to this line of questioning. Finally Judge Kaufman took the matter under advisement and the next morning ruled that Chambers' life story could be "self-serving" since it could not be corroborated. Consequently Murphy was asked to sketch the tale lightly rather than etching it in stone as had been his wont.

The story told by Chambers about the critical Hiss period was much the same as the one he had testified to dozens of times before: the apartment loan, the Oriental rug, Colonel Bykov, the purloined documents, his break from the Party in 1938 (he had originally told HUAC and the FBI it was in 1937, which was prior to the datelines on the Pumpkin Papers that were typed in the first four months of 1938), and the old Ford car. Chambers testified that Hiss never *gave* him the car; he merely borrowed it occasionally. New this time was a second automobile—also a Ford—which was to prove troublesome to Alger Hiss.

Chambers claimed that Hiss loaned him exactly $400 for a new car, which Mrs. Chambers purchased late in the year of 1937, after the deadline as specified in Count II of the indictment. At the trial, Murphy asked him about the size of the loan. "Not four hundred one dollars or three hundred ninety-nine but four hundred dollars?"

"Four hundred," Chambers said, "in cash."

The FBI was able to show that just before the purchase on November 19, 1937, Hiss had made a withdrawal that nearly emptied his account with the Riggs National Bank, leaving a meager $40.56 balance. Hiss countered that the $400 withdrawal was made to buy furniture in anticipation of the move to a duplex on Volta Place, a much larger domicile. The FBI searched the files of furniture stores in Washington and its environs but could find no records of any such purchases made eleven years before by the Hisses. Hiss contended that they had paid cash for the purchase, as evidenced by the large withdrawal. The FBI found the Volta Place house listed in the want ads of the Washington *Post* on Sunday, December 5, 1937, more than two weeks *after* Hiss's bank withdrawal. Hiss countered that he had a commitment from his rental agent, but, unfortunately for him, the agent testified that he had no authority to make such a commitment. Hiss replied that the agent was confused because the lease was signed on December 2, before the ad appeared in the *Post*.

The prosecutor was unwilling to contend that Hiss nearly emptied his savings account to buy an automobile for another man. Murphy merely presented this veneer of circumstantial but nevertheless compelling evidence: Hiss did indeed make a large withdrawal weeks before signing a lease, and four days after the withdrawal an impoverished Whittaker Chambers was at the wheel of a new Ford.* It was this car in which Chambers claimed he drove his family to Florida, effecting his break from the Communist Party.

Mrs. Chambers, however, expressed some confusion about this transaction. When questioned by Hiss's attorneys at the Baltimore libel deposition, she said, "Mother comes in there some place. I don't know. Mother did help us out at various times. She probably gave us the money for that."

The FBI papers reveal that although "sure" of the source of the loan, Chambers was less sure about the amount, despite his testimony under

---

*Writer Fred Cook and Attorney Ray Werchen discovered some shocking abnormalities involving this car transaction which they wrote about in *The Nation* on May 28, 1973. A photostatic copy of a single page of a ledger from the Schmidt Motor Company was introduced at both trials, and this "Exhibit 40" was accepted without challenge by the Hiss attorneys. According to the ledger, Esther Chambers purchased the new Ford on November 23, 1937. Her transaction is on the right-hand side of the page, while the left-hand side shows a purchase made by another customer on October 30—a three-week difference from one side of the page to the other. This would be a very unlikely time lapse in a busy car agency. The dates listed are in different styles with the left-hand entry reading "10/30" and Chambers' entry written as "11/23/37"—the year is added, in the opinion of the authors, "as if for emphasis." Another discrepancy is a significant difference in the folio numbers between the two transactions: Chambers' Ford is listed as "#335," while the left-hand entry is "265"—a difference of 70 numbers. In addition, Chambers' address is incorrect. Finally, the body style of the car is listed as a "For-door," which is a shorthand notation never used by Ford. The company called a four-door a "Fordor." From this and other evidence, Cook and Werchen conclude, ". . .The Chambers entry was probably made up and filled in on a blank section of this ledger page at a much later date." In other words, the purchase date appears to have been changed to correspond with Hiss's $400 withdrawal.

Murphy's lead. In an FBI summary of 184 pages written by Chambers after thirty-nine separate interviews with agents Spencer and Plant conducted between January and April 1946 (which the defense never saw until access was granted by the Freedom of Information Act), Chambers says, "I told him [Bykov] that I wanted to get the car immediately and that I would need about $500. Bykov said that he did not have that much money with him. . . . I subsequently asked Alger and Priscilla Hiss for this money and they agreed to loan it to me. It is my recollection that Priscilla Hiss told me that to get this money, it was necessary for her to close out her bank account. I further believe that this account was either in the main office of Riggs National Bank or at their DuPont Circle Branch. . . . As I have stated, it is my recollection that there was approximately $500. I gave this money to my wife to be used in the purchase of a new car." And again later: "Bykov had given me approximately $2,000 for my wages and rent for the photographic workshops [and] for the repayment of the $500 loan to Alger Hiss. . . . This money I kept and I did not repay the loan to Alger Hiss."

Why did Chambers amend the amount to $400 at the trial and exactly $400 at that? Hiss's Coram Nobis Petition speculates, "We assume that at some time between the date Chambers discussed the matter with his FBI interrogators and the date of his trial, Chambers discovered that the Hisses had withdrawn $400 from their bank in November 1937 and tailored his testimony to meet this fact."

Allen Weinstein is unpersuaded. He notes that Hiss's records at the Riggs National Bank in Washington were photographed by the FBI in February 1949. The originals were returned to the bank and "The records were not sent to New York where Chambers was then being interrogated by an agent of that FBI office, but remained in the Washington field office." It is curious that Professor Weinstein did not consider the possibility, but it is very likely that the FBI's Washington field office was in telephone contact with the New York office and that this information could have been transmitted that way.

When it was Stryker's turn to cross-examine Chambers, he leaped out of his chair as if he couldn't stand a moment more of these lies. Beginning with the words "Mr. Chambers, do you know what an oath is?" Stryker elicited from the witness a host of admissions of lies, major and minor perjuries, and examples of aberrant behavior: the fact that he suppressed evidence before the grand jury, producing the damaging packet from the dumbwaiter shaft only *after* the libel suit. (In Chambers' mind this omission was done to protect his ex-friend Hiss. He called it his "Christian duty . . . not to hurt him any more than possible"); the fact that after breaking with the Communist Party and going into hiding in Florida, he slept by day and watched "through the night with a gun or revolver within easy reach"; the fact that he had written a pornographic poem, "Tandaradei," which appeared in Samuel Roth's quarterly *Two Worlds*. The reader will recall that it was Roth who knew Chambers as Crosley, but Hiss's attorneys felt his publishing ventures were so dubious that his testimony might prove detrimental to the case and did not call him as a witness.

Although tepid by today's standards, the poem "Tandaradei" was not in 1949. In the courtroom Stryker claimed the poem offended his tender sensibilities, so he asked the author, Whittaker Chambers, to read it. Chambers' recitation couldn't be heard, however, so the court reporter did the poetry reading with so little warmth it quickly defused the bombshell that Stryker had been hoping for. One stanza read:

> And as you draw your limbs like a pale
> Effulgence around me, I must have them drawn into me,—as you fail
> And begin to leave me. You shall be a hand thrust.
> Into my flesh; your hand thrust into me impales.
> My flesh forever on yours, driven in through the bodycrust.

Stryker pried from Chambers other admissions of deviant behavior—"a tactic," Alistair Cooke wrote, which was "clearly designed to build up a plausible body of facts that might suggest an emotionally pathological background." He elicited from Chambers the fact that he had been charged with stealing books from the library, the fact that as a lad of seventeen he ran off to New Orleans and lived in a "dive" with drunks and a prostitute called "One-eyed Annie," so named for the cataract over one eye. Stryker asked what became of Chambers' grandmother who had lived in the family home in Lynbrook, Long Island. "She went insane," Chambers said quietly.

Stryker was a grim reaper. He dredged up the suicide of Chambers' younger brother, Richard, an act that left Whittaker Chambers immobile for weeks after. Stryker tried to equate immobility with paralysis but met with little success. In his book, Chambers described brother Richard as "unlike me," an outgoing, popular, "uncomplicated" youth who had been on the track and baseball teams in high school. Leaving Colgate College before finishing his freshman year, Richard came home disillusioned for unexplained reasons and turned with a vengeance to the bottle. He became morbid, talked often of death and tried to enlist his older brother in a suicide pact. When the offer was refused, Richard said, "You're a coward, Bro."

Later Chambers saved him from a suicide attempt. Richard thanked him with the oath "You're a bastard, Bro. You stopped me this time, but I'll do it yet." At his mother's request, Chambers stayed up through the nights watching over his younger brother until Richard married and moved into an apartment of his own. Shortly thereafter, Richard was dead at his own hand. He left his family this sad epitaph: "We are hopeless. We are gentle people. We are too gentle to face the world."

Under Stryker's prodding, Chambers next admitted cohabiting with a woman, Ida Dailes, in his mother's home. His mother, Chambers had once said, condoned this illicit relationship "because she had lost one son and did not want to lose another." Stryker asked Chambers if it were "a low-down, scoundrel thing for you to bring a woman, not your wife, in and foist her in your mother's home?" Murphy objected, and the judge ruled this matter was for the jury to decide.

The witness was by this time no stranger to testifying. Chambers did a good job for the six days he was on the stand. Only his eyes betrayed any uneasiness, and it was not their expression but more their direction as he stared off into space, avoiding the eyes of the defense attorney, the defendant, and the jury. Alistair Cooke wrote of his performance:

> To the end, this bulky, pale man with the expressionless, translucent eyes had told what he knew in the manner of one long resigned to a life of profound error and disillusion and the hope that perhaps a little peace and quiet before the end came. He had sat there for six days, but at the end he appeared no more tired than when he had come in, as if he had passed beyond tiredness years ago and turned abjectness almost into a social attitude. . . .

After Whittaker Chambers stepped down, the prosecutor paraded witnesses by at a dizzying pace, such as Barbara Morse from Columbia University, who said Priscilla passed a typing test while a student there in 1927, proving that she at least had the skill to type the documents. Also testifying was Dr. Meyer Schapiro, likewise from Columbia, who said he purchased four rugs for Chambers from the Massachusetts Importing Company in New York City. Schapiro claimed that he received the rugs on December 29, 1936, and sent them to Washington a few days later to a man named Silverman. The government also introduced into evidence a delivery ticket from the Massachusetts Importing

Company showing the purchase price as $876.71 as well as the date of delivery to Schapiro as December 29, 1936. Since this would have meant the shipping to Washington occurred "a few days later," in 1937 (after the January 1, 1937, deadline as specified in Count II), it is shocking that the Hiss attorneys chose not to cross-examine this witness.*

As Dr. Schapiro was probably the most important witness addressing himself to Count II, the most important one for Count I was Ramos Feehan, the FBI's document expert who said the typewriter used to type Hiss family correspondence *also* typed the Baltimore Exhibits produced by Chambers. Again, there was no cross-examination of Feehan, and this later turned out to be a genuine tactical mistake.

Murphy also called Esther Chambers to the stand to testify on the second count of the indictment—i.e., that Hiss never saw Chambers after 1936. Whittaker Chambers' wife was a small, rather severe-looking woman simply dressed in a plain gray suit. On the stand, Mrs. Chambers was nervous, especially during Lloyd Stryker's cross-examination, and Stryker did little to put her at ease.

In the beginning, the prosecutor wondered if the "first and only wife" of Whittaker Chambers lived and worked on a farm. "I milk eighteen cows and take care of some forty head of cattle, plus some chickens. I guess that's all," she said softly. Under Murphy's prodding, Mrs. Chambers proceeded to describe the many meetings she had with the Hisses, especially Priscilla. Over the two or three years they were friends, there were mutual visits to the zoo and to parks with the children, trips on the Potomac, a shared anniversary and holidays together, including a New Year's Eve celebration at the Hisses' Volta Place home at the end of 1936. Mr. Murphy gently pointed out that the Hisses didn't move there until the end of 1937. Then it must have been the last day of 1937, amended Esther Chambers, in a voice that rarely registered above a whisper. Prosecutor Murphy would often ask her to speak up, which she could manage only for a few lines before becoming inaudible once again. During Murphy's examination, Judge Kaufman stepped down from the bench and sat in the clerk's chair so he could better hear what Mrs. Chambers was saying.

When his turn came, Stryker wondered why Mrs. Chambers had

---

*Hiss said the rug was in his home more than six months before this date. Clydie Catlett, the Hiss maid, also testified to this fact. It is conceivable that there was another rug purchased and delivered months prior to the four rugs shipped by Schapiro, but the defense did not raise this issue.

testified in this courtroom and also in a deposition given in Baltimore that the New Year's Eve party took place in 1936 at the Volta Place house since the Hisses then still resided at 30th Street. To this, Mrs. Chambers, after several long pauses and false starts, concluded the New Year's celebration was probably at 30th Street and the party at Volta Place was perhaps a housewarming. Although she had been confused by the dates in this convoluted story, she had less trouble with the decor of the Hiss residences at 30th Street and Volta Place.* In his summation, Murphy, taking some license, was to say of her recollection, "She was either there or she was psychic."

As a preface to his cross-examination of Mrs. Chambers, the defense attorney reviewed the crimes and perjuries of Whittaker Chambers' life—a theme the jury was becoming well acquainted with. Asked if she was thoroughly familiar with her husband's philosophy, his connections with the Communist Party, and his commitment to a criminal conspiracy to overthrow the government by force and violence, Esther Chambers said she shared it with him. She lovingly described her husband as a "decent citizen, a great man."

Stryker inquired whether their daughter also had an alias during the period that the Chambers traveled under a variety of names. Mrs. Chambers admitted calling their daughter Ellen "Ursula Breen." "Did that bother your conscience?" he asked. Mrs. Chambers answered, " It worried me."

Nearing the end of the government's case, Murphy introduced the forty-seven stolen documents and two rolls of microfilm showing eight documents into evidence. At last on display for the world to see were the papers that committee chairman Karl Mundt described as of "such startling and significant importance and [they] reveal such a vast network of Communist espionage in the State Department, that they far exceed anything yet brought before the Committee in its ten-year history."

It took nearly two days for the documents to be introduced into evidence. The meticulous Mr. Murphy read them aloud as Walter Anderson of the record branch of the State Department monitored the

---

*The Hisses resided at 30th Street from July 1 to December 29, 1937, and at Volta Place from the end of 1937 to 1943. Murphy was attempting to prove that if Esther Chambers knew the decor of these two houses—especially the Volta Place house—she and presumably her husband saw Hiss after the January 1, 1937, deadline as specified in Count II of the indictment.

recitation for accuracy. Also part of the production were two prosecutors' assistants who turned over five-foot photographic enlargments in timely fashion with Murphy's words for the benefit of the jury.

Although to see and hear "secret" documents was titillating in the beginning, after a morning of the task, Stryker, for one, could take no more. He interrupted the prosecutor to offer this compromise: "I have a suggestion that may be acceptable to Mr. Murphy and perhaps the court. With the next—I think there are thirty-seven exhibits yet to be taken up—if Mr. Murphy will be kind enough to sit down with us after court . . . I think we could very promptly determine, and I expect there will be no objections, and they can go in as a group and we can get on. We have spent the whole morning . . . and we are down to Exhibit Ten." Murphy objected and was allowed to continue. Toward the end of the second day, Judge Kaufman himself could take no more, so at his suggestion the last few were accepted with no reading at all.

The Baltimore documents copied by typewriter were primarily incoming cables to the State Department from American consulates and embassies. There were accounts of diplomatic conversations, especially with representatives of what were soon to be the Axis Powers: Germany, Italy, Austria, and Japan. Several discussed Hitler's pressuring of Austria's Chancellor,* while the cables from the Far East concerned Japanese troop and supply movements in the Sino-Japanese War. The microfilm (Pumpkin Papers) were primarily an internal file of trade agreements from the State Department; two cables for the Secretary of State and cables from Hiss's chief, Assistant Secretary of State Sayre; also memoranda from the State Department; four handwritten documents which Miss Eunice Lincoln, a State Department secretary, identified as being in the handwriting of Alger Hiss (Stryker conceded this); and a lengthy aide-memoire discussing a trade agreement between America and Germany and written in German.

Some of the microfilmed documents even had Hiss's initials on them. Commenting on this fact, John Chabot Smith, a recent Hiss biographer, posed this rhetorical question: "Whoever heard of a spy carefully putting his initials on a stolen document . . . so that it could be easily

---

*The reference above is to Kurt Schuschnigg, who was Chancellor of Austria in the immediate pre-World War II era. Schuschnigg prospered under the protection of Mussolini, whom he skillfully played off against Hitler. When he lost the Italian dictator's support, he rapidly encountered difficulty with the Nazis and was imprisoned by them at the start of the war.

traced to him if it fell in the wrong hands?" Also as noted by Smith, the set of trade agreements was incomplete. "It showed all the work done on the subject in the Trade Agreement Section of the State Department, where Julian Wadleigh worked, but it didn't match the set of documents on the same subject in the office of Assistant Secretary of State Francis Sayre, where Hiss worked." Miss Lincoln and Walter Anderson, also from the State Department, testified that many of the documents had never been routed through Sayre's office where the defendant worked, and Anderson said the dispatches, cables, and strictly confidential reports were distributed to some fifteen agencies. And although records were kept on where the documents went, none was kept on who returned them.

The government did call another suspect in this matter, Julian Wadleigh, to the stand. The former State Department employee—now unemployed—looked disreputable. His blue coat was wrinkled, his shirt was soiled, his tie hung well below his belt, his socks, unsupported by garters, rolled around his ankles, and he sported a four-inch-high shock of tight brown curls which today might be called an "Afro." Following Mr. Murphy's lead, the minister's son traced his personal history: his move abroad as a child, a B.S. from London University, an M.A. from Oxford, and further postgraduate work at the University of Chicago. Although, Wadleigh said, he never actually joined the Party, he did agree to "collaborate" with the Communists. He admitted taking unauthorized documents from his *own* desk and giving them to a man known as David Carpenter, who had the same role as Chambers in another apparatus. When Carpenter was not around, he gave the pilfered documents to Chambers, who used the name Carl Carlson.

When it was Stryker's turn, Wadleigh maintained he never removed documents from *other* offices. "I never did anything so foolish," he said with conviction, as if the very thought was beneath contempt. Stryker wondered if he believed in such general tenets of the Communist Party as lying. "I would hardly call that a tenet," Wadleigh said pedantically in his slight English accent which Stryker disparaged as an "Oxford accent." "I would call it a procedure," corrected Wadleigh.

"I am not going into semantics," Stryker thundered back. "I didn't go to Oxford." He rephrased the question so that even a man with Wadleigh's education might understand: "If it were necessary to tell a lie, was that a policy that met with your concurrence?" Wadleigh could concur with that.

Later Stryker got another witness, Charles Darlington, a former assistant chief of the Trade Agreements Section of the State Department, to say that when he returned to his office after lunch, he would occasionally find Wadleigh reading a document off his desk. He had a "well-developed curiosity," Darlington said, "in a lot of things that were going on." He also recalled seeing Wadleigh alone in Alger Hiss's office but said, "I never gave any particular thought to that."

Following his initial refusal to testify before HUAC where he took the Fifth Amendment, Wadleigh was to become a most cooperative subject for HUAC, the FBI, and the grand jury. To an agent interviewer he said he was "genuinely amazed" when Chambers named Alger Hiss as a member of his Communist apparatus when the case broke. In fact, Wadleigh and Chambers often discussed fellow members and travelers of the Party. The FBI files say of this particular session, "He [Wadleigh] described Hiss as an individual whom he believed to be the genuine liberal, but with a definite conservative approach. He stated that even after reading the early accounts [of the case] in the paper, he was disposed not to believe that Hiss was guilty of underground espionage involvement." Wadleigh, the confessed spy, was never prosecuted, and perhaps a deal had been worked out between Wadleigh and the prosecutor.

Wadleigh was important to the prosecution because, other than Chambers, he was the only one to testify to the existence of an espionage network, although he could not say for sure that Hiss was a member. Wadleigh was also important to the defense, for they tried to deflect some suspicion on him. In reality, however, Wadleigh's participation was not central to the documents that Chambers produced in Baltimore. Wadleigh did admit to stealing over 400 documents to be turned over to the Russians, and it certainly is possible that he took some from other offices in the State Department, including that of Alger Hiss; still, he could not have stolen *all* of those that appeared in Chambers' damaging packet for reasons identified below.

The Baltimore documents covered the period from January 5 to April 1, 1938. Chambers wrote that the selection of material in his so-called life preserver "was not aimed at any individual. There was much of the Hiss material because he was the most productive source. . . . There was no material from Julian Wadleigh because, in the spring of 1938 [actually Wadleigh left on March 9], he was out of the country on a diplomatic mission to Turkey, though I believe he had returned by the time I broke." Wadleigh testified that a couple of weeks before he left, he was instructed by Chambers, "Don't deliver any more documents for the time being."

Three of the five rolls of microfilm and a section of one of the typed documents were not introduced at the trial. People were led to believe they were still too confidential despite the passage of eleven years. The deleted section of the typed document was on display at the second trial, however. It discussed possible French and Russian alignment with Germany before World War II, scarcely a topic of major concern in 1949.

Away from the klieg lights, no one was willing to speculate on the intelligence value of the documents or rolls of film. All the experts would say was that since the cables had been sent in code (and presumably picked up by the Russians), the Russians having been supplied the decoded messages had, no doubt, broken the code. Thus, it followed that they would have access to all subsequent messages sent in this code.

It took more than a quarter of a century for the country to discover just how confidential the three suppressed rolls of microfilm were. Richard M. Nixon had described the entire cache as "conclusive proof of the greatest treason conspiracy in this nation's history."* On July 31, 1975, the government, complying with the Freedom of Information Act of 1966, released copies of the film. Of the three suppressed rolls, one was overexposed and blank, while two others contained barely legible copies of Navy Department documents written by Rear Admiral A. B. Cook which discussed such innocuous matters as fire extinguishers, chest parachutes, and life rafts.

Alger Hiss was quick to call a press conference the day after their release. The former State Department official said on August 1, 1975, "I could not possibly have seen those memos. They certainly are useless for espionage." Hiss also charged that these particular Navy documents

---

*Nixon's sense of the meaning of the term *conspiracy* became far more refined as a consequence of his experience over the years. He himself was named an unindicted conspirator in a "conspiracy involving deceit, craft, trickery, and dishonest means" by the Watergate grand jury in 1974, and later undertook at some length to instruct television interviewer David Frost in the finer nuances of conspiratorial behavior in May of 1977, saying in vintage "Nixonese," "Let's get it clear what a conspiracy actually is. . . . If a cover-up is for the purpose of covering up criminal activities, it is illegal. If, however, a cover-up . . . is for a motive, that is not criminal, that is something else again."

were never confidential; in fact, on issue, they were displayed on open shelves in the Bureau of Standards library and available to the public. Most of Hiss's charge has been confirmed. Columnist I. F. Stone checked with the information chief at the Bureau of Standards and wrote of his findings on the Op Ed page of the New York *Times* on April 1, 1976: "Access to them was unrestricted," Stone was told, but no one there could recall "just how they were displayed." Stone concluded, "When Harry S. Truman, a week after the midnight scene at the pumpkin, dared to call the Committee's inquiry 'a red herring,' its chairman, Karl Mundt, challenged the president 'to authorize publication of all the documentary evidence the Committee had.' It is a pity now that Truman did not accept the dare."

Before Lloyd Stryker put his client on the witness stand, he endeavored to show the jury just what type of man they were being called on to judge. The tactic was to be greatness, or at least "innocence by association," as Nixon had once so cynically described it. Stryker began by reading a deposition from Illinois Governor Adlai Stevenson, who was too busy with state business to make a personal appearance. Stevenson, who had known Hiss since his tenure at the Agricultural Adjustment Agency in 1933 and later worked with Hiss at the State Department, characterized Alger Hiss's reputation for integrity, loyalty, and veracity as "good." Murphy countered this deposition with a cross-interrogatory where the governor admitted that he had never been in the home of Alger Hiss, nor had he ever heard prior to 1948 that Hiss was being called a Communist or sympathizer or that he had unlawfully removed documents from the State Department.

If the courtroom thought the governor of Illinois would be difficult to top, they were wrong. "I will call Mr. Justice Frankfurter," Stryker announced proudly. As the diminutive Supreme Court Justice took the stand, Judge Kaufman stepped down from the bench to shake his hand in what appeared to be a spontaneous reaction. Rarely, if ever, had a witness been accorded such a welcome. When order was restored, Judge Frankfurter said that while on the faculty of Harvard Law School, he was the one who selected Alger Hiss to be the law clerk of Justice Oliver Wendell Holmes. He described Hiss as the best qualified for this important and prestigious position. Speaking of Hiss's integrity, veracity, and loyalty, he characterized them as "excellent." Justice Frankfurter said he never heard Hiss's reputation "called into question."

Mr. Murphy was not shaken by the appearance of so important a personage as Justice Frankfurter and was willing to cross swords with him over the goodness of Hiss's character. There was this byplay:

MURPHY: *Didn't you hear in 1944 that it [Hiss's reputation] wasn't so good?*
FRANKFURTER: *I think Judge Jerome Frank had a difference of opinion with Mr. Hiss in the Department of Agriculture which I heard contemporaneously and did not bear on questions of loyalty or integrity.*
MURPHY: *It didn't, Judge?*
FRANKFURTER: *Not as far as my memory goes.*
MURPHY: *But you remember talking to Judge Frank about it?*
FRANKFURTER: *No, I remember his talking to me.*
MURPHY: *Then I assume that you talked to him when he talked to you?*
FRANKFURTER: *Well, let us not fence. All I meant to say was—*
MURPHY: *Well, you were the one that started fencing with me, weren't you, Judge?*
FRANKFURTER: *I am trying to answer as carefully as I can with due regard to your responsibility and mine and the jury's. . . .*

Then Justice Frankfurter admitted that he had a "vague memory" of having heard of an altercation in the Department of Agriculture between the lawyers and the non-lawyers.* Although he could not remember what the difference of opinion was about, he was sure it in no way "affected [Hiss's] loyalty to this country or involved the slightest betrayal of this country."

From one Supreme Court Justice, Stryker went on to another, Stanley Reed. Hiss had been on his legal staff when Reed was Solicitor General of the United States from 1935–1938. Justice Reed said the defendant had a good reputation which, as far as he knew, had never been questioned. The prosecutor's effort to show that Judge Frank had questioned Alger Hiss's loyalty and integrity was somewhat shaken when he asked Mr. Reed who recommended Hiss for his staff. "Judge Jerome Frank," said Reed.

How did the defense attorney and prosecutor react to the witnesses

---

*This was a policy matter concerning the treatment of tenant farmers on cotton plantations. The non-legal Cotton Section of the Department condoned the action of plantation owners who fired tenant farmers for political action, such as union membership. The legal section took the opposite stand. The opinion, as written by Hiss, stated that political action did not make a tenant a "nuisance" and such activists could not be evicted from plantations. The non-legal section won on this matter of policy, and the legal section was disbanded. By this time, however, Hiss was on loan to the Nye Committee. This matter is discussed further in Chapter 10.

called to this point? Alistair Cooke framed well Mr. Murphy's and Mr. Stryker's thinly veiled disdain for each other's attesters. "Where Mr. Stryker had adopted an attitude of nauseated contempt for Chambers, Mrs. Chambers, and Wadleigh, Mr. Murphy conveyed the subtler imputation that the Hiss witnesses were sentimentally united in a rather snobbish plot to prove that the defendant was altogether too charming a type, too scholarly a lawyer, and too devoted a husband to be capable of associating with such low characters as the Chamberses and the Wadleighs of this world."

Before Mr. Stryker would turn to the main course, Alger Hiss, he had one more appetizer to serve up. This was the literary critic Malcolm Cowley,* who had once been editor of the *New Republic*. Although never actually joining the Communist Party, he registered as a Communist for voting purposes in the 1932 and 1936 elections. No one could quite understand what he was doing on the witness stand until he began his story.

Around December 13, 1940, Cowley received a phone call at his Connecticut home from Whittaker Chambers' secretary. She said that Mr. Chambers was preparing an article for *Time* to be entitled "People Who Jumped Off the Moscow Express." The article was to be about authors such as Cowley, Waldo Frank, Lewis Mumford, and Granville Hicks, all of whom had once been Communists but had later changed their allegiance. The secretary asked if Mr. Cowley cared to make a statement. Cowley said since this was such a sensitive matter, he preferred to meet the writer in person. This was arranged, and he met Chambers for lunch at the Hotel New Weston in New York City.

In the course of their conversation, Chambers mentioned several alleged Communists still working in the government. Cowley did not then know the name Alger Hiss, so he could not remember if his name came up. Cowley said there was one name mentioned by Chambers that he would not care to repeat even today. Stryker implored him to do just that and promised profusely that such a reference would in no way reflect on this man's venerable career in government service. "It was Francis B. Sayre," Cowley said with much reservation.

Mr. Sayre was Woodrow Wilson's son-in-law, Alger Hiss's former chief in the State Department, and at that very moment the U.S.

---

*Cowley, a well-known American expatriate, had edited the much praised *The Portable Faulkner* for Viking Press in 1946.

representative on the United Nations Trusteeship Council. According to Cowley, Chambers labeled Sayre as "The head of the Communist apparatus in the State Department." Judge Kaufman interrupted at one point to say that this testimony was being given merely to point out inconsistencies in Chambers' statements and that the charge should not be misconstrued as truth. The judge's caveat was not what Stryker needed at this point. Rather than pointing out inconsistencies, the attorney hoped to show that no man, no matter how well respected, was safe from the accusations of a disturbed man like Whittaker Chambers.

"That shocked me," Cowley opined about this particular allegation, but the prosecutor, not interested in Cowley's opinions, objected. So Stryker led him onto other things. Asked to describe Chambers' appearance at this luncheon, Cowley said, "Chambers looked as if he had slept on a park bench the night before. His clothes were old, unpressed and rather dirty. His linen was not clean. He would never look me in the eyes, but kept glancing suspiciously around the restaurant." When Stryker asked whether Chambers had commented about anything while furtively glancing at the midday crowd, Cowley answered, "He said something to the effect that we were surrounded by spies, traitors, and conspiracies."

Cowley said that he told Chambers he was glad he had never actually joined the Party because, in his opinion, "All former Communists had been warped by their experience, that they felt the loss of something and could be likened to a bunch of defrocked priests." Chambers had disagreed, Cowley said. Chambers claimed that he was glad he had become a member because it gave him an opportunity "to learn their methods, and I am going to use their methods against them."

This statement about tactics was what Stryker had been waiting for. It showed that years after Chambers left the Communist Party—he told the committee he made the break in 1937 and the court in 1938—he was still willing to use their methods, such as lying, for his own purposes. In his summation, Stryker jumped with a vengeance on Chambers' words about tactics. He described Chambers as "a man that for twelve years or so was an enemy of the republic that we love, a blasphemer of Christ, a disbeliever of God, with no respect for matrimony or motherhood." Then having sufficiently warmed to his subject, Stryker continued that Chambers, having told the committee he broke from the Party in 1937 and disavowed "Stalin's tactics," told Malcolm Cowley in 1940 that he intended to use such tactics again.

Cowley, on the basis of this conversation, concluded that Chambers was crazy. The luncheon was so disturbing to him that immediately upon his return to Connecticut, he made notes on the conversation. These notes, entitled "Counter Revolutionary," were read to the jury.

One person, reportedly well known to the Chamberses, was Hiss's stepson, Tim Hobson. It is important to note that Hobson did not testify at either trial. Hobson remembered the Chambers family. He recalled their brief stay at his family's P Street home while the Chamberses waited for their furniture to use in the apartment subleased from his stepfather. During this visit, Hobson was sick in bed, and Esther Chambers painted a portrait of him to cheer him up. No one in the family cared much for the painting, and it was later thrown out.

Also during the time period that Chambers claimed to have visited the Hiss home "every seven or ten days" to pick up the stolen documents and typewritten copies on "transmission day," young Hobson was bedridden with a broken leg following an automobile accident. His bedroom at the Volta Place house overlooked the front door, and with little else to occupy himself with while convalescing, Hobson was well aware of the traffic to and from the house. Alger Hiss has stated categorically to the authors that his stepson could have told the jury that Whittaker Chambers never came to this particular house.

Before the trial, Hobson, then in his early twenties, was living on the West Coast, physically as well as emotionally out of the family. Hiss said his trial reunited the family. "It brought Tim back a sense of family affection and loyalty. He offered to testify. . . ." But Hobson at this time was a homosexual, and his stepfather was loath for this to become known. Hiss feared if it did, it would make the headlines and typecast Tim sexually forever.

To testify "would have hurt him," Hiss said. "It would have ruined his life. It would come out. We knew the FBI had interrogated all his homosexual friends. He was very definitely gay at that time. . . . He was encouraged in this direction as a way of getting quick release from the Navy." A confidential source for the FBI said one theory is that Chambers' intimate knowledge of the Hiss family and his access to the typewriter stemmed from the fact he carried on a sexual relationship with Tim Hobson. In the informant's words, "Chambers, an admitted pervert, was in Hiss's household having relations with Hiss's stepson."

Chambers was aware that Hobson was a troubled boy. He told the FBI that Tim was a rather pathetic child. According to Chambers, Alger and Priscilla were greatly devoted to Tim but were very worried about his lack of masculinity. He recalled that the relationship between Hobson and Hiss, although correct, was very cold. Chambers told the FBI that Hobson could corroborate his story of closeness to the Hiss family if properly approached and under the right circumstances.

The authors asked Hiss if it were possible such a relationship could have existed between his stepson and Whittaker Chambers. "I don't believe it," Hiss said, understandably upset. "Timmy even agreed to take Pentothal" [to prove his truthfulness] Hiss added that Dr. Meyer Zeligs (author of the book *Friendship and Fratricide*) had reviewed the Pentothal tapes and that "there is absolutely no evidence" to support that allegation.

For four weeks at the trial, as the web of accusations wound ever tighter around him, Alger Hiss sat calmly, yet attentively. A foreign observer commented on his demeanor. "If he was innocent, this serenity could be only the deep well of security in a character of great strength and purity. In a guilty man, certainly, his detachment would be pathological in the extreme."

When Hiss took the witness stand it was apparent that this was the moment he had been waiting for. Following Mr. Stryker's lead, Hiss testified that he had never belonged to the Communist Party, nor was he a fellow traveler; never in his life had he furnished unauthorized documents to Whittaker Chambers; and the testimony he gave to the grand jury was accurate. These preliminaries disposed of, they turned to Hiss's imposing *curriculum vitae*. This took some time, because there was no dearth of academic honors or impressive positions in the public and private sectors: Phi Beta Kappa at Johns Hopkins and the *Law Review* at Harvard; the position with a prestigious Choate law firm in Boston; his government appointments; and his contribution to the United Nations Conference in San Francisco, where Hiss was secretary-general.

Then they played point-counterpoint as Stryker sounded Chambers' accusations and Hiss parried. Chambers said he visited Hiss every week or ten days. Hiss said that except for the two- or three-night stay while Chambers and his family were waiting for their furniture before moving into the apartment sublet from him, he hardly ever saw Chambers. Hiss thought he saw Chambers no more than a dozen times prior to the confrontation at the Hotel Commodore. Hiss said there were no parties together and certainly no New Year's Eve celebration at the Volta Place house or any other place.

There was no rent either for the sublease, Hiss said. Hiss claimed he occasionally made small loans to Chambers, and these too were never repaid. Hiss finally became convinced that Chambers had no intention of repaying him. He confronted Chambers, saying that he would forget about the money, but "any further contacts had best be discontinued." After this altercation, which allegedly occurred in June 1936, Hiss testified he never saw Chambers again until the committee brought them together twelve years later. According to Hiss, their entire relationship, which began in late 1934 or early 1935, spanned just eighteen months.

Hiss testified that the $400 withdrawal from the bank was not made to buy a car for Chambers; it was to buy more furniture for the new house, which was larger. A table, bureau, and a few Hitchcock chairs were some of the items purchased; in fact, they still had the chairs purchased on that occasion. Stryker asked if he knew a Colonel Bykov "or *Bekov*"; the defense attorney tacked on this mispronouncement each time he said it to show his disdain for Chambers and his story. "No," Hiss responded; he did not know Bykov. He did receive an Oriental rug from Chambers, who said it was from a "wealthy patron." Hiss assumed that it was given to square accounts for the rent and the loans. In Hiss's mind, it did not.

More important, however, was the fact that it was Hiss, not Chambers, who mentioned this gift to the House committee. The delay in filing the libel suit after Chambers took his dare and made his accusations in a public forum beyond the shelter of the House committee was because Hiss's attorney was in London at that time. As soon as the latter returned, the suit was filed. It was Hiss who instructed his counsel to find the typewriter and offered to let the FBI examine it. And it was Hiss who willingly turned over to the FBI typed family correspondence so they could make their fateful comparison.* "Were these the actions of a guilty man?" Stryker seemed to be saying.

Before Stryker gave Hiss to the prosecutor, they voiced this coda:

> "Mr. Hiss," said Stryker, "you have entered your formal and solemn plea of not guilty to the charges against you, have you not?"
>
> "I have."
>
> "And in truth and in fact you are not guilty?"
>
> "I am not guilty," Hiss said solemnly.

---

*It is most important to note that Hiss did not have to introduce the typewriter nor turn over typed correspondence to the FBI. The rules against self-incrimination do not give the prosecution subpoena power over the defendant's possessions.

Then it was Murphy's turn. The prosecutor had the benefit of Nixon's help on the matter of cross-examining Hiss. In June 1949, in the midst of the first trial, Murphy received a letter from Victor Lasky, a right-wing journalist who had J. Edgar Hoover's personal seal of approval. ("He has written anti-Communist articles," wrote Hoover.) The Lasky letter to Prosecutor Murphy included this apology: "He [Nixon] hopes you don't resent his interest and I assured him you [Murphy] are not that kind of guy." Portions of the letter, which followed a two-hour telephone conversation between Lasky and Nixon, read:

> As you probably realize, Dick has a heck of a lot at stake in the outcome. Anyway, I got a couple of things which he thought you should like to know, based on his many dealings with our boy, Alger, in the House Committee. First off, about the "good" impression Alger made [probably a reference to Hiss's direct testimony]. He made a similar impression upon the House group the first time he appeared in public. But Dick's sure when you begin hammering away at him, at his inconsistencies, that impression will disappear. Dick, who is a lawyer, feels strongly Alger should be kept under cross at least three days, if possible. This of course is a difficult problem but Dick believes it's worth boring the jury rather than let Alger get off the stand with his exterior veneer unshaken. You recall from the hearing of August 25, Alger was pretty discredited. But Nixon recalls that a number of people thought in the middle of the afternoon that Nixon was keeping Alger too long and was going into too great and tedious detail. In the end, however, it paid off.

Then Nixon and Lasky as well, who thereby appointed himself to the prosecution team ("You're not the kind of guy who resents if [sic] I make myself 'assistant prosecutor,'" he wrote to Murphy in explaining such arrogance), offered sundry other techniques for discrediting Hiss.

Mr. Murphy and Mr. Hiss opened by bickering over the number of times Hiss had seen Chambers. Murphy felt the number was sixteen, whereas Hiss believed it was more like ten or eleven. So Murphy decided they would review the record together. Finally Hiss said he was counting the two- or three-day P Street visit by the Chamberses as one ("Well, I was not counting the separate motions in and out of a room," said Hiss with some sarcasm). "That would throw the tabulation off," Murphy allowed.

That perjury disposed of, they next discussed Chambers' sublease of the 28th Street apartment. Murphy asked, "Mr. Witness—" the reference he used for Alger Hiss throughout the trial—"who paid the utility bills?" Hiss said they were included in the rent. Murphy asked what if Chambers made long-distance calls? "You'd have *ran* after him?"

Murphy asked ungrammatically. "I would have *run* after him," Hiss punctiliously corrected. Murphy was angry. "You didn't mean to correct me that way, did you?" "Not at all," Hiss said. "It was answered in my normal speech."

Then they turned to the typewriter. Hiss had told the FBI that he remembered the typewriter at the Volta Place house, which, if true, could have meant that the typewriter was still in his possession in 1938 when the Baltimore documents were typed. Murphy wondered whether that was his impression today. "It certainly is not," Hiss said adamantly. "My *knowledge* today is that we gave the typewriter to the Catletts at the time when we moved from 30th Street to Volta Place in December 1937. . . . I know [that] from what the Catletts have told us. . . ."

Clydie Catlett had been the Hiss maid and cook for several years. Hiss occasionally employed her two sons, Mike and Perry, for odd jobs. The typewriter was given to them in partial payment for helping the family to move, but the Catletts had trouble pinpointing for *which* move or *when* (i.e., before, during or after) they received the typewriter, which was the kind of help Hiss so desperately needed. The move to Volta Place occurred just one week before the earliest Baltimore document, so if Hiss could prove he gave the typewriter away at the time of the move or shortly thereafter, he was on much firmer legal grounds. As Hiss told Murphy, "One does not carry an old typewriter to a new house only to give it away." Murphy objected to this quip, and it was stricken from the record.

The testimony of the Catletts came immediately after that of Justices Reed and Frankfurter. The New York *Times* reporter was struck by this strange juxtaposition of characters. He wrote, "Following the two legal experts on the witness stand were the Hisses' maid and her son whose testimony showed a scant grasp of legal procedures. While they were on, the jury and spectators laughed more than they had on any day since the trial began. . . ."

Prosecutor Murphy made short shrift of Mike Catlett. The twenty-seven-year-old man became so frustrated with Murphy's legal machinations that he said in frustration, "It's been so long ago." Murphy and Mike Catlett had this colloquy:

MURPHY: *I don't want to confuse you. Do you believe me when I say that?*
[*Mike Catlett shakes his head no.*]

MURPHY: *You don't believe me? You believe I do want to confuse you, do you?*

CATLETT: *Well, you would need to be a good friend of me if you didn't.*

MURPHY: *But you won't believe me now when I tell you I don't want to confuse, when I tell you honestly, I don't want to confuse you.*

CATLETT: *You know. I was told, I mean—*

MURPHY: *No.*

CATLETT: *When I was brought up young.*

MURPHY: *What were you told when you were brought up young?*

CATLETT: *To believe a whole lot of things like God, and about fellow like you, I mean—*

JUDGE KAUFMAN: *All right now. We will not get into a discussion of that!*

Mike Catlett's brother Perry had previously told the FBI that the Hisses "could have lived on Volta Place for several months before they gave it [the typewriter] to me." Murphy, recalling these words when Perry Catlett took the stand, asked him if the Hisses could have lived on Volta Place for several months before receiving the typewriter. "They could have," he said, and Hiss's case began to crumble with these three words.

Perry Catlett also said the typewriter was in "pretty bad condition" and needed repair. Upon receiving it, he took it to a repair shop on K Street and Connecticut Avenue in Washington. Mr. Murphy was only too pleased to point out to Perry Catlett and the jury that "the Woodstock repair shop at Connecticut and K did not come into existence until September of 1938." He wondered, "Would that cause you to fix the time after September when you took it there?" These facts notwithstanding, Catlett couldn't remember precisely when he took the machine for repair.

On the second day of Hiss's cross-examination Murphy made reference to Whittaker Chambers' brother's suicide. He asked Hiss if there was a like experience in his own background. Apparently what was good for the goose was not good for the gander, because Lloyd Stryker, who had asked Chambers about his brother's suicide, objected to this line of questioning. Judge Kaufman sustained the objection, saying that this episode had no bearing on the "credibility of this witness in the slightest."*

---

*Murphy was referring to the fact that both Hiss's father and sister had committed suicide. His intent was clear here: If both men could be portrayed as coming from flawed backgrounds, their physical and biographical differences could thereby be diminished.

Before the defense concluded its case, Stryker had a few witnesses left in his retinue, the most important being Priscilla Hiss. The first part of her testimony was confined to discrediting the testimony of Mrs. Chambers about decorating details of the Volta Place house. This was necessary because the Hisses moved there just prior to the January 1, 1937, witching hour—important as a part of count two of the indictment. Mrs. Hiss said the living room was green; Mrs. Chambers testified that it was pink. The dining room and kitchen were in the front of the house; Mrs. Chambers said in the rear. According to Mrs. Hiss, there was wallpaper in the dining room; Mrs. Chambers said it was paneled. That disposed of, Mrs. Hiss corroborated various parts of her husband's story, such as the typewriter going to the Catletts when he said it had and the $400 withdrawal being used to purchase furniture for the new house.

When it was Mr. Murphy's turn to cross-examine, he got Priscilla Hiss to admit that she probably had the requisite skills to type the documents, although she "certainly [did] not" do so. Murphy took her back to the early 1930s, when the Hisses were living in New York City on Central Park West. He inquired if she were at that time a member of the Socialist Party. She did not think so. Mr. Murphy then introduced the registry of the Board of Elections in 1932, which showed after her name the affiliation "Soc." Mrs. Hiss said that was merely an indication of whom she intended to vote for—Norman Thomas—but she was not a party member. "Mrs. Hiss," said Murphy, "don't you know that the records of the Socialist Party, Morningside Branch, list you as a member?" No, she did not know that.

When it was again the defense's turn and before building to the expected crescendo of psychiatric testimony on the sanity of the accuser, Stryker called to the stand a boyhood chum of Chambers to show how deep-seated was his abnormal and antisocial behavior. Chambers was described as a sloppy dresser and often in need of a haircut in high school. While walking home from school, he had been known to walk through a stream with his shoes and socks on, saying it cooled his feet. In high school, Chambers had written a class prophecy which the principal found offensive and consequently made Chambers write a different version. When the time came to read the speech, he read the first version.

At the first trial, this turned out to be the most learned opinion on Chambers' mental health, because before the psychiatrist, Dr. Carl Binger, could answer a single question, Mr. Murphy objected and Judge

Kaufman ruled it was for the jury to pass on Chambers' sanity. Finally there were the rebuttals and the surrebuttals, and then Mr. Stryker said the defense rested. Mr. Murphy immediately agreed that "the Government rests," and only the summaries stood between the jury and Mr. Hiss's fate.

The loquacious Mr. Stryker took almost five hours stretching over a two-day period. Addressing the jury at various times as "citizen judges of the facts" and "soldiers of justice," he thanked them for their kind and careful attention during the long weeks of the trial. These pleasantries out of the way, Stryker attacked the entire case, calling it "an outrage." He had special words for the second count of the indictment, calling it "absurd and preposterous," as if it were a crime to see someone after January 1, 1937. Actually the second count, he said, was telescoped into count one. "We are trying one question here and only one. Did Mr. Hiss *furnish, transmit, and deliver* to Chambers restricted documents in February and March 1938?"

It is the burden of the prosecutor to prove "beyond a reasonable doubt" that the defendant did those things. What is reasonable doubt? posed Stryker. "It is merely a doubt based on reason—that's all." He said that if Judge Kaufman didn't see it this way, the jurists would no doubt correct him. "There is only one man in the whole world who says Mr. Hiss furnished documents to him," said Stryker, "and that man is Whittaker Chambers, at least, I believe that is the name he is going under now."

Then harkening back to the performance of the FBI's document expert, who placed five-by-four-foot photographic enlargements of the typewriter keys on an easel in front of the jury, Stryker expressed the wish that the FBI would provide a similar service for the defense. He wanted large portraits of the defendant and the accuser and also a five-foot sign quoting the prosecutor's words from his opening, "If you don't believe Chambers, the Government has no case."

Who is the accuser? he asked. "A man who believes in nothing . . . not God, not man, not the sanctity of marriage or motherhood, not in himself. I can't think of any decent thing that he has not shown himself against. . . . Roguery, deception and criminality have marked this man Chambers as if with an iron. . . ." Chambers joined the Communist Party, embraced their tenets of "lying, stealing . . . street fighting . . . and to destroy the United States by any and all means." Here Stryker paused, searched the eyes of the jury for a moment, and then said, "If I didn't know anything else about a man, just that . . . I would not believe him if the FBI erected a stack of Bibles as high as this building,'' a structure that towered thirty-two stories over Foley Square.

Lloyd Stryker broke off his sermon on one point and said with anger, ''I wish that one part of the audience would not indulge in open smirks to the jury—that gentleman there with the earphone.'' The man with the hearing aid was William Bullitt, a former U.S. Solicitor General in 1912 and 1913. At this time he was a trustee of the Carnegie Endowment—Hiss's former employers. Mr. Bullitt, while convalescing after an illness the year before, had written a pamphlet critical of Alger Hiss. Bullitt said that John Foster Dulles suggested that Hiss voluntarily resign to relieve the Endowment of any embarrassment. Someone lied, Bullitt had concluded (Hiss said he had not been asked to resign); was it Dulles or Hiss? The pamphlet was republished during the trial in the New York *World Telegram*, and Mr. Bullitt endeavored to distribute it outside the courtroom. Stryker wanted him cited for contempt of court, but the judge, although lamenting some of the more scurrilous press attacks generating from this trial, didn't agree that Bullitt should be singled out.

Starting again, Stryker suggested Julian Wadleigh was the thief in the State Department. He could have taken the handwritten notes from Hiss's wastepaper basket or off his desk. That was his style, ''this miserably abject specimen of humanity.''

What about the microfilm? Chambers testified ''time and again'' while under oath that he couldn't remember who had given him these documents. What about the typewritten documents? The ''now famous typewriter'' was found by the defense, and it was the defense who offered to let the FBI examine it. ''But they never accepted our invitation ever.'' Instead they brought their experts and easels and large photographs and said the documents were typed ''on the same typewriter, period.'' And where did they get the documents to make their fancy comparisons? Why, from Mr. Hiss, that's who. ''You have heard the testimony of Mike and Perry Catlett—undoubtedly very ignorant colored boys but honest—that they were given the typewriter before the documents were typed. The corroboration is perfect and complete,'' said the attorney, reaching a bit. If Mike and Perry Catlett were confused about some of the details, remember that the conduct of the FBI toward the Catletts was ''close to oppression.''

Who can you believe? ''If there was ever a man in the world who has established a finer character than Alger Hiss, I don't know where that happened.'' You have watched Mr. Hiss on the stand. He didn't have to do that. He could have remained silent ''through his long, long bitter ordeal." Stryker doubted there was "any jury in the world that could sleep with their consciences and say that they believed beyond a reasonable doubt that Chambers was a truthful man."

He warned Prosecutor Murphy—a man for whom he had "nothing but personal goodwill"—to bring witnesses along when he cleared off his desk in the courtroom at the completion of the trial. Be certain that you "have left none of your handwriting around the table," he said in reference to the pilfering habits of the prosecutor's witness, Chambers. The courtroom erupted in laughter. Mr. Stryker then concluded that if he had done anything they [the jury] didn't like or offended them in any way, "hold it against me, not Alger Hiss. Alger Hiss, this long nightmare is drawing to a close. Rest well. Your case, your life, your liberty are in good hands. Thank you, ladies and gentlemen."

Mr. Murphy also exchanged pleasantries with the jury and thanked them for their "exhibition of courage," as evidenced by their refusal to have their picture taken by the newspaper people. "This is a real jury," he commented, for such a scrupulous act of conscience. Then Mr. Murphy spoke on the subject of reasonable doubt in order to correct some erroneous impressions left by the defense attorney. It would be unreasonable, Murphy said, to decide not to believe a man because you didn't like the way he combed his hair. That isn't reasonable doubt. "The doubt in a criminal case is the doubt that exists in your minds after you have applied reason."

Now let's apply reason to the immutable witnesses in this case: the typewriter and the documents. You have heard that all but one of the State Department documents were typed on a machine belonging to Alger Hiss. "Only one inference can be drawn from the uncontradicted facts—only one—and that is that defendant, that smart, intelligent American-born man gave them to Chambers." Don't be misled by the appearances of the accuser and the defendant. "Hiss is handsome, clean-cut and intelligent, while Chambers is short and fat and once had bad teeth. Mrs. Chambers is plain and demure. Mrs. Hiss is demure and attractive, intelligent to boot. Very intelligent." But look beyond such surface details.

When you scratch the surface of Alger Hiss, you'll find a "traitor, a Judas Iscariot." Judas Iscariot had a "fairly good reputation. . . . He got so close to his boss that he stabbed him. . . . And then Benedict Arnold. He came from a fine family. . . . And what happened? He is made major general, and he sold out West Point." Benedict Arnold didn't get caught, but if he had, "He could have called George Washington as a reputation witness.'' Speaking of character witnesses, Murphy advised the jury to forget those you heard here. The typewriter and the documents are the facts in this matter.

Stryker had labeled Whittaker Chambers a ''moral leper.'' Well, if he is a moral leper, what is Mr. Hiss? They were ''bosom pal[s].'' Alger Hiss is a traitor to his country, a Benedict Arnold and Judas Iscariot, Murphy repeated, then added a new name to the group, ''a Judge Manton''—a judge of the Circuit Court of Appeals who served nineteen months in prison for judicial corruption. ''Right here in this building,'' said Murphy, speaking of the Manton case, ''twelve jurors like yourselves tried a man from high places, from the United States Court of Appeals. Someone has said that roses that fester stink more than weeds, and I say that a brilliant man like this man who betrayed his trust stinks and under that smiling face his heart is black and cancerous.''

Murphy asked about Chambers' motive for risking his $30,000-a-year job. The defense thought it was because he still owed Hiss $135 in loans. ''I wondered—good God, that can't be the motive.'' The defense also speculated that it was politically expedient for Chambers to attack a Democrat of Hiss's stature in an election year. Might get him a job in the new Republican Administration. Murphy said, ''Judge Kaufman gets fifteen thousand dollars. Do you know what members of the [President's] Cabinet get? . . . Nothing like thirty thousand dollars. So that can't be it. Mr. Stryker can't prove a motive because there wasn't one. There are only facts, like the documents,'' he said, patting them almost lovingly.

Turning to the old Ford, the prosecutor reminded the jury how Hiss testified that he threw in the automobile ''to clinch the rental agreement.'' Then there was the apartment sublet. There was no written agreement made with a guy he didn't know from Adam. ''Can you imagine being forty-four before meeting that type of character—a landlord who was not concerned with wanting the rent in advance?'' Hiss gave Chambers the car, a year later even though ''the guy gyped him a little bit in between,'' because after all Hiss gave his word. What happened to that old Ford? It went to Cherner Motor Company, but unfortunately Judge Kaufman ''would not let me prove what happened after that.''

Let's look at the Oriental rug. You heard Dr. Schapiro testify about the purchase. In evidence is the bill of sale, and here, said Murphy, holding up a piece of paper, ''is the check of Dr. Schapiro.'' Hiss said he received a rug from Chambers, still has it, even, "but he has no idea why he [Chambers] gave it to him."

You remember their refrain? Hiss claims he has been aboveboard. Helped the FBI and the authorities in every way possible. Turned over to them typed family correspondence, told the FBI about the typewriter they found, and told the committee about the rug, as if he who yells for the cops *first* must be innocent. Where did they expect the evidence to go? Just disappear?* Speaking of the typewriter, Hiss told the FBI either he or his wife turned it over to a secondhand typewriter dealer in Georgetown. "They thereby eliminated all other cities in the United States," and yet Hiss claims he wanted to help the FBI. And Mrs. Hiss testified before the grand jury that their maid Clydie Catlett was dead. That was helpful too; "just eliminate her from the list of people to see."

It was Perry Catlett who couldn't remember when he received the typewriter, remembered it as before, during or after "some moving." He took it to a repair shop, "and that is true. . . . He did take it to some place . . . in 1939" because that is when the office came into existence. So then the defense parried and suggested it was another repair shop, down the street on K. "Well, we checked that, and you heard the witness [from the FBI] say that the shop didn't come into existence until May 1938. The defense could take their choice. They had that typewriter when the Baltimore exhibits were typed and gave it away *after* Whittaker Chambers broke from the Party."

When faced with the damaging repair-shop testimony, posed Mr. Murphy, how did the defense attorney handle it? "Well, he started on the FBI. You know, there's an old saying, when you haven't the facts on your side, then you knock the District Attorney's head off. That's changed now. It's open season on the FBI. . . . It's the liberal approach." Mr. Murphy, for one, was having none of that. "If one juror thinks the FBI was unfair" in their handling of the Catletts, "acquit this man."

The prosecutor concluded by saying that on May 31 the jurors had given the court their oath. "Today is the day. I ask you as a representative of the United States government to come back and put the lie in that man's [Hiss's] face."

---

*Murphy was clearly wrong here. Hiss did not have to introduce this evidence as discussed in footnote on page 134.

It was over. After listening to testimony for five weeks, hearing 803,750 words about the guilt or innocence of Alger Hiss, the jury was about to receive the case. Judge Kaufman read his charge, and now it was in their hands.

When the jury retired from the courtroom on July 7, the whole confusing tale had been thrashed out and, as it turned out, inconclusively once again. On and off during the next day, there were unofficial intimations that the jury had no clear idea about a decision. They returned wearily to the courtroom late on the night of July 8, 1949, at 8:55 P.M., and Judge Kaufman asked the question: "Is the jury still deadlocked?" The foreman replied, "Yes, sir it is. [It is] impossible to reach a verdict." "Well," said Kaufman, "that leaves me no alternative but to discharge the jury. . . . You are discharged with the thanks of the court." The United States Attorney immediately announced the government's determination to petition for a new trial, and five or six jurors stayed around to shake the hand of Mr. Murphy, who was to say about the jury, which was deadlocked eight to four for conviction, "By the way the jury's split, righteousness appears to have been on the side of the government two to one."

Reporters found the behavior of the jury on completion of the trial to be very strange indeed. Several jurors were in the press room and, according to Alistair Cooke, were "sounding off as freely as a revival meeting." When the jury first retired, one man demanded an immediate vote for a Hiss acquittal. Three of the jurors were angered by this proposal and thereafter were noted to be for conviction. When the jury returned that night, the vote was eight for conviction and four for acquittal, which turned out to be the final tally.

There was no voting change through all the next day. The eight for conviction based their position on the documents and the typing samples. The reluctant four found it impossible to accept Chambers' doubtful credibility. In the waning hours of deliberation, the jury finally took all of the typewriter exhibits and tried to determine whether they could have been typed by the same person—i.e., whether there was a personal typing style. This extraordinary scientific presumption—one which neither the prosecution nor the defense was willing to consider—brought no resolution. Although they looked for personal characteristics in the typing specimens, there was no agreement on the results. Interestingly, the eight for conviction found enough similarities between the purloined documents and Mrs. Hiss's letters to vote for conviction; the resistant four did not find those same similarities. The division was reportedly irreconcilable, and that matter required the second trial.

The FBI, for its own reasons, also interviewed four jurors who voted for conviction. The weeks of listening to testimony, the hundreds of opinions and the very few uncontested facts they were called on to digest, and then the long hours of negotiation with no resolution left the jurors exhausted but with a strong aftertaste of antipathy too for those on the other side of the guilty–innocent fence. Exasperated by their inability to reach a verdict, they were all too happy to vent their spleen to the press and the FBI.

The four jurors described two of their brethren, whose names are deleted from the FBI files, as "The most stupid individuals that they had ever encountered; that at no time had they had a complete grasp of the case; the technical aspects of the typewriter and the documents made absolutely no impression on them. They did not have the mental capacity to absorb them."

Life was especially difficult for the four jurors who voted for Hiss's acquittal. On July 10, 1949, a story appeared in the Chicago *Tribune* stating that the FBI was conducting an investigation of the four who voted for acquittal. To requests by the Associated Press for confirmation of the story, the FBI offices in Washington and New York had no comment.

After the trial the FBI received a phone call from Mrs. Arthur L. Pawliger, wife of one of the jurors. She complained that since the trial, the family had received more than a dozen phone calls that were threatening or molesting in nature. The first caller said, "Your husband is a Communist, and he is one of the four. And he is going to get his." Mrs. Pawliger had a suspicion that one of the jurors in the case made the call. The other calls had a "get-the-hell-out-of-the-country tone," she said. Another juror, Louis Hill, received a postcard that, not by coincidence, was written in red ink. "You are one of the four bums. Where did you get the name Hill? Drop dead or go to Russia." A third, Louise A. Torian, said during the third or fourth week of the trial an attempt was made to "enter my apartment." She too received a note after the trial ended which said, "You red . . . we will trap you soon and that will be your end. So you are a traitor." The letter was signed Carlos K.

The jurors interviewed by the FBI approved of Prosecutor Murphy's performance. "His handling of the case was excellent," they said. Lloyd Stryker faired less well in the opinion poll. They described him as "too old'' and a practitioner of the ''tub-pumping [sic], blood-and-tears method,'' which to them was ''passé.'' Especially unpalatable were the aspersions Mr. Stryker cast upon the FBI during his summation when he described their treatment of the Catletts as ''close to oppression.'' This inspired Mr. Murphy's impassioned defense of the FBI and his ultimatum to acquit Hiss if the jurors thought the Bureau treated the Catletts unfairly.*

The jurors said that Stryker's attack on the FBI caused irreparable damage. One juror stood up during deliberations and made a speech condemning it, and the women alternates also made their displeasure known. Even those jurors who voted for acquittal were offended by Mr. Stryker's remarks.

The eight for conviction and the two alternates found Hiss unbelievable as a witness. They felt his testimony was in ''too apple-pie order.'' Bothersome also was what the jurors characterized as Hiss's ingratiating acting and reacting for their benefit. Two of these, Thomas C. Bryan and Robert W. Pitman, who impressed the agents who interviewed them as showing ''above average ability'' and being ''keen observers of what went on in the courtroom,'' said Hiss's habit of ''bending over to lift his leg by the lower calf to cross his legs was annoying to them.'' They thought Hiss did that to attract their attention, and once they turned to him he smiled at them. They also said that when Hiss corrected Mr. Murphy's grammar when the latter used the wrong verb, ''it had a very noticeable reaction among all jurors.''

Many years later we asked Hiss what would he do differently if he had the trial to do over again. Hiss, who had by that time read most of the FBI files, said:

> I know what the jurors who reported to the FBI said. I would have been more restrained and less at ease. After all as a lawyer, I felt very happy in court. That is where I wanted to be. Maybe I was patronizing. I'm not at all surprised. This, of course, would annoy them. . . . But I know I'm arrogant. I know I have a great deal of what Christians—I'm an Episcopalian—call ''false pride.'' And I would try to seek a greater humility than I normally have. . . .

---

*Alger Hiss had this to say about Prosecutor Murphy's ultimatum: ''He was leading from strength. He knew the average middle-class American regarded the FBI as sacrosanct.''

The twelve jurors agreed to ignore the testimony of the character witnesses, such as Adlai Stevenson and Justices Frankfurter and Reed. "All jurors agreed that the character witnesses were not qualified to pass on Alger's character, and they did not confine their testimony to the period in question," the FBI papers read. Pitman and Bryan agreed that if the defense had called a neighbor who knew Hiss during the period in question, "it would have been more helpful than all the 'big shots' he called."

Not only was the testimony of the Supreme Court Justices not helpful to Hiss, but as the legal process ground on it became a tremendous liability. When Hiss was found guilty at the second trial and the case was finally appealed to the Supreme Court, the court's decision was four to two against hearing the case. Three Justices disqualified themselves: Reed, Frankfurter, and Tom C. Clark, who was Attorney General during the trials. Three votes would have been enough to grant the petition for a hearing.

Hiss was asked about this ironic turn of events. "I wasn't thinking of appeals," he said. "I was sure we would win. . . . It never occurred to me [that we might have to appeal]. We did not call them [the Justices] a second time—although they were already disqualified. . . . I didn't want people to be hurt by being close to me."

When the trial results were made known, Congressman Nixon was once again discomfited. He joined three other members of the Congress in a statement that accused Judge Kaufman, who was appointed by President Truman, of bias for Hiss and demanded that his fitness be investigated. Nixon said at that time that the judge's "prejudice for the defense and against the prosecution was so obvious and apparent that the jury's 8 to 4 vote for conviction frankly came as a surprise to me." He accused Kaufman of refusing, for political reasons, to permit two witnesses* to testify and said, "I think the entire Truman administration

*Hede Massing, who claimed she knew Hiss as a Communist, was one, and William Rosen, who allegedly purchased Hiss's old Ford from the Cherner Motor Company, was the other. As Hiss was being tried for perjury involving espionage, *not* Communist Party membership, Judge Kaufman ruled against Mrs. Massing's appearance because it involved a collateral issue. Kaufman also decided to forego the appearance of Rosen, whom the FBI labeled a Communist. Rosen had taken the Fifth Amendment each time he was called on to testify. Recognizing the judicial precedent that a witness who takes this constitutional privilege can adversely reflect on the defendant, the judge decided to forego Rosen's appearance.

was extremely anxious that nothing bad happen to Mr. Hiss. Members of the administration feared that an adverse verdict would prove that there was a great deal of foundation to all the reports of Communist infiltration into the government during New Deal days."

Nixon complimented Mr. Murphy's handling of the case, saying he "did a great job against great odds. I mean no disparagement of him when I say that it might be wise—considering the importance of the case—to appoint a special prosecutor to work with him. I think Mr. Murphy might welcome such assistance." These were Nixon's sentiments then; his affection for the office of the special prosecutor was to diminish over the years.

A brief five weeks had elapsed since Judge Kaufman had opened the trial on a warm day at the end of May in the year 1949. And after all the arguments were over, there was no comforting resolution for any of the concerned parties. In four months the energetic Murphy would be back at front center of the court, but Stryker would be gone for good. Gone also would be Judge Kaufman, but the cast of characters would remain much the same. Hiss, personal virtue and belief in the system of justice somewhat shaken, would now insist on a slightly more psychiatrically oriented approach that would ultimately make little difference. Chambers, whose expectations were so ambiguous and whose testimony integrated all contradictions, direct or implied, would reappear rumpled as ever. And last and certainly not least in the permanent cast would be the typewriter and the documents—those silent sentinels that came finally to stand for immutable evidence of Hiss's perfidy.

CHAPTER 7

If the American people understood the real character of Alger Hiss, they would boil him in oil.

—RICHARD NIXON, as reported by Whittaker Chambers

Just before the second trial, Walter Winchell wrote that the Hiss attorneys would subpoena interoffice cables and memoranda from *Time* magazine files. Winchell reported in his September 12, 1949, column that one lengthy cable sent by foreign-news editor John Osborne would provide drama in the courtroom by attacking the "credibility of accuser Whittaker Chambers."

When the cable was sent, Chambers was acting foreign-news editor filling in for Osborne, who was on leave in Europe. According to Winchell, Osborne, after analyzing an issue of the magazine produced by his stand-in, cabled *Time*'s editor-in-chief Henry R. Luce to charge that Chambers had changed "*Time* from a 'news magazine' to a 'fiction magazine.'"

The FBI interviewed Chambers following Winchell's reference. Chambers told the FBI agents that his problems with Osborne were caused more by interoffice politics than by international politics—although he had seen stacks of the *Daily Worker* in Osborne's office. In editing this section in Osborne's absence, Chambers adopted a lighter tone in order to make it "more attractive and interesting for readers," read the FBI files summarizing the interview. According to Chambers, his superiors at *Time* apparently preferred his style of editing to that of Osborne's and to avoid "demoting him [Osborne] or hurting his feelings," they created a new responsibility for him, the International Section.

Chambers knew of two articles in particular to which Osborne objected. Chambers himself had written one describing a Winston Churchill interview in Rome. The tone was "light," said Chambers, and the article was friendly to Britain's leader. The second article was about a Communist revolution in Czechoslovakia, written by former *Time* correspondent William Walton, who had parachuted into Normandy with American troops during World War II. Walton sent back a thirteen-page cable describing the people's revolution there which Chambers refused to print, believing the story was "unsubstantiated and inaccurate." Chambers further related to the FBI that the cable division was the home of the "majority of Communists" at the magazine. Chambers' refusal to print Walton's article, he said, "caused these Communists to think and speak unkindly of me."

Then, apparently warming to the subject, Chambers attempted to cast political aspersions against others he alleged were Communists in Mr. Luce's Time-Life empire, some of whom had gone on to first-chair positions in the community of letters. To the FBI—and the material is in the FBI files—he maintained that in 1939 there was a Communist Party cell at the magazine which put out a house organ known as *High Time*. Members of the cell, Chambers claimed, included John Hersey, who became the Pulitzer Prize-winning author of *Hiroshima* and *A Bell for Adano;* John McManus, who wrote the radio and cinema section; and John Osborne himself. Osborne later wrote five books on Richard M. Nixon's Presidency as well as a weekly political column for the *New Republic*.

While Chambers did not tell the FBI that the foreign-news section was pro-Communist, he did tell the FBI that the section was especially "pro-Russian" and "pro-left," and that as soon as he came into authority there he changed the tone; it was this which caused the animosity. Chambers related how *Time*'s foreign correspondents established an *ad hoc* group in opposition to him. They were known as the "Round Robin," and Chambers claimed they sent disparaging telegrams about him to Henry Luce. Members of this group included Theodore White, later the well-known author of the estimable Making of a President series of books; Richard Lauterbach, who was to write *These Are the Russians* and *Danger from the East;* Charles Wertenbacker, who wrote *Invasion;* and John Scott, author of *Behind the Urals* and *Democracy Is Not Enough.*

Between trials, Chambers accepted an invitation from his friend Ralph de Toledano, national-affairs writer for *Newsweek,* to appear on that magazine's television program. Chambers not only wished to make the appearance but suggested that while on the air he be monitored by a lie detector while being questioned about the case. It should be remembered that Alger Hiss declined the committee's offer to take such a test, describing it as not scientifically accurate.

As *Newsweek* was a direct competitor of *Time*—Chambers' former employer—the people at *Newsweek* were much amused by the idea of his appearance and liked the format he proposed. Before appearing, however, Chambers asked the FBI to react to the proposal. The agents he talked to said Chambers welcomed the opportunity to appear, saying that he described it as a great idea "from the standpoint of a newspaperman." The FBI summary continued: "Chambers felt it would be a dramatic success as well as helpful to the Government's case because he feels the program would arouse public sentiment against Hiss."

The agency, having assumed a proprietary interest toward their "witness," advised him against such an appearance. According to the Bureau's files, Chambers fought them on this point. The FBI record read:

> He sees no objection to telling the truth under the proposed auspices and is inclined to insist on his appearance as planned. Chambers sees no objection to the Government taking the initiative and reviewing the questions to be asked prior to the telecast. And at one point in the discussion, Chambers critically stated that it might be best to take the attitude, "the less the Government knows about this thing, the better." It is unknown whether he meant from the effect on the public, or the defense or whether he might possibly be inferring that the Government was infringing on his rights as a private individual.

Finally, and after a long conversation with Prosecutor Murphy, who pointed out the reasons why this appearance might "severely hinder the

Government's case at the retrial," Chambers acquiesced and canceled his television appearance.

Between trials Hiss decided he stood a better chance if the second trial was held in Vermont, where he had been a summer resident for several years. Believing that the New York City press had prejudiced any chance he had for a fair trial, his attorneys petitioned for a change of venue. "The press count [survey] said that this was the center of the Hearst press and the Scripps-Howard press and the hostility," said Hiss. "Vermont was a nice quiet place.* I had a good argument since I had maintained a residence there in the summer. I was as much a resident of it as I was of New York where I had just gone to live."

Confident that the change of venue would be granted—"My problem was one of overweening optimism"—Hiss decided to change attorneys, thinking that the histrionic Stryker "would not play well to a staid Vermont jury." Hiss also chose a new attorney because he wasn't sure, he said, if Stryker "could stand the strain of a second trial because he was slightly apoplectic anyway."

Lloyd Paul Stryker had invested a great deal of time and effort and charged a token $20,000 for the more than half a year he and two of his assistants spent on the case. This was a laudable gesture because Stryker was accustomed to charging a very high fee for his services. The sacrifice was not made out of friendship for Hiss, either. Hiss said, "I'd never known him before. He did it out of belief in my innocence." Stryker was willing to see the matter through to completion, and although he was not optimistic about winning the case, he was cheerful about winning enough support to result in a hung jury the next time around. Hiss later commented that Stryker, not surprisingly, seemed relieved when he went elsewhere for legal counsel.

In Stryker's place Hiss chose Claude B. Cross, whom he felt a rural-state jury would find easier to identify with. He was a short, matter-of-fact man, with a round, pudgy face and a full belly which pressed against his vest. It is striking in a case in which so much attention was paid to personal appearance—Hiss and Chambers, Stryker and Murphy—that no singular picture of Cross emerges anywhere. When Hiss's change-of-

---

*The defense conducted a survey to be certain that Vermont was indeed "a nice quiet place." Vermonters from Rutland, Burlington, and Brattleboro were questioned about how much they knew and what they had concluded about the Hiss case.

venue petition was denied,* it was the quiet Claude Cross who would defend him in cosmopolitan New York City.

Hiss's legal fees over the years 1948–1950 amounted to about $100,000 (approximately $250,000 in current economic terms). Hiss was responsible for his legal expenses, unlike Chambers, whose legal fees from the libel suit were handled by his employer, Henry Luce, who viewed an attack on one of his editors as an attack on the magazine itself. Hiss raised $20,000 on his own, was given a like amount by his brother, Donald, and secured $70,000 from the Hiss Defense Fund, an *ad hoc* group of friends and sympathizers. Many of his attorneys were friends, and some, like William Marbury, who handled the libel suit, did not even submit a bill.

Hiss gave instructions to his friend from Harvard Law School, Richard H. Field, who was treasurer of the Defense Fund, that he should be extremely careful who was solicited for contributions. Hiss wished no grandiose effort like that mounted by the supporters of Sacco and Vanzetti and, above all else, wished to avoid any political taint—especially Communist Party support. Field followed these instructions carefully.

However, the FBI and the House committee were interested in who was contributing to the Defense Fund. The following directive was sent from Washington FBI headquarters to the FBI office in New York City:

> Your files on Alger Hiss should be thoroughly reviewed for the purposes of identifying any financial contributor to the Hiss defense. During the review, you should also look for any confidential informants or contacts. . . . You may have informants through whom you could safely check the bank accounts of Priscilla and Alger Hiss and the Hiss attorney.

The FBI's search was fruitful. From a confidential informant they received the following information:

---

*Hiss's attorneys contended under Rule 21(a) that "the publicity which has been given this case . . . has been of such unprecedented volume and in some respects of such extraordinary virulence that the defendant cannot obtain a fair and impartial trial [in New York]." In response to the petition Prosecutor Murphy surveyed the press coverage of the trial. In 1949 there were eight major daily newspapers in the New York area. Murphy found that, of 470 stories written about the trial, 68.5 percent were completely factual, 8.3 percent pro-Hiss, 6.1 percent anti-Chambers, and 17.5 percent pro-and-anti Judge Kaufman. District Judge Alfred C. Coxe ruled against the motion, concluding: "I find nothing in the papers submitted on the present motion to indicate that there exists in this district any such prejudice."

> The Hiss attorneys are being paid from what is known as the Alger Hiss Defense Fund, to which prominent people have been making $3,000 contributions. This source stated that Eugene Meyer, former owner of the *Washington Post* . . . and Dean Acheson, Secretary of State, had made a similar contribution through a member of Acheson's law firm. This source stated that the House Un-American Activities Committee is endeavoring to prove these contributions and intends to subpoena bank officials to produce any cancelled checks to support these contributions.

Not surprisingly, when Senator Joe McCarthy took the lead in the anti-Communist crusade—which began not by coincidence less than one month after Hiss's conviction—he too showed an interest in the contributors to the Hiss Defense Fund. An article entitled "Guilt by Association," which appeared in the Washington *Post* on March 10, 1951, written by Carl Marquis Childs, said that Senator McCarthy's supporters had obtained a list of contributors to the Hiss Defense Fund. Childs predicted that if the Senator followed his usual methods, he would bring the names of those who contributed into the headlines with the intent of doing as much harm as possible to all concerned.*

Alger Hiss's second trial began on November 17, 1949, and lasted until January 21, 1950. The Honorable Henry W. Goddard was the judge, and the prosecution provided essentially the same team as at the first trial; the central figure was, once again, the redoubtable Thomas Murphy. Claude B. Cross had a single theme from beginning to end. It was that Chambers had, through the years, kept a "productive file of private lives and incriminating documents" in order to use them on other people if he ever got into serious trouble. He pointed out that Chambers had testified, "I make it my business to know all I can about people." Mr. Murphy, now fully experienced with the trial issues, continued to hammer at the circumstantial weight of the typewriter and the purloined documents.

The machine that had been produced by the Hiss side came, as in the first trial, to haunt him. Chambers, the confessed spy, admitted that he personally had disposed of a typewriter which he felt might incriminate him by leaving it on a streetcar or subway train in New York City and

---

*Such headlines did appear: One named William Putnam Bundy, administrative assistant of the CIA who was also Dean Acheson's son-in-law, as a contributor of $400. Also listed for a large contribution was Eugene Meyer of the Washington *Post*, who was the father of that newspaper's present publisher, Katherine Graham.

walking away from it forever. Hiss, who denied knowing anything about his typewriter (he could not even type) and who would really have had excellent reason if Chambers' allegations were true for disposing of his typewriter, could never seem to manage the task.

How like Hiss to have been unable to give away such a weapon, and how like the Hiss forces painstakingly to recover it at great final cost to their own cause! The path the typewriter took is not crucial; what is central is whether the Hisses had the typewriter at the time Chambers alleged that Mrs. Hiss copied stolen documents for him. That is the true significance of the great typewriter search. If the Hisses disposed of the typewriter in advance of the early 1938 dates on the incriminating documents, it was unlikely that they could be charged with participating in the plot. If, however, the typewriter could be connected to the Hisses through the first few months of 1938, their defense was notably weakened. This, at least, was how the whole matter appeared in 1949.

This effort to find a yea or nay answer soon came once again in the second trial to resemble a comic opera. It all came to naught, for it was impossible *at that time* in the maelstrom of conflicting testimony to make a definitive finding on the matter of dates, and that was what was needed to save Hiss. In reviewing this whole episode, The Earl Jowitt concluded with admirable English detachment, "Indeed, perhaps it was inevitable that it should be vague since the witnesses were trying to recall events which took place some twelve years before the date of their testimony."

The recent release of the FBI papers on the case produces information that is mind-boggling—i.e., the FBI itself seriously questioned whether the typewriter the Hiss forces produced, and which the government experts utilized to such advantage, was even genealogically the typewriter that all conceded it to be. This last matter—of crucial importance to both parties—will be discussed in Chapter 8.

There were still the documents visible every day in the courtroom, and Mr. Murphy devoted many hours to the process of reading them aloud, line by line. A reporter commented that Murphy could "make every word sound like a casual tolling bell." When it came time to make the summation arguments, Cross had still been unable to provide for an alternate thief in the State Department. The existence of the documents prove, he said, that there was a thief in the department. "I don't know who it was, but it was the person who stole these documents and gave them to Chambers." He was also unable to solve the typewriter conundrum. The typing could have been done by a "stooge" or by Chambers himself, who might have been able to gain use of the instrument at some early point in its peregrinations.

When Murphy's final turn came, he was to make much fun of this defense explanation:

> What probably happened, Mr. Cross testified, is that somebody, not Chambers—he's too smart, but one of his conspirators, one of his confederates (those are good names . . .)—he went up to the Volta Place house and asked innocent Clydie Catlett: "I'm the repair man. Where's the machine?" I can just see it now. It's terrific. You can have this guy coming with a Woodstock hat on—"Woodstock Repair"—saying: "I'm the repair man to fix the typewriter." Then Clydie says, "Well, which one do you want? The Remington, the Royal, the L. C. Smith . . .?" "No, we want the Woodstock." "Oh, that's over in my boy's house over at P Street."
>
> And the next scene: It is in the middle of one of these dances. [The Catletts had kept the typewriter in a spare room in which they occasionally danced.] And you see Chambers sneaking in at night, mingling with the dancers, and then typing, typing the stuff, holding the State Department document in one hand. Oh, Mr. Cross, you got better than that.

Still, when all of Mr. Murphy's fun and games were over, he was not able to prove that Mrs. Hiss had sat at the typewriter and reproduced the documents. And so there remained in the courtroom the "guilty" typewriter and the purloined documents, and no one—neither the prosecution nor the defense—had been willing to risk the possible argument that there is a distinguishable typing style as there are distinguishable elements in handwriting.

In an effort to corroborate Whittaker Chambers' testimony that he saw Hiss after January 1, 1937, or that at least a closer intimacy existed between them than Hiss was willing to admit, the FBI concentrated on finding two maids who worked for the Chamberses when they resided in Baltimore. Addressing himself to the incongruous position that a Communist should hire a maid in the first place, Chambers wrote seriously in his book, " . . . a servant is a worker like any other, and workers live by working." There was also the sticky matter of wages for the domestic of a Communist. "To pay more might cause comment and even trouble," wrote Chambers, who characterized a maid's wages then as "shockingly low." He continued, "I decided that we would pay a little more." Still in keeping with his egalitarian view of society, Chambers insisted that the maid take her meals at the table with the family.

The FBI was able to locate two of the maids, Edith Murray and Evelyn Matley. Edith Murray was used as a surprise witness for the prosecution on the last day of the second trial. She was rather unflappable, and her testimony was important because she was the only witness who claimed to have seen Hiss and Chambers together. And Murphy was quite taken by her contribution to the government's case, referring to her at various times as "that lovely girl" and also "the essence of simplicity and truthfulness."

Mrs. Murray had worked in two homes of Whittaker Chambers', although she knew him as Lloyd Cantwell, the alias he was using at the time. She could recall four visits by Priscilla Hiss and one by Alger. According to Mrs. Murray, Priscilla once stayed overnight at the Baltimore house to take care of the Chamberses' baby daughter, while Mrs. Cantwell (Chambers), again pregnant, went to New York to consult with her physician. Mrs. Murray also testified that one night just before leaving at the end of her work day she let both Alger and Priscilla Hiss into the house.

Since Mrs. Murray was a surprise witness, presented during the rebuttal phase of the trial, the defense could do little to contradict her testimony. During the cross-examination Mr. Cross asked how she could so readily identify Alger Hiss when she had seen him only once, years before, and then for just a moment. Edith Murray commented with a laugh on the "difference of the two couples," which apparently was a reference to the disparity in dress and demeanor between the Hisses and the Chamberses. She also said her employers "never had no company at all, only those two."

When the FBI first located Mrs. Murray, they showed the maid a portrait Esther Chambers had painted of her to jog her memory. This pleased her, and the next day she was taken to Chambers' farm in Westminster, where she enjoyed a reunion with the family she had not seen in more than a dozen years. Also during one of her early interviews by the FBI she was shown a photograph of Priscilla Hiss. Hiss's attorney asked her to relate what she had said about the picture. In the courtroom she recalled her comments: "It looks like someone that I know. It looked like—I thought maybe it was an actress or something." Shown a picture of Alger Hiss, she was unable to make a positive identification.

That was not precisely what she said to the FBI, however. Her signed statement on September 28, 1949, read in regard to Priscilla Hiss, "The agents have shown me a photograph and have told me that it is a photograph of Priscilla Hiss. The name is not familiar, and I do not recall Mrs. Cantwell ever introducing me to a lady by that name, but I think the photograph might be a picture of the lady from Washington."

Regarding Alger Hiss, she told the FBI, "The agents have shown me a photograph of a person, they have told me it is Alger Hiss and the photograph looks something like the slender man who accompanied the lady from Washington on the above visit to the Cantwells'." Note in both signed statements that not only is this a less-than-positive identification, but Edith Murray said that the FBI told her who it was she was identifying (Priscilla and Alger) before asking her to make the identifications.

The FBI files read, "In an effort to preclude any embarrassment on the part of the government," Edith Murray was brought to the courtroom on the first day of the second trial. She saw Alger and Priscilla Hiss surrounded by their attorneys, news reporters, and photographers. The FBI files say this about the surreptitious "show and tell": "Edith Murray made a positive identification of Priscilla Hiss and a tentative identification of Alger Hiss."

Interviewed recently on the subject of this witness, Hiss, commenting on the question "You know Edith Murray saw you and your wife in the hallway before the trial?", said, "Oh, absolutely! She had plenty of time to identify me and my wife. And she said she had seen us in newsreels. I think she was suggestible. I think she had come to know Chambers at some time. And had known the children. They took her to Chambers' farm. I think she was a kindly woman feeling some loyalty to her employer. I think she was suggestible, which is an old trick with prosecutors."

Yet Mr. Murphy, standing at some distance from the truth and Mr. Hiss's conclusion, said in regard to the government's handling of this witness in his summary, "Everything was done to avoid any criticism of prompting. She did not know what it was all about."

Usually in a case like this, one witness is not permitted to hear the testimony of another, but the psychiatrist, Dr. Carl Binger, was granted special dispensation to listen to Chambers' testimony at the first trial. The doctor suggested questions that the defense attorney might ask Chambers in order to help in his psychological evaluation. Dr. Binger further undertook an extensive analysis of Chambers' writing, which included his student work, translations from the 1920s and '30s, and his work at *Time* magazine—a monumental task for which, as a friend of Hiss's he neither asked for nor received any remuneration.

The prosecutor in both trials, Thomas F. Murphy, objected strenuously to a psychiatrist being called to testify for the defense on Chambers' mental health, just as he objected to the psychiatrist's presence in the courtroom for the six days Chambers was on the stand. The judge at the first trial allowed Stryker to do Dr. Binger's bidding, but after Mr. Murphy's objections and the submission of briefs by both sides, he would not allow the psychiatrist to testify in the trial itself.

Before this ruling, however, Mr. Stryker managed in a roundabout manner to tally some psychological points. Beginning with the words "Now, Doctor, assume that the following are true . . ." the attorney was able to summarize in forty-five minutes some of the more aberrant chapters in Chambers' strange life. Alistair Cooke described this technique well: "There followed a roll-call of conjunctive clauses, lining up like a battalion of deserters, each pointing a shabby finger at the life and character of Whittaker Chambers."

After Stryker's long-winded assumptions—for which he was not lacking for material—he finally came to the heart of one of the longer sentences ever sounded in a courtroom or anywhere else for that matter: "Now, Doctor Binger," he said, "assuming the facts as stated in the question be true, and taking into account your observations of Chambers on the witness stand, and your knowledge of his writing and translations, have you as a psychiatrist an opinion within the bounds of reasonable certainty as to the mental condition of Whittaker Chambers?"

The long-silent Murphy erupted in a fire of objections, saying he thought the judge had indicated he would not permit this sort of testimony. Moreover, he contended that Stryker's question, rather than being a question, was a summary of the case before the time for such summaries. Also it was not by a "long shot" a complete summary either, Murphy opined. He characterized this defense tactic as "a grave injustice." This time Judge Samuel H. Kaufman ruled for the prosecution. He ordered the jury to ignore Stryker's summary, "because," he said, "the record is sufficiently clear for the jury, using its experience in life, to appraise the testimony of all witnesses who have appeared in this court."

A bit late perhaps, since a jury's memory is not at the beck and call of a judge, but ultimately the prosecutor won the first battle against psychiatric testimony. From Dr. Binger's questions and Stryker's summary, the prosecutor was able to infer the prospective nature of the testimony. Between trials he studied the subject and consulted with other experts, and when the decision regarding expert testimony went against him at the second trial, Murphy was able to match wits with the psychiatrist on his own terms.

This was the first federal case where a psychiatrist testified on the credibility of a witness. Judge Henry Goddard, recognizing the precedent being set, noted that prior to this, a psychiatrist could not testify as an expert witness, but this ruling was made in 1921, "before the value of psychiatry had been recognized." The judge ruled, "evidence of insanity . . . affect[s] credibility."

Dr. Binger, feeling his explorations of Chambers' writings and observations of him at the trial were sufficient to make a diagnosis, concluded that Chambers showed a psychopathic personality, one symptom of which was pathological lying. Binger testified that people like Chambers "have a conviction of the truth and validity of their own imaginations of their own fantasies without respect to outer reality; so that they play a part in life, play a role . . . and on the basis of such imaginations, they will claim friendships which were nonexistent, just as they will make accusations which have no basis in fact. . . ."

When his turn came, Murphy was ready, and his cross-examination of Dr. Binger lasted three days. Time after time he went for Binger's heart in an assault which Cooke characterized as the "first public trial run of the common man's resistance to psychiatry." Hiss described the prosecutor's cross-examination in this matter as "a particularly savage attack. He [Murphy] had nothing to fear. A cross-examiner always has to worry that if he is too savage, he will provoke sympathy for the witness. [But the] . . . jury didn't like psychoanalysis."

In his cross-examination of Dr. Binger, the prosecutor wondered if it were possible that another psychiatrist might come to a different conclusion about Chambers' mental health. Dr. Binger agreed this was possible. After the psychiatrist stepped down, the defense called another physician, who agreed with Binger's analysis. This was Henry Murray, who had developed a personality test using clues from a subject's writing. Dr. Murray, a former director of the Harvard Psychological Clinic, found in Chambers' writing "A higher proportion of images of disintegration and destruction, filth and dirt, decay and decomposition and death than in any writings I have ever examined." Like Dr. Binger, he concluded that Chambers was a psychopathic personality.

Hiss had been warned by his attorneys that psychiatric testimony could very well prove a liability. Hiss commented on his attorneys' resistance, "They said, 'The jury doesn't like psychoanalysis.' I said, 'I don't care. I want the record complete. This is an area where you don't have any explanation. The only explanation is a psychiatric one, and I insist on it.' ''

Richard M. Nixon, gone by this time but unwilling to be forgotten, had some political ammunition to help in the "search and destroy" mission of Dr. Binger, if it were needed. Nixon phoned the FBI with information that Dr. Binger was once called before the Massachusetts State Medical Association for what the FBI papers cryptically describe as "unethical charges growing out of a book he published." The Bureau also received a call from Nixon's mentor in these matters, Father John Cronin, who told them Dr. Binger had a "slight" Communist record.

In addition, a confidential source for the FBI said a New York *Times* article described Dr. Binger as "making pro-Communist remarks in England." The informant, no champion of psychoanalysis either, said that Binger was a Freudian, and therefore was "anti-God in his beliefs." The *Times* article referred to reads in part, "In an interview after a speech, the doctor [Binger] pointed to the bugaboo of Communism, which he said was now spreading a state of neurotic anxiety throughout the United States."

When Judge Goddard charged the jury at the end of the two lawyers' summations, he began by reading the "counts" under which Hiss was being tried: (1) that Hiss had committed perjury by "falsely testifying" before the grand jury that he "did not turn any confidential documents of the State Department or copies of confidential documents over to Whittaker Chambers or any other unauthorized person" and (2) that the times Hiss saw Chambers after January 1, 1937, were the very times he was charged with turning over the documents, and the place of transmission was "the defendant's Volta Place residence."

Judge Goddard warned the jurors that it was not necessary for Hiss to prove how or from where Chambers had received the documents. It was the government's responsibility to prove beyond a reasonable doubt that Hiss or his wife did deliver them. He further warned them that testimony about Hiss's good character should be carefully weighed. "It may be that those with whom he had come in contact previously have been misled and that he did not reveal to them his real character or acts."

Judge Goddard similarly cautioned that defense testimony by two psychiatrists that Chambers was suffering from "a mental disorder which would tend to reduce his credibility" was to be carefully weighed by the jury, who had to answer the primary question: Was Chambers lying when he said he received the stolen documents from Hiss?

Finally, Judge Goddard summarized the evidence. The jury had seen the documents, the Woodstock typewriter, and the letters that Mrs. Hiss had typed on it on other occasions. "It is the contention of the government that this," the judge said, indicating the defense's typewriter on display, "is the typewriter upon which [the] Baltimore exhibits were typed."* The jury was to weigh that testimony; they might accept it or reject it in its entirety. And so, after a few more brief procedural remarks, the jury left the room at 3:10 on the afternoon of January 20, 1950.

They were back almost immediately. Less than two hours had elapsed, and they had moved promptly to the same old issues which confounded all observers. Among some other odds and ends, the jury wanted to have all of Mrs. Hiss's testimony read to them, as well as the confusing Catlett family recollections, and lastly they wanted a reprise of all the material relating to the Hiss move to the Volta Place residence.

Judge Goddard pleasantly demurred. "It would take about five days to accomplish those tasks," he advised. An hour later the jury was back with a shorter request; this time it was still the documents which were needed, but only Mrs. Hiss's statements on how and when she disposed of the troubled typewriter and the Catlett recollections of how they came to possess it. After some discussion by the contesting sides, it was finally agreed to "isolate" the above testimony, and it was read to the jurors late that evening. The next morning the jury requested that Judge Goddard once again define the concept of reasonable doubt and that of corroborative evidence.

By midafternoon the vigil was over and the jurors filed back into the room. The foreman, a widow from the Bronx, came to her feet and announced rapidly, "Guilty on the first count and guilty on the second." Hiss's attorney asked that the jury be polled. They were and the result remained unchanged. Mr. Murphy moved that Hiss be now remanded to jail or at least have his bail raised as was the case with all convicted persons. Judge Goddard's response was pleasant once again, but firm. "I think not, Mr. Murphy."

On the following Wednesday, Alger Hiss appeared before the judge, this time for sentencing. Mr. Cross moved for a new trial on eleven counts, and his motion was denied. He then asked for an "arrest

---

*The government witness, Ramos Feehan, did not say this. Carefully ignoring the defense's Woodstock, he testified that the typewriter which typed the Hiss letters also typed the Baltimore exhibits. This is discussed extensively in Chapter 8.

judgment," which would permit Hiss to go free on the grounds that he had, in Cross's words, "suffered the worst punishment, of mind and heart, that he could undergo." Cross continued the sorry tale:

> What little savings he has had were gone long before the conclusion of the first trial. He has borrowed heavily on notes still unpaid. And the second trial has been financed by loyal friends and loyal Americans who believed and still believe in his innocence.

Judge Goddard listened patiently but was unmoved. This was not a proper case for an arrest judgment. "There should be a warning to all that a crime of this character may not be committed with impunity. The defendant will stand up."

Both Hiss and Cross came to their feet, but Cross spoke first. He asked if Hiss could be allowed to speak a few words before sentence was pronounced. Judge Goddard nodded, and Hiss came forward to the bench and in a firm voice said:

> I would like to thank your honor for the opportunity again to deny the charges that have been made against me. I want only to add that in the future the full facts of how Whittaker Chambers was able to carry out forgery by typewriter will be disclosed. Thank you, sir.

A moment later, the sentence was pronounced. Hiss was to serve two concurrent terms of five years on the two counts and was temporarily paroled into the custody of his attorneys pending the results of his expected appeal.

For Chambers, the good news came by phone from one of the national wire services. He had been severely disenchanted by his treatment at the first trial. He felt that he had no personal relationship with Thomas Murphy and that Lloyd Paul Stryker, Hiss's attorney, was allowed to "spin and flail like a dervish in the courtroom, while Judge Kaufman snapped 'denied' to most of the government's motions." His worst moments came on the morning when Mr. Stryker began to cross-examine him and bit by bit led Chambers to admit to lying in taking his oath for government employment in the WPA, in falsifying his academic background in the same application, and in talking the authorities at Columbia into allowing him back into the university after he had been expelled.

Chambers flatly admitted stealing books from both the public library as well as the university library, using many names, lying to State

Department officials when he first went to them with the Hiss story and also when he applied for a passport under a false name and concealing the truth about the spy ring from the FBI, the House committee, and the grand jury. The sad saga finally concluded with Chambers' admission that he had willfully concealed from his own attorney the full information about the finding of the documents in Brooklyn.

Mr. Stryker was becoming wildly excited with each new admission and saw the perjury score rising very high. In contrast to his own subjective impressions, Chambers had looked remarkably cool to observers. His replies were low-keyed and without defensiveness; even when Stryker pointed out that there were certain conspicuous contradictions with regard to dates and places (no small matter in that trial) between Chambers' testimony there and that given before, Chambers answered quietly that his testimony in the past had been correct as far as it went but that now he was remembering things that he had not recalled before. Hot on the scent, Stryker accused Chambers of erroneously referring to a Hiss address on Dent Place when actually Hiss lived on Volta Place. When Chambers denied it, he was shown the printed record and without rancor replied, "The printed record is wrong."

The record was "wrong" in other places as well, but the defense had no way of knowing this at that time. Before the indictment, Hiss's attorneys filed the usual discovery motion. They asked the prosecutor to release to them "all written statements and affidavits, whether signed or not, made at any time by Whittaker Chambers to the Department of Justice, the Federal Bureau of Investigation . . . concerning any matter relevant to the issue in this action." The motion was denied at that time.

At the first trial, Hiss's attorney asked Chambers about his various interviews with the FBI. He ambiguously admitted making "three signed statements." Asked to whom they were made, Chambers responded, "They were made to an agent in Baltimore on December 3, I believe. . . . And two were made, I believe, in New York, also in the month of December 1948, I believe." The next day at his cross-examination he was asked, "Prior to that [December 1948 interview and the resultant signed statement] had you made previous statements?" "I had made no signed statements" was the witness' evasive response.

Stryker requested to see the three signed statements, believing they would be helpful during his cross-examination of the witness. Judge Kaufman demurred, however, saying, "I will direct the government to supply me with the grand jury minutes and with the statements to which you refer and if there is any inconsistent statements made in prior testimony, those matters, and only those matters, will be exhibited to counsel for the defendant."

Despite Chambers' specific mention of seeing the FBI three times, including December 1948, the prosecution supplied the judge with only two statements: May 14, 1942, and June 26, 1945. It was not a complete record, even by Chambers' ambiguous tally. Nevertheless, Judge Kaufman found significant inconsistencies there. The most important was the date Chambers defected from the Communist Party. The judge said, "One of the crucial things in this case is the testimony of Chambers as to when he left the Communist Party. If he made the statement once, there might be the possibility of misunderstanding, but here in this statement of May 14, 1942, Mr. Whittaker Chambers advises that he was a member of the Communist Party from 1924 until the spring of 1937, at which time he ceased connection with the Party."

In fact, Chambers made that statement numerous times prior to the discovery of the Pumpkin Papers. He told HUAC on August 30, 1948, that he broke from the Party in 1937. He gave the same information on six occasions to FBI Agent Spencer, although the defense never learned of this interview until recently. Summarizing this matter, Spencer wrote on March 26, 1946:

> . . . His actual knowledge of Hiss' activities concerned the period shortly preceding 1937 and he was unable to elaborate on any information concerning Hiss' connection with the Communist Party or Communist front organizations other than what he reported at the time he was interviewed on March 13, 1942, and again on May 10, 1945. He recalled that after 1937 . . . he had lost all contact with Alger Hiss and the only information that he has concerning him is that which appeared recently in the various newspapers which attempted to attach him in some way to the Communist Party.

Such information would certainly have imperiled Chambers' charges because, according to his own testimony, offered on numerous occasions, he would have been out of the Party for almost a year before securing the allegedly stolen Hiss documents, which bore datelines of the first four months of 1938.

Judge Kaufman asked the prosecutor if Chambers had made any other statements to the FBI prior to December 1948—the date he had previously mentioned. Mr. Murphy responded, "My recollection is, Your Honor, that he gave three after these two you just read which were written statements, but in 1948, commencing in December, he gave three written statements." "Is that correct?" Murphy asked his assistant, Thomas Donegan. "That is correct," Donegan replied. Judge Kaufman asked to see these statements, and over Murphy's objections they were delivered. The judge found no inconsistencies in this batch, however. Ostensibly, the same process was repeated at the second trial.

The recently released FBI files show that there were indeed other statements made by Chambers, but not signed, although Chambers had no qualms about affixing his name to these documents. Among the unsigned statements was the 184-page history, written by Chambers and referred to in the previous chapter in regard to the $400 loan. There was also the description of Chambers' homosexual activities. None of these was signed—a very atypical practice—because neither Murphy nor Donegan, who himself was a former FBI agent, wanted Chambers to sign them.

A note in the FBI files explains this:

> Donegan pointed out that since Chambers would be a government witness and a friendly one, no material benefit could be gained by him signing this statement. He pointed out that on the other hand, if he did sign it, this fact might be brought out during the course of the trial, and although the statement might not be actually presented to the jury, there was a possibility that the Judge might allow the defense attorneys to read the statement, which would probably result in some complications. It was pointed out to Donegan that Chambers has already signed three rather brief statements, but he indicated that he does not believe that these statements will cause any conflict if they are introduced, in view of the fact that they are brief and are concerned with specific matters. He pointed out that the statement in question, of course, is very lengthy and if the defense attorneys got their hands on it, they might use some of the material therein to at least cloud the issue.
>
> Donegan suggested that although he is aware that this is contrary to the general Bureau custom, he felt that if the Bureau actually wanted the statement signed that arrangements could be made whereby Chambers could set his signature to this statement subsequent to the trial.

Allen Weinstein argues that Chambers did not sign the interviews because the FBI feared his homosexual confession might destroy the case. After Chambers' confession, Weinstein described the "shock and consternation that spread among FBI and Justice Department officials . . . when they learned that their star witness had confessed to many homosexual involvements during the same years he had been meeting with Hiss." In summarizing the above FBI memo, Weinstein writes, "The agent, apparently alluding to the homosexuality issue and perhaps to other information helpful to the defense, explained that, if Chambers did sign it, the fact of the signed statement's existence might be brought out during the course of the trial. . . ."

Weinstein is confused about this matter. It is not an "agent" who argued persuasively that Chambers not sign his statements, but United States Attorney Donegan, no doubt with the blessing of Prosecutor Tom Murphy. Rather than worrying about Chambers' confession of homosexuality, the prosecution team, the FBI, and the Justice Department most certainly had much larger concerns.

The government must have feared that the very foundation of their case would be shaken if the information about the date of his Party departure and the confusion about the amount of money involved in the car loan were made known to the jury. In serious doubt would be such key pieces of evidence as the microfilm, typewritten documents, and the handwritten notes, because by Chambers own admission he had left the Communist Party months before the source material was circulating in the State Department.

Such information would have been exculpatory, conclude the Hiss attorneys in their *coram nobis* petition, and with good reason. Not only did the government hide these interviews but, when specifically asked about them, denied their very existence.

Between the two trials, Mr. Murphy had visited Chambers at his Maryland farm, and a better relationship developed between the two men. Murphy had been transformed by his previous experience, and now Chambers saw in him a real defender.

> He understood the case, not only as a problem in law. He understood it in its fullest religious, moral, human and historical meaning. I saw that he had in him one of the rarest of human seeds—the faculty for growth, combined with a faculty almost as rare—a singular magnanimity of spirit. Into me, battered and gray of mood after a year of private struggle and public mauling, he infused new heart, not only because of what he was, but because he was the first man from the Government who said to me in effect: "I understand." I needed no more.

When Mr. Cross succeeded Stryker as the Hiss counsel, Chambers had an easier time; the badly furrowed ground had been gone over unprofitably before, and the thin tissue of lies and religious mysticism no longer seemed quite so central in the case. Throughout both trials Chambers also continued to draw support from Richard Nixon. According to Chambers, the Nixons frequently were at the farm, and Nixon even occasionally brought his own parents to see the Chamberses. The Chamberses' children came to love Nixon and even called him "Nixie."

Chambers wrote:

> To them, he is always "Nixie," the kind and the good, about whom they will tolerate no nonsense. His somewhat martial Quakerism sometimes amused and always heartened me. I have a vivid picture of him, in the blackest hour of the Hiss case, standing by the barn and saying in his quietly savage way (he is the kindest of men): "If the American people understood the real character of Alger Hiss, they would boil him in oil."

Without Nixon it is hard to know what might have become of Chambers, and without Chambers it is hard to know what might have become of Nixon. Chambers was openly grateful for that support but apparently had difficulty with Nixon's political ambitions. Writing to Buckley in 1960, he complained that he and Nixon had nothing to say to each other any more and described Nixon as "personally inadequate" for the Presidency.

The afternoon the Hiss verdict was announced brought a second phone call, one of those encounters that seemed to come to Chambers gratuitously at eventful moments in his life. This time it was an elderly man who said, "I know your telephone will be ringing every minute now, and I had to reach you first. I had to say 'God bless you, God bless you, oh, God bless you!'" and then hung up.

Five days after the verdict was in, Mr. Nixon made a request for a "special order" and addressed the House of Representatives. Even with Hiss's conviction, Nixon was still vitriolic. He was furious with the results of the first trial, where the jury had been unable to reach a verdict. In consort with Representatives Frank B. Keefe of Wisconsin and William K. Macy of New York, Nixon attacked Judge Kaufman's fitness. The *Congressional Record* gives evidence that the passage of time had scarcely diminished their "many misgivings as to the manner the judicial proceedings were conducted" as well as to their deep antipathy to editorial support given Kaufman "by the Communist Press in the city of New York" and by the Washington *Post*, which turned out to be both a longtime and significant enemy of Nixon's.

Nixon's brush was broad as he painted a list of enemies of the republic which resembled nothing save a *Who's Who* of the Democratic Administration. These individuals were mentioned, according to Nixon, as examples of how effectively the conspiracy was concealed and "how far it was able to reach into high places in our government" to obtain apologists for its members:

> The President had referred to the case as a "red herring" and did so even after the indictment. The Secretary of State, Mr. Acheson, before his confirmation, declared his friendship for Mr. Hiss and the implication of his declaration was that he had faith in his innocence. Two justices of the Supreme Court, Mr. Frankfurter and Mr. Reed, in an unprecedented action, appeared as character witnesses for Mr. Hiss. Judge Kaufman, who presided at the trial, stepped off the bench and shook hands with these defense witnesses, one of many of his actions during the trial in which he showed his obvious bias for the defendant. The wife of the former President of the United States, Mrs. Roosevelt, on several occasions during the two trials, publicly defended Mr. Hiss in her news columns. Among the high government officials who testified in his behalf were Mr. Philip Jessup, then President Truman's ambassador at large in Europe, and now the architect of our far-eastern policy; the Governor of Illinois, Mr. Stevenson; Judge Wyzski, of the United States District Court, Boston; Judge Magruder, chief justice of the circuit court of appeals, Boston; and Francis B. Sayre, Assistant Secretary of State.

That much quoted *Congressional Record* (January 26, 1950) showed Mr. Nixon in his most extravagant posturing. For a well-trained lawyer, Nixon took remarkable umbrage at those who invoke the Fifth Constitutional Amendment for self-protection purposes.* Speaking in advance of the McCarthy era, Nixon laid this time-honored Constitutional safeguard to rest in his strange statement:

> You will note that on several occasions I have pointed out that many of those who were named as being members of espionage rings, when they came before the committee or before the grand jury and were asked about their activities, answered by saying, "I refuse to answer the question on the ground that my answer I give to the question might tend to incriminate me." Let me say on that point, I do not see how a no answer—"No"—to a question

---

*On another occasion at a HUAC hearing, Nixon said to a witness who took this Constitutional privilege, "It is pretty clear, I think, that you are not using the defense of the Fifth Amendment because you are innocent."

as to whether a person engaged in espionage activities could incriminate anybody.

Nixon continued this frank summary with an expression of nostalgia for the Canadians (who were themselves involved in an espionage case), who did not allow individuals to "hide behind the self-incriminating plea." He said, "The Canadians who were so charged [with espionage] were compelled to answer questions, and that is one reason for the great success of the Canadian investigations." Continuing the House speech, Nixon blamed Hiss for being responsible for the increase of Communists in the world. In a feat of magnificent overstatement and in reference to Hiss's participation in world congresses, he stated:

> In other words, in 1944, before Dumbarton Oaks, Teheran, Yalta, and Potsdam, the odds were 9 to 1 in our favor. Today, since those conferences, the odds are 5 to 3 against us.

Nixon concluded somberly that the tragedy was that Hiss came from a good family, was a graduate of one of our best schools, had been awarded highest honors in government, but sadly "had found the Communist ideology more attractive than American democracy."

Nixon spoke in favor of McCarthy's renomination in 1952, telling the Wisconsin Republican Convention that "a fair investigation" of McCarthy's charges "will never be made until the Republican Party comes into power." McCarthy, in grateful return, supported Nixon strongly in the presidential campaign of 1952. Although Eisenhower made little secret of his distaste for the Wisconsin Senator, the latter charged, "The Communists know Nixon's election will be a body blow to the Communist conspiracy." Later when Nixon was in serious trouble over the discovery of the "special fund" during the 1952 election campaign which led to his famous "Checkers Speech," McCarthy added, "The left-wing crowd hates Nixon because of his conviction of Alger Hiss, the man for whom Adlai Stevenson testified."

Nixon, not unmindful of certain ideologic similarities between himself and McCarthy, said to biographer Earl Mazo in a most revealing moment:

> My feeling is that the McCarthy thing was a tragedy. I think he was really a casualty in the great struggle of our times, as Hiss was a casualty on the other side. Both deeply believed in the cause they represented. The reason

McCarthy became a casualty was because in dealing with this conspiracy it takes not only almost infinite skill, but also patience, judgment, coolness . . . it takes all this plus dedication and courage and hard work. He had the last three qualities in abundance. But in the other respects he was erratic. To an extent he destroyed himself, but he was also destroyed because of the very character of the force he was fighting. This does not imply that all those who were against McCarthy were Communists. On the contrary, many people just as honest as he was opposed him because they felt his methods hurt the cause to which they were dedicated. It is important to remember that when fighting Communists in the United States, domestic Communism, it isn't enough to be right on the merits. You have got to bend over backwards to be fair in tactics because those who oppose you are smart enough never to attack you frontally. They always direct their fire at how you do it, rather than what you do. The reason I succeeded in the Hiss case when McCarthy failed in some others is because I was right on the facts and I did not give the opposition any target to shoot at on the tactics.

Nixon's recollections of this period formed the first chapter of his book *Six Crises.* Nixon began and concluded his book with a question from his then fifteen-year-old daughter Tricia, who wanted to know "What was the Hiss case?" Nixon chose to answer young Tricia with a quote from Whittaker Chambers that the situation which involved Hiss and himself was not just human tragedy but rather a "tragedy of history . . . in which the two irreconcilable faiths of our time, Communism and Freedom, came to grips in the persons of two conscious and resolute men."

For Nixon, political capital gained notwithstanding, the Hiss conviction provided a pleasant and symmetrical ending:

Because through that case, a guilty man was sent to prison who otherwise would have remained free; a truthful man was vindicated who otherwise would have been condemned as a liar, and the nation acquired a better understanding, vital to its security, of the strategy and tactics of the Communist conspiracy at home and abroad.

While experience with Judge Kaufman at the first trial brought to Chambers a sense of disenchantment, Hiss came to feel that Judge Goddard at the second trial was "lenient with the prosecution witnesses and brusque with the defense witnesses." Nonetheless, the second trial seemed to Hiss and to his attorney to be going along very well. To the Hiss team, Thomas Murphy appeared discouraged as the trial ended, and Murphy stated with irritation when the jury requested a rereading of testimony that if this jury disagreed, that would end the case as far as he was concerned.

During the lunch break, while waiting for the jury to return, a reporter from the New York *Times* approached Hiss and told him that he had just canvassed the working press who had covered the trial and asked whether Hiss would be interested in hearing "their verdict." The reporter said the consensus was "acquittal in five minutes." When the jury foreman announced the verdict, Hiss was bitterly disappointed. Nonetheless, he remained confident of obtaining a new trial and said, "My spirits and confidence remain high. Knowing my innocence of the charges, I was, as I still am, certain of eventual vindication."

During the following months, Hiss worked diligently with his attorneys in preparing for an appeal. The trial record stretched to some 4,000 pages, and it was necessary to reduce the entire matter to 125 pages. The appeal was to be based on presumed improprieties in Judge Goddard's charge to the jury at the conclusion of the trial as well as to the claim that a grievous number of erroneous rulings were made during the trial. The Appellate Court denied the appeal in December 1950, but the Hiss group felt that the court's misconceptions about the case were so grave that a rehearing should be immediately scheduled. That request was denied in January 1951.

Despite their recognition that the Supreme Court limits its reviews to cases (1) involving an important Constitutional question, (2) involving decisions by two of the appellate courts in different parts of the country which are in disagreement, and (3) involving legal issues which may affect numerous other cases or individuals, a decision was made to ask the highest court in the land to pass on the merits of the Hiss conviction. The Supreme Court refused to review the case, and Hiss finally went to prison on March 22, 1951.

Even while in prison, Hiss continued to take as active a role as possible in the effort to uncover new evidence. Mr. Hiss's attorney throughout this period was Chester Lane, a former classmate of his at the Harvard Law School. Experts in typewriter construction, in paper chemistry, in the analysis of documents, and in metallurgy were employed to examine that still paramount issue of the typewriter and the documents. From these experts the Hiss team discovered that "forgery by typewriter" was possible and that the typewriter they found and introduced into evidence at the trials showed signs of having been "deliberately altered."

This and other evidence bearing on the condition of the Baltimore documents when found in the Levine bathroom (the envelope was unevenly stained and the documents themselves were not uniformly aged, suggesting that they had been placed in the envelope at different times) were argued before Judge Goddard. This time Hiss was so optimistic that he asked fellow prisoners if there was anything he could do for them on the outside. Once again, however, the judge ruled that no proof had been submitted that would support a finding by a jury that the typewriter received in evidence at the trial "was constructed by or for Chambers or that the typewriter was not the original Hiss machine."

The Hiss forces moved for another appeal, and this request was unfortunately heard by the same three judges who had ruled against them three years earlier. That second effort was similarly denied in July 1952, and the case once again was moved to the Supreme Court for review. On April 27, 1953, the Supreme Court declined to do so.

Hiss wrote shortly thereafter:

> The ordeal of fighting false charges has disrupted my life and has brought pain to me and to my family. But nothing can take away the satisfaction of having had a part in government programs in which I strongly believed. I feel deep satisfaction that I took part in the creative efforts of the New Deal and in the formation of the United Nations. . . . The democratic ideals which motivated me in government service continue to shape my outlook on life.

Years after with the release of the FBI's papers on the case, it was discovered that there were other grounds for appeal. Like Hubert James, the jury foreman at the first trial who in Murphy's mind was not as clean as "Caesar's wife," two jurors at the second trial also had idiosyncratic taints. An FBI office memorandum reads:

> ASAC [Assistant Special Agent in Charge] Belmont of New York called today—and advised that two members of the jury in this case have relatives working in the Bureau. Mrs. Beatrice Link, one of the jurors, is the mother of Beatrice Link Granville, a former clerical employee of the New York office, who resigned to get married. Additionally, Mrs. Link is also related, very distantly, to Marcella Judge, who is presently employed in a clerical capacity in the New York office. The husband of Mrs. Luis (Georgy) Brunty (another juror) is the second cousin of Special Agent Fred Baukham, who is assigned to the New York office.

Recognizing that such relationships had the government facing a potential mistrial, the author of the memorandum finished with this warning: "Mr. Belmont stated that this information has been made available to Murphy and Donegan, who are handling the prosecution of this case, who expressed appreciation upon receiving it, and requested *that it be kept quiet*" (authors' emphasis).

In prison, Hiss worked as a stock clerk. In the words of another prisoner, "He sought no favors, asked for no soft jobs." He was released on November 27, 1954, after serving three years and eight months, the adjusted maximum time of his five-year sentence. Returning home, he began to work on his book, *In the Court of Public Opinion.* Although Doubleday and Company had asked for first option on the book, they were disappointed in it and refused to publish it. Ultimately, Alfred Knopf concluded that the book deserved publication and bought it out. It was also a disappointment to many of his friends, for the book was written, according to Hiss, "as a lawyer's brief."

> I'm not going to write an autobiography and I'm not going to write about my experiences in prison, because I hold certain strong views about privacy. I see no reason why my personal life should be compared with my case.

As a consequence, the book did little to explain to the public the persona behind the Hiss revealed in the case. The book had one unexpected consequence: Its publication led to Hiss's first post-prison job. He went to work for a small manufacturing firm in New York City and shortly became the executive assistant to its president. The business was too small to support two executives, and Hiss was soon at liberty. Unfortunately, the above dilemma was complicated by a growing estrangement between Hiss and his wife that resulted in their separation.

Straitened financial circumstances forced Hiss to move to a cold-water flat, and after a lengthy period of unemployment he took a job in 1960 as a stationery salesman and has made what could be called a mildly successful career in this field. To this day, he has remained essentially unchanged in his expectations that some inquiry may some day unearth new evidence and still finds it difficult to criticize either Thomas Murphy ("Murphy had a job to do and he just went ahead and did it") or Judge Goddard. Hiss seemed no closer to an understanding of Chambers' motivation, and when asked by author Meyer Zeligs to think about the whole way he had dealt with Chambers' accusations against him in 1948, Hiss replied:

> While I was in prison in Lewisburg, I had much time to reflect about the entire

case. I kept thinking, would I do this over again? There was no moral alternative. Though, in part, my behavior was motivated by real anger, that is, that someone I had befriended would turn on me like this—I would defend my name and honor in the same way if I had to do it over again, despite the tragic consequences of my actions.

Chambers was deeply troubled with the impending appearance of Hiss's book. He received several requests to review it, all of which he rejected. Finally he attempted to work out a deal with one editor to prepare an article on the book *without* reading it and with all proceeds going to Hungarian refugee relief. Charitable contribution notwithstanding, the editor wasn't interested in a book review by a writer who was incapable of reading the book, and nothing came of this project.

Just before the publication of Hiss's book in May 1957, Whittaker Chambers was again interviewed by the FBI. Commenting on Hiss's book, he said in a session summarized in the FBI files:

That [he had heard] the material contained in the book was so libelous that he and Vice President Nixon would be forced to sue. Chambers stated that he had always felt that there was the possibility "the other side" would attempt to get to Vice President Nixon through Chambers. Chambers expressed the opinion that Vice President Nixon would not comment on any book written by Hiss, and certainly Chambers is not disposed to furnish any information about any such book. Chambers did say that to avoid any trouble, he was considering leaving home a few days around May 7, 1957 until such time that the present notoriety ran its course.

Chambers felt that the Hiss case had turned both himself and his wife into "old people." He never returned to work in journalism and, apart from an occasional column he contributed to William Buckley's journal, *The National Review,* he remained quietly on his Maryland farm. Following the publication of *Witness,* which was a Book of the Month Club selection and garnered for Chambers a good deal of money, he grew increasingly critical of his writing and began obsessively to rewrite materials and finally to burn them in despair. He wrote in his book of his concern for his son. The language was still mystical and religious—political writer I. F. Stone described Chambers as the "*Time-Life* Dostoevski"—and it is difficult to penetrate to the real meaning of this passage:

I want him to know, in that dark, continuous struggle, that it is by his soul, and his soul alone, that he may sometimes glimpse, if only roughly, the hour

> of the night and his direction in it. I want him to understand, when he lifts up his eyes, that against the range of space and cold his soul, and his soul alone is life for which, in the morning and the evening, he gives thanks to God to Whom it ties him. I want him to know that it is his soul, and his soul alone, that makes it possible for him to bear, without dying of his own mortality, the faint light of Hercules' fifty thousand suns.

Chambers had spent the years immediately after the trial writing *Witness*, and he concluded it on a very solemn note, as if his life were over and the end at hand:

> In a world grown older and colder, my wife and I have no dearer wish for ourselves—when our time shall come, when our children shall be grown, when the witness that was laid on us shall have lost its meaning because our whole world will have borne a more terrible witness or it will no longer exist.

Chambers was indeed still the "witness," and he assumed that role when he made speeches. Once in an address to a Rotarian–Kiwanis meeting near his home, he repeatedly spoke of himself in the third person as "the Prodigal Son." Chambers described a phone call from a Baltimore newspaper which asked him for his opinion of Senator McCarthy. He replied, "What do I think of Senator McCarthy? I hand my mantle to him!"

Although the mantle was freely given by Chambers and presumedly freely accepted by McCarthy, this relationship rapidly foundered. McCarthy sought an appointment at Chambers' farm to seek his advice on how to advance McCarthy's opposition to Eisenhower's proposed appointment of Charles "Chip" Bohlen as ambassador to the U.S.S.R. McCarthy's deviousness troubled Chambers, as did the vagueness of his charge—that Bohlen consorted with known "leftists"—and this seemed to confirm Chambers' feeling that he was being used by the Senator.

Phone calls from the press the next morning verified Chambers' fears. McCarthy had alerted the Washington press corps to his visit to the Westminster farm and intimated that Chambers had given him important information about Bohlen's "un-American" activities. Thus Chambers was forced to make a statement that no such incriminating information existed as far as he was aware but stated nevertheless that he didn't believe Bohlen was worthy of the ambassadorship. Chambers later wrote to a friend that failure to declare himself openly and directly in opposition to the Senator on this matter would mark him as McCarthy's accomplice in the eyes of the world.

Years after Hiss's conviction, the government saw fit to maintain its proprietary interest in Whittaker Chambers. The FBI was especially concerned with the contents of his book. Before publication, Chambers submitted the manuscript and then a set of the galleys to the Bureau for their approval in a literary liaison that was unknown even to his publishers at Random House.

Describing the nature of this relationship, the FBI files read:

> It was Chambers' thought that any suggestions that the Bureau might care to make with regard to this book could be made from the Bureau's review of the "raw copy" already furnished to the Bureau, and any corrections necessitated by these suggestions could be effected by him during his review of the temporary galleys.* Chambers has reiterated his desire to cooperate fully with the Bureau in connection with . . . the book . . . but pointed out that this was somewhat restricted by the standard practices and procedures of the publisher.

As revealed in the FBI papers, there was something extraordinary in the Bureau's near priestly influence on the thoughts, writings, and actions of many of those connected with this case. In keeping with Malcolm Cowley's parable "All former Communists had been warped by their experience, that they felt the loss of something and could be likened to a bunch of defrocked priests," men like Chambers, Wadleigh, and others openly welcomed the FBI into their lives. It was not unlike a religious conversion.

Almost to the end of his life, Chambers sought Bureau approval of his writings and actions and confessed to them all his sins, great and small. Wadleigh was another former Communist who became a convert to the

---

*At least one change was desired by the FBI and effected by Chambers. From the FBI files came this give-credit-where-credit-is-due critique:

> It is noted that in the chapter where Chambers gives credit for the successful conclusion of the Hiss case, Chambers gives credit to Prosecutor Murphy, Senator Nixon, and the men of the FBI, but he does not give any credit to Donegan, who very capably handled the grand jury which indicted Hiss. It is suggested that . . . we point out to Chambers that Donegan handled this matter before the grand jury and had exceptionally good background to present it capably and fairly because of his previous long experience with the FBI.

Chambers acquiesced. Tom Donegan, one of the FBI's favored graduates, indeed is mentioned for his role in the case. This sentence comes from Chambers' book: "Those were the forces—Thomas Murphy, Richard Nixon, and the men of the FBI—who, together with the two grand juries and *Tom Donegan* [authors' italics] . . . finally won the Hiss case for the nation."

FBI's "church." The files say that "out of anxiety," Wadleigh "has made a habit of contacting [FBI] agent [Bert] Zander regarding everything he is doing." In fact, Wadleigh in his fervor became such an annoyance that the Bureau stopped giving him any encouragement. In a sense, then, he was excommunicated.

Similarly, the FBI's cultivation and reliance on confidential informants was like nothing save a priest-confessor relationship. The Bureau certainly encouraged this type of communion and made good use of the information it obtained to secure the conviction of Alger Hiss. For many, the cleansing act of confession to such an authority was reward enough; in other cases absolution was granted in more tangible forms such as freedom from prosecution, a privilege which both Chambers and Wadleigh enjoyed.

Presidents came and Presidents went, but there was always FBI Director J. Edgar Hoover. Although several Presidents considered removing Hoover, none had the courage to act against him. For an organization so able to trade on its Godlike image, the FBI dearly protected their God and also itself as agent-angels of justice. To support its image, the FBI maintained a public-relations arm that was omnipresent. Under these circumstances the request for FBI acknowledgement in Chambers' book came as no surprise.

Shortly after the publication of *Witness,* Chambers severed the umbilical cord linking him to the FBI and also ended the relationship Senator Joe McCarthy was attempting to establish with him. From the FBI files comes this termination order of Esther Chambers:

> On March 31st, 1953, Mrs. Whittaker Chambers stated that she is of the strong opinion that Special Agents of this Bureau will never again be able to transact any official business with Whittaker Chambers, even a quiet, personal interview. . . . Mrs. Chambers had advised that visits by Special Agents upset both herself and her husband. She asserted that this is true merely because they associate the Special Agents with the past Alger Hiss trials which together with the Congressional hearings in which Mr. Chambers had participated were, Mrs. Chambers felt, the direct cause of his heart attack. She added that the recent visit of Senator McCarthy to their farm in Westminster, Maryland, had seriously upset the entire family and had put them in a position where Mr. Chambers was forced to make a statement to the press through her.

An article written by Fred Rodell, professor of law at Yale, entitled "Was Alger Hiss Framed?" is instructive not just for Mr. Rodell's affirmative conclusions but also for the FBI's reaction to the article. Rodell wrote in *Progressive* magazine that Alger Hiss's conviction was won with highly suspect and possibly manufactured evidence. Rodell commented that regardless of Alger Hiss's guilt or innocence, "I was convinced that there was something malodorous, to put it mildly, about certain FBI activities in connection with the case."

Rodell went on to describe the roadblocks set up by the FBI to impede Chester Lane, Hiss's appeal attorney who worked on while his client was imprisoned. Mr. Lane said of his difficulties, "We search for records, the FBI has them; we ask questions, the FBI will not let people talk to us. When we ask people to certify information in files they have shown us, they must consult counsel, and we hear no more from them." Rodell also echoes Lane's charge that the FBI was so interested in his investigation of the typewriter that they kept him under surveillance during the entire phase.

The following section comes from the FBI's summary of the Rodell article:

> The article has no basis. It can only be concluded that it is an unwarranted and vicious attack on the Bureau. The Bureau knows that Lane's allegations against it which are reiterated and implemented by Rodell are absolutely unfounded. If it were not for the fact that the Hiss motion is presently pending before Judge Goddard, it is believed that we should take aggressive steps to put Mr. Rodell and the *Progessive* magazine in their place.

Other writers were used perhaps unwittingly as public-relations men for the FBI. One was Bert Andrews, Nixon's reporter friend, who signed a contract with Appleton-Century-Crofts to write a book on the Hiss case. The publisher requested that a psychoanalyst provide an interpretation at the end of each chapter.

Andrews phoned the FBI and asked for information on the politics and character of Dr. Gregory Zilboorg, a psychiatrist suggested by his publisher. Zilboorg came highly recommended, Andrews noted, as he had treated such prominent figures as Marshall Field, the Chicago department-store magnate who was also publisher of the liberal newspaper *PM*. Zilboorg was described by Andrews, according to the FBI papers, as being "Russian-born but is a valid, White Russian," and as far as Andrews knew, "Zilboorg was not a publicity seeker and has a clean record from a subversive standpoint."

The FBI agent, L. B. Nichols, who talked to Andrews declined to pass judgment on the character of Dr. Zilboorg but after this conversation he speculated to his superior, Clyde Tolson, that Andrews might be a good source to get the FBI's version of the Hiss–Chambers story across to the nation. Nichols wrote, "I frankly think that on this we will have to find some way whereby we can get across to Andrews in an informal manner, of course, our side in order that we can assure against unfavorable comment." Nichols added that Andrews might be the right reporter for the job because he "was quite outspoken in denouncing Hiss—however, this might have been purely for effect. He [Andrews] stated that Hiss was goaded into filing a suit by the Washington *Post* when their ringing editorial suggested this course of action, and, after a lapse of time, the Board of Directors of the Carnegie Foundation began to ask questions as to why he didn't [file suit]." Bert Andrews died before finishing his book. The project was completed by his son, Peter Andrews, and contained no comments by Dr. Zilboorg or any other psychiatrist. It was published in 1962 by R. B. Luce, not the original publisher.

Not surprisingly, Alger Hiss concluded that there was something self-serving in the FBI's search for evidence in his case. Hiss commented that the Bureau—much like the HUAC, which incidentally had its own private problems with the FBI—employed a verdict-first, find-the-corroborating-testimony-second *modus operandi.* Hiss said to the authors, "There are documents [in the FBI files] which say to agents that your job is to support Chambers' story."

Asked to assess responsibility for his tragic life, Hiss was unwilling to blame either Richard Nixon—that his attack was politically motivated—or Whittaker Chambers—that he was mentally ill. Nonetheless, Hiss bore great antipathy for FBI Director J. Edgar Hoover. In Hiss's mind, Hoover was the "real villain," Hiss said, "only in the sense that he directed and pressured his men to get any kind of evidence that would support Chambers' stories. . . . He saw the chance for making this what it became, a *cause célèbre,* if the FBI won it. . . . So he simply wanted a victory."

It is somewhat difficult to believe that Hoover should bear the major responsibility for Hiss's fall from grace. Chambers and Nixon come off relatively well while the FBI chief becomes the culprit. This kind of rationalization does not surprise those who know Hiss. To this day he seems to be unwilling or unable to respond with feeling to the Chambers–Nixon link.

Notwithstanding, Nixon's Watergate *contretemps* did, in interesting fashion, revive public interest in Hiss. When interviewed by the Los Angeles *Times* in February 1973, Hiss commented wryly on the fact that he had received a fund-raising letter the previous summer from the Committee for the Reelection of the President addressed "Dear Fellow American." Hiss's book on the events of 1948 to 1950 was reissued, and the Los Angeles *Times* commented that even at sixty-eight, Hiss retained "the lean good looks that have bothered so many people, including Whittaker Chambers, who wrote: 'Quite unconsciously for the most part, [people] want to believe me wrong and wicked. It corresponds to a deep need in them. . . . The parts were miscast.'" Hiss spoke of Chambers, but still there was no fire:

> Why does a sniper shoot down one man and not another? Why will someone kill a total stranger who resembles his father? Why did Bremer shoot Wallace? . . . Motivation baffles me. It was always one of the most tormenting aspects of the case. . . .
>
> My feelings about him now are no more friendly than they were at the time. Asked about the value of the secret documents he is said to have passed to Chambers, he said: They had somewhat less security value than the Ellsberg documents. . . .

The Los Angeles *Times* reporter wrote: "The dispassionate quality, in both the man and his book, is disturbing."

The interview was a follow-up to Hiss's appearance at a meeting of the Soçiety of Magazine Writers in New York. At that meeting he spoke of three possible theories which connected the Woodstock typewriter with the Baltimore documents, but it was said that no one any longer understood the issues.

When asked when he would be cleared, Hiss remained convinced the record would be changed, but it was an event that would no longer take place in his lifetime but certainly would in his son's lifetime. The release of confidential FBI documents on the case has changed Hiss's caution. Moreover, internal scandals in the FBI and the CIA giving evidence of their willingness to collude with each other and with others in high government office to protect "national security," has brought about an increase in public cynicism about their probity in investigative matters.

What became of Whittaker Chambers when the furor had at last died down? During the period that followed Hiss's conviction, Chambers suffered a coronary from which he slowly recovered. Hiss's release from jail distressed him, and he wrote to William Buckley in a letter published much later in *Esquire:*

> Alger came out more fiercely than ever I had expected. . . . His strength is not what it was. But that it exists at all is stunning. Every time that, in the name of truth, he asserts his innocence, he strikes at truth, utters a slander against me, and compounds his guilt of several orders. . . . It is this which squirts into my morale a little jet of paralyzing poison.

Despite his illness, Chambers continued to work hard on his farm, looking after his herd of fine cows and other assorted animals. His insistence on maintaining an arduous schedule troubled his physician, who felt that Chambers was determined to die with his boots on. A fire broke out in the attic of the large Chambers' house, and they went to live with neighbors until the damage was repaired. They never moved back into the house and lived instead in a much smaller and rustic second house which they had lived in when they first acquired the farm in 1937. Chambers became increasingly reclusive and found all travel to be fatiguing. Another coronary occurred, and when he sought somewhat later to change his living pattern by a trip to Europe, he once again became ill and had to return home suddenly.

When he recovered from this illness, Chambers did a very unusual thing. He enrolled at the age of fifty-eight in a small college near the farm and spent the next two years going daily to class and doing his homework each evening. He was a science major and achieved good grades.

During this time Nixon's political fortunes had declined with his defeat by John F. Kennedy in the presidential election of 1960. Concerned with how Nixon was dealing with adversity, Chambers wrote to him in February 1961:

> It seems possible that we may not meet again—I mean at all. . . . I do not believe for a moment that because you have been cruelly checked in the employment of what is best in you, what is most yourself, that that check is final. It cannot be. . . . Great character always precludes a sense of comedown.

A month after his only son's marriage in June 1961, Chambers suffered a final coronary attack and died. The death was not publicly announced until seventy-two hours after it had occurred, and this fact, plus the story that Mrs. Chambers had taken an overdose of barbiturates on finding Chambers dead and had to be hospitalized as a consequence, revived the public's flagging interest in the case.

The New York *Times* went immediately to the other two principals. The contrast between their reactions was illuminating:

"His book *Witness*," said Nixon, "is the most penetrating analysis of the true nature of deadly appeal of Communism produced in this generation. It should be required reading by every American who is concerned by the threat of Communism."

Mr. Hiss declined comment on the death of Whittaker Chambers. He has steadfastly vowed confidence in his "eventual vindication." He charged Chambers "fabricated the evidence against me."

CHAPTER 8

The typewriters are always the key. . . . We built one in the Hiss case.

—RICHARD NIXON talking to Charles Colson as reported by John Dean

Several questions need to be answered about the ill-fated Hiss typewriter, described by Richard Nixon as "the key witness in the case," which was found by the defense forces and which did as much as any piece of evidence to bring about Hiss's eventual conviction. For example, was the typewriter prominently on display at both trials, an old desk-model Woodstock bearing the serial number N230099, the same typewriter that had been in the Hiss home? If this were not the Hiss typewriter, did the FBI know it? And if it were a fake, who was responsible for this fabrication?

Previously confidential papers, released under the Freedom of Information Act, and other documents and reports related to the Watergate scandal enlighten us on the questions that have gone begging for answers for nearly three decades. In the first place, the Woodstock 230099 could not have been the Hiss machine. It was manufactured in 1929, while Hiss's typewriter could be traced back to 1927. Secondly, the FBI was well aware of this discrepancy and its importance and went to great lengths to keep this information from coming out at the trials. It is even conceivable that the FBI passed off this doctored machine on an unsuspecting Hiss defense team. Thirdly, the FBI's famous crime laboratory did not have much facility in interpreting typewriter evidence. Finally, the documents reveal that the FBI made good use of Horace Schmahl, a double agent who reported to the Bureau while working for Hiss, primarily on the typewriter phase of the investigation.

The Woodstock belonging to the Hiss family had originally been owned by Priscilla Hiss's late father, Thomas Fansler, who used it in a small insurance office he ran with Harry Martin from 1927–1930. In an FBI interview, Martin said they bought a new Woodstock typewriter at the inception of the partnership. The business was dissolved three years later when Fansler retired. It is agreed by all the principals that the latter took the typewriter and a roll-top desk and gave the typewriter to his daughter. Priscilla used it for occasional correspondence and also for a book she wrote and published in 1934 on the fine arts in corroboration with her sister-in-law.

The big Woodstock proved to be a poor typewriter for Priscilla's purposes, as the keys jammed, the ribbon didn't advance properly, and it showed other maladies common to old manual typewriters. As a consequence, she wrote much of the book in longhand. The machine was simply more trouble than it was worth, and ultimately the Hisses disposed of it. Alger Hiss felt that his wife had given the Woodstock to a secondhand typewriter establishment in Georgetown. More than a decade after the event, Priscilla was unable to recall exactly what happened to it despite being asked about it several times by both the FBI and the House committee. Hiss's stepson, Tim Hobson, had a vague recollection that the typewriter was given to the Catlett family. Hiss related this to his attorneys but not to the grand jury or the FBI.

After Alger Hiss's indictment, the son of Hiss's former maid called Donald Hiss and told him the FBI had been asking him about the typewriter. Raymond Sylvester "Mike" Catlett said he knew where it was but hadn't told the agents. Mike Catlett remembered he and his brother Perry receiving the typewriter as a gift when the Hisses moved from either P Street to 30th or from 30th Street to Volta Place. Both moves were prior to the earliest Baltimore document,* but fixing the precise time of the gift proved difficult for Catlett, who became befuddled during the prosecutor's sharp cross-examination. This testimony proved costly to Alger Hiss.

A Southern black like Mike Catlett was a stranger in a strange land in front of the all-white jury in the federal courtroom in New York City. Catlett's incongruity was magnified by the fact that he followed closely on the heels of Supreme Court Justice Felix Frankfurter, who had testified for the good character of Alger Hiss. But Catlett was not about to be cowed by any lawyerly machinations. The cross-examination opened with Prosecutor Murphy inquiring if he might call him "Mike." "My name is Raymond Sylvester Catlett," he replied with some passion. This angry pride was maintained during the rest of his testimony.

News observers were struck by this strange juxtaposition of witnesses. One wrote, "Justice Frankfurter's performance had been a legal charade put on for the benefit of the peasant layman, but in the afternoon watching the graveled and outraged Mike Catlett, it was the layman who felt as learned as the nine Supreme Court judges rolled into one."

In response to the prosecutor's effort to pinpoint the date of the typewriter gift, Catlett erupted in frustration, saying that Murphy was getting him "all balled up." It went rapidly downhill from there. The nadir came when Catlett said the FBI had offered him "$200 or more" for the typewriter. The agent he accused—not exactly a disinterested party—was only too happy to deny this as well as defense attorney Stryker's charge that the FBI had bullied Mike and his brother Perry. In the defense attorney's redirect examination, Stryker characterized the Catletts as playing out of their league with the FBI, calling them "those poor little colored boys," a remark that elicited little sympathy from the jury. The entire episode was not helpful to either Catlett's credibility or Alger Hiss's case.

Although the typewriter proved no longer to be where Catlett remembered it, he was able to establish the first link in the chain of possession that the defense team followed. Donald Hiss, Mike Catlett, defense attorney Edward McLean, and a black attorney, Charles Houston, who

*The name given to those papers produced by Chambers in the libel suit hearing in Baltimore, which were ultimately turned over to the Justice Department.

acted as an ambassador to the suspicious black community, tracked the Woodstock through its tangled web of ownership and, presumably scooping thirty-six FBI agents in the process, recovered it on August 16, 1949, for $15 from a night watchman, Ira Lockey.

Even this was no simple matter. When Catlett first went to Lockey's house, he said he saw several FBI agents there and fled. When he went there on another occasion, Lockey refused to talk to him. Next, Lockey said he knew nothing about a typewriter despite Catlett's offer of $50 for its recovery. Finally, when the defense team accompanied Catlett to Lockey's house, the typewriter had materialized, as if by magic, and Lockey seemed only too happy to turn it over for $15, far less than the previously mentioned sum. He said he had received the typewriter in partial payment for a move he did for a man named Marlow. When Lockey first saw the typewriter, it was outside Marlow's home, weathered in the elements and obviously in bad shape. Lockey found it nearly inoperable and gave it to his daughter, Margaret, who felt the machine was beyond repair and rarely used it. Given this checkered history, the machine seemed in surprisingly good health upon inspection.

Besides looking for the typewriter, the Bureau scoured the country for any correspondence typed on the Woodstock by the Hiss family. Even former cub scout chums of Alger Hiss's stepson, who was now in his twenties, were subjects of FBI attention. After an exhaustive search where anyone who ever had anything to do with any of the Hisses was asked for such correspondence, four communications* were found by the Hiss family and furnished to the FBI. They came to be known as the "Hiss standards" at the trials. The date of the latest standard was May 31, 1937, typed seven months before the earliest Baltimore document.

The Bureau interviewed everyone they could find who was connected with the Fansler–Martin insurance office, including the secretary who used the Woodstock. Although they were vague on some of the details after all these years, all agreed that (1) the typewriter was purchased in

---

*One of the Hiss standards was entitled, "Descriptions of Physical Characteristics of Tim Hobson"; the second was a letter addressed to a Mr. Heillegrist, Director of Admission at the University of Maryland; the third was a letter typed by Priscilla's sister, Daisy Fansler, and addressed to a Miss Hellings; and the fourth was a president's report to the alumni of Bryn Mawr College typed by Priscilla, who was then president of the alumni association. The FBI expert, Feehan, testified that all of the sixty-four Baltimore documents, except No. 10, which was typed on a Royal typewriter, matched the four Hiss standards.

1927, (2) that there was never more than one Woodstock in the office, and (3) that they never traded it on a new machine. The FBI also found Thomas Grady, the man who sold the Woodstock to Fansler–Martin. Grady said the partnership had just formed when he sold them the typewriter. This had to be in 1927 because Grady resigned his position with Woodstock on December 3, 1927, and never again sold a typewriter. The FBI interviewed Martin and Grady on at least two other occasions. Neither was ever to change his testimony.

The FBI then went to the Woodstock factory near Chicago to get the appropriate serial numbers for typewriters manufactured and sold around the year 1927. Naturally there was a time lag between the date of manufacture and the time the typewriter actually reached the market place. A dealer's manual showed that the first typewriter produced in 1926 bore serial number 145,000; 1927: 159,300; 1928: 177,100; 1929: 204,500; and 1930: 246,500.

Armed with this information, the agents searched for Woodstock typewriters with serial numbers "less than 177,000," as the files read, which corresponds to typewriters manufactured in 1928 or earlier. They were able to find dozens of old Woodstocks but *none* which matched the type characteristics of the Hiss standards or the Baltimore documents. Understandably, the agents had no interest in later-model typewriters, such as the defense's Woodstock 230099, which, according to the factory's records, was probably manufactured around 1929.

FBI agents learned that the defense had possession of the long-lost typewriter in the course of an interview with Hiss's former maid, Clydie Catlett, on May 14, 1949. That the Goliaths of the investigation world were bested by the ragtag forces of the defense was not exactly welcome news. There was much posturing to avoid the wrath of J. Edgar Hoover. Within twenty-four hours the FBI traced the machine to Ira Lockey, who showed them McLean's bill of sale, which noted the serial number 230099. Long-winded memoranda were sent to the Director explaining why the search had concentrated on typewriters with serial numbers much lower than 230099.

An immediate investigation into the history of Woodstock 230099 was undertaken by the Bureau. Thomas Donegan, special assistant to the Attorney General, who assisted Prosecutor Murphy in the case, requested in the course of this investigation: " . . . that all persons being interviewed re Woodstock typewriter phase of this investigation should be informed that they should regard such interviews as strictly confidential as are any other interviews conducted by the Bureau. Do not, however, advise persons interviewed that they should not inform Hiss or his attorneys that they were contacted by FBI as it would be embarrassing at trial if it developed that such instructions were given by Bureau."

The FBI went back to the factory to try to find the history of this particular Woodstock typewriter. There were no records on the sale of 230099 nor any data on any subsequent service by the factory. The Bureau did, however, manage to focus the date of its manufacture a little sharper. The FBI's files read, " . . . serial number 230099 was placed on a machine between March, 1929, when serial number 220,000 was used, and August, 1930, when serial number 265,000 was used." As for when the typewriter might have reached the market, " . . . sale of a machine with serial number 230098, one digit lower than the machine in question, has been located. It occurred in Philadelphia, Pennsylvania on September 21, 1932." This was two years *after* the Fansler–Martin partnership ended and five years after their typewriter was allegedly purchased.

The conclusion is inescapable. Unless all witnesses lied and the factory records are wrong, Woodstock 230099, the typewriter produced by the defense which proved to be so damning to Alger Hiss, was not the Fansler–Martin machine, nor by extension the Hiss typewriter. But the defense was unaware of this at the time of both trials (and for many years later as well).

Defense investigator Horace Schmahl followed the same trail blazed by the FBI to Harry Martin, the surviving member of the Fansler–Martin partnership. Martin told Schmahl the same story he had previously told the Bureau: The partnership was formed in 1927; upon its inception, a Woodstock typewriter was purchased from the company's representative, Thomas Grady; this was the only typewriter used in the office. Martin further suggested that Schmahl contact Fansler–Martin's parent office, Northwestern Life Insurance Company, for typewritten letters which might be useful for comparison purposes.

When Schmahl phoned the main office in Milwaukee, he was told all such specimens had been turned over to the FBI. On December 9, 1949, a representative from Northwestern Life called the FBI advising them of Schmahl's interest. Schmahl also called the Woodstock office, which refused to cooperate with the defense. They, too, advised the FBI of Schmahl's investigation. The private detective was unable to locate Thomas Grady.

Not only did the wandering path of ownership of 230099 seem plausible to the defense, but when they took the typewriter to their own experts, the Hiss group was told that it now had in its possession the very machine that typed both the Baltimore documents and the Hiss standards. Since they had searched for and found the typewriter at no small cost in time and manpower, the opinion of the experts must have been shocking to the Hiss camp. A compelling question, never satisfactorily answered, is why would Hiss authorize such a costly search if he knew it to be the "guilty" machine?

Nevertheless, the defense did announce its "find" and also introduced the typewriter into evidence at both trials. Why? As Hiss said recently, "Our records showed that we got rid of the typewriter before the dates on the documents. Therefore, it is really like a case where someone says X is guilty because the car driven in a murder was his. It doesn't prove who's driving it. Particularly if you can show it was out of his possession at the time of the crime." Hiss curtly dismissed the question of whether it was a tactical error to introduce into evidence a typewriter which he knew had been identified with the incriminating Baltimore documents: "I was seeking to bring forward all the facts before the court."

The defense found their typewriter on April 16, 1949. The first trial convened one and a half months later, and the court impounded the machine on July 8 at the end of the trial. This allowed the defense expert less than forty-five days to inspect the Woodstock, and consequently only a cursory examination was performed. The only physical anomaly found was its surprisingly good condition.

With the application of a little oil, it worked very well. This belied the Woodstock's known history. When Hiss gave the typewriter away, it was "in poor condition. . . . some of the keys stuck and the roller had to be turned directly by hand." In the courtroom Mike Catlett said that "It was broke" when he received it. "The keys would jam up on you, and it wouldn't work good. You couldn't do any typing hardly with it." Catlett also said there was a knob missing, and the ribbon didn't advance correctly, which was why it was taken to a repair shop. When Ira Lockey first saw the machine, it was outdoors in the rain.

Several years later, in 1952, a more important irregularity was found. Dr. Daniel Norman, an expert in the physical and chemical analyses of metals, said that Woodstock 230099 "is not a machine which has worn normally since leaving the factory but shows positive signs of having been deliberately altered, and that many of its type are replacements of the originals and have been deliberately shaped." Such information led Hiss to make his last appeal for a retrial, but the three appellate judges found his arguments unconvincing.

The defense introduced Woodstock 230099 into evidence at both trials. Several witnesses for the defense identified it as the typewriter that had been in the Hiss home, but the government knew its history was wrong; *it was manufactured too late to have been the Fansler–Martin machine.*

John Lowenthal, a professor at Rutgers Law School, wrote in *The Nation* (June 26, 1976) that the "Government faced the tactical problem . . . [by] *appearing* [Lowenthal's emphasis] to accept the machine as the Hiss Woodstock, but without putting any witness on the stand to testify to that effect under oath." At the first trial, in fact, the government wouldn't even admit that the typewriter in question was a Woodstock. Carefully choosing his words, the government's document examiner, Ramos C. Feehan, said that the Hiss standards and the Baltimore documents were typed on the *same* typewriter, but declined to identify which make of machine.

Lowenthal, an old family friend of the Hisses who had worked on the Hiss appeal while a law student at Columbia, noted that in summing up the government's case at the second trial, Prosecutor Murphy subtly changed Feehan's carefully worded testimony, saying, "They [the documents] were typed on that [indicating the defense Woodstock] machine. Our man [Feehan] said it was." Lowenthal pointed out that Feehan had said no such thing, but Murphy's summation "was only a lawyer's argument . . . [not] made under oath." Judge Henry Goddard was misled by Murphy's sleight of tongue, for he said, "It is the contention of the Government that this [#230099] is the typewriter upon which [the] Baltimore exhibits . . . were typed."

Hiss may have been misled by his own experts when he was told 230099 typed both the Hiss standards and the Baltimore documents, but in addition to the time constraints, the defense had trouble in attracting any experts to their side in such a highly publicized case where the FBI's famous crime lab would be their competition. The FBI had the resources, talent, and, most important, the public relations to "prove" whatever they wanted. The Hiss case was not an attractive prospect for the best men in the field.

If the Hiss experts were confused, the government's witnesses came up with the same conclusions: that Woodstock 230099 typed both the Baltimore documents and the Hiss standards—although they said nothing like that at the trials. One such reference comes from J. Edgar Hoover, dated May 25, 1949. It says that Woodstock 230099 " . . . has been identified by the FBI laboratory as being the machine which was used to type documents . . . known as the 'Baltimore' documents." This is a fascinating admission since it comes five days before the opening of Hiss's first trial. The Hiss lawyers would not even introduce the typewriter into evidence until after Feehan, the government's typewriter expert, gave his trial testimony.

It is possible that Hoover was in error here, but in the same directive he says, "Under no circumstances during the course of the above-requested investigation should the *fact* [authors' italics] be disclosed that the typewriter bearing serial 5N 230099 has been identified as the machine used to type the 'Baltimore documents.'" Could it be that the FBI had examined 230099 even before the defense "discovered" it and then planted it with Ira Lockey with instructions to allow the defense to buy it, which it did for the bargain price of $15 rather than the $50 that Mike Catlett has previously mentioned? This, however, is mere speculation, because there is no such conspiracy plot proven in the 40,000 pages thus far released under the Freedom of Information Act, which, however, doesn't prove that such papers do not or did not exist.

Between trials, Prosecutor Murphy was granted an *ex parte* order allowing him to take samples from the defense's Woodstock, then in the custody of the court clerk. Twenty-one pages of type and ten carbons were taken from Woodstock 230099. What was the conclusion of the FBI's document expert? The report from the laboratory dated October 27, 1949, says, "It was concluded that the typewriter with serial 230099 which was used to type the specimen designated above [21 pages of type and 10 carbons] was the typewriter that was used to type the evidence below." The evidence listed below was the Baltimore documents and the Hiss standards.

When defense investigator Horace Schmahl called Northwestern Mutual Life for typed correspondence from Fansler–Martin, he was told that all such communications had been turned over to the FBI. The FBI's laboratory, however, was unable to correlate these twenty-one-typed pages with the Baltimore documents. An FBI office memorandum dated June 9, 1949, with the heading "an analysis of information . . . concerning the Woodstock typewriter featured in this case," reads "Typewriter specimens . . . from Martin–Fansler written in 1927 have been definitely found non-identical with the 'Chambers' documents." There were no samples available for 1928. "A specimen bearing the date of June 29, 1929," continues the report, "has been determined by the FBI Laboratory as being non-identical. However, a specimen dated July 8, 1929 . . . on the letterhead of the Martin–Fansler partnership has been determined by the FBI Laboratory to have been prepared on a Woodstock typewriter of the same general style on which the 'Chambers' documents were prepared."

Similar, yes, but identical? No. The FBI explained their inability to make this important correlation by saying in the same report, "It is felt that in the intervening years between 1929 and 1938, characteristics could have been developed by normal use and wear which might account for changes in typewritten characteristics."

The inability of the government's expert, Feehan, to perform this critical correlation leads to some surprising conclusions: (1) the typewriter that typed the Baltimore documents was not Hiss's typewriter—it was another Woodstock; it could have been Chambers', the FBI's, or anyone else's for that matter; (2) the government experts were either sloppy in their examination of the typewriter and documents or didn't have the tools or knowledge to do the job right; (3) typewriter characters change over time [this the FBI admitted in the last quote] in an unpredictable fashion that could make criminal laboratory correlations tenuous; and (4) there was another typewriter in the insurance office, but former Fansler–Martin employees and Harry Martin deny this.

Despite Martin's steadfast refusal to change his testimony about there being a second typewriter, the FBI became much attracted to this thesis. Noting that a specimen dated July 8, 1929, was prepared on a typewriter "of the same general style" as the typewriter that typed Chambers' documents, the Bureau speculated that sometime between then and June 29 (when a "non-identical" letter appears), a second typewriter was purchased. Hoover wrote, "There are at least indications that the Woodstock typewriter in question (230099) was purchased on or shortly before July 8, 1929."

Assuming for a moment that there was a second typewriter, it could not have been 230099. Representatives of attorney Chester Lane, who after Hiss's conviction was working for a new trial, interviewed Joseph Schmitt, head of the Woodstock factory and custodian of the records. From this interview, Lane concluded, "On the basis of the use of serial number 220,000 in April, 1929, and the monthly production statistics set forth . . . above, it appears that Woodstock . . . 230099 was manufactured during the *latter part* [authors' italics] of July or in August, 1929." Therefore, it could not have been sold on July 8, 1929. Schmitt, however, refused to sign this affidavit, although he admitted to the FBI that it was "substantially" correct. Schmitt also mentioned to the FBI that there was a time delay between manufacturing of the machine and its packing, shipping, and selling. This was at least a couple of months, and when inventories were high, as they were in 1929, it was as long as eighteen months. This was borne out by the fact that Woodstock 230098—one digit lower than the defense's Woodstock and similarly manufactured in 1929—was not sold until the end of 1932.

Such discrepancies were disturbing to the FBI, and some of their attempts to reconcile this matter began to resemble a comic opera. On May 25, for example, Hoover directed the Milwaukee office " . . . to interview Thomas Grady to obtain an explanation as to how he could sell a machine which was manufactured in 1929 to the Fansler–Martin partnership in 1927." Grady would not budge. In an office memorandum from the Washington office of the FBI, dated June 9, 1949, it was decided that this discrepancy of dates is evidence of "Grady's unreliability," a characterization also given to him by Harry Martin.

Allen Weinstein seems much taken with this description of the typewriter salesman. He wrote, " . . . Thomas Grady entertained no doubts about the date [of the sale]. Nor without opening himself to a charge of larceny could he have." Weinstein suspected that Grady stole a typewriter from his employer in Philadelphia and then sold it after he terminated his employment on December 3, 1927. This theory still does not explain how he managed in 1927 to steal and sell a typewriter which was manufactured in 1929. Moreover, according to the FBI files (January 11, 1949), which Weinstein presumably had read, a report made by agent James L. Kirkland says, "John Carow, manager of the Philadelphia agency for whom Grady worked during all of 1927, and for a number of years thereafter, has advised that there were no inventory shortages prior to 1933. This would eliminate the possibility that Grady stole a typewriter while in the employ of Woodstock.

It wasn't until after Hiss's conviction and after his appeal was rejected that his attorney, Chester Lane, began to question the authenticity of the typewriter. Lane thought it strange that the government expert had never linked the trial Woodstock to Chambers' documents or to the Hiss standards. As mentioned earlier, Lane sent representatives to interview Schmitt, boss of the now defunct Woodstock company, to find out when Woodstock 230099 was manufactured. From this interview Lane concluded that the typewriter was manufactured in late July or early August.

Remembering that Horace Schmahl, the defense investigator, had learned from Harry Martin that the one and only Woodstock typewriter was purchased by the Fansler–Martin agency in 1927, Lane contacted Martin, who refused to talk with him without FBI permission. Lane went back to Schmahl, who refused to sign an affidavit outlining the results of his investigation done years before.

Following now what was becoming a well-worn path, Lane contacted Northwestern Mutual Life, which refused to provide him with samples of typing from the Fansler–Martin partnership. The company finally relented and agreed to release some letters to Donald Doud, a document expert hired by the defense. It is interesting that Doud's conclusion was the same as Feehan's (although the defense never knew what Feehan's conclusion was until very recently). Doud told Lane that "the Fansler–Martin office acquired a second Woodstock machine between the period of June 29, 1929, and July 8, 1929." Of course Martin had said that wasn't so; but neither typewriter could have been Woodstock 230099, for that machine was not manufactured until late July or early August of that year.

When Lane petitioned for a new trial on the basis of newly discovered evidence, one of his strongest arguments was that 230099 was not the defense's Woodstock. The government responded to this argument by contending that Feehan never said it was. (In fact, at the first trial, he had testified even before the defense had introduced their Woodstock into evidence.) Feehan testified only that the Baltimore documents were typed on the same machine as the Hiss standards. Myles Lane, who argued the case for the government, said, " . . . No identity with Ex. UUU [the defense Woodstock] was attempted or needed." This was not true—an identity was attempted, and the conclusion was that 230099 typed the Baltimore documents—but the Hiss team didn't learn this until the release of the FBI papers. Lane also said in grandiose fashion, ". . . Even assuming the government's argument that the trial exhibit was a fabricated machine and not the Hiss machine, the soundness and completeness of the government's evidence is not affected one iota."

Long forgotten, however, was Prosecutor Murphy's summary argument at the end of the second trial. The Baltimore documents, he said, "were typed on that machine [indicating the defense Woodstock]; our man [Feehan] said it was." Their man, Feehan, had cleverly said no such thing. Also forgotten was Judge Goddard's echo of these misleading words: "It is the contention of the Government that this is the typewriter upon which the Baltimore Exhibits. . . were typed."

Richard Nixon, with his usual air of certitude, wrote about the infallibility of typewriter evidence in his book *Six Crises:*

> A typewriter has one characteristic in common with a fingerprint; everyone is different, and it is impossible to make an exact duplicate unless the same machine is used.

This assertion will be examined shortly, but what is obvious from the FBI papers is that the experts could not interpret typewriter evidence with much facility. Even if all fingerprints are different, there is no practical purpose in crime detection if the crime laboratory is unable to tell one fingerprint from another. This seemed to be the state of the art in typewriter and document interpretation back in the year 1950.*

There was more subterfuge on the part of the government involving typewriter evidence. At the first trial, FBI document examiner Feehan said that in comparing the Hiss standards to the Baltimore documents, ten typewriter characters were used in looking for "similarities" as well as "variations" in the documents. Raymond A. Werchen and Fred J. Cook, writing in *The Nation* (May 28, 1973), noted that at the second trial, "There was no mention of 'variations.' . . . Feehan's second trial testimony made it appear to judge and jury that the characteristics he had found and compared in the documents were absolute and unvarying with no exceptions whatsoever. This has to be considered little short of deception."

It was not until after Alger Hiss's conviction that Chester Lane challenged Feehan because he compared ten characters rather than all of them. In an affidavit answering Lane's challenge, Feehan changed his testimony, saying, "I examined each and every character of typewriting

---

*An Associated Press story in April 1977 leaves the impression that things have not changed much in the ensuing quarter century. Evidence produced by federal, state, and local crime labs seems to lose its sacred-cow status with the results of a test conducted by the Forensic Science Foundation on 240 such laboratories. The conclusion was that these crime laboratories "often do poor work and sometimes are wrong about the evidence against defendants in criminal trials."

A spokesman for the foundation commented that it was *possible* the FBI's laboratory was involved in the test, but there is no way to tell how they scored. Although typewriter evidence was not part of the study, such basic correlations as comparing bloodstains saw some labs score as low as 40 out of a perfect score of 100. According to the story, "The report . . . said that many crime laboratories operated by state, local, or federal governments made mistakes and overlooked stains, firearms, glass, paint, soil, and important features in comparing blood [and] other items used as evidence."

appearing on the questioned and known documents." He said he concentrated on only ten characters to avoid confusion for his presentation to the jury. Cook and Werchen conclude that such a task was "patently impossible." They calculated that it would have taken Feehan "6,400 hours or 800 eight-hour days with no recess for lunch" to examine the "93,267" characters in the manner that he said he did.

Hiss, commenting on the alleged infallibility of typewriter evidence, said recently, "Now the real trickery into which we tumbled was the mystique about typewriter evidence, that a typewriter can't be forged." The defense's own expert told them that typewriter evidence can't lie, and so at both trials Feehan's words were accepted as gospel. It is sad to note that Feehan wasn't even cross-examined.

In 1949—and years before as well—typewriter evidence could indeed be forged. Intelligence operatives had been doing this since before World War II, and FBI Director J. Edgar Hoover was well aware of this capability; in fact, he personally visited an establishment in Canada where this work was being done. Typewriter forgery was a small part of a massive intelligence operation set up by Prime Minister Churchill and President Roosevelt to counter the Nazi war machine. This worldwide intelligence agency, which included some of the best men from either country as well as underworld figures, was run by Sir William Stephenson, whose code name was Intrepid. Called Camp X, it was situated on the Canadian shore of Lake Ontario, 300 miles from New York City. William Stevenson (no relation to the book's hero) writes in *A Man Called Intrepid* that the site "was chosen in part because it could be reached easily by the FBI agents." It could not be located in the continental United States, for we were officially "neutral" at this time.

Describing the operatives culled for this intelligence work, he writes:

> The experts came from all levels of society. Some were men and women of . . . distinction. Others were safe-crackers, forgers and professional bank robbers whose expertise could not be duplicated by legitimate entrepreneurs. . . . [One was] a typewriter manufacturer [who] could duplicate any patented machine in the world. . . . I was associated with an industrial chemist and two ruffians who could reproduce faultlessly the imprint of any typewriter on earth.

The forgery section was designated Station M, and one job they performed was to duplicate a typewritten letter, including the type characters, the letterhead, ink, and paper. It was so perfect, Stevenson reports, that "it caused the removal of certain key pro-Nazis in South America."*

Although unaware of the forgery factory in Canada, Alger Hiss's attorneys received information some time after Hiss's conviction that typewriters could be forged. To prove that "forgery by typewriter" was possible, the appeal team set out to find an expert who could construct a duplicate of Woodstock 230099, which, according to Chester Lane's affidavit for appeal, had to be done "without seeing the typewriter in evidence [which was impounded by the court] . . . but working simply from sample documents typed on that machine." The defense asked a typewriter expert, Martin Tytel, to manufacture such a machine. It was no easy task. Tytel searched about for the Woodstock type that had become obsolete in 1928. Fortunately, he was able to locate such a type in the shop of another expert, Adam Kunze.

For $7,500 the Hiss team got a typewriter that was described by Hiss, perhaps a bit enthusiastically, as "completely successful." Documents typed on 230099 were coded, as were duplicates typed on the forged typewriter. The Hiss team challenged the FBI's experts to distinguish between them. Hiss said the FBI refused this offer. "If you read the FBI documents carefully," he said, "they were afraid they couldn't tell the difference. . . . The FBI documents show they couldn't."

Commenting recently on how unpersuasive the original typewriter evidence was, Hiss said, "We now know, particularly from the Stevenson book, that the FBI knew about this [typewriter forgery] because Hoover himself went to Stephenson's laboratory where they were making fake typewriters. But in our case, it didn't even take any elaborate forgery trick because the testimony is so flimsy."

To the question "Who might have perpetrated the forgery?" Hiss said that it was not beyond Chambers to use an old Woodstock, any old Woodstock. Although Hiss offered one theory that Chambers stole the Hiss Woodstock and substituted another, possibly 230099, he doesn't place much store in this.

---

*If this intelligence operation has a James Bond quality about it, it is interesting to find that 007's creator, Ian Fleming, worked at Camp X in naval intelligence. He was a commander known as 17F and showed a special fascination for intelligence gadgets, which he was to use with such facility in the Bond series. Stevenson reports that Fleming showed Sir William the manuscript of his first James Bond thriller, *Casino Royale*. "It will never sell, Ian," Stephenson predicted. "Truth is always less believable. . . ." Apparently Sir William was better at intelligence than at judging the marketability of literature.

I don't think it has to be as fancy as that. He [Chambers] had been in our house. He knew the kind of typewriter we had. He was a man who was alert to typewriters; he used them all the time. All he had to do was get another old Woodstock. I think that was the one he left on the subway train. You remember, he said he had a typewriter he disposed of.

The story Hiss referred to involves testimony by Whittaker Chambers at the second trial about a typewriter he abandoned on a subway in New York City. Asked by Prosecutor Murphy whether he ever owned a typewriter during the critical period of his story, Chambers admitted that he owned a Remington portable from "roughly 1934 until some time in 1940." He testified that the typewriter had been the gift of a man named Uhlrich, who, he said, was head of the "first Soviet apparatus to which I was attached."

The next day Hiss's attorney, Claude Cross, picked up the thread of Chambers' testimony, but beyond eliciting a few additional details of the story, Cross could do nothing with it. It is sad that this intriguing admission by Chambers was not examined with much enthusiasm by the defense attorney. The relevant testimony follows below:

CROSS: *Did you dispose of it [the typewriter] at some time?*
CHAMBERS: *I did.*
CROSS: *When?*
CHAMBERS: *Either in 1940 or thereabouts.*
CROSS: *How did you dispose of it?*
CHAMBERS: *I left it on a streetcar or on an elevated train.*
CROSS: *Where?*
CHAMBERS: *In New York City.*
CROSS: *Where were you living at the time?*
CHAMBERS: *I was living on my farm in Westminster, Maryland.*
CROSS: *Well, is it fair to say that you came up by train with this typewriter and got on either the New York City subway or the elevated or a streetcar in New York City and then left the car or subway or elevated and left the typewriter behind?*
CHAMBERS: *That is quite fair.*
CROSS: *Now was the reason for disposing of that typewriter in that manner because it reminded you of the past and you wanted to dispose of it?*
CHAMBERS: *That is right.*
CROSS: *And you wanted to dispose of it in such a way that it could not be traced to you, didn't you?*
CHAMBERS: *No, I do not think it could have been traced to me.*
CROSS: *Well, you knew that when you disposed of it in this manner.*
CHAMBERS: *That was not the thought in my mind. I wanted to get rid of it.*

CROSS: *You knew that at that time, didn't you?*
CHAMBERS: *I certainly must have known it, yes.*
CROSS: *And this abandonment of this typewriter was planned before you left your farm in Westminster, wasn't it?*
CHAMBERS: *I believe it was.*
CROSS: *In just this manner?*
CHAMBERS: *Perhaps. I am not sure about that.*

Hiss was reminded that Chambers said he got rid of a Remington, not a Woodstock. Hiss responded, "He not only said it was a Remington, but he said it was a portable. But I don't accept Chambers' statement. Psychologically, why did he get rid of any typewriter? He didn't say it could incriminate him; he said he did not want to be reminded of the past."

Zeligs was intrigued by Chambers' strange confession. He wrote, "Like the arsonist who returns to the fire," Chambers was compelled to tell his adversary, Alger Hiss, and the court as well that he had once felt the need to divest himself of a typewriter. "In his mind," Zeligs continued, "his past life is interchangeable with the symbol of his guilt—the typewriter." Dr. Zeligs also wondered if there was a less symbolic meaning in this act. "If an examination of his . . . typewriter would expose his past actions, it is not just an innocuous fetish but an incriminating piece of hard evidence that he had need to get rid of," proposed the psychiatrist.

It is now known that the FBI had a spy in the Hiss camp. An intelligence operative named Horace Schmahl was utilized as a private investigator by the defense to do much of the leg work during the pretrial period beginning in October 1948 until early 1949. As it turned out, Mr. Schmahl was more of a "public" investigator, as he shared all the information uncovered while working for the defense with the FBI and Prosecutor Murphy. Before the indictment, the Hiss forces were fully cooperating with the Bureau in an effort to show they had nothing to hide. Such collaboration was supposed to have ceased when the indictment was handed down on December 15, 1948. However, Schmahl continued to inform the FBI what the Hiss forces were doing. The FBI papers are full of references to Horace Schmahl and the aid he gave them.

Using private investigators to do routine research and interviewing is a common practice of law firms. It saves the more valuable—and expensive—time of the attorneys. Schmahl entered the picture after Hiss filed his libel suit against Chambers in the fall of 1948. The defense turned to him after attorney Edward McLean's regular investigator, John "Steve" Broady, decided the case was too small for his personal attention. In his stead, he suggested two of his employees: Horace Schmahl and Harold Bretnall.

Schmahl, who was born in Germany and then worked as a private detective and in U.S. intelligence for twenty years, spent most of his time on the typewriter phase of the investigation specifically looking for typed documents produced on the Hiss machine. Hiss, although he had some vague reservations about the man—Schmahl was short, stocky, tough-looking, and spoke in a heavy German accent—described his initial work as "very good." Schmahl managed to interview Chambers' mother, as well as Harry Martin. As reported in the FBI's files, Schmahl won Martin's confidence by telling him and his attorney that Schmahl himself doubted Hiss's innocence. Schmahl said there were inaccuracies in Hiss's story about the typewriter and about other key pieces of evidence. A teletype summarizing the Schmahl-Martin interview reads, "Schmahl did state that if Hiss were proven wrong on 'one more thing,' his firm would withdraw from the case." According to Hiss, this was later denied by Schmahl's superior, Steve Broady, who added that he personally was much impressed by the defendant.

Shortly thereafter, the working relationship between the private detective and the Hiss side deteriorated, and Schmahl ultimately refused to sign an affidavit outlining the results of his investigation, citing the fact that his firm hadn't been paid for the work.

In the course of his investigation, Schmahl also discovered that Esther Chambers made a fraudulent application for credit by posing as the wife of one Jay Chambers, who had been a teacher at the Park School since 1935 and was then employed in the Treasury Department. Schmahl advised McLean about this but never produced any documentary proof.

McLean did instruct him on January 21, 1949, to follow up on this matter. There was no reason to, for Schmahl knew the entire story and even had photostatic copies of the credit application, although he neglected to tell this to his client or to supply him with copies of the application. Instead, he turned this information and the photostatic copy over to the FBI. Without documentation, the defense elected not to ask Mrs. Chambers about this matter.

The FBI's relationship with Schmahl continued at least until June 1949. On June 1 an FBI memorandum notes, "During the past several weeks, Schmahl has had telephonic contacts with special agents James P. Lee and D. V. Shannon of this office [New York] in reference to the Hiss–Chambers case.'' When Schmahl reported to agents that he had information on an assassination attempt on union leader Walter Reuther, which took place in Detroit on April 20, 1948, the FBI became wary of their informant. Reviewing their file on Schmahl, the Bureau concluded that no further interviews be held, as it might place the Bureau in a compromising position. The Bureau's conclusions about the investigator's character are instructive. An FBI memorandum dated June 1, 1949, says:

> . . . Schmahl has an unsavory reputation in New York City as a private investigator and translator and among other things . . . he has claimed to be ''cooperating with the FBI'' while conducting investigations in the past. Additional references in the files reveal that subject has come to the attention of this office periodically in the past several years. In 1947, he attempted to volunteer his services in connection with Communist Party matters, although no action was taken because of his previous unethical practices.

Prosecutor Murphy, however, was neither bothered by Schmahl's ''unsavory reputation'' nor worried about ''compromising'' his own integrity. When the FBI lost interest in Schmahl in June, he began surreptitiously meeting with Murphy. The private investigator met the prosecutor at least twice (June 5 and 6, 1949) in the midst of the first trial. From these meetings Murphy learned that the defense had acquired an old Woodstock typewriter for purposes of comparison from a firm headed by Adam Kunze. Murphy asked Hiss about this particular typewriter during his cross-examination and again mentioned it during his summation to the jury.

The Hiss attorneys contend in the *coram nobis* petition that there was in Murphy's file a six-page document entitled ''Outline of Investigation'' and dated December 28, 1948, a time when Schmahl was still employed by the defense. The outline delineates, according to the *coram nobis* petition, ''defense tactics and investigative programs in connection with the typewriter.''

After a guilty verdict was returned at the second trial, attorney Chester Lane asked Hiss if he had reason to suspect Schmahl of being a double agent. Although Lane did not say how he knew, he said he was sure of this charge. Hiss remembered that for the first month or so

Schmahl had "been very active and seemingly useful." Shortly thereafter, however, Schmahl informed the defense attorneys that he had contacts with the House Un-American Activities Committee and could get access to some of their files. Hiss rejected this offer. "We were playing it straight," Hiss said, "and were not interested in that kind of underhanded operation. I said, 'No. Small thanks but no thanks.'"

Schmahl now lives in Fort Lauderdale, Florida, where with his son, a marine architect, he runs a boat-building and repair business. Hiss said, "I've been told he has a violent temper. Still, I've also been told that he would be ready to talk if approached properly." The authors of this volume were markedly unsuccessful in getting Schmahl to talk, no matter how "properly" we approached him.

According to Hiss, who has read some pages from the CIA files on Schmahl, the latter was implicated in the Galindez affair.* Jesus de Galindez was a Dominican Republic politician and scholar who disappeared from Columbia University several years ago. Morris Ernst, the well-known liberal attorney, was hired by the Dominican Republic to investigate the matter for a fee of $500,000 and was reportedly unable to find any Dominican government involvement in Galindez' disappearance. Hiss felt that this was a "whitewash" by Ernst, evidence of the fact that there was some softening in Ernst's liberal attitudes near the end of his life.

More importantly, Galindez has never been found, although newspapers report that a man thought to be sick but also described as possibly drugged was carried out of a building at Columbia University and put into an ambulance. Hiss described Schmahl as the driver of the ambulance.

Since these are serious charges, Hiss was asked who had told him of the connection between Schmahl and the Galindez affair, and he said:

> Fred Cook wrote an article for *Look* magazine on the Galindez disappearance. It was cancelled because of objections by Washington. In the course of

---

*In August 1977 Hiss went to court in an attempt to gain access to the complete CIA file on Schmahl. The file, which was alleged to contain some ninety documents, was denied Hiss in November of the same year on the ground that Schmahl's right to privacy outweighed Hiss's interest in an old case.

that research, he [Cook] discovered the connection between Schmahl and the Galindez kidnapping.*

It is not entirely surprising that the CIA, like the FBI, has been less than enthusiastic about releasing their papers dealing with Horace Schmahl, citing the catch-all phrase "national security." From the few Hiss has read, he learned that Schmahl, who left his native Germany in 1929 at age twenty-one, had come under suspicion during World War II as being a Nazi sympathizer. He was interviewed about this matter by the U.S. Army. In this interview Schmahl denied the charge. However, he did admit knowing certain pro-Nazis in America, such as Fritz Kuhn (leader of the German-American Bund) and Adam Kunze. It should be remembered that Martin Tytel, the expert who made the copy of Woodstock 230099 for the Hiss side, located much of the type he used in the reconstruction in the shop owned by Schmahl's acquaintance, Adam Kunze. Also the defense purchased a second Woodstock typewriter from Kunze.

---

*Former newspaper reporter, now writer, Fred J. Cook spent ten years researching the Galindez disappearance. He discovered that the Dominican Republic exile had written a pamphlet entitled *Babes, Bananas, and Bloodshed* which was highly critical of Rafael Trujillo, the Dominican Republic dictator. Trujillo is reported to have said after reading the article, "This bandit will eat the paper I hold in my hands." On March 12, 1956, Jesus de Galindez disappeared after giving a lecture on Latin American politics at Columbia. Cook discovered he was drugged and then transported in an ambulance to Amityville, Long Island, and from there by plane to the Caribbean island. The pilot of this plane was a young American, Gerry Murphy, who was later murdered in the Dominican Republic. *Life* magazine in 1957 was the first to make the connection between Murphy and the Galindez kidnapping.

It is a matter of record that a month before Galindez was kidnapped, Horace Schmahl chartered a plane, piloted by the same Gerry Murphy. It later came out in a trial that Schmahl shared his private investigation office in New York with Joe Frank, a former CIA and FBI agent. Frank had worked as bodyguard for Trujillo when he visited the U.S. The plane Murphy used to fly Galindez to the Dominican Republic was chartered in the name of Joe Frank.

The Justice Department was not very enthusiastic about prosecuting anyone in this matter—although the FBI had done an exhaustive investigation of Galindez' disappearance. Eventually Frank was indicted by a Washington grand jury on a technicality—that he failed to register as an agent of a dictator. At the trial, Frank's secretary testified that he told her he was hired by the dictator to keep an eye on Galindez and that Trujillo was billed for these and other services. Although two important witnesses died just before the case went to trial, Frank spent eight months in prison.

Schmahl has refused requests for interviews from both Cook and the present authors. Of the investigation of Trujillo conducted by Morris Ernst, Cook wrote, "It can be taken for granted as a simple rule of life that when a man spends $500,000 to have himself investigated, he does not expect to be branded as a kidnapper and murderer."

It was mentioned earlier that Steve Broady put two of his private detectives, Schmahl and Harold Bretnall, on the Hiss investigation. More than a dozen years later, Bretnall happened into the offices of Chester Lane, Hiss's appeal attorney. There he spied the Hiss file, which contained the courtroom testimony. "I know all about that," Bretnall was reported to have said. "Hiss was framed."

The attorney to whom Bretnall made this remark expressed great interest in hearing more, and Bretnall related that Schmahl's friend, Adam Kunze, fabricated the typewriter that helped to convict Hiss. Kunze was dead by then and this could not be verified, and Schmahl was not talking to anyone about such matters. The only confirmation of the story was the fact that Bretnall knew precisely where Kunze's shop was located. Unfortunately, before the Hiss attorneys could learn more from Bretnall, he was killed in an automobile accident.

Despite an exhaustive search, the FBI continued to claim they were unable to locate the Hiss typewriter. This was a surprising admission because the Bureau—deeply committed to good public relations—was not wont to air its failures publicly. Over the years there have been many rumors that the FBI was actually "playing possum." One of the first such mentions that the Bureau's search for the Hiss typewriter was more fruitful than they were willing to admit to appeared in Walter Winchell's column on December 18, 1948, about four months before the defense found Woodstock 230099. Winchell wrote, "On unimpeachable authority, the FBI has cracked the Chambers–Hiss case." The column continued that "despite headline-hunting ploys by the members of the House Committee [HUAC], the FBI will get the credit for breaking the case. Through the work of the G-men, and their scientific laboratory, the typewriter involved in the copying of stolen documents has been identified."

Winchell was most friendly with J. Edgar Hoover, a relationship writer Herman Klurfeld, who ghosted hundreds of columns for Winchell, characterized this way in his biography of the popular commentator and columnist: "J. Edgar Hoover, who adored publicity, courted reporters in general and Winchell in particular because Walter had the largest audience." This is not to say that Hoover was Winchell's "unimpeachable authority" for this particular item, but if the reporter doubted its veracity, he could easily have checked it with the director of the FBI.

The second reference to the fact that the FBI found the typewriter appeared in the December 30, 1951, final report of the House Committee on Un-American Activities. Entitled *The Shameful Years—Thirty Years of Soviet Espionage in the United States,* it read:

> The Committee wished to commend the Federal Bureau of Investigation for its work in bringing this case to a successful conclusion when all odds were against it. The location of the typewriter and certain pieces of other evidence needed during the trial of the case were amazing.

In 1962, reference to the FBI typewriter wizardry was heard again, this time by none other than Richard Nixon in his book *Six Crises.* The first chapter is devoted to the Hiss case, and with the number of subsequent references to it in the Watergate tapes, the Hiss case must have been Nixon's favorite crisis. Nixon writes, "On December 13 [1948] FBI agents found the typewriter." This is one week prior to the Winchell story and four months prior to when the defense found their typewriter on April 16, 1949. Another reference from *Six Crises* reads:

> On December 15, the critical last day [of the grand jury's term], an expert from the FBI typed exact copies of the incriminating documents on the old Woodstock machine and had them flown up to New York as exhibits for members of the Grand Jury to see.

The hue and cry that followed on the revelation of these passages in Nixon's book caused him some anguish. As recounted earlier, public-relations men soon attributed the confusion to a researcher's error. Hiss was amused by Nixon's blunder and his handling of the situation:

> And here this is supposed to be the case that he [Nixon] knew so much about. He had been the fearless prosecutor, yet he missed that. And you know there have been several official statements by the committee that the FBI found the typewriter. . . . We have a letter from Congressman McDowell [John R. McDowell of Pennsylvania] . . . saying the FBI found the typewriter. Many references. Then from Nixon—getting back to him—comes the tape [Watergate tape] from February 27, 1973, where he says "we found the typewriter." Those dumb associates he had and he says *we* found the typewriter.

The tape that Hiss refers to involves a conversation between President Nixon and his Chief Counsel John Dean following a 70-0 Senate vote to establish a committee to investigate the break-in at the Watergate. As the heat became uncomfortable in the presidential kitchen, it was the

Hiss case that was on Nixon's mind. The President is speaking to Dean in this passage:

> When you talk to Kleindienst [Attorney General]—because I have raised this [inaudible] thing with him on the Hiss case—he has forgotten, I suppose. Go back and read the first chapter of *Six Crises*. But I know, as I said that was espionage against the nation, not against the party. FBI, Hoover, himself, who's a friend of mine, said, "I am sorry I have been ordered not to cooperate with you," and they didn't give us one [adjective omitted] thing. I conducted that investigation with two [characterization omitted] committee investigators—that stupid—they were tenacious. We got it done. Then we worked that thing. We then got the evidence, we got the typewriter, we got the Pumpkin Papers.

Weinstein offered a contradictory interpretation of this tape. Using a later transcription prepared by the House Judiciary staff—an effort Weinstein characterized as "made by more objective transcribers using superior equipment" and later as "generally acknowledged to be more accurate than Nixon's doctored White House rendition"—he described Nixon as "babbling." "But we broke that thing . . . without any help. The FBI then got the evidence which eventually— See, we got Piper who— We got the, the, the, oh, the Pumpkin Papers, for instance. We, we got all of that ourselves. . . . The FBI did not cooperate." Weinstein concluded, "There is no mention of HUAC finding the Woodstock typewriter."

Although professing initially to be troubled by the non-word *Piper*, Weinstein's historical research presumably led him through many turns in the road to the fact that William Marbury, onetime Hiss attorney, later joined the firm of Piper and Marbury. The material may be read another way, however. Linguists, psychologists, poets, and even lovers are aware of the effects of emotion on pronunciation and the way we condense and transliterate under pressure. When it is read in this light, Weinstein's bent for minutiae had led him astray, for "Piper" may more realistically be a contraction of "typewriter." If the taped transcript is reviewed this way, it says precisely what the less "objective transcribers" wrote—i.e., "We [HUAC] got the typewriter."

There has long been a conspiracy theory to explain the confusing typewriter evidence. It has been suggested that the FBI manufactured or doctored the typewriter and then allowed the defense to find it, thereby neatly effecting Hiss's downfall. The conspiracy theory was given new impetus in 1976 with the publication of John R. Dean's book *Blind Ambition.* The New York *Post,* whose front page that summer was full of the sexual exploits of Ohio Congressman Wayne Hayes, dropped this carnal tale for the story of the Hiss typewriter as told by the former chief counsel to the President. The front page read: "DEAN: NIXON ADMITTED FAKING HISS TYPEWRITER."

The passage describes President Nixon discussing Dita Beard and the International Telephone and Telegraph Corporation scandal with Charles Colson, special counsel to the President. Dita Beard, an ITT lobbyist, typed a memorandum to her supervisor saying that the way to get the Justice Department to drop their antitrust suit with ITT was to contribute $400,000 to the GOP convention scheduled for San Diego.*

When columnist Jack Anderson reproduced the memo, the land rang with alarum. Nixon is reported to have told Dean that the way to handle the situation was to have the memo declared a forgery. Dean related how the typed memo reminded the President of the Hiss case. According to him, Nixon said to Charles Colson, "The typewriters are always the key. . . . We built one in the Hiss case."

John Dean is certain that President Nixon made that statement, but he isn't exactly sure what it means. "The man best able to answer it is Richard Nixon," he said in a television interview. Dean described how the President "seemed to enjoy" reliving his memories of this case with the younger members of his staff. It could be nothing more than the President striking the tough prosecutor pose to impress his staff, or it could mean, as it says, that the FBI—alone or in collusion with Schmahl's friend Kunze or even the unlikely possibility of working with HUAC, for whom there was little love lost—manufactured evidence to convict Hiss. Nixon, now in exile in San Clemente, has refused our—and everyone else's—request for an unrenumerated interview.

Without an answer from Nixon or the release of more pages from the FBI and CIA files, it is dangerous to make categorical statements, but this much is certain: The typewriter that sat in the courtroom and seemed so integrally connected to Hiss's fate was a false witness. It was not his typewriter. The FBI knew this, and the worst assumption is that the agency had a typewriter constructed to match the Hiss standards and then simply typed the incriminating documents on that machine. This

---

*The 1972 Republican convention site was later shifted to Miami after a public outcry.

typewriter could then have been turned over to Ira Lockey, who was told to let the defense find it.

On the other hand, the best that can be said about the FBI and the typewriter is that they kept the information that Woodstock 230099 was not the machine locked safely in their files, where it would have remained forever had not the Freedom of Information Act required the Bureau to divulge it. Even then the FBI struggled to keep its records closed.

This is perhaps a good time to recall Prosecutor Murphy's summation to the jury: "If you think that any bit of evidence in this case, any bit, material, or immaterial, was manufactured, conceived or suborned by the FBI, acquit this man. The FBI ought to be told by you that they can't tamper with witnesses and evidence." Murphy's words convey a horrifying cynicism when read alongside the information in the FBI's own files—information Murphy was well aware of.

CHAPTER 9

[The difference between Truman and Nixon is] . . . the difference between a warmhearted fellow who plows into a fist fight for the sheer joy of it and the grim-minded fellow who charges in with a pipe wrench. —Washington *Post*

*M*uch has been made of the straitened circumstances of Nixon's childhood which reputedly contributed to his ambitious character, of his Quaker upbringing which formed and hardened his attitude toward personal and public morality, and of his love for America and its public institutions which has come to dominate his political life. Yet, closer examination of the Nixon background fails to mark his early years as especially impoverished. The Nixons were middle-class poor, but so was much of the country.

They were devout Quakers, and Nixon's father even taught Sunday school. It was generally conceded in Whittier, California, that Hannah Milhous married beneath her station, but the family seems in no way remarkable. Richard Nixon experienced what has generally been described as a rather lonely childhood.

A young cousin described him as so fastidious that he refused to ride the school bus "because other children didn't smell good." It has been commented that Richard was his mother's favorite and that she frequently expressed the wish that he would become a minister or a classically trained musician. He seemed to be determined to please her. His mother described him as a "very serious child," adding, "Most boys go through a mischievous period; then they grow up and think they know all the answers. Well, none of these things happened to Richard. He was very mature even when he was five or six years old. . . . He was thoughtful and serious. He always carried such a weight." Later, Nixon was to describe his mother as a "saint," saying in his East Room farewell speech in August 1974, "Yes, she will have no books written about her, but she is a saint."

Another relative, Jessamyn West, the novelist, in picturing him as polite, tidy, and earnest, said, "He wasn't a little boy that you wanted to pick up and hug. It didn't strike me that he wanted to be hugged." Years later, Nixon commented to David Frost, "I'm not a very lovable man."

Richard's relations with his father were of a somewhat different variety. Frank Nixon was a teasing, passionate man whose conversion to the Quaker religion was a whimsical one. He became a Quaker to marry a Quaker: Hannah Milhous. This man who never finished the sixth grade was a strict disciplinarian, and at least one Nixon biographer (Bela Kornitzer) pictured Frank Nixon as "tough, opinionated, capricious, argumentative, and unpredictable." With his son's rise in the party he became a rabid Republican. When Nixon was selected as Eisenhower's running mate, he commented with pride, "Truman hates Dick like a rattlesnake." Nixon himself offered the following assessment of his father: "If he wanted something, he wanted it at once. He had a hot temper, and I learned early that the only way to deal with him was to abide by the rules he laid down. Otherwise, I would have felt . . . the ruler or the strap as my brothers did."

Nixon came to recognize that he had to be very convincing to escape the wrath of his father, and so Nixon became quite convincing. Fawn Brodie, a UCLA historian who is writing a Nixon biography, believes that this might very well have been the origin of Nixon's lifelong faith in the stratagem of lying.

From family recollections it is clear that Richard was successful in this effort, for he was rarely punished. And in his farewell speech on the last day of his public career Nixon characterized his father as "a great man. He didn't consider himself that way. You know what he was, he was a streetcar motorman first, and then he was a farmer, and then he had a lemon ranch—it was the poorest lemon ranch in California, I can assure you. He sold it before they found oil on it. And then he was a grocer, but he was a great man. Because he did his job, and every job counts up to the hilt regardless of what happens." Historians have pointed out that even in Nixon's final hours of public service he was making a questionable statement. There was never any oil on his father's lemon ranch.

There were two chronically ill children in the family, and both ultimately died of tuberculosis. There were numerous separations. On one occasion Mrs. Nixon moved to Arizona with one of her ill children, leaving the others at home with their father for three years. On one such occasion Richard expressed his loneliness in a letter to his mother beginning with the salutation "My dear Master" and signing it "Your good dog." Family legend has it that when young Richard was eleven, he was so outraged by the Teapot Dome scandal of 1924 that he pledged himself to become "a lawyer who can't be bought."

In high school he was a successful politician and debater. These traits were also highly visible during his collegiate career at Whittier College in his own hometown. Once again he was active politically, and although his debating skills were high, he was an indifferent athlete. He was remembered as being markedly fastidious, and he owned two tuxedos while at college. After college he attended law school at Duke University. He graduated high in his class and told the dean at an employment interview that he was attracted to the idea of a position with the FBI. Dean H. Claude Horack tried to discourage him: "I said I thought he was too good a man for that. He [Nixon] said that he didn't know why, but he was attracted to it."*

Despite his strong credentials, Nixon returned to Whittier, where he

---

*In 1937 Nixon did apply for a job with the FBI. He was never informed of the result. While Vice President, Nixon asked J. Edgar Hoover what became of his application. Hoover promised he would check his files and find out. Nixon, who related this story while President, said, "I don't know whether this part of the story is true or not—although Mr. Hoover always tells the truth—but nevertheless, he said I had been approved as an agent of the FBI except for one thing: Congress did not appropriate the necessary funds requested for the Bureau in 1937." Nixon, having just received from the FBI Director the gold badge of a special agent with the FBI's motto "Fidelity, Bravery, Integrity" engraved on it, continued: "I just want to say in Mr. Hoover's presence and in [Attorney General] Mitchell's presence that that will never happen again."

began a career in tax law. By 1940 he had married Pat Ryan (in fact, he proposed marriage on the night he met her) and two years later enlisted in the Navy, where he remained until 1945. His election as a Congressman, Senator, and Vice President followed in rapid succession. After he was defeated twice—once by Kennedy in the presidential contest in 1960 and two years later by Pat Brown in the gubernatorial race in California—Nixon's political career seemed over. History demonstrates that this judgment was a premature one, but even at the time the record of achievement was a large one.

His sympathetic biographer, Earl Mazo, who was then political correspondent for the New York *Herald Tribune*, wrote in 1959:

> For Richard Nixon, the end is power—specifically the incomparable power of the Presidency. He moved toward it in a spectacular, meteoric career. Congressman at 33. Important Congressman at 35. Senator at 37. Vice-President at 39. Only two-term Republican Vice-President at 43.

There would be even more striking achievements: election to the Presidency in 1968 and an awesome landslide re-election victory in 1972 which nearly decimated the Democratic Party. Nixon's long interest in foreign affairs brought a U.S. rapprochement with Red China and, even more surprising, a significant detente with the U.S.S.R. Visitors to the Eastern world in 1972 brought the strange news that the Russians favored Mr. Nixon's candidacy over the presumedly more friendly one of the liberal George McGovern.

Along the way Nixon suffered a series of rebuffs which at times seemed seriously to cloud his political future. His first movement onto the real national scene as Vice Presidential candidate in 1952 brought a severe setback when a story surfaced about a contribution of $18,000 by a group of California businessmen which was used to pay Nixon's out-of-pocket expenses as a Senator.

Theodore White reported that Eisenhower had selected Nixon as his running mate through default. A Republican party official said at that time, "We took Dick Nixon not because he was right wing or left wing but because we were tired, and he came from California." When the story of the secret fund came out, Eisenhower was quite upset. He was campaigning for a government that was to be "as clean as a hound's tooth," and Nixon's gaffe seemed to stop the campaign almost before it started. Thomas Dewey, always Nixon's *bête noir*, called him personally to say that he had been at a dinner party with Eisenhower, and seven of the nine men present felt that Nixon should resign.

The story broke when Nixon was campaigning in the Central Valley of California. At first he decided to answer "No comment" to requests for information, but when someone shouted as the train was about to pull out of Marysville, "Tell us about the eighteen thousand dollars," his early resolve was broken. The response was straightforward and the defense was a familiar one. "That did it," wrote Nixon. "Despite all of our plans to ignore the attack, I could not see myself running away from a bunch of hecklers. I wheeled around and shouted, 'Hold the train!'" Nixon then explained to the crowd:

> You folks know the work that I did investigating Communists in the United States. Ever since I have done that work the Communists and the left wingers have been fighting me with every possible smear. When I received the nomination for the Vice Presidency I was warned that if I continued to attack the Communists in this government they would continue to smear me. And believe me you can expect that they will continue to do so. They started it yesterday. They have tried to say that I had taken $18,000 for my personal use.

Notwithstanding the boldness of Nixon's response, the Washington *Post* and the New York *Herald Tribune* called editorially for his resignation from the ticket, and when Eisenhower phoned Nixon and reportedly requested that he step down, Nixon announced his determination to make a public explanation. Nixon then made the famous "Checkers" speech on television, an appearance that cost $75,000 to explain away an $18,000 fund. On that program, maudlin, disjointed, effective, and interlarded with references to his cocker spaniel dog Checkers, a gift from a supporter in Texas ("The kids, like all kids, love the dog . . . [and] regardless of what they say about it, we are going to keep him"), Pat Nixon's coat ("A good Republican cloth coat, but I always tell her she'd look good in anything"), and Alger Hiss ("I can say to this great television and radio audience that I have no apologies to the American people for my part in putting Alger Hiss where he is"), Nixon closed with this plea for public support:

> I know that you are wondering whether I am going to stay on the Republican ticket or resign. Let me say this. I don't believe that I ought to quit. I'm not a quitter. And incidentally Pat is not a quitter. After all her name was Ryan, and she was born on St. Patrick's day, and you know the Irish never quit. But the

> decision, my friends, is not mine. I would do nothing that would harm the possibility of Dwight Eisenhower to become President of the United States. . . . Let them [the Republican National Committee] decide whether my position on the ticket will help or hurt. And I am going to ask you to help them decide. Wire or write to the Republican National Committee whether you think I should stay or get off.

Despite Nixon's baring his soul and incidentally what he termed his "complete financial history" before a national television audience, there was no quick decision from Eisenhower, who said he would decide after he talked to Nixon personally. Nixon was enraged by the decision:

> For the first time in almost a week of tremendous tension, I really blew my stack. "What more can he possibly want from me?" . . . He was being completely unreasonable. I had been prepared for a verdict. I was expecting a decisive answer. I didn't believe I could take any more of the suspense and tension of the past week.

When the two came face to face in Wheeling, West Virginia, Eisenhower, with tears in his eyes, shook his hand and said simply, "You're my boy," and Nixon broke into tears. Although the two ran successfully, there was little closeness between them, and Nixon was lonely and morose throughout the next four years. Once he said of the President:

> Despite his great capacity for friendliness, he also had a quality of reserve which, at least subconsciously, tended to make a visitor feel like a junior officer coming in to see the commanding general.

Eisenhower paid him compliments, but they were always carefully couched in terms that clearly conveyed Nixon's junior status. He was a "comer," "a splendid example of the younger men we want in government," or "a very likable person." When the President suffered a coronary, Nixon was circumspect in avoiding any semblance of interest in presidential succession. As a consequence, Nixon appeared hurt when, on Eisenhower's return to the White House, the President made no special mention of Nixon's work:

> He [Eisenhower] thanked us all for our "perfect" performance during his absence. But to my knowledge, he did not thank anyone personally. He felt that all of us, no matter how hard we worked, were merely doing our duty, what was expected of us under the circumstances. . . . But after this most difficult assignment of all—treading the tightrope during his convalescence

from the heart attack—there was no personal thank you. Nor was one needed or expected. After all, we both realized that I had only done what a Vice President should do when the President is ill.

Their relationship became even more complicated when Eisenhower confessed to Nixon in 1955 that his poll ratings were "most disappointing." Eisenhower thought that a "crash program" might be necessary to build Nixon up, and there was considerable talk throughout the Republican party about the need to take Nixon off the national ticket and give him a cabinet post. Nixon reacted with anger to this idea, feeling that it would provide a clear sign to the public that "Nixon had been dumped." He had been able to rationalize away the earlier difficulty with Eisenhower over the fund because the latter had not really known him at that time, but this later challenge came after Eisenhower "had had an opportunity to evaluate my work over the past three years." He finally sought one of his rare audiences with the President and told him that he would not want to force his way onto the ticket against Eisenhower's wishes.

They ran again and were re-elected, and again Nixon lapsed into obscurity. When Nixon was nominated for the Presidency at the 1960 Republican Convention, Eisenhower had already left the hall and belatedly congratulated him on "at last being free to speak freely and frankly in expressing your own views." Eisenhower made it clear throughout the campaign that it was vital that Nixon establish his own identity and took a very minor part in the electioneering. The low note came in late August 1960 when Eisenhower was asked at a press conference to describe an idea of Nixon's that had contributed to a major policy decision. His response—"If you give me a week I might think of one"—was to prove a considerable liability for Nixon. It was all of a piece with a remark that James David Barber reported Nixon made to a friend when he was left on the front lawn of Eisenhower's Gettysburg estate as the President escorted some of his old military friends into the house: "Do you know," Nixon said, "he's never asked me into that house yet."

However, in Miami in 1968, as Nixon stood at the podium to accept his party's nomination for the nation's highest office, he paid a remarkable tribute to his former mentor. Nixon said, "General Eisenhower lies critically ill in Walter Reed Hospital tonight. I have talked, however, with Mrs. Eisenhower on the telephone. She tells me that his heart is with us. And she says there is nothing that he lives more for and there is nothing that would lift him more than for us to win in November."

Finally, like Pat O'Brien in the movie *Knute Rockne—All American,* Nixon shouted, "Let's win this one for old Ike!" Perhaps Nixon himself was aware of the strangely theatrical nature of the statement, for it was excised from the text of the speech when it was later reprinted in the book of his speeches.

All things taken into consideration, Nixon ran well in 1960, and the election provided his archrival, John F. Kennedy, the smallest plurality in memory. Nixon retired to the practice of law in Los Angeles, and although he professed little interest in political life, he was soon to be found at the eye of the hurricane in the state of California, where he sought to unseat Pat Brown for the governorship in 1962.

That campaign was notable for two features. Nixon was assembling a new group, largely from southern California and made up of old supporters from the business world, including a soupçon of advertising men and a number of others who later came to constitute the Nixon team. It was also in this campaign that he ran into trouble once again with newspapermen and with dirty tricks. Richard Tuck, a Democratic Party strategist and no mean trickster himself, has written of that election:

> They were all there then: Klein, Haldeman, Ziegler, Chapin, even Segretti and Kalmbach. Scurrilous mailings appeared all over the state. Governor Pat Brown was soft on the Communists (whatever happened to them?). Doctored photographs showed the Governor bowing to an Asian Communist leader and with his arm around Harry Bridges, the well-known Communist. When Brown sued to stop the mailing, the "pranksters" involved in the fakes directly (or indirectly through their participation in the campaign) were revealed: Klein, Haldeman, Ziegler, Chapin, Segretti (even Segretti), and Kalmbach. (John Mitchell hadn't joined the team yet.)

Nixon had long been associated in the public mind with aggressive political campaigns. By his own admission he ran a "fighting, rocking, socking" effort against his Democratic opponent, Jerry Voorhis, in his first California try for the House in 1946. Similar tactics led to the defeat of Helen Gahagan Douglas in the senatorial race of 1950. Here the Nixon team produced their "Pink Sheet," supposedly delineating the dubious political activities and affiliations of the Representative they dubbed the "Pink Lady." Those who opposed his candidacy in 1952 were similarly treated to short shrift. Columnist Jack Anderson has written of the late Drew Pearson's being threatened by a "Red smear" if he persisted in opposing Nixon's continuation on the Republican ticket after the knowledge of the fund became public. The message came directly to Anderson, who was then Pearson's assistant, via a phone call from William Rogers speaking from Nixon's campaign train.

Nixon's assault on Adlai Stevenson in the campaigns of the 1950s brought forth from the Democratic candidate the characterization of "Tricky Dick" who lived in "Nixonland . . . [a place of] slander and scare, sly innuendo, poison pen, the anonymous phone call and hustling, pushing, shoving—the land of smash and grab and anything to win." The gubernatorial campaign in California brought a resounding defeat to Nixon, but it also revealed the immutable signs of the Nixon political style which came to plague and finally fell him a dozen years later.

A suit by the California State Democratic Party led to a finding in 1962 that the Nixon for Governor Committee contributed $70,000 to a group it had organized under the misleading name "Committee for the Preservation of the Democratic Party in California." The California Superior Court found that Nixon and H. R. Haldeman approved a circular addressed to registered Democrats which misled recipients about a poll reflecting adversely and falsely on candidates endorsed by the California Democratic Council. The court further held that the postcard poll was "revised, amended, and finally approved by Mr. Nixon personally." The results were ostensibly distributed under the imprimatur of "the voice of the rank and file Democrat," while in fact the cards were mailed by the Nixon for Governor Finance Committee. Nixon was also censured for the distribution of a misleading pamphlet called "The Communist Dynasty in California" and for a similarly falsified leaflet about Pat Brown by a nonexistent organization called "The Citizens' Fact Finding Committee."

Nixon had closed his campaign with a statewide television address charging that he (rather than Brown) was the victim of a "malicious smear campaign." He had run a very strange race, and his experience in the rarefied international atmosphere of Washington caused him to appear alienated from the mundane affairs of the state house in Sacramento. His speeches were dotted with vague promises to "clean up the mess in Sacramento" or "to throw the Communists out of the Golden State," proposals which had a strangely national sound to them. Brown was reputedly unfit for continuation in office since he had "proved himself incapable of dealing with the Communist threat within our borders" by his failure to introduce a single item of antisubversive legislation in four years.

Nixon was uncertain about entering the gubernatorial race in 1962; he also expressed the same ambivalence in 1968 before deciding to run for the Presidency. The polls showed him running comfortably ahead of Governor Pat Brown, but Nixon still feared he would lose the election and thus become a proven loser in the public's eye. Such a reputation might work against him when bigger and better political things came along.

Two men, however, were instrumental in helping him to make up his mind about the California race, Nixon reports in *RN: The Memoirs of Richard Nixon*, as much to acknowledge their strong influence over him as to imply they shared in this decision which, in hindsight, nearly ended his political life. The first was Dwight D. Eisenhower. At the Eldorado Country Club near Palm Springs, the former President instructed Nixon, "It has been my experience that when a man is asked by a majority of the leaders of his party to take an assignment, he must do so or risk losing their support in the future. If you don't run and the Republican candidate loses, you will be blamed for it, and you will be through as a national political leader."

The other man whose words weighed heavily on Nixon was none other than Whittaker Chambers. Nixon wrote that while agonizing over the election decision, he learned of Chambers' death. Describing his feeling at this moment, Nixon commented, "I knew that Chambers had not been well, but he had survived so much in his life that I suppose I had come to think of him as indestructible. Now he was dead." That night Nixon reread a letter Chambers wrote when the former Vice President had just returned to California to begin the practice of law. Chambers said:

> It seems possible that we may not meet again—I mean at all. So forgive me if I say here a few things which, otherwise, I should not presume to say.
>
> You have decades ahead of you. Almost from the first day we met (Think, it is already 12 years ago) I sensed in you some quality, deep-going, difficult to identify in the world's glib way, but good, and meaningful for you and multitudes of others. I do not believe for a moment that because you have been cruelly checked in the employment of what is best in you, what is most yourself, that that check is final. It cannot be. . . .
>
> You have years in which to serve. Service is your life. You must serve. You must, therefore, have a base from which to serve.
>
> Some tell me that there are reasons why you should not presently run for governor of California. Others tell me that you would almost certainly carry the state. I simply do not know the facts. But if it is at all feasible, I, for what it is worth, strongly urge you to consider this.

Nixon finally concluded, "My inclination was against making the race. But Eisenhower's advice and the pressure brought to bear by . . . Whittaker Chambers' letter and by the importunings of many close friends began to tip the balance in favor of a decision to run."

Mark Harris, the American novelist, teacher, and essayist, was hired by *Life* to cover Nixon during the 1962 gubernatorial campaign. After several weeks of listening to repetitive speeches and informal chats by Nixon with newsmen, centering largely on baseball and how to order tomatoes in a restaurant, Harris tried to get some impression of Nixon's intellectual make-up by asking him to identify three terms which appeared often in the candidate's speeches—*indoctrination, individualism,* and *indecision.* He carefully planned an approach to Nixon that would ask him to "comment, compare, and contrast" his use of the three words.

When Nixon entered a correspondents' dinner that he had promised to attend, Harris leaped to his feet and followed Nixon around the room as the latter exchanged the usual pleasantries with those assembled. Stepping in front of the candidate, Harris said, "Sir, I want to ask you about three words you've been using." Nixon replied, "Oh, yes" and started to move on as if the conversation was over. Harris moved with him, saying, "At San Diego this noon you equated education with *indoctrination. . . .*"* The following conversation transpired:

> "I'm speaking of *indoctrination* against Communism," he [Nixon] said.
>
> "You also favor *individualism.* But isn't *indoctrination* the antithesis of *individualism?*"
>
> "Read up on the Hiss case," he said, his voice confidently raised. He attempted again to resume his motion, depending upon me to understand that his resumption of motion implied that my question was answered.
>
> "I'm speaking about Communism, too," I said, holding my place. "You say you'll prevent Communists from speaking on college campuses. . . ."
>
> "I didn't say Socialists," he said.
>
> "Aren't you meddling in my classroom now? What about *my* individualism?"
>
> "I said tax-supported colleges only, not Socialists, only Communists. . . ."
>
> "I don't care *who,*" I said. "I don't care if it's the Ku Klux Klan."

---

*Nixon had said, "What are schools for if not for indoctrination against Communism?"

"You should," he said, pointing his finger at me, and with this I fell back, and he passed.

Harris spent much time trying to decipher the instruction "You should." Was Nixon speaking of a moral duty or the idea that the Klan was as evil as Communism, or was the finale an admonition that each man should look after himself? Harris was dismayed by the last alternative, for he felt that, if true, it conveyed Nixon's inability to understand the cardinal democratic idea that we achieve freedom "only by the perilous course of granting it to our enemies."

In an effort to settle the matter, Harris once again approached Nixon at the Oakland airport the next morning. The following dialogue occurred:

"Sir, last night our conversation was unfortunately interrupted just as we were discussing *indoctrination* and *individualism*. . . .

"Oh, yes," he said.

"And I've been thinking about *indecision,* too, my third word, and banning Communists from campuses. . . ."

"From tax-supported college campuses. . . ."

"Yes, and to teach my students about Communism my method would be *indecision,* letting my students arrive at good ideas themselves, not through my *indoctrinating* them.

"Remember, I didn't say Socialists," he said. "I said Communists."

"I don't care *who,*" I said. "The point is that you're telling me how to run my classroom, it's out of your jurisdiction, you haven't the facts, it's like your telling me the other night that we need a bigger auditorium. . . ."

"My goodness," said Mr. Nixon, "you're not comparing Communism to auditoriums, are you?"

Again Nixon walked away, but this time the annoyed Harris followed and shouted at Nixon's back, "See, see, you say you'll answer questions but you walk away; you walked away last night." Nixon turned back and, pointing a finger, shouted back, "Don't point, don't point, keep your hands down!"

From these and other experiences Harris concluded that Nixon suffered from something called "low intellect." Their relationship was doomed from the beginning when Harris introduced himself to Nixon at the latter's home in Beverly Hills by identifying himself as a professor at

San Francisco State College. After a moment's pause, as if he were trying to visualize the campus, Nixon responded brightly, "You need a bigger auditorium up there." Since an auditorium was, by Harris' calculation, among the least needed things on that campus, he recognized the perfunctory, political nature of the response. His summation of Nixon is understandably a scathing yet strangely prophetic one:

> For the truth of Mr. Nixon was really so very simple the student of American civilization should have recognized it sooner: Mr. Nixon was not wicked or evil or malevolent except to the extent that these defects of character arise from low intellect.
>
> Mr. Nixon lacked the self-awareness which intellect produces. He did not know, though he had almost touched upon it at Pomona, that the failure of his language was directly traceable to his lacking a reason to be against those he was against. Empty of response, he walked away, or he told you to put your hands down.*

Finally, Harris faulted Nixon for his inability to distinguish between the pulls of public policy and private success and for his failure to realize that private success might coexist with public loss. He described Nixon as an example of a particular kind of American success:

> His reason for running for governor was the hope of his own success, and he believed success was enough because nobody had ever told him otherwise. He possessed no farther vision of the end. Nobody had ever warned him that the enunciation of a world view toward the end of a private ambition was a torture of credibility which the English language would never contain. Under sufficient, prolonged exposure his reason was bound by stages to fade and vanish.

The morning after Nixon's defeat in California he was interviewed at what was then commonly regarded as the nadir of his political career. There was an unfortunate confrontation with the press that seemed to confirm Harris' deprecatory evaluation of Nixon's motivation. In

---

*Nixon's self-image was markedly different. In Emile de Antonio's documentary film *Milhous,* journalist Jules Witcover commented, "On several occasions, Mr. Nixon described himself to me as an intellectual. Called himself, at one point, 'the egg-head of the Republican Party.' He compared himself to Adlai Stevenson. . . . He told me one time he'd rather be teaching in some school, like Oxford, and writing two or three books a year."

despair, he reached out for that special hallmark of his first success, the watershed event in his political life, the Hiss case, and linked it to another familiar preoccupation, the liberal newspapers and their reporters, saying:

> Now that all the members of the press are so delighted that I have lost . . . I believe Governor Brown has a heart, even though he believes I do not. I believe he is a good American, even though he feels I am not. . . . You gentlemen didn't report it, but I am proud that I did that. . . . And our 100,000 volunteer workers I was proud of. I think they did a magnificent job. I only wish they could have gotten out a few more votes in the key precincts, but because they didn't Mr. Brown has won and I have lost the election. . . . I don't say this with any bitterness. . . . I don't say this with any sadness. . . . And as I leave the press, all I can say is this:
>
> For sixteen years, ever since the Hiss case, you've had a lot of—a lot of fun—that you've had an opportunity to attack me and I think I've given as good as I've taken. . . . And I can only say thank God for television and radio for keeping the newspapers a little more honest. . . . Just think how much you're going to be missing. You won't have Nixon to kick around any more. . . . They [the press] have a right and a responsibility, if they're against a candidate, to give him the shaft, but also recognize if they give him the shaft, put one lonely reporter on the campaign who will report what the candidate says now and then.

The defeat in California, coupled with the earlier disappointment on the national scene two years earlier, seemed to many to wrap up the Nixon political career. When it was all over, Nixon moved to New York and joined a prestigious law firm that included William P. Rogers, formerly chief counsel for the Senate Investigating Subcommittee and later Attorney General of the United States in the Eisenhower Administration. Later, Nixon joined forces with a firm headed by John Mitchell, who was to serve as Attorney General in the first Nixon Administration. After a few years of obscurity, Nixon began bit by bit to put his political house back in order. *Esquire* magazine reported that he refused to commit himself to Goldwater until the last moment of the 1964 convention because he believed that he might seize control of the Republican Party once again should the Senator from Arizona fail to secure the nomination. He said in regard to his presidential aspirations, "My interest in the nomination is only in the event that the convention is unable to settle on a candidate and feels I would be the best man to unite the party." Nixon spoke all over the country in the next three years; he had always been a party regular and he had many "chits" to call back in.

There was little real Republican party opposition to his nomination in 1968, and his campaign against the tired Democratic warrior, Hubert Humphrey, scarcely taxed the picture of the new and cool Nixon. The only cloud on the horizon was the surprisingly strong late finish by the Minnesotan. Nixon barely had time to bring out the familiar heavy artillery. One of his best shots went: "Hubert is a loyal American. Hubert is against Communists. Hubert is for peace. Hubert is a good speaker. Hubert is a very plausible man; he's a very pleasant man. He is a good campaigner. But Hubert is a sincere, dedicated *radical!*" (author's italics). Moreover, it was Humphrey who was trying unsuccessfully to disassociate himself from the Vietnam War policies of Lyndon Johnson, while Nixon cleverly refused to discuss the war for fear of imperiling the possibility of peace. He made it abundantly clear that if Johnson's efforts failed, he, Nixon, could bring the war rapidly to an end.

When it was all over and after spending some $24,000,000 to become President, twice as much as any candidate had ever spent, Nixon appointed a cabinet which he described as "the greatest in history." In this speech he paid a special tribute to Attorney General John N. Mitchell as "more than one of the country's best lawyers. . . . I have learned to know him over the past few years as a man of superb judgment, a man who knows how to pick people and lead them and to inspire them with a quiet confidence and poise and dignity."

Throughout his first term Nixon remained preoccupied with what had come to be regarded as his special interest, foreign affairs. He traveled widely as old foreign alliances were re-evaluated and new ones were being forged. At home he was defeated by the Senate on two Supreme Court nominees, G. Harrold Carswell and Clement F. Haynsworth, and responded with predictable anger. He maintained that the Senate action sorely assaulted the "preservation of the traditional constitutional relationship between the President and the Congress" and implied that the President was indeed "the one person entrusted by the Constitution with the right of appointment." Senatorial friends and foes alike took large exception to this construction, and soon the President tried to fashion a Northern liberal–Southern conservative confrontation. He called a sudden press conference and stated:

As long as the Senate is constituted the way it is today, I will not nominate another Southerner and let him be subjected to the kind of malicious character assassination accorded both Judges Haynsworth and Carswell. . . . My next nominee will be from outside the South and he will fulfill the criteria of a strict constructionist with judicial experience from either a Federal bench or a state appeals court.

I understand the bitter feeling of millions of Americans who live in the South about the act of regional discrimination that took place in the Senate yesterday. They have my assurance that the day will come when men like Carswell* and Haynsworth can and will sit on the high court.

When his anger finally diminished, Nixon set about methodically to find a man who was eminently acceptable to most factions, and Harry A. Blackmun was approved easily by the Senate.

As the conflict in Southeast Asia rapidly turned worse, student dissent increased geometrically. Never popular with the young, Nixon's posture in the Vietnam War continued to reap a harvest of locusts, and his sudden decision to invade Cambodia, apparently against the counsel of his advisers, provoked an immense outcry on college campuses. Unmoved by it all, Nixon persisted in speaking out against "humiliation and defeat" while escalating the war and simultaneously promising an increased rate of military withdrawal from Vietnam. The New York *Times* headlined an editorial "Military Hallucination Again," and the Senate Foreign Relations Committee condemned the President for "conducting a constitutionally unauthorized presidential war in Indochina." U.S.S.R. Premier Kosygin pointedly asked:

What is the value of international agreements which the United States is or intends to be a party to if it so unceremoniously violates its obligations? It is impossible not to give serious thought to the fact that President Nixon's practical steps in the field of foreign policy are fundamentally at variance with

---

*This was not the last the world would hear about Nixon's Supreme Court nominee, G. Harrold Carswell, who went down in a 51–45 Senate vote for reasons ranging from a mediocre judicial record to making racist statements. On June 27, 1976, Carswell was arrested in Tallahassee, Florida, for "battery and attempting to commit unnatural and lascivious acts."

Florida State Attorney Harry Morrison said Carswell, then an attorney and bankruptcy referee, approached a man in the restroom of a shopping center. The restroom was considered a homosexual meeting place and was under police surveillance. The two men left in Carswell's car and parked in a wooded area. There Carswell "actually and intentionally" touched the other man, George Greene, a police undercover officer, who arrested him. When the news broke, Carswell, married and the father of four, entered a Florida hospital for what was termed "nervous exhaustion and depression." Later the charge was changed to simple battery.

those declarations and assurances that he repeatedly made both before assuming the Presidency and when he was already in the White House.

Even the tragedies at Kent State and Jackson State failed to touch the President, whose comments at the news of the students' deaths ("[this] should remind us all once again that when dissent turns to violence it invites tragedy") inflamed some of his friends, notably Republican Senator Robert Dole, who was then national party chairman and was later to run for Vice President on the Ford ticket, and Secretary of the Interior Walter Hickel. Both men were ultimately cashiered from their posts by the President, who said philosophically:

> I knew the stakes that were involved. I knew the division that would be caused in this country. I also knew the problems internationally. I knew the military risks. . . . I made this decision. I believe it was the right decision. I believe it will work out. If it doesn't, then I'm to blame. . . .

When the national explosion came, Nixon rapidly backed off. The Cambodian chapter would soon be over; the incursion into that country would be limited to twenty-one miles; and there would be a "full" investigation of the killings on the two college campuses. The war went on nonetheless. We were still at the point of victory, each new bombing would break the back of the enemy, the pacification program was an immense success, the Vietcong's will to resist was about to break, the bombs that fell on the Hanoi hospital were from the enemy's own antiaircraft guns, and we would never leave that alien land until Thieu's government was secure and all of the POWs were safely home.

In Washington, Nixon marched with the same somber step to his own drummer. Advisers did not advise, cabinet officers rarely met with the chief executive, the President vetoed bills without warning Republican legislators who had sponsored them at the behest of cabinet members, and in a thousand different ways, Nixon was determined to show the country that he alone was responsible for what was taking place. Executive officers approved by the Senate were rapidly shunted aside by a series of presidential appointees from the West Coast, and newspapers soon began to describe the Nixon Administration as being managed by Orange County (California) Republicans. Arthur Burns had been replaced by the ambidextrous Daniel P. Moynihan, who was succeeded by John Ehrlichman, who finally spoke for Nixon domestically. His old law partner William Rogers, now Secretary of State, served as an inarticulate apologist for foreign policy, which Rogers seemed not fully to understand, while Henry Kissinger traveled the world negotiating the on-and-off peace. John Mitchell, another former law partner, had a large piece of the President's ear as Attorney General, and bright young men and a few not so young or bright men, many from California, either lawyers or public-relations specialists like Herb Klein, Ron Ziegler, Jeb Magruder, John Dean, Gordon Strachan, Herb Porter, Hugh Sloan, and Bob Haldeman, kept the President away from the distractions of everyday political life.

Late in his first term Nixon took the unprecedented step of giving a private interview to the New York *Times*. It followed close on the heels of an interview that appeared in the *Sunday Telegraph* of London in which the President explained how his parents' fierce adherence "to what is now deprecatingly referred to as 'Puritan ethics'" shaped his own strong sense of individualism and opposition to the New Deal. In the *Times* interview, Nixon saw himself as one who had kept faith with the 1940s while the nation's once farsighted liberal establishment was running out on its principles "toward isolation and weakness."

By late 1971 he regarded himself as an internationalist, railing against those who spread "disillusionment" about the war in Vietnam. He viewed his interest in Southeast Asia as a healthy antidote to the Munichlike appeasement of Hitler, and in a statement that must have made some of his supporters shudder and some of his opponents blink in wonder, the President delivered himself of an amazing *coup de grâce:* "The nation's fundamental disease is the psychology of the cold war, our obsession with power, our assumption that the great problems that glare upon us so hideously from every corner of the horizon can be solved by force." Pointing to his impending meetings with Peking and Moscow, Nixon offered the thought that even "many Communist leaders gratefully recognize that the United States wants nothing for itself save the chance for everyone to live and let live."

The war did not end, and when Nixon ran again in 1972, he asked his old friend John Mitchell to step down from the nation's highest legal office and chair the Committee to Re-elect the President. The Democrats finally put forward George McGovern, whose star-crossed candidacy was immediately marred by revelations of an emotionally troubled running mate and whose campaign limped from crisis to crisis.

The campaign itself was a curious one. Senator Edmund Muskie, who public-opinion polls showed to be leading Nixon in popularity early in 1972, presumably lost the nomination in his own section of the country, where he was laid low by some vicious rumors about his feelings for ethnic minorities—in this instance, the French Canadians—and some scurrilous references to his wife by a Manchester, New Hampshire, newspaper publisher, William Loeb. Muskie was so enraged that he wept as he stood outside Loeb's office on a snowy day answering the charges, and the widely printed picture that resulted from that less than classic confrontation served to discredit his stability in the eyes of many of the country's citizens.*

Other Democratic candidates suffered similar mishaps. Party regulars in California received letters on official Humphrey stationery misrepresenting the Senator's position on some critical issues, and Floridians received mail reputedly signed by Muskie accusing both Henry Jackson and Humphrey of sexual misbehavior. At the same time, a former CIA agent and then White House employee named E. Howard Hunt undertook a project to forge cables linking John F. Kennedy with the assassination of Ngo Dinh Diem.

With the blessing of Charles Colson, special counsel to the President, Hunt simulated two cables involving President Kennedy with the assassination of the Vietnamese Premier. However, he was unable to secure from the Secret Service the original typewriter used for such correspondence—"too sensitive" was the reason according to Colson—and the effort was, in Hunt's words, "convincing to the reader, though not—as I had warned Colson—invulnerable to technical examination." The text of these forged cables was provided to *Life* magazine's Pulitzer Prize-winning journalist Bill Lampert, who saw in them a major scoop but neglected to ask if the cables were authentic.

The magazine was interested in running the story but first wisely asked for copies of the originals, presumably to judge their authenticity. Since this could not be done, the story was never printed, and all of the

---

*At the direction of Attorney General John Mitchell, G. Gordon Liddy helped set up an intelligence organization for the 1972 campaign. At least two Nixon men infiltrated the Muskie organization: John Buckley, an Office of Economic Opportunity employee, photographed mail and other campaign documents, and Tom Gregory reported on contributions and other finances. E. Howard Hunt, who, like Liddy, would become well known for his part in the Watergate break-in, personally hired Gregory and was the contact man for Buckley.

material was locked in Hunt's safe in his White House office.* When he and the plumbers unit were later arrested for breaking into the Watergate complex, these cables, along with other incriminating evidence, were seized by John Dean and illegally destroyed by acting FBI Director L. Patrick Gray.

The Republicans, running easily in front in the presidential campaign in 1972 with the same well-tried team, were becoming preoccupied with the Pentagon Papers. This matter came to public consciousness on June 13, 1971, when the New York *Times* began publication of a lengthy document which in its totality comprised forty-seven volumes and some 7,000 pages reproduced from a government copy loaned to the Rand Corporation in Santa Monica, California.

The government immediately began efforts to enjoin the distribution of its secret papers. Initial success in this endeavor brought about a fascinating "rash response" as newspapers in different parts of the country successively took up the publication responsibility as others were being prohibited by court injunction from reprinting sections of the report. Those responsible for the "theft" were identified as Daniel Ellsberg and Anthony Russo. The first was a former Johnson Administration official who had become disaffected from the war, and the latter was a Rand Corporation associate who shared Ellsberg's animus.

The President was understandably upset that documents classified as top secret were now coming into the public's hands, dealing as they did "with military and diplomatic moves in a war that was still going on." What troubled Nixon even more was that the *Times* had published material which he said was not even in the forty-seven-volume study, thus raising in his mind "serious questions about what and how much

---

*Hunt first met Nixon in a Washington restaurant after the young politician had given an address to a meeting of former FBI agents. Hunt approached Nixon's table, introduced himself, and congratulated Nixon on his role in the Hiss case. Nixon was flattered and invited Hunt and his wife to join Pat and himself. According to Hunt, politics was the topic of conversation. Later they were to meet again in Montevideo, Uruguay, when Nixon was Vice President and making his South American tour and Hunt was CIA chief of station.

The CIA agent was favorably impressed with the Vice President but was unimpressed with his chief, President Eisenhower, whom he had once served as an interpreter. Hunt wrote in his book *Undercover*, "From several . . . contacts with the President [Eisenhower] I found there was considerable disparity between the public figure of a broadly grinning Ike and the private man, who struck me as petulant and autocratic toward his staff. This experience considerably dimmed my enthusiasm for the President but convinced me that Vice-President Nixon would be a more than worthy successor."

might have been taken." Finally, he spoke of his real fear that "there was every reason to believe this was a security leak of unprecedented proportion."

Although the Department of Justice fought hard, the injunctions were soon set aside, and the *Times* brought out the Pentagon Papers in book form. There was little question that the government was not going to back down on the matter, and it sought and succeeded in obtaining indictments to try Ellsberg and Russo in Los Angeles County. The long-term Administration antipathy toward newspapers was fanned by this episode, and Mr. Agnew and others became even more outspoken in condemning the liberal press, which in this era had come more and more to include television commentators. One of the Administration's young men, Clay T. Whitehead, who was acting as director of the Office of Telecommunications Policy, made such injudicious statements at a news conference about the consequences of press independence that he was forced to stage another such conference one week later in order to recant many of his earlier comments.

The aforementioned E. Howard Hunt had joined the Nixon White House staff following Daniel Ellsberg's theft of the Pentagon Papers. He was recruited by Charles Colson, who, like Hunt, was a Brown University graduate, to work on the so-called "Pentagon Papers project." Hunt writes of this work, "In and around Colson's offices, the question was often asked: Why did Ellsberg do it? What was his motivation? Was he a madman, martyr, the tool of a foreign power or that most unlikely of possibilities—an idealist and an altruist?"

To answer such questions, a psychiatric profile of Dr. Ellsberg was prepared by a CIA psychiatrist. It was "superficial," Hunt wrote, "and perhaps prepared by a partisan of Ellsberg's." So Hunt and G. Gordon Liddy, head of the White House's Special Investigative Unit, and four Cuban exiles whom Hunt knew from his work during the Bay of Pigs invasion burglarized Ellsberg's psychiatrist's office in Beverly Hills looking for his psychiatric profile. These same six men, along with the addition of an electronics expert, James McCord, were later to break into the Watergate complex in Washington. These two events came to be known by reporters as "Watergate West and Watergate East." However, the real Watergate entry became public knowledge immediately, while the West Coast effort surfaced for the first time months later during the Ellsberg–Russo trial.

Describing the rationale of the President's men for this and other surreptitious and illegal activities, Hunt commented, "If Daniel Ellsberg viewed the legitimate government of the United States as 'criminal,' then we perceived him as treasonist and alien. We were deeply concerned by extremist elements: the yippies, hippies, and zippies, the mob, the SDS and the movement—all groupings of counterculture directed by the countergovernment, whose purpose seemed clearly aimed at the destruction of our traditional institutions they could not hope to eliminate through elective process."

Summer unofficially began in Washington on June 18, 1972, with the announcement that some men had been arrested early that morning at the Democratic campaign headquarters in the Watergate complex. Bit by bit their identities emerged, but the White House continued to refer to the story as a "silly and bumbling effort by a few misguided men acting entirely on their own." It soon developed that some of the men were employees of the Committee to Re-elect the President (CREEP), and one, E. Howard Hunt, even had an office in the White House complex. However, there were clear intimations that he had no official duties in the White House at the time of the robbery.

None of those arrested was willing to talk, but the circumstantial evidence pointed to efforts to "bug" Democratic party phones for strategic reasons and to collect evidence at the headquarters that might prove useful in the impending campaign. The White House was happy to point out that it had no official connection with the committee, and one insider dismissed the whole thing with an airy statement that "the Watergate was strictly a Keystone Kops, 1701 [Pennsylvania Avenue, the headquarters of the Committee to Re-elect the President] operation." More contemplative apologists argued that such "nonsense" as the Watergate break-in was almost part and parcel of the American political tradition, and if one only had time to look, the Democrats could be proven to have performed equally heinous tasks.

The Washington *Post* had assigned two young reporters, Robert Woodward and Carl Bernstein, to a continuing investigation of the case, and that paper became the source of a number of stories which created the growing impression that there was more than a simple exercise in "summer madness" in the Watergate affair. After ten weeks had passed, Nixon held a press conference coincidental with a large celebrity cocktail party at his San Clemente estate and made this fateful statement:

Mr. Dean [the presidential counsel] has conducted a complete investiga-

> tion. . . . I can say categorically that his investigation indicates that no one in the White House staff, no one in this administration presently employed was involved in this very bizarre incident.*

About this time it became public knowledge that the legendary John Mitchell, the bellwether of the Nixon presidential campaigns, was encountering rough matrimonial seas. After a particularly colorful episode which brought his wife, Martha, into difficulty at a Newport Beach, California, motel that unhappy summer, it was reported that he wished to step down as director of the committee in order to spend more time with his wife and daughter. Mr. Nixon was strangely philosophical at losing such a valuable aide but remarked that he certainly understood the forces that made Mitchell's resignation necessary. Mitchell's assistant, Clark MacGregor, was named to replace him, and the Nixon campaign sailed merrily on.

Woodward and Bernstein continued to pick methodically away at the threads of what the Republicans were calling jocularly in their gentlemen's-club manner "the Watergate caper" (as if it were all good fun and only grubby outsiders would take the matter seriously). The evidence of growing responsibility that might cross Pennsylvania Avenue to the White House made MacGregor angry, and he countered with a testy attack on the *Post* delivered on October 16: "Using innuendo, third-person hearsay, unsubstantiated charges, anonymous sources, and huge scare headline, the *Post* has maliciously sought to give the appearance of a direct connection between the White House and the Watergate, a charge which the *Post* knows—and a half dozen investigations have found—to be false."

The *Post* answered softly:

> The Watergate bugging is part of a much broader campaign of political espionage and sabotage that is basic strategy of the President's 1972 campaign.

In reference to the touchy issue of presidential responsibility, the newspaper responded: "All we need to do is to quote the President when he said, 'When I am the candidate, I run the campaign.'"

All of the above seemed to make little difference to the voters, and Mr. Nixon was one of the nation's easiest victors as he won all but the

---

*In Dean's book, *Blind Ambition,* he admits that no such "complete" investigation was made, and his tentative conclusions were at odds with Nixon's summary.

state of Massachusetts and the District of Columbia in defeating George McGovern in November 1972. The victory was clear evidence of Nixon's careful reading of the public pulse. The entire nation could accept a warriorlike Quaker in preference to a Christian missionary liberal, wrote Howard Stein. Although both candidates promised to root out evil, McGovern's moralism apparently frightened the electorate, who feared that he really meant business. The Democrat's evangelical stance convinced many that he would not merely combat the symptoms but intended to revitalize the society so that evil no longer existed.

Retrospectively, what the electorate got in this election was a man with a stern veneer and a corrupt core—i.e., a President who got away with as much as possible while righteously punishing those who got away with too much too openly. It is not surprising that the Nixon era was later characterized as providing a national climate in which sins of commission were made slyly, secretly, or vicariously, while the exhortations for decency were made in the piety of public places. Playwright Arthur Miller compared the morality of Nixon's campaign with public reaction to the revelation that Tom Eagleton had undergone psychiatric treatment. Miller concluded that the people "could trust a man who is corrupt but could not trust a man who knew despair."

The Watergate unfortunates went to trial, and none offered any defense in a District of Columbia courtroom presided over by a judge who owed his appointment to Mr. Nixon. Judge John Sirica was a fabled law-and-order man who was known colloquially as "Maximum John" by unhappy prisoners who had felt his sinews in cases where they received the longest possible sentences for their offenses.

The only exotic event of the fall and winter was the death of E. Howard Hunt's wife in an airplane crash in Chicago. The post-crash investigation revealed that she was carrying many thousands of dollars in $100 bills in her purse. Hunt said the money was from their savings and was for an investment in a hotel-management company, although he failed to explain why the money was in cash. The media speculated that the money was a political payoff for the burglar's silence.

All seemed in usual order until L. Patrick Gray, Mr. Nixon's replacement for the late FBI Director J. Edgar Hoover, came to appear before the Senate for confirmation. Immediately there were stories that John Dean, presidential counsel, had been allowed to sit in on the interrogation of White House personnel who were being seen in connection with the Watergate affair. Gray admitted that he had permitted this and also stated that he knew that Mr. Dean had private interviews with those individuals after Gray had left the White House. To compound the matter even more, Gray had allowed Dean access to the FBI files on the Watergate investigation and had also destroyed some files that Dean and Ehrlichman had instructed him should never "see the light of day."

When that day of testimony before the Senate confirmation committee was over, it was clear that Mr. Gray was *persona non grata* with the Nixon Administration and hardly much better off with the full Senate. It was but a matter of a day or two before he withdrew his name from consideration. A month before, James W. McCord, Jr., the electronics expert of the Watergate seven who had been convicted and was waiting sentence, wrote to Judge Sirica and offered to "tell the truth" about Watergate. McCord, recruited especially for this event and thus not a dedicated team player, stated that the White House was involved "up to and including [the man in] the Oval Office" and that Magruder, Mitchell, Ehrlichman, Dean, Haldeman, and many others shared guilty knowledge of either the planning, execution, or cover-up.

McCord's tales of the offer of money to the defendants and executive clemency after a year to all who went to jail without testifying about the true facts in the case brought the Ervin Senate committee, which had been despondently picking away at the edges of an evanescent rumor, into a position of central authority. There had been innuendos of Justice Department inactivity for almost a year, and the McCord charges tended to lend substance to the feeling that no one in the Administration was willing to take on an active investigation of White House involvement in the messy affair.

Throughout this period the colorful Henry Kissinger had been on the shuttle between Washington, Paris, and Saigon, and a shaky cease-fire had finally been negotiated. Although its terms were seemingly unenforceable and the U.S. bombing continued for many months, at least the bulk of American POWs were released and flown back to the West Coast. As each group descended from the aircraft, the highest-ranking officer offered a eulogy to President Nixon, whose manifold talents and devotion to their cause had made their releases possible. The similarity of the statements brought inquiries from reporters, who learned that the Defense Department either had or had not—depending on whom one talked to—suggested a possible planeside statement to the senior officer aboard.

Nonetheless, there seemed little question that the prisoners were solidly grateful to their President, and when Nixon told them at a party he held for them at the White House (save for those who were either divorced or in the process) that the "job of the Executive is to keep from the people what they ought not to know and to keep from the press what they ought not to print," he was roundly applauded. This was consistent with his post-election comment that most of the American people "wanted to be treated like children," and he still exuded confidence that no one of any consequence was really likely to take the Watergate break-in as anything other than a foolish mistake.

Mr. Gray, the deposed aspirant to J. Edgar Hoover's throne, did not take his loss in the expected fashion. Newspaper reports that he had been warned off the Watergate matter by White House aides continued to proliferate, and soon the name of Richard Helms, former director of the CIA, began to be heard in the rapidly expanding affair. The CIA had been asked by Hunt and Gordon Liddy to provide support for their intelligence efforts, and when Helms grew suspicious of their stories he called John Ehrlichman, who assured him that their actions had White House approval. Helms finally grew unhappy with the pair and refused to cooperate with them. He found himself suddenly asked to resign from the CIA in order to become Ambassador to Iran.

The rejected L. Patrick Gray now reported that he was asked by the White House to move cautiously in his investigation of the Watergate affair because it might involve CIA contacts in Mexico. Although Mr. Helms was described as saying that he had assured White House aides Haldeman and Ehrlichman that the CIA had no involvement in Mexico at that time, Gray was nonetheless warned to proceed slowly by the pair lest his investigation endanger "national security."

Somewhat later in the game, when it came John Dean's turn to testify before the Ervin committee, he recalled Nixon instructing him to advise L. Patrick Gray to read Nixon's chapter on the Hiss affair in *Six Crises* in order to place the FBI role in cooperating with the executive branch into proper perspective. When the furor began about executive privilege, Dean testified that he advised Nixon that his current stand was at variance with that he had taken during the Hiss affair. Apparently surprised, the President asked Dean to reread the earlier statement and report back to him. Almost as an afterthought, Dean stated that he never talked with Nixon on the subject again.

As the Ervin committee freed itself from its early lethargy, a New York grand jury considering a case involving a prime Republican Party contributor named Robert Vesco began to investigate the relationship between his secret campaign contribution of $200,000 and a charge that was then pending before the Securities Exchange Commission which accused Vesco and others of bilking investors of some $224,000,000. When those hearings were over, two of Mr. Nixon's most prized lieutenants were indicted—John Mitchell, formerly the highest law-enforcement official in the land, and Maurice Stans, who had stepped down as Secretary of Commerce to manage finances for the Re-election Committee. The charges were briefly stated, but the tone was ominous—"conspiracy, obstruction of justice, and perjury."

The man Nixon thought to displace in the 1964 election spoke up promptly. The party was being torn asunder, Senator Goldwater stated; Mr. Nixon must come forth and speak plainly and openly of the Watergate affair before all was lost. The President's spokesman, Ronald Ziegler, continued to vent his fast-diminishing spleen on reporters. There had been Mr. Dean's investigation, the White House was then and was now clear, Mr. Nixon was being treated poorly by the liberal press, who had been after him since 1948, and, finally, the President would have nothing more to say about the scurrilous rumors.

In direct counterpoint, Mr. Gray was testifying to a Congressional committee that he had warned the President as early as July 6, 1972, that the men around him were misusing both the FBI and the CIA. Even more directly, Gray was said to have told Mr. Nixon that "You are being mortally wounded by the men around you." Gray portrayed Nixon as asking for proof, and this does present a strange vision of the former Hiss "prosecutor" asking that he be convinced that something bad was happening to him personally.

The Ellsberg trial began in Los Angeles, and a long series of famous Americans testified to the pros and cons of the case. Members of the Kennedy Administration were prominent supporters of the Ellsberg–Russo duo even though the release of the papers seemed directly to indict the Kennedy and Johnson era for foolish naïveté and self-deception; the generals and officials of the Nixon Administration were prosecution witnesses even though Nixon came off relatively clean in the reprise.

Throughout the early days of the trial the defense kept trying to bring into evidence the fact that Ellsberg's phone had been tapped and the indictment was illegal as a consequence. The government was slow in responding to Judge Matthew Byrne's orders to produce its tapes of Ellsberg's conversations, and the judge was becoming angry at the dilatory tactics. Suddenly, a report that Hunt and Liddy had broken into Ellsberg's psychiatrist's office in search of incriminating information—

an effort that was reportedly approved once again by White House officials—stunned the public. Then came Judge Byrne's statement that he had two meetings with White House officers and one with the President personally to discuss the possibility of his accepting the FBI directorship—all taking place while the Ellsberg trial was going on.

Judge Byrne rapidly came to the conclusion that the government's behavior in the Ellsberg–Russo affair left him no alternative but to declare a mistrial, and in his decision he made it abundantly clear that he could not conceive that the case could be retried under existing conditions. Writing in the New York *Times,* Tom Wicker placed the blame directly on the President: "Nixon himself compounded the follies of the Justice Department and his own staff in the Ellsberg prosecution by making approaches to the trial judge that, as a lawyer, he should have known were improper, appeared improper, and risked a mistrial or judicial reversal."

Throughout the late winter, Mr. Nixon remained above the gathering clouds. He refused to speak out on the Watergate conflagration and began to argue that no one around him could be expected to testify. In a startling extension of the doctrine of executive privilege, he sent his new Attorney General, Richard Kleindienst, to testify before the Senate that the President could blanket in all two and a half million federal employees and order them not to testify before the Congress should he choose to. According to Kleindienst, the President had the right to determine what Congress could or could not hear from federal employees. In response to questions as to what remedies the above course of action would leave the Congress, the Attorney General blithely suggested two: (1) cut off funds to the executive branch, or (2) either impeach the President or defeat him at the next election. (Of course, in his second term, Nixon would not be subject to a "next election.") Once again, Tom Wicker spoke out plainly:

> The thought is chilling. For by now it is clear that these Nixon men are not merely trying to cover up whatever responsibility they may have for the Watergate affair. They are the same men who have gone to unprecedented lengths to seize the power of the purse from Congress, who are conducting unauthorized war in Cambodia in contradiction of the President's own pledges, who are trying to make it a felony to disclose almost any foreign policy or national defense information and another felony to publish it.
>
> Until thwarted in the Supreme Court, these same men claimed the unlimited right to wiretap and bug anyone they accused of domestic subversion,

and imposed the first prior restraint on publication in American history. Is there any limit to the raw and unchecked power they seek?

As the spring of 1973 approached, a number of other names were being frequently heard in connection with the Watergate investigation. Two of the most prominent were John Mitchell, already indicted in the Vesco matter, and his protégé, John Dean, who was now being described as a central figure in the attempts to cover up White House complicity in the Watergate. The Ervin committee, officially known as the Senate Select Committee on Presidential Campaign Activities, had already begun to discuss the necessity of calling White House officials to testify, and the President had immediately refused. The Kleindienst statement was the obvious reply to the possible request. Increased speculation about the probable involvement of Ehrlichman and Haldeman on the one hand and Jeb Stuart Magruder on the other brought a further White House modification—that those in daily contact with the President would respond to written questions which the President regarded as proper but would not submit to cross-examination under any circumstances. There were finally rumors that one of the President's men, John Dean, would be "allowed" to testify under the clear understanding that this "once-and-for-all" appearance would suffice for the entire staff.

Pressure continued to build, and on March 15, 1973, Mr. Nixon held his first announced press conference of the year. He began with the important announcement that David Bruce would head the U.S. mission to Peking, but once that by now perfunctory item was out of the way, the first question concerned the President's attitude toward allowing John Dean to testify.

Mr. Nixon was quick to reply that Mr. Dean would not be allowed to testify. The whole matter reminded him of his difficulty with the Truman Administration in the Hiss days when the House committee had asked for the FBI report on Hiss:

> And Mr. Truman, the day we started our investigation, issued an executive order in which he ordered everybody in the executive department to refuse to cooperate with the committee under any circumstance. The FBI refused all information. We got no report from the Department of Justice, and we had to go forward and break the case ourselves. . . . I would like to say, incidentally, that I talked to Mr. Hoover at that time. It was with reluctance that he did not turn over that information, reluctance because he felt that the information, the investigation they had conducted, was very pertinent to what the committee was doing.

> Now, I thought that decision was wrong and so when this administration has come in, I have always insisted that we should cooperate with the members of the Congress and with the committees of the Congress and that is why we have furnished information but, however, I am not going to have the counsel to the President of the United States testify in a formal session before the Congress. However, Mr. Dean will furnish information when any of it is requested, provided it is pertinent to the investigation.

In response to further questions that day, Nixon pointed out that his Administration had always kindly offered the corpus of Henry Kissinger when there were foreign-policy matters under Congressional consideration and that

> I am very proud of the fact that in this administration we have been more forthcoming in terms of the relationship between the executive, the White House, and the Congress, than any administration in my memory. We have not drawn a curtain down and said that there could be no information furnished by members of the White House staff because of their special relationship to the President. All we have said is that it must be under certain circumstances, certain guidelines that do not impinge upon or impair the separation of powers that are so essential to the survival of our system.

Finally, Nixon denied any concern about the story that his personal attorney, Herbert Kalmbach, had any part in the employment of one Donald Segretti, a young man whose name kept cropping up in connection with unbelievable tales about employing political spies and mercenaries to disrupt the Democratic campaign. The President was glad for the opportunity once again to express his complete confidence in those "White House people" around him and was not going to go into the matter any further.

When the session was over, it was evident that the challenge had been laid down by Nixon: if Dean were to be subpoenaed, the President would welcome it, for "perhaps this is the time to have the highest court in this land make a definitive decision with regard to this matter." Far in the distance and understandably unmentioned was Nixon's personal position during the Hiss case that "any such order of the President [Truman] can be questioned by the Congress as to whether or not that order is justified on its merits."

As rumors of high-level involvement proliferated, there were some defenders of the Administration. Spiro Agnew, after a long and uneasy silence (supposedly caused by advice to cut loose from Mr. Nixon while he still had time), spoke out in defense of the President, but the old fire was gone. He appeared strangely subdued, as if he had lost his own moorings in these strange goings-on. The once hostile Ronald Ziegler, soon to be promoted to the position of White House Chief of Communications as Herbert Klein suddenly disappeared from sight, admitted that all his previous Watergate disclaimers were now "inoperative" (a statement happily seized on as a wonderful example of Nixonian "newspeak") and publicly apologized to the Washington *Post* for his earlier deprecations. That certainly was a red-letter day for the press corps, since Mr. Nixon himself appeared at the White House press room and made the surprising announcement that he hoped that he was worthy of their trust and that he encouraged them to continue to "give me hell" when he erred.

John Mitchell and California Governor Ronald Reagan were on another reaction track. Mitchell, who was by now indicted in the Vesco case and in whose office at the Justice Department the early Watergate plans allegedly had been discussed, wanted it to be understood that he was in danger of being made a scapegoat for others' wrongdoings. He had a clear conscience, he said, and continued with the mind-boggling comment, "I've never stolen any money. The only thing I did was to try to get the President re-elected. I never did anything mentally or morally wrong."* In a bitter editorial, the New York *Times* concluded:

> Clearly in Mr. Mitchell's mind—as in the minds of others in the President's palace guard—there existed one set of laws for the common people and another for themselves as the governing elite. The hot cash in the committee's safes and briefcases, after all, was not stolen money, and those who used it unscrupulously considered their cause sufficient justification of their questionable actions. As Mr. Mitchell phrased it, the only thing he did was "to try to get the President re-elected." And so, the former Attorney General rests his moral case and his conscience.

Reagan similarly protested that those Republicans involved in Watergate were not "criminals"; in their heart, they suffered from a surfeit of loyalty for their President and their country. Even Kissinger in a press conference devoted to complications in the Vietnam cease-fire asked for

---

*Speaking later from the federal prison in Alabama where he was serving a thirty-month to eight-year term, the former Attorney General had a change of mind. In a successful effort to get his sentence reduced, he finally said on October 4, 1977, "No set of circumstances, whatever they might be, could ever again cause me to perform such actions or lead me to commit such deeds."

"compassion" for the unfortunates associated with the by now famous illegal entry. The whole matter was neatly summed up in a cartoon by the irrepressible Jules Feiffer, who pictured Nixon in a discussion with himself about the scandalous affair:

> I do not say Watergate was not illegal. It was! . . . But I say it is a body blow to the whole American system to say it was criminal. . . . First of all the perpetrators held respected and sensitive jobs in the highest branch of government. . . . Now I know some people would call that criminal. I don't. . . . Next, they are white, come from good homes and held impressive track records in private enterprise. . . . Now I know some people would call that criminal. I don't. . . . Next, their acts were not directed at personal gain or mob violence. Not at all! . . . Their acts, overzealous perhaps, were directed at perpetuating four more years of peace with honor and law with order. . . . Now I know some people would call that criminal. I don't. . . . No, Watergate was not criminal. Daniel Ellsberg, Dr. Spock, Chicago in '68 were criminal. . . . Watergate was self-defense.

Public suspicion was shifting to the fact that immense sums of money had been collected by the Committee to Re-elect the President and that even after a free-spending campaign, over $5,000,000 still remained in party coffers. There were even rumors that some of the money had been used to secure the silence of the Watergate conspirators. Most difficult of all for the White House to deal with was the fast-growing impression that Mr. Dean's "complete" investigation was perhaps not very complete at all and possibly not even an investigation in the commonly accepted sense of that term.

Under much pressure, Mr. Nixon took to the air waves to make another statement on Watergate on May 22, 1973. In a fascinatingly cynical move, radio station WBAI in New York City followed its transmission with a repeat of the 1952 Checkers speech. The station's news editor was quoted as saying, "The similarities were amazing." It was Nixon's fourth effort to come to peace with the Watergate equation.

His first attempt was the simple declarative statement at the outset in the summer of 1972 that he "wanted all the facts to come out" and that he "was going to get to the bottom of the whole thing." At that time Nixon refused to answer any further questions and seemed to hope that the trouble would spontaneously disappear. This was followed by the unfortunate reference to the Dean "investigation" at San Clemente in August of that year which for Nixon seemed to lay the matter to rest for good. The Gray affair brought him back to the television cameras on April 17, 1973, when he seemed less than candid about what had taken place in his Administration but admitted to the possibility that he might have been misled by some of his own loyal aides.

The May 22 statement was by far the longest and the most revealing. There were a number of confessions made during that speech—there were wiretaps, intelligence plans of such magnitude that they cracked even the hardened sensibilities of J. Edgar Hoover and also general encouragement of burglaries and the like—but "none of these took place with my [Nixon] specific approval or knowledge. To the extent that I may have in any way contributed to the climate in which they took place, I did not intend to; to the extent that I failed to prevent them, I should have been more diligent."

And then came the argument that echoed resoundingly down the long years to Jerry Voorhis, Helen Gahagan Douglas, Alger Hiss, Adlai Stevenson, and the others who offended him; the argument that was linked similarly to all of the related issues beginning with his wish for the U.S. to intervene in the French Indochina War at Dien Bien Phu and which still dominated his reluctance finally to give up in Cambodia; the argument that everything Nixon did or failed to do (in this instance at Watergate) was directly motivated by his concern for "national security." The explanation grievously offended James Reston of the New York *Times*, who charged that this was a favorite Nixon tactic under pressure. He wrote:

> It is the main theme of his political life. Whenever he has been charged with dubious political or executive decisions, he has always justified them on the ground that, right or wrong, they were done in the name of "national security." . . .
>
> And the tragedy is that more crimes and brutalities have been done in the name of "national security" in this country in the last quarter century than in the name of anything else, and Mr. Nixon is still falling back on this excuse, as he has done throughout his long career. . . .
>
> But this is precisely what he is doing. He is failing the inquest. By his own testimony, he has created an atmosphere of fear, suspicion, and hostility in the White House, which has infected not only the Haldemans and the Ehrlichmans and the Mitchells but all the other minor characters in the tragedy.

By the end of April 1973, the White House team was being decimated by retirements, forced and otherwise. Haldeman and Ehrlichman were in the otherwise group; they "resigned," much to the President's regret, and were publicly commemorated as "two of the finest public servants it has ever been my pleasure to know." John Dean stood in the forefront of the fired group; his perfidy was sufficiently extreme so that the President could make it known that he had asked for his counsel's resignation. Almost immediately Attorney General Kleindienst (reportedly never very popular with the President) found that his friendship with some of those likely to be in legal trouble was of such magnitude that he could not effectively function and stepped down precipitously as Attorney General.

Jeb Stuart Magruder, about whom it was being said that he had agreed to tell all he knew about Watergate, similarly resigned as Director of Policy Planning at Commerce, and Egil Krogh, another believed implicated, was relieved of his responsibilities at the Transportation Department. G. Bradford Cook, chairman of the Securities and Exchange Commission, also voluntarily left government service when complaints developed over his handling of a commission action against the omnipresent New Jersey financier Robert L. Vesco. The whole matter was capped by one of the most amazing games of musical chairs in cabinet history when Elliot Richardson moved from the secretariat of HEW to that of the Department of Defense and then to the Attorney Generalship in a brief five-month period.

The price for Richardson's confirmation as Kleindienst's successor was the appointment of a prosecutor for Watergate who would be independent of the Justice Department, and Mr. Nixon gave ground in grudging fashion on that issue. When the Senate finally showed its teeth, Richardson conceded more readily than his chief and promised complete independence for the man finally selected, Archibald Cox, Jr., a distinguished member of the Harvard Law School faculty.

As a succession of young men appeared before the Ervin committee either to confess and/or deny their part in the Watergate planning or its cover-up, the White House withheld comment. This was true even when Maurice Stans, the silver-haired money wizard of the Nixon forces, admitted giving Gordon Liddy $199,000 without inquiring what use Liddy might have for that astounding sum. Only when John Dean came before the committee was there any overt sign that people at 1600 Pennsylvania Avenue were listening. A group of questions were rushed to the committee to be put to Dean, whose testimony provided an alleged link to Mr. Nixon's role in at least the cover-up phase of the Watergate investigation. The questions were at first described as official, then as nonofficial, and finally, in what was fast becoming characteristic of the whole affair, as not representing Mr. Nixon's position at all although the latter had been "briefed" on their contents.

When Senator Daniel Inouye, Democrat of Hawaii, agreed to serve as the questioner (all Republican members refused to do so), the White House brief turned out to be a series of charges against Dean, which he answered with straightforward denials. Dean's testimony was damaging to Nixon in at least two areas. The former presidential counsel charged that Nixon had congratulated him on his success in keeping the White House out of the investigation. Even less palatable to the President was Dean's contention that Nixon told him that the figure of $1,000,000 to secure the silence of the Watergate seven did not seem to be an unreasonable amount.

Throughout it all, Nixon was described as unperturbed. Deputy Press Secretary Gerald Warren said the President was "in a very good mood" and that he still "stood behind that [May 22 speech in which Nixon denied any involvement in either the cover-up or offers of executive clemency] statement." Warren continued solemnly, "That statement is going to stand. That statement is on the record."

At the same time, attention was being drawn to Nixon's real-estate ventures, notably the two vacation White Houses at Key Biscayne and San Clemente. Persistent rumors that the Committee to Re-elect the President had "invested" some of its left-over capital in those two estates brought denials from the President. Nixon was angered over continued press skepticism regarding presidential disclaimers that Teamsters Union campaign funds were involved in the real-estate matter and were also related to executive clemency granted to James Hoffa. The White House first claimed that money for the improvements came largely from contributions from friends. It was easy to identify the bulky outline of Bebe Rebozo on the Florida scene, but a new name, Robert Abplanalp, came into view as a co-owner of the San Clemente property.

The White House's first assay of the improvements made to the California estate listed the amount involved as $39,525 and noted virtuously that the sum was spent for "security purposes" at the request of the Secret Service. Soon, however, the General Services Administration boosted this small figure to $703,367 for San Clemente and added another $579,907 for "improvements" at Key Biscayne. The White House held to its story of "national security" until the GSA produced records indicating that a roof had been retiled and a lawn sprinkler system, a beach cabana, and a pool heater had been installed along with landscaping and electrical rewiring which consumed almost a quarter of a million of taxpayers' money.

The White House was insistent that the expenses had been "requested" by the Secret Service, which benefited most from the outlay. Cited as an example was the fact that the men guarding the President were able to remain warm and snug in the new cabana. The Secret Service, tight-lipped as ever, replied that it had "acquiesced" in many of the improvements but had "requested" only a small portion of the total involved. Newspapers complained about public payment for private gain, but Ronald Ziegler described the President as "appalled" by what Ziegler described as "a malicious and persistent effort to suggest wrongdoing."

In response to questions, Ziegler insisted that the White House "had given an accurate, precise and factual" account of the financing of the President's vacation homes. Ziegler angrily concluded, "We do not intend to issue any further statement on these problems." Viewing Ziegler's performance over the years, Dan Rather, White House correspondent for CBS, was moved to say:

> He is a recorded announcement, a classic young J. Walter Thompson junior executive. He is a decent, uncomplicated fellow. He walks in every morning and reads a script. That's what he's paid to do. The difference between Ziegler and previous secretaries is that he doesn't know much about what's going on.

Although the President was generally silent about Watergate after his May 22 statement, his elder daughter, Julie Nixon Eisenhower, was moving into the forefront as a family spokesman. When she and her husband found the atmosphere toward the President and Watergate too frivolous at the Washington Radio and Television Correspondents' Dinner, she burst into tears and protested the "tastelessness" of the proceedings. A spokesman for the group sponsoring the dinner explained later that the White House had called two days before the event and said that Julie and David would like to attend despite the warning that some barbed humor could be anticipated.

On her twenty-fifth birthday, the same young lady, nothing daunted, held a press conference at the San Clemente Inn and discussed a family dinner at which the President, playing the devil's advocate, asked the family's views on whether he should resign. The family was very much opposed to this possibility, for it felt the President had much good work left to do, and soon Gerald Warren was back before the press corps reporting that the President's dinner conversation was not terribly serious. David Eisenhower, by now a briefly experienced sports writer for a Philadelphia newspaper, also called a conference to chide his new colleagues for not being fair to his father-in-law in the Watergate matter. The young Eisenhower informed his confreres to pay more attention to foreign affairs, an area where Mr. Nixon was hitting the ball very well indeed.

None of these family tidings could cover up what was taking place in the country at large. The news of the break-in into Ellsberg's psychiatrist's office, the wiretapping of reporters, the list of enemies of the government who were to be harassed by the Internal Revenue Service,* the effort to incriminate John F. Kennedy in Diem's death, the secret police activities of the Liddy–Hunt plumbers group, the attempts to subvert the FBI and the CIA, the talking to Judge Byrne during the Ellsberg trial and the illegal intelligence plan that angered J. Edgar Hoover, the talk of bribes for the silence of the Watergate seven and allegations of executive clemency—this was the pox that would not go away. Nixon entertained Leonid Brezhnev, and there were many pictures of champagne being happily consumed. The Ervin committee cancelled its hearing for that week at the request of Senators Mike Mansfield and Hugh Scott, who felt the President's possible embarrassment might imperil the U.S.–Russian accord.

Throughout it all, Charles Colson, another former Nixon counsel, continued to insist Nixon knew nothing of the Watergate break-in as if innocence in the matter of the June 17 affair was overbearing proof of innocence in all that transpired thereafter. Mr. Colson's philosophic approach was best exemplified by two statements: (1) that a polygraph had established his own innocence in all Watergate matters (there was no evident willingness on Colson's part to produce the results for

---

*The omnipresent Tom Wicker, who found himself holding high rank on the "enemies list," complained angrily in the New York *Times:*

> They [the "enemies list"] are sad. They are sad because they show that even great power could not make of Mr. Nixon and his aides anything but small fearful men. They are sad because they disclose a great nation being led by men unworthy of her and her history. They are sad because they represent so graphically, for so many people, the last crumbling of illusion—the final evidence that there is nothing magical or ennobling about the President, nothing about American power that makes it less corrupting than any other brand of power.

examination) and (2) that as late as March 21, 1973, the President was personally complaining to Colson that "he was not being told the truth."

Even more devastating was the strange picture of Nixon that emerged from the Dean testimony. The President was portrayed as being almost paranoid about demonstrators and demonstrations. Dean stated:

> I was made aware of the President's strong feelings about even the smallest of demonstrations during the late winter of 1971 when the President happened to look out the windows of the residence of the White House and saw a lone man with a large 10-foot sign stretched out in front of Lafayette Park. Mr. Higby called me to his office to tell me of the President's displeasure with the sign in the park and told me that Mr. Haldeman [Higby's boss] said the sign had to come down. . . .
>
> When I came out of Mr. Higby's office, I ran into Mr. Dwight Chapin [Nixon's appointments secretary], who said that he was going to get some "thugs" to remove that man from Lafayette Park. He said it would take him a few hours, but they could do the job.

Dean said he dissuaded Chapin from taking that action and offered to go out himself and personally negotiate with the lone demonstrator. The ploy was successful; Dean was able to persuade the placard-carrying citizen that he would soon be in trouble if he didn't move out of Nixon's sight, and the day was saved without the intervention of the "thugs."

Dean also commented on Nixon's displeasure on his arrival in Akron, Ohio, for a ceremony at the Football Hall of Fame. Across the street from Nixon's motel room were flag-waving Vietcong sympathizers:

> The President, after seeing the demonstrators, told the Secret Service agent beside him, in some rather blunt synonyms, to get the demonstrators out of there. The word was passed, but the demonstrators wouldn't be moved—much to the distress of the advance men who were responsible for the presidential trip. . . . Any means—legal or illegal—were authorized by Mr. Haldeman to deal with demonstrators when the President was traveling or appearing some place.

Even at home Dean said that the President created "a climate of excessive concern over the political impact of demonstrators . . . had an insatiable appetite for political intelligence [which] culminated with the creation of a covert intelligence operation. . . . The strong feelings that the President and his staff had toward antiwar demonstrators . . . permeated much of the White House. . . . The White House was continually seeking intelligence information about demonstration leaders and their supporters that would either discredit them personally or indicate that the demonstration was in fact sponsored by some foreign enemy."

Dean concluded his report with the summarizing statement:

> . . . the information regarding demonstrators—or rather lack of information showing connections between the demonstration leaders and foreign government or major political figures—was often reported to a disbelieving and complaining White House staff.

Perhaps this was what the President was referring to in his candid statement on May 22 that he could have unwittingly contributed to the creation of an "atmosphere" at the White House that led to the Watergate affair, but such admissions and small insights were almost lost sight of as the Watergate revelations cascaded on.

When the Ervin committee spoke of inviting the President to testify before them in order to dispel the deep public suspicion that was developing, Mr. Nixon was staunchly opposed to this notion. Moreover, he would not provide the committee with any presidential documents which might illuminate some of the charges made against his Administration. He said:

> I have concluded that if I were to testify before the committee irreparable damage would be done to the constitutional principle of separation of powers.

Pointing out that he had agreed to permit "unrestricted testimony" of present and former White House staff members before the committee, Nixon ruled out access to his papers in order to preserve "the indispensable principle of confidentiality of presidential papers." He continued:

> Formulation of sound public policy requires that the President and his personal staff be able to communicate among themselves in complete candor, and that their tentative judgments, their exploration of alternatives, and their frank comments on issues and personalities at home and abroad remain confidential. . . . Arguments can and have been made for the identification and perusal by the President or his counsel of selected documents for possible release to the committees or their staffs.
>
> But such a course, I have concluded, would inevitably result in the attrition and the eventual destruction of the indispensable principle of confidentiality of presidential papers.

Much to the surprise of historians, the President at this time invoked the name of Harry S. Truman, formerly a target for much of Nixon's venom. Now, Truman's move in refusing to testify before Nixon's old committee, the House Un-American Activities Committee, in 1953 was cited as an excellent example of constitutional law. Nixon pointed out that Truman had declined to appear before the committee eighteen months after he had left office "on the ground that the separation of powers forbade his appearance" and that Truman's decision "was not challenged by the Congress."* Mr. Nixon was now happy to be associated with his old enemy, whom he had attacked four months before in retrospective fashion for his failure to allow an FBI report involving the Hiss affair to be released to the House Un-American Activities Committee in 1948. At that time Mr. Nixon's remark placed him at some distance from Truman's position: "The Truman order cannot stand from a constitutional standpoint in the merits of the case."

The Ervin committee pressed on, and its new witness was John Mitchell, the former chief of law and order in the Nixon Administration. Previous testimony had clearly implicated Mitchell in both the planning and the cover-up of the Watergate affair and had also revealed a White House plot to attach the Watergate blame to the former Attorney General. Throughout the preceding week there were continued reports that the so-called "strong man" of the Nixon Administration would say little that might possibly implicate his former chief, and his appearance confirmed this assumption.

It has been commented that Mitchell resembles a Marine Corps officer, and he quickly established a manly ambience. Mitchell made no introductory statement (John Dean's had consumed a full day) and began by figuratively rolling up his sleeves and explaining to an apparently surprised Ervin committee what it was like to run a large governmental-military-business enterprise. Although his memory was patchy about things he was uncomfortable with, Mitchell was quick to recall and disclaim events that implicated either himself or the President in the Watergate and its aftermath. Had he testified before another Congressional committee that he had clearly kept out of the election campaign before stepping down from the Attorney Generalship—well, Senator, that was at another time, and he now recalled the events with new

---

*Mr. Truman had then declined an invitation by Congressman Harold Himmel Velde to testify regarding the continuation in government employment of Harry Dexter White, who figured largely in the Hiss–Chambers hearings in 1948.

clarity. He had indeed "made major decisions" about the Nixon campaign while still Attorney General because the President had asked him "to keep an eye on the campaign committee, to see that they didn't get out of line there." No one was visibly concerned with such contradictory sworn testimony, Mr. Mitchell least of all.

When Senator Herman Talmadge of Georgia asked him whether Magruder or Dean were guilty of perjury in testifying falsely about his involvement in Watergate, he was disdainful of such legal talk. He was a lawyer, he said; he knew what the legal definition of perjury was. But for himself, Mitchell was more comfortable with the idea that those who had testified falsely about him were simply subject to "confusion." His speech was dotted with profanity; "damn" and "hell" were heard continuously, and his gruff, masculine charm was hard to resist.

Mitchell's technical language was equally interesting, and such obscure terms as "time frame" and "subject matter" (the word "subject" was never used alone; every "subject" also had "matter") were heard in profusion. Mitchell had the knack of never answering a question in the same context it was asked (when Committee Counsel Samuel Dash asked whether Haldeman and Ehrlichman were, to his personal knowledge, also trying "to keep the lid on"—which was Mitchell's oft-repeated statement—the Watergate mess, Mitchell answered, "I assumed, Mr. Dash, that they shared my concern"). The difference between an "assumption" which was such an important matter two weeks before when Dean testified and "personal knowledge" fell by the wayside, as did many other substantive issues. One had the feeling that the questioners were operating on AC while Mitchell was on a DC band. He took liberties with the Senators, responding curtly to such inquiries as "Did you see the document personally?" with the rejoinder "Why don't you get the FBI to examine it for my fingerprints?" This abashed even the most sympathetic Republican Senator, Edward J. Gurney of Florida, who laughed uneasily and responded, "I'll pass your suggestion on to the FBI."

When it was all said and done, the most telling moment came when Herman Talmadge led Mitchell down the primrose path with a series of assertions that described the contradictions between Mitchell's behavior as the former chief of law enforcement in the land and his role in the Watergate cover-up. Talmadge, speaking in a broad Georgian accent, asked Mitchell, "Why, when you discovered the crimes, the conspiracy, the cover-ups were committed, didn't you walk in and tell him [Nixon] the truth?" Mitchell, all Western range-rider now, allowed as how maybe he "had been wrong, it occurred to me that it was the best thing to do, to just keep the lid on through the election."

Talmadge pressed on again, asking whether the election was more important than honesty, than duty, or honor. Mitchell would not budge. In a fashion that reminded one of an experienced father lecturing an idealistic and immature child, he offered the final thrust: If he told Mr. Nixon what he, Mitchell, had discovered, it would "affect him in the election." "Was that so bad?" Talmadge asked. Then it came out. Mitchell answered that Nixon's election was "so much more important than what was available on the other side [McGovern] that I am prepared to comfortably place all of my behavior within that context." He frankly acknowledged that his statement might sound "expedient," but that's how it was.

The audience sat silent. Here was the new honesty, the means justified by the ends, loyalty to Nixon the overriding consideration in any and all decisions. This was a different kind of law and order. The Presidency had to be saved for Nixon irrespective of cost. Mitchell was the perfect Mr. Outside to Nixon's Mr. Inside. He was straightforwardly manipulative, cynically devoted to Mr. Nixon's welfare as the highest public good, and the possessor of a very elastic concept of justice. It began to seem likely that Mr. Nixon's piety was sustained by an angry band of gunslingers who kept their chief pure.

The contrast between this aspect of the Senate hearing and the Hiss affair was suffocating. Mitchell casually offered up "not to my recollection" or "I don't recall that" dozens of times over the days about events which reportedly occurred a brief three or four months before, while Hiss went to jail because he recalled in faulty fashion something removed from the present by more than a dozen years. Nobody on the committee seemed courageous enough to speak seriously of perjury in 1973, which was, of course, the major issue in 1948.*

Only at the very end of the interrogation did the word finally emerge, and then it came from the mouth of Sam Dash, the placid majority counsel. Throughout the long two and a half days Mitchell had held firmly to the idea that Nixon's election was far more important than most ordinary law-and-order considerations, that its very importance made it possible to accept the notion that crimes could indeed be committed in

---

*This was the case throughout the Senate investigations. However, perjury did play a role later on in the convictions of some involved in the Watergate affair.

order to achieve the election goal. When pressed, Mitchell would only exclude "treason and other high crimes" from his bag of tricks. During the final cross-examination, Dash asked flatly, "Would you call perjury the kind of crime that could be committed in order to re-elect the President?" "Mine or someone else's, Mr. Dash?" asked Mitchell. "Yours," said Dash. After the longest pause of the day, Mitchell, face red and now looking every bit his age, answered, "I have to give long thought to that, Mr. Dash." And to make sure that the full meaning of that brief confrontation was made clear to Mitchell, Dash then said with skepticism hanging from each word, "Mr. Mitchell, is there any reason to believe in your testimony before this committee . . . ?" And then almost before the startled Mitchell could begin to answer, "I resent that statement very much, Mr. Dash," Dash rapped out, "No further questions, Mr. Chairman."

The next witness, fatherly, gray-haired presidential counsel Richard Moore, told of how the President turned contemplative about his stewardship of the government as the Watergate events came crashing in on him in thc spring of 1973. During a private conference on May 8, the President was quoted as telling Moore: "I have racked my brain, I have searched my mind. Were there any clues I should have seen that should have tipped me off?" The President went on to say, according to Moore, that he wondered whether he should have noticed such clues despite the many other matters competing for his attention. This colloquy roused Senator Ervin to an enraged interrogation of Moore as Ervin successively recounted item by item all of the stories which had appeared in the news media in the first two months after Watergate and demanded that the easily terrified Mr. Moore advise him whether any American who could "read a newspaper, listen to a radio, or watch TV" could not have known that the President's closest friends and advisers were involved in the scandal. The lionesque Ervin seemed scarcely mollified by Moore's quaking response: "Obviously something was rotten, but I thought the 'rottens' had been removed."

And soon the "rottens" were falling from the tree like overripe apples. Before it was all over, confessions, resignations, threats of prosecution, and even several jail sentences decimated the Nixon ranks. Clamor for the President's impeachment and/or resignation began in his own party as a Congressional committee to decide his fate began its hearings. When Garner Cline, who was temporarily serving as Judiciary Committee Clerk during the debate, read in an unaccented and clear voice the following statement which prefaced the debate, it surely ranked as one of the most horrifying statements of that or any other year:

> RESOLVED, that Richard M. Nixon, President of the United States, is impeached for high crimes and misdemeanors and that the following articles of impeachment be exhibited to the Senate. . . .

In a rare and gratifying expression of governmental probity, the House Judiciary Committee voted that there was sufficient evidence to conduct an impeachment proceeding against Mr. Nixon in the Senate, and rumors of presidential resignation swept the Capitol. Talk of new "bombshells" was heard, and on the evening of August 9 Mr. Nixon publicly resigned from the office of the Presidency. Tapes which offered irrefutable additional evidence that the President had known of Watergate from its very beginning and had actually interfered with the investigation were identified as the precipitating factor; still, all Congressional support had finally evaporated, and the country had finally had enough. After a tearful farewell to his legions, Nixon flew off to San Clemente, the first American President to be forced to resign from office.

What had happened to public morality, where was the law-and-order Administration, where was the public outcry at the deceptions and the deceits? How could a man like Nixon who spoke so virtuously end up in such public disorder? More importantly, was there some defect in Nixon's character which led to this sad state of national affairs? These questions will be addressed in the final chapters.

CHAPTER 10

We come into the world with sealed orders, and our job in life is to try and find out how to live and what those sealed orders may be.
—SÖREN KIERKEGAARD

Alger Hiss was born in Baltimore in November 1904, the fourth child and second son in a family of two sisters and three brothers. Both his parents came from substantial families who could trace their roots to the Chesapeake Bay city's beginnings in the middle of the eighteenth century. Hiss's great-great-grandfather was born in Germany in 1729. Upon immigrating to America, he changed his name from Hesse to Hiss. Valentine Hiss married well, became a successful farmer, and owned property in and around Baltimore.

Hiss's father, Charles Alger Hiss, was born during the Civil War to a family of six children. He was a handsome man whose distinguishing feature was large ears, which both Alger and his younger brother, Donald, would inherit. He married a middle-class girl, Mary Lavinia Hughes, known as "Minnie," who had attended Maryland State Teachers College. After marriage at age twenty-four, Charles entered the business world and shortly thereafter joined Daniel Miller and Company, importers of dry goods. Minnie Hiss was active in Baltimore society, belonging to several fashionable clubs, and took this role much more seriously than her maternal duties until she died in 1958 at age ninety-one. Charles did well in business and soon became an executive and stockholder in his firm. Although never rich, the Hiss family was well known and prominent in civic life. The family often drove about town in a horse-drawn carriage, in which Alger's mother took much pride.

In 1895 Charles's older brother John died suddenly of a heart attack at age thirty-three, leaving a widow and six children. For the Hisses, there existed a strong sense of family, and when a relative was in need, aid was given—and accepted—as a matter of course. Consequently, Charles assumed the financial and emotional support of his late brother's family. Although the families kept separate residences, brother John's six children and Charles's eventual five were together much of the time.

Charles also helped his wife's favorite younger brother secure a job with his employer. In the beginning, at least, brother-in-law Albert Hughes distinguished himself and rose to treasurer at Daniel Miller and Company. However, in a complicated financial maneuver, the young man failed to fulfill some monetary obligation which was part of a joint arrangement. Charles was forced to sell his stock in the company to make good Hughes's debt, and then Charles himself resigned.

This was in 1907, the year of the Great Panic—one of the country's worst economic periods. After several months of unemployment, complicated by the fact that his wife had just had another baby (Donald), Charles Hiss began to experience melancholia and hypochondriasis. His older brother, George Hiss, stepped in and offered him a partnership in a cotton mill in North Carolina. Charles's depression lifted until he visited the mill and realized that rural North Carolina was nothing like cosmopolitan Baltimore. When he proposed the move to North Carolina to his wife, Minnie answered, "Leave Baltimore? Leave my home and carriage? Never!" Charles's depression returned, worse this time than ever, and one night in despair he sent his wife off to summon the doctor. When Minnie returned to tell him the doctor was on his way and dinner was ready, she found her husband had cut his throat with a razor.

Alger Hiss was then under three years of age. The cause of his father's death was kept from him until he inadvertently heard some gossip about it from neighbors many years later. At that time, Alger angrily confronted his revered older brother, Bosley, who produced the true story of their father's demise.

The father's memory had been glorified by the family in a protective gesture, and Hiss experienced a profound sense of disillusionment on learning the truth. Feeling that his father had "let the family down," he determined to try to restore the Hisses' good name. Among his many substitute father figures was Supreme Court Justice Benjamin Cardoza, whose own father had been a dishonored justice of the New York State Supreme Court. It was well known that Cardoza had vowed early in his life to erase the blemish on his family record by exemplary conduct. Hiss was conscious that his identification with the jurist was based on similarities in their pasts. Others who filled this void for Hiss were brother Bosley, Felix Frankfurter, his professor at Harvard, and Supreme Court Justice Oliver Wendell Holmes, Jr., for whom he would work on graduation from law school.

As a child, Alger was what was conventionally known as a "good boy." Family legend has him as "most obedient, always listening to reason and giving the other fellow an opportunity to speak first." His mother would often entreat him to be a brave boy. Indeed Hiss was a brave child, a trait he magnified as a man to the point of bravado. This proved costly in his legal battles years later when he staunchly refused to believe he was ever in any real danger. As Hiss commented to his son Tony in the latter's book *Laughing Last,* "I certainly didn't feel any fear [when called a Communist]. Maybe that's partly a phobia against fear. Dr. Rubinfine, my analyst, says I have a phobia against fear and don't get afraid when I should get afraid. I'm also cocky and always optimistic."

There was about $100,000 in insurance money left after Hiss's father's funeral. Mrs. Hiss continued her life as an active club woman, but her role as a mother was even more perfunctory. Her major determination was to see her daughters established in wealthy marriages and her sons in successful careers. To this end, there were music lessons for young Alger from a woman across the street and German lessons from Fräulein Hogendorf. Minnie wished for her children to display their talents and be nice to important people.

Alger, however, was disturbed by his mother's aggressiveness. He has remarked that from the earliest time he can remember, he knew it was necessary to resist her will. He said in explanation: "I now have an attitude, so conditioned as to seem second nature, that exhibitionism is, to say the least, bad taste. In others, it bores me or annoys me except where it is so evidently pathological as to arouse sympathy or curiosity. . . . It just isn't my idea of the civilized man. My admiration (and emulation) are reserved for British understatement and restraint, for absence of display." Hiss remembers himself as showing great self-control as a youth.

As he was brought up in a fatherless home by an absentee mother, the major influence in his life was women, such as his late father's sister Lila, who often read aloud to Alger from the Bible or chivalrous tales like *King Arthur's Knights of the Round Table*. Such readings are among his earliest and most pleasurable memories, and he continued to find pleasure in this activity as a man. While clerking for Justice Holmes, Hiss often read aloud to the octogenarian. Their reading list ranged from literature to improve one's mind to what they referred to as "a little murder."

Alger also became attached to his brother Bosley, who was five years his senior. Bosley was a very colorful young man who turned out to be a ne'er-do-well, first living with and later marrying a wealthy interior decorator from Rye, New York, some twenty years older than himself. For a time he worked as a police and court reporter for the Baltimore *Sun* and lived the life of a boulevardier. However, Bosley contracted a kidney disease apparently induced by drinking, and his condition rapidly deteriorated. On Alger's graduation from Johns Hopkins in 1926, he went to Margaret Owen's home in Rye to nurse his brother until he was forced to leave for Cambridge and law school. Bosley died shortly thereafter at the age of twenty-seven. His death profoundly depressed his younger brother.

Hiss was just sixteen when he graduated from high school. Minnie decided he was too young and, at 116 pounds, too small to attend college, so she enrolled him in a prep school outside Boston, near his newly married sister, Mary Ann Emerson. After his high-school post-graduate year, Hiss entered Johns Hopkins—brother Bosley's alma mater. He selected Johns Hopkins because scholarships were available, and the campus, situated in Baltimore, allowed him to live at home. In a conscious effort to eclipse the long shadow of Bosley, Alger distinguished himself there. By the time he graduated, he had served as president of the student council, columnist on the school paper, ROTC cadet commander and a member of many honor societies—and was even an enthusiastic member of a drinking club. He showed a flair for drama and, having reached his full height of six feet one inch, ran (according to his own report) a very slow quarter mile on the track team. Hiss graduated from Johns Hopkins as a Phi Beta Kappa.

He spent the summer in Europe after his sophomore year in 1924. On the boat crossing the Atlantic he met a dainty blond Philadelphia girl named Priscilla Fansler, who had just graduated from Bryn Mawr and was a year older than he. He was smitten by her rare combination of brains and delicate beauty, and she became his first romantic attachment. Hiss spent some time with her in London both before and after going off to France, where he visited museums and cathedrals and took a bicycle trip through Normandy with a college friend. On returning to the States, Hiss corresponded with Priscilla, who was then doing graduate work at Yale in English literature.

Priscilla came to Baltimore during Easter of 1925 ostensibly to see another of her beaus, Dudley Diggers, whom she knew from Yale. While riding to a party in a car with several of Dudley's friends, including Hiss, on whose lap she happened to be sitting, she announced that she was marrying a divorced man, Thayer Hobson, whom she had met while at school in New Haven. Hiss, who at that very moment thought she might be taking his affection more seriously, was disappointed. Following this visit by Priscilla, Dudley Diggers' mother told Minnie Hiss that Priscilla was "boy crazy."

Hiss was admitted to Harvard Law School in the fall of 1926 and earned a scholarship the following winter. At Harvard, he came to know Professor Felix Frankfurter, who would later recommend him as Justice Holmes's secretary and also for his position in government with the Agricultural Adjustment Administration. Hiss often visited the professor and his wife, Marion, who was then editing the Sacco and Vanzetti letters. Frankfurter, too, was interested in this case, feeling it was a tragic miscarriage of justice.* Ironically, the Hiss case was to become as important an event in the political and social history of the country as the Sacco and Vanzetti case.

At Harvard, Hiss was once again a superior student and, as a mark of special achievement, was on the editorial board of the *Harvard Law Review*—the most prestigious achievement of a student's law-school career. The only dark note at this time was occasioned by another family

---

*Nicola Sacco and Bartolomeo Vanzetti were two Italian immigrants executed for a payroll robbery and murder of two guards in South Braintree, Massachusetts. Their guilt or innocence has been a subject of debate for fifty years. In July 1977 Massachusetts Governor Michael S. Dukakis issued a proclamation removing from the names Sacco and Vanzetti "any stigma or disgrace."

tragedy. A month before his graduation in 1929, Hiss received news that his sister, Mary Ann Emerson, had committed suicide.*

Her husband, Eliot Emerson, a successful Boston stockbroker, was financially ruined by the economic depression following the First World War. The Emersons were spared bankruptcy only by financial aid from her family. Alger Hiss, then eighteen, contributed $2,500 to this end. Such sudden poverty deeply depressed Alger's older sister. There were numerous domestic quarrels about money and occasional separations. Eventually, Mary was treated at a sanitarium. Finally, after a midnight quarrel with her husband, Mary Ann Emerson drank a caustic cleanser and died.

Although Hiss was very close to his sister and saw her almost every weekend since she and her husband lived near Cambridge, he seemed totally unprepared for and shocked by this event. Hiss later commented that after a profound initial upset, his effort to find the reason for the suicide left him confident that it was "a sudden irrational act without prior warning." He was much relieved by this discovery, as it left him without guilt that he should have perceived her depression and perhaps have been in a position to prevent the suicide.†

In his senior year at Harvard, a law-school friend mentioned that he had two extra tickets to see Wagner's *Parsifal* in New York City and invited Hiss to bring a date and join him. Hiss thought of Priscilla, who he had learned was recently divorced and was supposedly living a fast life in New York. Priscilla was at this time a graduate student at Columbia, completing the M.A. degree in English she had begun at Yale. Her academic life had been interrupted by her marriage and the birth a year later of a son, Timothy Hobson. Married less than four years, Thayer Hobson was granted a Mexican divorce from Priscilla so he could take a third wife, Laura Zametkin, who as Laura Hobson would write *Gentleman's Agreement,* the 1947 best-selling novel. Before returning to school, Priscilla worked as an office manager for Henry

---

*The prosecutor at the first trial attempted to question Hiss about the suicides of his sister and father. Judge Kaufman sustained the defense attorney's objection to this line of questioning, saying it had no bearing on the "credibility of this witness [Hiss] in the slightest."

†Meyer Zeligs, the psychiatrist who spent many hours talking to Hiss, concluded that Hiss's "lack of awareness and willingness to accept such a poorly rationalized explanation" were indeed significant in a personality sense and relate to his earlier reactions to his father's suicide.

Luce, a friend and classmate of her ex-husband who had just started a new magazine called *Time*.

Hiss invited Priscilla to come with him to the opera; she invited him to stay at her apartment, and Hiss made ready for what he expected to be his first truly illicit weekend. He even purchased a diaphragm toward this end. When he arrived in New York and announced his intentions, Priscilla chided him, calling him a "virgin boy," and told him a woman had to be fitted for a diaphragm. There was also another man staying in her apartment who looked to Hiss like no stranger at all. Priscilla did attend the opera with Hiss, however.

When Hiss graduated *cum laude* from law school in 1929, he was selected by Felix Frankfurter to serve as secretary to Supreme Court Justice Oliver Wendell Holmes, Jr. During his last free summer before joining the jurist, Hiss went off to Europe with younger brother Donald. That summer his thoughts often turned to Priscilla, and there was occasional correspondence between them.

This was not a good period for Priscilla, neither physically nor emotionally. She had become pregnant by her lover (the comfortable young man of recent notice), who then refused to marry her, saying his wife also was pregnant. Priscilla was forced to have an illegal abortion.*

Emotionally, this was not a very good time for Hiss either. His sister had just taken her life, and this act occurred just two and a half years after the loss of his beloved brother Bosley. Hiss was a very lonely young man. When his boat arrived back in the States, Priscilla was there on the dock to meet him.

After his return Priscilla went into the hospital for an operation to correct a uterine disorder that had been troubling her for years. Hiss spent much time with her at the hospital during her convalescence. They got on well, and he decided he could not live without her. Priscilla decided she wouldn't live with him without the benefit of marriage.

Shortly after beginning his tenure with the Justice, Alger and Priscilla were married. Holmes's secretary, however, was not supposed to be married, but somehow Hiss was unaware of this stricture. When he discovered his error, he rushed to tell the judge what had happened and to apologize. Forgiveness was forthcoming, and Justice Holmes offered

---

*A recent Hiss biographer, John Chabot Smith, felt Hiss's desire to hide from the jury the fact of his wife's abortion made him appear covert at the trial. Smith writes, "Many things that happened at the Hiss perjury trials were influenced by Hiss's knowledge of this secret [the abortion] and the way he protected it."

him two weeks off for a honeymoon. Hiss refused and asked instead for a signed copy of Holmes's collected speeches. The judge wrote in the book, "To Alger Hiss." Hiss wanted more, however, so Holmes added, "*Et Ux,*" Latin shorthand for "and wife."

At their wedding, the Hiss family was represented only by brother Donald. On the wedding day, December 11, 1929, his mother, possibly because of Mrs. Diggers' aspersion about Priscilla's moral fiber, wired Alger from Texas, saying, "DO NOT TAKE THIS FATAL STEP." Hiss showed the cable to his new bride, who became very angry at her mother-in-law for writing it and at Hiss for showing it to her.

Priscilla's son Timmy was two when she remarried, the same age Alger was when his father took his life. Hiss took his role as a father very seriously, being a model if perhaps impersonal father. In a custom quite rare for this time, Timmy referred to his stepfather by his first name and to his mother by the nickname "Prossy."

After his year as a clerk for Justice Holmes was over, Hiss secured a position with the Boston law firm of Choate, Hall and Stewart. He expected to make the practice of private law his life's work. In 1932, however, Priscilla, who missed New York, returned there to collaborate on a book with her sister-in-law, Roberta Fansler, on the fine arts on college campuses, written for the Carnegie Foundation. Alger commuted to New York to be with her and his stepson on weekends until he completed his casework for the Boston firm. Then he moved to Manhattan and went to work for the law firm of Cotton, Franklin, Wright, and Gordon, where he did antitrust work, involving such corporations as RCA, General Electric, and AT&T.

The Depression, which had begun in 1929, was much more evident in New York than it had been in Boston. Along Riverside Drive, hard by the Hudson River, were miles of shacks and wooden and even cardboard boxes providing temporary shelter to the thousands of unemployed. The area was disparagingly known as "Hooverville" for President Herbert Hoover, who simplistically concluded that the solution to the complex problems of the Depression was for the private sector to put the unemployed back to work.

Relative to others, the Hisses were living very well, but the grim lot of the have-nots was disturbing to them. On many streets in New York City were soup kitchens run by such charitable missions as the Salvation Army where long lines of hungry men stood waiting for a free meal. There was one such kitchen near the Hiss apartment on Claremont Avenue near Columbia University run by the Socialist Party. In sympathy with the plight of the unemployed, Priscilla made sandwiches at this feeding station and also attended some Socialist Party meetings, occasionally in the company of Alger and on one occasion in the company of her brother, Tom Fansler, who was then unemployed. Objecting to the speaker referring to those assembled as "comrade," Tom Fansler stood up and shouted, "Don't call me 'comrade!'" He was pulled back to his chair by his embarrassed sister.

In 1932 Priscilla registered as a Socialist for voting purposes—although she claims she never actually joined the party. During this time Alger became a member of the International Juridical Association (IJA), which put out a publication modeled after the *Harvard Law Review*. This organization was later to appear on the FBI's subversive list.

The good living that Hiss was making from his work at the law firm and the fact that the Hisses were living together again and did not have to maintain separate residences in different cities allowed them to move to a much nicer apartment on Central Park West. At night Hiss made a public-service contribution with his writings about sharecroppers for the IJA, while Priscilla was busy raising a child, writing her book, and doing charitable work. All things considered, it was a good time for the newly married Hisses.

Hiss was content with his work in private practice. In fact, he had already turned down an appointment in the Hoover Administration in the Justice Department. However, when he was twenty-eight years old and the Roosevelt era was beginning, Hiss received a call from Jerome Frank, an attorney who was working in the Department of Agriculture. Frank was then recruiting lawyers for his staff. Hiss had been recommended to him by Felix Frankfurter. At first, Hiss declined the offer, saying he had just joined the law firm in New York and was not in a position to leave them so soon. Shortly thereafter he received a cable from Professor Frankfurter, who implored him to answer the call of his government on the basis of a "national emergency." Hiss did just that, leaving his wife and stepson in New York and taking up quarters at the Racquet Club in Washington, D.C., in May 1933.

At the Agriculture Department, Hiss was surrounded by other bright young men who were filled with the same enthusiasm he had for his new work. He knew several of his co-workers before coming to Washington. These men were given a chance to solve some of the problems they had merely argued about while attending prestigious Ivy League colleges. There were such men as Nathan Witt, Henry Collins, John Apt, Charles

Kramer, and Lee Pressman. These five men, along with Hiss and his brother Donald and sundry others, were later named by Whittaker Chambers as being members of his Communist apparatus in Washington.

Indeed some were admitted Communists, such as Lee Pressman, an intellectual Hiss knew from the *Law Review*. When Pressman became counsel to the CIO, he often found the Communists' alliance with labor helpful to the union's cause. Also, before World War II, the Communists were—admittedly on an "on and off" basis—one group actively resisting Hitler. This was one of the party's biggest attractions for many Jews, like Pressman, and other liberal thinkers. As prosecutor Tom Murphy was so accurately to frame it a dozen years later, "There were people who felt that the advance of Nazism and Fascism . . . was being stemmed or stopped by nobody but the Russians. . . . So you can see how a person of Chambers' intellect . . . could become . . . involved with that type of thinking of these foreign philosophies."

Hiss worked in the Agricultural Adjustment Administration as assistant general counsel to Jerome Frank. Since food and the other agricultural products are basic commodities, one of Roosevelt's priorities was to guarantee that the farmers received a fair price for the fruits of their labor. This was done by artificially manipulating the laws of supply and demand. Some farmers were paid *not* to grow—or not to grow as much of—certain crops, which decreased the overall supply and provided a better price for those farmers who were growing the crop.

Hiss was assigned to the Cotton Section, where as an attorney he made sure things were done according to law. Roosevelt's planned economy worked much the same way he hoped it would with farmers—or at least landowners—receiving better prices for their crops. There was an inequity, however, in this system of parity. Sharecroppers—the South's answer to the abolition of slavery—who did much of the work on the larger farms and plantations saw little or nothing of such government funds. Since 40 percent less land was in cultivation, fewer sharecroppers were needed to plant cotton, and those who joined political associations to protect themselves from greedy landowners were often labeled "nuisances" and evicted from the lands they worked.

It was a bitter struggle between the landowners and tenants, and Hiss sided with the latter group; in fact, he wrote the legal opinion that delineated their rights. Eventually, Secretary of Agriculture Henry Wallace—later to be the champion of the common man—decided in favor of the landowners. This signaled the end of the legal division where Hiss worked. Jerome Frank, who sided with Hiss in this matter, was fired and the division he headed abolished.

At the time of the shakeup, Hiss was on loan to the Nye Committee, a Senate committee headed by Senator Gerald P. Nye, which was investigating the munitions trade. The Nye Committee was established in 1934 at a time when Hitler was rearming Germany in direct defiance of the Treaty of Versailles, which ended World War I. With wonderful capitalistic logic, the Nye Committee reasoned that the way to end wars was to take the profit out of them. In the course of their investigation, the committee discovered that several American companies were selling products that could be easily converted to military use in Hitler's Germany. The Communists found the Nye hearings of more than passing interest to their propaganda machine and followed the work of the Nye Committee closely.

It was during Hiss's work here that the man he would finally know as Whittaker Chambers entered his life. According to Hiss, "George Crosley" told him he was a freelance writer of magazine articles, then working on a story on the Nye Committee, which he hoped to sell to *The American Magazine*. He wished to see the transcripts of the Nye Committee's hearings and also wished to ask Hiss some questions about them and the committee's work. Such requests were normal and were Hiss's responsibility. He cooperated with Crosley, whom he came to like. Crosley was a good raconteur—although Hiss took exception to his stories of sexual conquests—and he reminded Hiss of his brother Bosley.

There was an occasional lunch together. Later, when Crosley told Hiss he would be staying around Washington for a few more months to complete his articles and would need a place to live, Hiss, who was moving to a new apartment, reportedly offered the writer his old apartment on 28th Street. According to Hiss, the lease on the place ran for another three months and, besides, the rent had already been paid. Crosley could pay him the rent when he sold his articles. The story Chambers told about their association was much different. Chambers was there, he said, to set up a Communist apparatus, joined by men like Hiss, who would rise to high positions in government and then influence policy in keeping with Communist ideology. On the way up, Chambers added only after Hiss filed the libel suit, these men would turn over secret documents to the Soviets.

In 1944 Hiss joined the State Department and served with distinction at the Dumbarton Oaks Conference. This conference was among the Big Three Powers—the United States, Britain, and Russia—and was held to discuss proposals for world peace and security just prior to the end of World War II. From this conference came the groundwork for the United Nations.

The UN was further discussed at Yalta, a Russian resort on the Black Sea, where Hiss was again an administrator for the U.S. delegation. Hiss similarly served as Secretary General of the San Francisco Conference, where the UN charter was signed in 1945. It was Hiss who brought the charter back to Washington to present to President Truman. Hiss's picture appeared in *Life* magazine, taken as he disembarked from an Army plane carrying the UN charter, which was fitted with its own parachute and a forwarding address in the event of a crash. Hiss himself was provided with no such safety equipment. For his important but certainly not pivotal contribution to the UN, the right-wing John Birch Society called the United Nations "The house that Hiss built."

When his United Nations work was completed, it seemed to Hiss to be a good time to leave government service. He had never planned to work in the public sector, and after joining the government in the "national emergency," he hardly expected to stay there as long as he did. Hiss decided to rejoin his old law firm, Choate, Hall and Stewart in Boston. He advised his superiors in the State Department of his plans. The Secretary, Edward Stettinius, asked him to stay on for a while longer, at least until the next session of the United Nations in London, scheduled for January 1946. Hiss agreed to do so.

Around this time he received a telephone call from the naval commandant in New York, who told him that his stepson, then in the Navy V-12 program, was in serious difficulty. Tim Hobson had claimed that he was a homosexual, said the officer, and wanted an immediate discharge from the Navy. Deeply concerned and somewhat shocked by this report, Hiss managed to get young Hobson transferred to a naval hospital on Long Island, where he underwent psychiatric treatment. Shortly thereafter Hobson was discharged from the service and departed for the West Coast, breaking all ties with his family. In California he worked in television for a time, and a few years later he returned to New York to become a television set designer.

When his stepfather's name appeared in the headlines in 1948, Hobson contacted Hiss and said he would be available to testify at the trial. He said he could contradict Chambers' claim of numerous late-night visits to the family home in Georgetown. As recounted earlier, Hiss refused his stepson's generous offer because he feared such an appearance might jeopardize Hobson's future. Hiss told his attorneys, "I'd rather go to jail than let the boy testify." There are a few who conclude that this is precisely what happened: In keeping Hobson off the witness stand in order to avoid public knowledge of his homosexuality, Hiss went to prison.

Eventually, Hobson went to college at New York University, working at night at Bellevue Hospital to put himself through. Hiss was unable to help him financially at this time because he was in jail, and his resources were depleted by the two trials. Although a wealthy man, Tim's true father, Thayer Hobson, was unsympathetic to his son's homosexuality and refused to help at all. On graduation, Hobson went to medical school in Switzerland. Today he practices family medicine on the West Coast, is married and the father of four children.

Hiss's wish for an end to his government life took a strange turn at this time. John Foster Dulles, having become chairman of the board of the Carnegie Endowment for World Peace, offered him the position as president of the Endowment. Hiss had worked with Dulles at the San Francisco conference and hadn't much liked him. Dulles, a longtime Republican, was appointed an adviser to the session planning the UN in order to give the American delegation a bipartisan look. Adlai Stevenson, the eventual Democratic standard-bearer, was in charge of press relations. Fearing this delicate and important mission might be scuttled prematurely by sensational news stories, Stevenson told the press as little as possible. Stories were leaked, however, and what appeared in the papers seemed to be consonant with the Republican view of the world. Hiss concluded that Dulles was the source of these leaks and was angry about it. In considering Dulles' offer, Hiss decided Dulles' chicanery was politically motivated and had little to do with his character, and he accepted the position.

On returning from London in the summer of 1945 at the conclusion of the UN meetings, Hiss learned from Secretary of State James Byrnes that he was in danger of being called a Communist by two Congressmen. Byrnes advised Hiss to call J. Edgar Hoover immediately. Hiss did so and was interviewed by one of the director's lieutenants. Hiss remembered the interview as being rather "perfunctory." Byrnes recalled either being told over the phone or reading a report to the effect that the FBI had given Hiss a "clean bill of health." In spite of clearance from the FBI Byrnes continued to have doubts about Alger Hiss.

This was not the first time that Hiss fell under suspicion; many people in government—especially New Dealers—were being called Communists by someone or other. Stories about Hiss were first heard in 1939, even before Whittaker Chambers told his story to Isaac Don Levine. A report had come to William Bullitt, American ambassador to France, from Premier Edouard Daladier saying that French intelligence listed Alger and Donald Hiss as Soviet agents. This report was fairly well known in the State Department, as was the fact that Igor Gouzenko, a code clerk in the Soviet Embassy in Ottawa, had stated that he had been informed that the Soviets had an agent who "was an assistant to the then Secretary of State Edward R. Stettinius." There was no question that the finger pointed directly to Hiss.

Even before Chambers became important in the anti-Communist crusade, the name Alger Hiss was so well known in this regard that the *Christian Science Monitor* could write in 1946, "More than one Congressman, whenever the subject of leftist activity in the State Department is mentioned, pulled out a list of suspects headed by Mr. Hiss."

Secretary of State Byrnes had a sister, Mrs. Leonora Fuller, who worked with Hiss, Lee Pressman, and John Abt at the Agricultural Adjustment Administration. (The three would later be called Communists by Chambers.) Mrs. Fuller told her brother that she considered these men to be "leftist thinkers." According to the FBI files, Byrnes came to believe that Alger Hiss was not "politically sound." The Secretary mentioned this to John Foster Dulles, chairman of the board of the Carnegie Endowment, when the organization was considering Hiss for its presidency. Byrnes told Dulles that he believed "Hiss's political thinking was entirely different from that of the persons who founded that body."

Others were telling Dulles that he was on the verge of hiring a Communist. One was Lawrence Davidow, who came to know Dulles in 1946. Davidow also became friendly with Ben Mandel, who, when they first met, was working for the security division of the State Department. Mandel was to become well known when he served as the chief investigator for HUAC during the Hiss–Chambers affair.

Davidow, in a conversation with Mandel, mentioned that he was personally acquainted with Dulles. Remembering this conversation, Mandel wrote to Davidow in November 1946 saying that Hiss was being considered by Dulles for the position of president of the Endowment. Mandel claimed there were people in Washington who could prove that Hiss was a Communist. Davidow passed this information on to John Foster Dulles. Shortly thereafter, the FBI interviewed Davidow in his home in Huntington Woods, Michigan. He told the agents that after telling Dulles about Hiss, "Mr. Dulles replied. . . that he knew Hiss personally and could vouch for his patriotism."

Despite the fact that Dulles said there was nothing to these rumors, Davidow supplied the information to radio commentator Fulton Lewis, Jr., who used it in a broadcast. After the program the FBI interviewed Ben Mandel, who said his source was Isaac Don Levine (the right-wing editor whose testimony before the House committee about Lawrence Duggan allegedly contributed to the latter's suicide). Mandel admitted he himself had no firsthand knowledge of Hiss's politics.

Mandel was a man not unlike Chambers; in fact, the two were friends. Mandel was born in New York City and taught public school there. In the 1920s he joined the U.S. Communist Party and worked as business manager for the *Daily Worker,* the same party organ Chambers wrote for. Chambers knew Mandel by his alias, Bert Miller. Chambers wrote in his book that it was Bert Miller who issued him his Communist Party identification card.

Following a Stalin purge in 1929, Mandel and many others were expelled from the Party. As a consequence, he became an avowed enemy of Communism. In this regard he would later write an anti-Communist book *(I Was a Soviet Worker)* and numerous pamphlets (e.g., "A Handbook for Americans: The Communist Party of the United States of America, What It Is, How It Works"). Learning of his knowledge of and animus for Communists, HUAC hired him as an investigator in 1939. He left the committee six years later to work in the State Department, where he was employed until 1947 before rejoining HUAC to assist them with the Alger Hiss case.

Mandel's stay at the State Department overlapped Hiss's. Hiss believes that Mandel may have been responsible for the theft of the State Department documents which Chambers attributed to him. In an interview Hiss said, "My theory is that Ben Mandel had something to do with it. Mandel worked for the committee. He worked for the State Department—although I didn't know it at the time—in their security division. He would have easy access to any kind of old documents he wanted." Mandel died in 1973 at the age of eighty-two, and Hiss's accusation remains unevaluated.

Hiss, not wishing to resign under a cloud of suspicion, remained in the government until the fall of 1946. While Hiss was still in Washington, Dulles received a phone call again impugning Hiss's reputation, this time from Alfred Kohlberg, an acquaintance of a trustee of the Carnegie

Endowment and the founder of *Plain Talk*. He told Dulles that Hiss was a Communist. Dulles asked for proof, and Kohlberg said he would get it. A short time later he wrote back saying that the only proof was in the "files of the FBI."

Kohlberg's source was Isaac Don Levine. Levine's source, of course, was Whittaker Chambers, who told him his story in 1939. It was Levine who took Chambers to see Assistant Secretary of State Adolf Berle—a meeting that triggered the Hiss case years later. When Kohlberg asked Levine for proof of Hiss's Party membership, the editor of *Plain Talk*, protecting his source, who didn't want his name mentioned, said there was no documentary proof other than in the FBI files.

Despite Kohlberg's inability to produce anything useful, Dulles had grave doubts about the man he had just employed. He checked with Kohlberg's friend, Republican Congressman Walter Judd of Minnesota, who was said to have proof of Hiss's Communist activities. Judd wrote back that he had such information. Dulles showed the letter to Hiss, who responded that there was no truth to it; he was a victim of guilt by association—i.e., such charges stemmed from his employment at the Agricultural Adjustment Administration, where he worked with several so-called radicals. Dulles dictated Hiss's response for his files, and then checked with the security division of the State Department, which said of Hiss, "[We are] convinced of his complete loyalty to the United States."

Speaking of the above conversation, Hiss testified to the House committee in August 1948: "He [Dulles] said he had heard reports that people had called me a Communist. I can only assume that he was satisfied there was nothing to report." Hiss continued, "When I accepted the post, I checked with Mr. Byrnes specifically to see whether he thought the issue had been laid to rest. It was his impression, as I recall it, that it had been entirely laid to rest."

Congressman Karl Mundt then noted that Hiss had not *volunteered* the information; Dulles initiated the conversation. Mundt was trying to make the point that Hiss hid this information from his prospective employer. There was this heated exchange:

MUNDT: *Did you bring it up or did Mr. Dulles?*
HISS: *Mr. Dulles called me up and I discussed it with him.*
MUNDT: *So it still stands for the record that you of your own volition did not bring this matter up with your prospective employers.*
HISS: *It stands for the record the way I testified.*
MUNDT: *That is the way you have testified.*

Others were calling Hiss a Communist at this time, but their source was that same recanted Communist, Whittaker Chambers. In 1946 Arthur Schlesinger, Jr., along with Barbara Kerr, a researcher with Time-Life, interviewed Chambers for an article on the Communist Party in America. Chambers told them the same story he had been telling for years. One name he mentioned was Alger Hiss. It was a dramatic story, and he told it well. Mrs. Kerr felt it was the truth because, she concluded, Chambers had no reason to lie. Besides, he seemed to be in fear of his life, as he insisted that no names be used in writing the account, and he requested that Mrs. Kerr remove her notes from her office lest they be stolen.

Barbara Kerr did not forget the name Alger Hiss. She had once been interviewed by him for a State Department job which she did not get. Hiss impressed her as being an "uptight pukka sahib type." At a New Year's Eve party in 1947, she told some people who were discussing loyalty oaths, "I don't see the point of loyalty oaths; they don't catch the dangerous people like Alger Hiss." The conversation broke off abruptly, and Mrs. Kerr, who had much to drink that night, realized too late that one of the men listening was Edward Miller, who worked in the State Department and was well acquainted with Hiss. Miller demanded an apology on the spot or else, he said, Hiss would sue for libel. Miller later told Donald Hiss about her comments, and Donald mentioned this occurrence to his brother. A few days later Miller phoned Mrs. Kerr, who said as far as she knew the story was true but promised not to mention it again.

Despite such warnings and Dulles' own doubts, Hiss went to work at the Carnegie Endowment. The Hisses, who now numbered four with the addition of Alger's and Priscilla's child, Anthony, moved to an apartment in New York on East 8th Street, where Priscilla still lives today. Hiss commuted uptown to his office, which was situated near Columbia University. Priscilla took a job teaching at the Dalton School, a progressive private institution on the upper East Side.

In June 1947 two FBI agents visited Hiss's office at the Endowment. He was asked if he was acquainted with any of a long list of people whose names they read. One of these names was Whittaker Chambers. Although Hiss said he knew no one by that name, he later said he remembered it because he knew a Bob Chambers at the AAA and a Tony Whittaker at Harvard. This was perhaps the most exotic event in Hiss's quiet existence in 1947. Much happened to Hiss, however, in the year that followed. His whole life came apart, for with the well-publicized HUAC hearings, the name *Alger Hiss* became a household word and the symbol of a generation.

Then followed the now familiar testimony by Chambers to HUAC, the stormy hearings, and finally the denouement at the New York grand jury. When Hiss's indictment was handed down by that investigative body on December 15, 1948, his confidence was not shaken in the least. He looked forward to the trial with unbounded optimism, a fact that surprised both his attorneys and friends. When the first trial resulted in a hung jury, Hiss was certain he would be vindicated at the second. When he was convicted there, he was sure he would win on appeal. After being released from prison in 1954, he believed in and worked tirelessly for his eventual vindication. Today he expects a writ of *coram nobis*—a very rare occurrence in the judicial system—finally to wipe the slate clean.

If Hiss suffered from boundless confidence in his ability to surmount life's obstacles, the same could not be said of his wife, who was horrified by what was happening to her family. The press and photographers surrounded their Greenwich Village apartment like vultures. The subway car taking them to the courthouse on Foley Square for the first day of the trial (May 31, 1949) looked as though it were under siege as flashbulbs exploded mercilessly about them.

Priscilla was unable to deal with the sudden notoriety. She approached the trial fearfully. From the moment Hiss's friend and libel-suit attorney Bill Marbury announced at a quiet supper that Alger was likely to be indicted, she was stricken with panic. She wondered aloud where the money would come from to pay for the trial, whether she would have to give up the apartment and automobile and whether their son Tony could continue in private school.

Hiss said Priscilla considered herself a Quaker. For Quakers there is a strong antilitigious tradition. She considered a trial an "impropriety," said Hiss, where one's troubles are aired in public. Hiss told his son Anthony, in the latter's book on his father, "No, I, myself, was not frightened. . . . But Prossy [Priscilla] did go into a type of collapse. I tried to impart courage to her. I didn't catch fear from her. . . . Once my old friend, Bill Marbury, said, 'Alger, you're going to be indicted,' there was this constant, I guess I'd call it 'free-floating' anxiety that was impossible to calm. If you eliminated one damned thing, talking rationally and reasonably, there was always another."

Before the trial, Hiss put in long days with his attorneys and then came home to Priscilla and spent long hours at night trying to quiet her fears. He would take long walks with her because she was convinced their apartment was "bugged." Not surprisingly, the Hiss lawyers found Priscilla a very difficult witness to work with. Hiss is certain Priscilla's diffidence with his attorneys and later on the witness stand was because of religious beliefs, but others, such as John Chabot Smith, believe that Priscilla's anxiety stemmed from the fact that she must have considered herself an unindicted co-conspirator because it was she who allegedly typed Chambers' incriminating documents. A guilty verdict against her husband was in a sense a confirmation of her own guilt. However, when Smith asked her about this, she said she had never considered the idea. This denial notwithstanding, one of the most persistent rumors heard during the trials was that Hiss was covering up for his wife, who was, if anything, alleged to be a much more rabid Communist than he was.

After his conviction and after his appeal was denied, Hiss went to see Austin McCormick, then head of the Osborne Association, the prisoner's aid society, to find out what he could expect in prison. When a friend asked if he wanted to see the penologist, Hiss's reply was classic: "Absolutely! If I were a nineteenth-century traveler about to set out for Arabia and somebody asked me if I'd like to see Doughty, I'd be very pleased." A meeting was arranged.

McCormick, like Hiss, had served in the New Deal. He had been the former deputy director of the Federal Prison Bureau. To get firsthand knowledge of the prison environment and the life of men he was to advise in his work with the Osborne Association, he voluntarily had himself incarcerated on several occasions.

At their meeting, the two men discussed where Hiss was likely to be imprisoned. Lewisburg was McCormick's guess because Danbury (Connecticut)—the federal penitentiary *nearest* Hiss's home and thus the logical choice—was considered a country club, and Hiss was regarded as too notorious a felon to serve what was held to be "easy time."* Atlanta was thought of as too rough an institution for political prisoners. McCormick was right. On March 22, 1952, Hiss went to the federal penitentiary in Lewisburg, Pennsylvania.

They also discussed Hiss's probable work detail. When Hiss expressed a desire to work in the infirmary, the penologist dissuaded him. He said Hiss would have easy access to drugs there. If he gave them to his fellow inmates, he could get in trouble with the authorities; however, if he refused, he would be unpopular with the prisoners. Hiss

---

*Watergate burglar, G. Gordon Liddy served part of his sentence at the federal penitentiary in Danbury.

suggested then that he could teach in the prison school. McCormick opined that would be unacceptable because someone would conclude that he was teaching Communism. Hiss suggested the library, but McCormick said that would be viewed as too easy a work detail. McCormick suggested Hiss work in the storeroom, which is what he did, sometimes unloading heavy sides of beef but often having the time and permission to read books from the prison library or the two magazines he was allowed to subscribe to: *The New Yorker* and *New Statesman*. The first book Hiss read in prison while being isolated for the first month in what was called "quarantine" was *Don Quixote*, a not surprising choice for someone of his temperament.

Hiss described his prison stay as not particularly difficult. It was little more than "cramping," he said, and he also claimed that he learned as much there as he did in his four years at college. He was friendly with the Italian inmates who ran the prison. Hiss had described his fellow inmates as "interesting." He was known to the younger men there as "Pop," and he played a lot of baseball during this time. In prison parlance he "kept his nose clean." Many of the prisoners had heard of him before he entered Lewisburg, and they looked forward to meeting him. In a way he was a celebrity. It was said that Hiss would talk easily to murderers and thieves alike. On his release, Hiss considered writing a book on etiquette for political prisoners but ultimately dropped the project. He served three years and eight months of a five-year sentence, actually a longer period of time than a less notorious prisoner was likely to serve. On November 27, 1954, he was met at the prison gates by his wife and their thirteen-year-old son. Hiss looked considerably thinner, and his face was more angular and somewhat harder in appearance. He was then two weeks past his fiftieth birthday, and the wide-eyed look of omnipotence was gone.

Priscilla looked older, too, having gained some weight in the ensuing years. While Hiss was incarcerated, Priscilla worked in the basement of the Doubleday Book Shop, located on Fifth Avenue in New York. Her take-home pay was $37 a week. She had some friends from her work at the bookstore, but her son, Anthony, said there were no special men friends. He also said when Hiss returned home and met Priscilla's new friends and they liked him, she dropped them.

Hiss stayed with Priscilla for five more years. She suggested the two of them leave New York and change their names to escape the notoriety. Hiss's parole officer also recommended that he change his name. While Priscilla was away at work and before Hiss started writing his book, *In the Court of Public Opinion,* he had much free time, which he particularly relished after nearly four years in jail. He was often seen about town in the company of an attractive foreign actress. Hiss said he never slept with her because he was a married man. Eventually, Priscilla became angry over the relationship, and Hiss, out of loyalty to his wife—a very important quality in their relationship—stopped seeing the actress.

When Anthony Hiss was a senior in prep school in Vermont, Priscilla, supposedly on the advice of her psychiatrist, asked her husband to move out of their apartment. He said he would not leave until their son's Christmas vacation from Putney School, but Hiss began sleeping in the living room. Once again he started seeing his actress friend. In January 1959 he moved out for good. By this time Anthony Hiss had been accepted at Harvard. Hiss tried to dissuade his son from going there, believing that Harvard had lost its edge and the pendulum of academic excellence had swung to the West Coast at Berkeley. But Anthony was not persuaded and chose to enroll at his father's alma mater.

Priscilla was later to become very bitter over Hiss's departure. She refused, and has refused to this day, to give him a divorce. Hiss stopped seeing the actress when she decided his notoriety was hurting her career. Moreover, she felt Hiss was too bossy. Today he lives with another woman in an apartment not far from the one he shared with his wife. Priscilla still lives in the same apartment on East 8th Street. The card at her doorbell reads, "MRS. ALGER HISS."

After Hiss completed his book, he began looking for work. His financial matters were mightily complicated by the fact that he was a convicted felon. As such, he was either undesirable to many employers or legally ineligible for many positions. Also with his conviction he had lost his license to practice law. To add to his difficulties, his book was neither a financial nor a critical success. Eventually, Hiss was offered a position with a talent agency which represented blacklisted artists. He was told the position as an executive assistant was his *if* he could learn shorthand and typing.

Hiss enrolled in business school. He was terrible at speedwriting and had even less success with typing. Commenting to his son, Tony, on his complete lack of success at the keyboard, he said, "I typed in unknown languages. My teacher gave me special sessions. She said, 'You've got a block about typing—maybe because of that damn case. . . .'"

Hiss finally got a job with a small manufacturing concern that imported and sold something called "Feathercombs." The owner of the company was aware of his financial situation and figured he could get Hiss cheap. He was right. Hiss went to work there as office manager for $6,000 a year—a huge comedown from the $20,000 he commanded at the Carnegie Endowment a number of years back. In his second year on the job the business began to fail. During this unhappy time, Hiss made the mistake of making allegations against his employer in front of other employees. For all practical purposes, his career with Feathercombs was over. On losing his job, Hiss had the first of two heart attacks.

He was unemployed during the year 1959, which was also the year he was to leave Priscilla. Again it was a lean time for Hiss, for, in addition to his other troubles, he had been denied his government pension of $61 per month. Congress had passed a bill that came to be known as the "Hiss Act," denying him the money.

During this time he happened to mention to a friend that he needed work. The friend, in turn, contacted another friend, Sam Chernoble, a Republican National Committeeman who ran a very successful printing business. Hiss got along well with Chernoble and learned the business quickly. He made a mildly successful career as a stationery salesman. Speaking about this, Hiss commented to the authors that his well-known name was an asset in his work. "I got in to see people who otherwise wouldn't have seen me," said Hiss. "It was an asset. I didn't always make the sale, but I don't think anyone I called up and said I'd like to see them as a salesman brushed me off. Maybe their interest was to have a story about me to tell at dinner. . . ."

In 1970 Hiss, in association with the American Civil Liberties Union, petitioned the government for the pension that had been denied him when he was released from prison. The judge who heard this matter was Roger Robb, a Nixon appointee. He ruled the so-called Hiss bill unconstitutional, and Hiss began belatedly to receive his pension. Ironically, when Congress was endeavoring to deny former President Nixon his pension, they were prevented from effecting this because of the Robb decision.

Alger Hiss was readmitted to the Massachusetts Bar in the summer of 1976. He was not exonerated of the charges for which he went to prison, however. In a unanimous decision, the judges of the Supreme Judicial Court of Massachusetts wrote: "We consider him to be guilty as charged." The basis for reinstatement was the fact that Hiss was considered to have paid his debt to society and had been rehabilitated thereby.

Hiss now describes himself as a legal consultant, and although still not licensed to practice in the state of New York, he works out of his New York apartment.

Debates about Hiss's guilt go on and on. As recent as April 1976 *Harper's* did a reprise of this very issue in an article provocatively entitled "The State of the Art of Alger Hiss." Feeling that this case was "tending toward final judgment," Philip Nobile, an author and editor, asked approximately 100 lawyers, journalists, and intellectuals ". . . If you had to pronounce Alger Hiss guilty or innocent, what would your verdict be?" With an approximately 50 percent return, the group split rather evenly, with some seventeen respondents feeling Hiss was guilty, fourteen voting him innocent, and six identifying themselves as being unable to make up their minds. What is strange is that liberals and conservatives are to be found in all three categories, and this was surely not the case in the late 1940s.

For Hiss, and understandably, the world is divided into two camps: pro-Hiss and con-Hiss. The New York *Times* seems to fall into the anti-forces. Hiss said in 1978, "Certainly the New York *Times* has clearly taken the position that they will have very little about the case, and let's call it in a pro-Hiss way, until after the court things [the *coram nobis* petition] get settled." And then Hiss added, "This is the same view the [TV] networks are taking."

There have been many books (at least thirteen) written on the case, and some authors get high marks while others fare less well with Hiss. Alistair Cooke, who was the first to approach the case in a cultural context, is no longer highly regarded; Hiss feels that Cooke's interest in the issues ended completely with the first trial. Moreover, Hiss cannot settle comfortably for Cooke's "generation on trial" motif. Although he sees himself as a victim of New Deal antipathy, Hiss still insists that there was nothing "personal" in what happened to him and devoted his own book to a dry legal appraisal of some paralyzingly human issues. Here it would seem that Hiss wants it both ways; he admits to mistreatment but not as a person. Rather, he is victimized because of his beliefs without regard to his actions. This is the position that Hiss has maintained for better than three decades, and it has provided him with the kind of necessary vacuum within which he can live because it reconciles all of the contradictions that so trouble other observers.

The Earl Jowitt, whose book *The Strange Case of Alger Hiss* was an effort to contrast Hiss's fate in America with his legal prospects in England and which came up with a favorable Hiss balance, understandably remains a Hiss favorite. Hiss had been annoyed by the fact that the manuscript was given by the publisher to Sidney Hook, then a philosophy professor at New York University, for review and that Hook, never sympathetic to Hiss, had suggested a number of factual revisions which dampened Doubleday's enthusiasm for the book.

Hiss's attitude toward Meyer Zeligs has been more ambivalent. According to Hiss, Zeligs had originally approached Hiss with the request that he cooperate in a book Zeligs was writing about Whittaker Chambers. When the book was expanded to include Hiss, purportedly at the request of Zeligs' publisher, Hiss reluctantly agreed to participate. Although Hiss described the book to the authors as an "embarrassing" one, he felt that Zeligs had accumulated the most complete file on the case. In the early 1970s, Zeligs decided to do a biography of Hiss and a psychobiography of Nixon (according to Hiss), and Zeligs and Hiss appeared together at a press conference in San Francisco in 1976. Although Hiss has remained cool to Zeligs' psychohistory effort, most of those individuals associated with Hiss's cause seem to be inordinately fond of *Friendship and Fratricide*. It is a book that was recommended to us a number of times as we spoke with Hiss sympathizers about the case.

There can be little question that John Chabot Smith's recent biography of Hiss, subtitled "The True Story," draws high marks from the Hiss circle. Smith is notably sympathetic to his subject, and the book is well written and has served to focus the revival of interest in the case. Smith, who once described Hiss as "a saintly man," did not have access to the FBI papers, and the "dated" quality of his book probably detracts from its merits.

Hiss has gone out of his way to avoid a public confrontation with Allen Weinstein, whose book *Perjury* sought to end all of the bickering by finding Hiss guilty. Hiss has stated privately that Weinstein "has done a disservice to public information" and also described Weinstein as "not unintelligent" but as "manipulative and biased."

Matthew Josephson, who once contemplated doing a biography of Hiss, wrote to Hiss, "When there is such mountainous material as in your case, if you have any kind of thesis you can seem to find something to support it somewhere." Perhaps this is why so many people have found so much to write about with regard to the case itself. And perhaps this is why Hiss, who has been so thoroughly examined over the years, is unusually sensitive to the issue of friends and enemies. There is no middle ground for Hiss. His very soul has been caught up in his political, social, and legal predicament; and all of his own disclaimers notwithstanding, he has been engaged for many years in a fight for his own life.

Hiss attempts to manage this by assuming an insouciance that somehow never seems completely to come off. Thus, he described his current life situation in hyperbole to *Life* reporter Thomas Moore, in the spring of 1972, as made up of "good wine, good food, Havana cigars—I have a depraved taste for them—and good thought go together." Although Weinstein found Hiss to be "relaxed, mellow, and amiable," it is hard for others who know him to miss the hard-driving, brittle, and angry qualities.

And not unlike the rest of the world, Hiss surrounds himself with friends and avoids those who are not fully in support of "the Case," as both he and Whittaker Chambers, as well, came to call it. The line between friends and foes is not always clearly demarcated, and Hiss not infrequently has been quick to pick up on even tentative apostates. Failure to understand the core issues or raising questions about some of the tightly held Hiss assumptions about the case bring clear evidence of impatience and even flashes of anger. Hiss has been known to brusquely turn away from an honestly curious interviewer with the statement "I'm pretty busy today; perhaps you ought to talk to someone else about that issue."

At such times one sees Hiss's sorely tried quality and his tired look, and it is almost possible to smell the odor of defeat. Then there is no mention of the good life, no wine and cigars and warm friends, just the sound of another name being registered somewhere as a member of the anti-Hiss forces.

The question of guilt aside—at least until the final chapter—the Hiss case surely marked the end of an era of innocence in the American body politic. We do not refer to the constitutional outrages of the McCarthy period; that linkage is so direct that it hardly requires further acknowledgement. Even more important was the impact of the trial and conviction on the progressive movement in the United States. It put to rest once and for all what Leslie Fiedler once called the implicit dogma of American liberalism: "that the man of goodwill is identical with the righteous man, and that the liberal is per se, the hero."

For many liberals, the sense of identification with Hiss was a strong one. Even a "softness" or sympathy toward Communism was not entirely retrograde. Most liberals had been "close"—some frighteningly close—to the Communist ideology.

Hiss's idealism was for this group a very strong element in their support of him. They recognized that reactionaries easily confuse idealism with Communism. Someone—Diana Trilling, we think—wrote that reactionaries assume that every idealist is a crypto-Communist. By analogy, Hiss was seen simply as a very large and symbolic victim of the above. For this liberal cadre Hiss had to be viewed as innocent, since his guilt threatened to undermine their own ideological agreement with him.

The Depression, the Spanish Civil War, the growth of European Fascism, the Russian fight against the German forces in World War II, all the central events in the third and fourth decades of the century tended to make the Russian political system attractive to some liberals. Franklin D. Roosevelt's decision to recognize the Soviet Union shortly after his inauguration brought him considerable support from the liberal movement, which was sorely tried by both the Red-baiting and isolationist tendencies of previous Administrations.

Still, Russia's attitude toward civil liberties within its own borders, its attack on Finland early in the war, and its flirtation with Nazism as illustrated by the Soviet–Nazi pact served to disenchant a large segment of the progressive movement in America.

Some who had avoided the flirtation with Russia found themselves with strange bedfellows in the Hiss affair. For these folk, Hiss seemed to epitomize the gullible man whose acts of indiscretion imperiled the *entire* liberal movement in the country. In this camp were some disillusioned former Communists and some who were originally sympathetic to the Communist position on certain ideological issues. Also here were found a few who throughout the years had felt that their vision of Russian despotism was confirmed by Hiss's seduction.

This entire group reacted sharply to the Hiss trials. Diana Trilling, writing in the *Partisan Review* shortly after Hiss's conviction, stated in somewhat reckless fashion, "Hiss, of course, has been found guilty not only of being a Communist, but of being a Communist spy." The first accusation was certainly not true; it was on this issue that Hiss tried to fight, and the government was wise to avoid the trap. What is evident is Diana Trilling's rage at the mess in which Hiss had left the liberal movement. She wrote:

> The only morality to a Communist is revolutionary morality, and according to revolutionary morality, Hiss performed a moral act because he was furthering the revolutionary goal. It is interesting to study why someone like Hiss who was bred by standards of bourgeois morality should have switched to so different a moral code; but such a study has only a coincidental relevance to

> his objective acts. What is immediately pertinent to his acts is his ideas. In lying and stealing Hiss took the fullest responsibility for his political ideas. He contemplated where his ideas might lead, and he was nevertheless willing to have these ideas and perform his acts. He really understood the reality of politics.

This statement should be compared with Richard Nixon's evaluation of Hiss's character on page 296. The conclusion is inescapable: Both sides of the political spectrum were outraged by Hiss's treachery. Diana Trilling and her husband, Lionel, bellwethers of the liberal anti-Communist movement, were and remained antagonistic to Alger Hiss throughout, and Lionel Trilling wrote in the *New York Review of Books* in the very twilight of his own life that Whittaker Chambers seemed to him to have been, in final assessment, an honorable man.

Leslie Fiedler provides another example, a better reasoned one, which illustrates the genteel feeling of anger that the Hiss trials engendered in the anti-Communist segment of the liberal movement. Fiedler constructed an attractive scenario which placed Chambers as a Third Period (Lenin's term for the last stage of imperialism) Communist, and Fiedler professed to believe that Chambers' name-changing, wandering, and stealing were standard operating procedures for rebel-intellectuals in this stage in the world revolution.

Hiss is similarly dealt with. Fiedler finds Hiss the perfect embodiment of the Popular Front Communist, those who emerged as Communism became respectable in the Roosevelt Administration. The Hiss hair cut, pressed suit, and urbane manner illuminated the growing rapprochement between the Administration and the Communist ideology as the New Deal moved American politics to the left. The fact that this movement opened university and governmental doors to Communists is the major point Fiedler makes: Men like Hiss walked into high position and thus could be easily corrupted by men like Chambers.

The theory is not a new one, but it still continues to intrigue many observers. Fiedler, a fine essayist and critic, is unfortunately a less successful novelist, and not surprisingly he is put off by Hiss's refusal to provide the elements necessary for the satisfactory ending to the tale. Hiss's failure to *confess* irritates Fiedler greatly, and Fiedler concludes:

> It is difficult to say what factor is most decisive in cutting Hiss off finally from the great privilege of confession; opportunism or perverted idealism, moral obtuseness or the habit of Machiavellianism—they are all inextricably intermingled.

In the end he failed all liberals, all who had, in some sense and at some time, shared his illusions (and who that calls himself a liberal is exempt?), all who demanded of him that he speak aloud a common recognition of complicity. And yet, perhaps they did not really want him to utter a confession; it would have been enough had he admitted a mistake rather than confessed a positive evil. Maybe, at the bottom of their hearts, they did not finally want him to admit anything, but preferred the chance he gave them to say: He is, we are, innocent.

The fact is that the damage done to the liberal cause by the Hiss case was in a sense indeed irrevocable, and for a significant number of people the signal element was to be found in the necessity to acknowledge that the worst enemy might well lie within. The liberal movement was badly split, and it has never completely healed over its scars. The confusion sown by the Hiss case marked the end of an age of innocence, not only for the progressive movement but for the country at large. For all intents and purposes, it was no longer possible to make those large character judgments endemic in an earlier era. After Hiss, there would have to be innocent liberals and guilty ones; after Chambers, there were good spies and bad ones; and after Nixon, finally, there were honest Presidents and dishonest ones.

The most recent reviewer of the Hiss–Chambers affair seemed to have fallen heir to many of the problems discussed in the foregoing pages. Allen Weinstein, a historian from Smith College, had reportedly been studying the Hiss case along with the Ethel and Julius Rosenberg trial as part of his interest in contemporary American history. Bit by bit that interest apparently shifted, and Weinstein began publishing more and more on the Hiss–Chambers struggle. His book, *Perjury,* which has attracted so many adherents and foes, brings a certain persuasive aura with it. To wit: It was begun in the belief that Hiss was innocent, but Weinstein reluctantly concludes that Hiss was guilty as charged. On all sides one hears that the book is strong as scholarship, long in length, dotted with thousands of footnotes, appendices and collateral sources, and heavy with new information.

Much has been made of Weinstein's conversion. *Time,* which apparently purchased the rights to delineate the final confrontation between Weinstein and Hiss, describes the scene with all the verisimilitude of Lt. Henry walking away from the hospital after Catherine's death in Hemingway's *A Farewell to Arms:* "After a while I went out and left the hospital and walked back to the hotel in the rain." In the *Time* review of Weinstein's book, Weinstein is quoted as saying to Hiss: "When I began working on this book four years ago, I thought I would be able to demonstrate your innocence, but, unfortunately, I have to tell you that I cannot; that my assumption was wrong." Hiss is reported to have replied, "I'm not surprised."

Weinstein told *Time's* senior correspondent, James Bell, "In the end, Chambers' version turned out to be truthful, and Hiss' version did not. Alger Hiss is a victim of the facts." What appears to have emerged after a half-dozen years is a book badly flawed by the need to arrive at a summary judgment—was Hiss innocent or guilty? It is somewhat uncommon to find a historian reviewing highly controverted evidence and ending up with a legal conclusion. Clearly, only courts can convict individuals, and Weinstein's judgment of Hiss's guilt seems overextended in any event. Moreover, a number of his most important sources of evidence have recanted.

To recall the instance cited earlier, Karel Kaplan, on the basis of eight years' work as an archivist for the Czech Communist Party Central Committee, was alleged by Weinstein to have confirmed Hiss's relationship with Noel Field in the U.S. Communist underground. This was a very crucial assertion, since all involved previously had been unable to link Hiss with Field, who had disappeared into Eastern Europe just as the Hiss case was beginning. If Field had confirmed "guilty knowledge" of Hiss, much of Hiss's reputation as a citizen of unimpeachable probity would have been diminished. However, Kaplan wrote later to another correspondent who raised the same inquiry: "N. Field's testimony, as far as I can remember, did not contain any facts or explicit statements that A. Hiss was delivering U.S. documents to the Soviet Union."

In another example, Sam Krieger, a former Communist, was identified by the historian as having used the alias of "Clarence Miller." He was, according to Weinstein, purportedly the man who proposed Chambers for membership in the Communist Party. These assertions Krieger admits. But, in an assertion vigorously denied by Krieger, Weinstein further wrote that Krieger was the same person as a murderer by the name of "Clarence Miller," who fled the country in order to escape prosecution and ended up in the Soviet Union. Weinstein identified Isaac Don Levine, a well-known anti-Communist writer, and Alden Whitman of the New York *Times* as his sources for this story. Warren Hinckle of the San Francisco *Chronicle* later called Levine, who denied the whole thing, saying, "It was Professor Weinstein who told *me* that this man Sam Krieger was Clarence Miller. I did not know Krieger. How could I identify him?" Similarly, Whitman told Hinckle that he was shocked to hear that Weinstein had identified Krieger as Miller. "It's utterly false that Sam Krieger is Clarence Miller," Whitman said. "I knew them both. Miller was red-haired and a foot taller than Krieger."

The above extended reprises represent only two of a number of such purported inaccuracies in the book. Some individuals such as Ella Winter, who was Lincoln Steffens' widow, denied Weinstein's attribution of a relationship between herself and Whittaker Chambers. Others, such as Maxim Lieber, refuted Weinstein's contention that Lieber identified Josef Peters as the head of the Communist underground in the U.S. Peters himself, who offered Weinstein no information at all, understandably bridled at Weinstein's suggestion that Peters' smile at a certain crucial Weinstein question—dealing with "open" versus "secret" Soviet functionaries—indicated "a certain awareness" of that second category. Paul Willert of Oxford Press disputed Weinstein's allegation that he "protected Chambers" by warning him of efforts against his life. Donald Hiss, faulted by Weinstein for not providing him with information on the finding of the typewriter, denied being asked any questions by the historian about the typewriter. Alden Whitman, supposedly the source for the report that Priscilla Hiss blew up at a Chicago dinner party and said she was tired of all the cover-up, said he "has no memory" of a story emanating from the 1968 dinner party. Sender Garlin, who was mentioned seven times in direct quotes in connection with Chambers having "gone underground" for "special work," insisted he never saw Weinstein nor made the statement to anyone else.

Even Philip Nobile, the indefatigable surveyor of pro-and-con Hiss attitudes, came to view Weinstein's scholarship with some reserve. Nobile, a close enough friend of Weinstein's to have been present at the twenty-first-floor meeting in the editorial office of Alfred A. Knopf when Hiss was delivered the *coup de grâce* by Weinstein, faults the historian for failing to show Hiss any of the documentary evidence which Weinstein professed to have in rich abundance, a procedure that Nobile suggested in an article in *The Nation* "any cub reporter would have done." Describing himself as "once Weinstein's friend," Nobile, who still believed Hiss to be guilty, felt that the Smith historian had "fiddled with the evidence." Nobile's conclusion was dismaying: "I would not be surprised if Weinstein's character betrayed history in *Perjury* where it suited his purposes."

It may well be that all of the above protestors were suspect anyway. Some were undoubtedly members of the current-day Baker Street Irregulars—the so-called "Hissniks"—who lie in wait for those who try to slay their hero. What was interesting was that Weinstein's scholarship offended some presumed friends, and Whitman stands as a ready example. Whitman is the now retired New York *Times* reporter who denied being a party to Priscilla Hiss's anti-Alger declarations. Having either loaned or given files on the Hiss–Chambers case to Weinstein on his own retirement, Whitman wished to review them and wrote to Weinstein offering to pay the necessary costs.

Whitman notes that Weinstein failed to respond at all, but Whitman was especially angered to discover that Weinstein had used Whitman's material without "my permission, consent, or knowledge." There was indeed an occasional attribution buried in the footnotes "by courtesy of Alden Whitman," but Whitman rejected the notion that he "collaborated or cooperated with Weinstein in the research or preparation of *Perjury*."

What was particularly troublesome was Weinstein's antihistorical response to all such complaints: He was directly quoted in the San Francisco *Chronicle* on May 4, 1978, as saying of the criticism, "Frankly, to me it's irrelevant." Moreover, his tapes of thousands of hours of interviews which contained the material to corroborate his assumptions were not available for verification. Weinstein wrote in the *New Republic* on May 13, 1978, that he would not meet with doubting Thomases: All with such interest could read his notes when they became available for public inspection in the Truman Library in 1979.

Weinstein clearly met with many people who said many things to him. However, the charges and countercharges which have surfaced after the book's publication led Victor Navasky to write in *The Nation:* "The target of *Perjury* is Alger Hiss and his claim of innocence, but its temporary victim is historical truth."

For the current authors, the worst confusion came in issues surrounding Weinstein's treatment of the typewriter evidence. He never seemed to have discovered that the typewriter seen at the trial was surely not the typewriter that produced the controverted documents. Since Weinstein had access to the same FBI material that we had, he either misunderstood—the FBI clearly enjoined its representatives from connecting that typewriter to the Hiss documents—or, to find Hiss guilty, he chose to de-emphasize the most important evidentiary item.

What we here refer to is the fact that the FBI knew that it did not find the correct typewriter—i.e., the one that typed the documents in Chambers' possession—and warned all associated with that machine to avoid identifying the courtroom typewriter in that manner. For Weinstein to slight this fact and to focus instead on the highly dubious evidence about Hiss's concealing the whereabouts of the typewriter indicates massive partisanship.

This was a careless analysis of what most observers agree was the most important item in the whole Hiss–Chambers evidence bag. Hiss did not go to jail for denying that he knew Chambers after January 1937, but he did go to jail because he failed to persuade his jurors that he had nothing to do with the typewritten documents in Chambers' possession. Thus, the whole case revolved around Hiss's connection to the correct typewriter, and it is here that Weinstein's errors were most egregious.

What, then, can be said of his book? If it was straight history, it was poor history. If it was a detective story based on partisan inferences, it ought to have been so labeled. Weinstein seemed either uninterested in or unable to understand the central elements in the puzzle—what was it that drove Richard Nixon (by far the most important figure in the case) to bring Alger Hiss to his sorry end, and what was there in Hiss that required him to cooperate in his own destruction? These are the human issues that we endeavor to address in the concluding chapters of this book.

CHAPTER 11

We have become . . . a nation of liars. . . . We accuse ourselves of everything, are forever under horrible indictments, on trial, and raving out the most improbable confessions. . . . Everything bad is done for the best of reasons. How can a man like Richard Nixon think ill of himself? His entire life was a perfect display of Saturday Evening Post *covers. He was honest, he had healthy thoughts, went to meeting three times each Sunday, worked his way through school, served his country, uncovered Communist plots. It is impossible that he should be impure.*

—SAUL BELLOW in
To Jerusalem and Back

Roger Kahn, a writer, was standing in the doorway of a restaurant with a friend, Vic Obeck, then athletic director at New York University. Richard Nixon, at that time out of office and practicing law in New

York, waved at the two men, neither of whom had ever met him before. On impulse, Kahn approached Nixon and asked if he might bring Obeck over and introduce him. Nixon quickly asked for a description of Obeck, and his NYU affiliation was offered. "What else should I know?" he asked urgently. Kahn added that Obeck had played guard with the old Brooklyn Dodgers professional football team in 1946 and was a good blocker. Now fully credentialed to the former Vice President, Obeck was invited over to the table. Kahn wrote:

> "Vic," Nixon began, "I'm really pleased to shake your hand and we were just wondering here why the city doesn't do more for NYU sports. They ought to build a new basketball arena.*
>
> Obeck beamed.
>
> "I remember you with the old Brooklyn football Dodgers," Nixon said.
>
> "Really?" Obeck was a political conservative and his eyes became so bright that I was afraid he might weep.
>
> "Yes, sir," Nixon said, "you were a great blocker. I'll never forget the way you blocked for Ace Parker."
>
> The bright eyes glazed. Ace Parker played in the 1930s. Obeck blocked for a passer named Glenn Dobbs.

Kahn's conclusion is sober: "In sports, as in Watergate, the blunderer gives himself away."

This is a book about lying and blundering. The latter quality is a human weakness which our central characters certainly share. Both fell from public grace in a shocking manner. Although the timing was dissimilar (Hiss at the middle of his career and Nixon close to the end), the sanctions have been much the same. David Frost's television interview with Nixon served only to underline the alienation of the former President, while the fact that Hiss can now practice law in Massachusetts but consults on legal matters from his apartment in New York City sounds a sorry ending to a public life that began so brightly.

---

*This sounds much like Nixon's statement to Mark Harris about San Francisco State University needing an auditorium. It also bears comparison in the sense of repetitive patterns with Nixon's behavior immediately after his father's death during the 1960 presidential campaign. Following a brief four-day mourning period, Nixon again resumed campaigning in upstate New York. At each stop, voice breaking and hands gripping the podium, Nixon would solemnly intone, "My father always said that Buffalo [or Syracuse or Rochester or Batavia or Schenectady] was his favorite city."

Lying is, of course, something else again. Hiss went to jail because he was convicted of lying to the New York grand jury, and Nixon was forced out of office when his own tapes revealed that he had lied in denying either knowledge of the Watergate break-in or culpability for the subsequent cover-up. It is curious that dates were important in each instance.

For Hiss it was January 1, 1937. Had he lied when he denied seeing Chambers after that date? The importance here was that documents bearing the dateline 1938 were alleged to have been passed from Hiss to Chambers, and it was Hiss's failure to convince two juries that he was finished with the accursed "witness" prior to this that cost him so heavily.

This entire issue of dates became a heavily controverted one and remains so even today. Through much of Chambers' early testimony, which portrayed Hiss as a misguided Communist occupying high government office, he dated his own break with the underground movement as occurring around the end of the year 1937. The break was occasioned, Chambers said, by the fact that he let God into his life. This newfound religiosity also caused him to cease his homosexual activities. The break from the Communist Party, usually described as the most painful of his adult life, brought about an immense personal reaction. Chambers purchased a gun and slept with it under his pillow and was finally forced to flee the capital area and hide in Florida with his wife and children.

Only after the thrust of Chambers' testimony changed and Hiss was portrayed as an espionage agent did Chambers reconsider the date of his party defection. It then became sometime in the spring of the next year—1938—most likely April 15, Chambers now reported. This coincided with his production of the Pumpkin Papers, which bore datelines of early 1938. Some have contended that the FBI produced the documents and transmitted them to Chambers and that it was this which then required a rewriting of the Chambers scenario. Those who take this position ask how Chambers could misdate such important events in his life that occurred simultaneously: finding God, giving up homosexuality, breaking from the Communist Party, and going into hiding in Florida with his family.

For Nixon, the crucial date was March 21, 1973. Since he claimed to be in the dark about Watergate affairs until that date, his complicity in the cover-up was, in the lexicon of the times, "deniable." Evidence that he had known of the affair, at least from the weekend of the entry, brought him to the end of the road. While such statements as "obstruction of justice" and "failure to perform constitutional duties" were associated with the termination of his Presidency, it was equally clear that the general public finally could not countenance another lie by Richard Nixon.

It is an impossible task to separate blundering from lying. It must be assumed that both are integrally connected—i.e., one either leads to the other, or at least creates the need for the other. In the material that follows, we will try to arrive at some judgments regarding how Nixon and Hiss fare in this final assay of their character.

James David Barber's book *Presidential Character* tells us something of Nixon's character in a psychoanecdotal sense, but Barber was forced finally to say, "There was a flood of data; the problem was in interpreting it. In 1972, there are still uncertainties as observers in search of the 'real Nixon' try to connect who he is with what he does."

A less scientific and yet more credible evocation is offered by Garry Wills, who said flatly that Nixon was at the mercy of his past without quite possessing it. Wills was not much taken with any of Nixon's predecessors either but felt that other Presidents "did have style, each his own, because they were marked with region and real likes; each had a usable background and personal tradition." He liked Kennedy for his direct and open interest in the political process, Eisenhower for his "exquisite courtesy," Truman for his "earthy majesty," and Johnson for his "frank and undisguised concern with political power."

But Wills found that Nixon, whose autobiographical statements depict him as the living fulfillment of the American dream, had lost his own identity in his pursuit of high political office, that his ambition and high station had taken more from Nixon than he received in return. Wills argued that the cost of Nixon's success was expressed in the loss of his "feel for region, the carriage and easy swing of identity, of defined style." In a remarkably predictive statement written in the late Sixties, Wills drew a long arrow:

> Nixon is President of the forgotten men, of those affluent displaced persons who howled at Wallace rallies, heartbroken, moneyed, without style—called by their moral code to Horatio Alger battle against the odds, then baffled by a cushy world admitting no heroics. These rugged "individuals" are tied together in live wires, endlessly multiplying, of compulsory technological interdependence. The Natty Bumppos of free enterprise did not want community in the first place; that would have been bad enough. But this mere entanglement, this parody of community, is even worse. Restless in such chains, our American frontiersman, grown of necessity flabby, feels he has

> betrayed a trust. And not only is he unfaithful, corrupted by affluence, by privilege without responsibility (supreme torture to a Calvinist); he must also stand by and watch while the code's heretics rake in the good things—the slothful and undeserving, unwed welfare-mothers, the hippies screwing in parks, priests breaking the code by praising unmade selves.

Still, there is another way to look at Nixon. One may conclude that Nixon's career has often been compromised by his need to dissimulate, to conceal his real problems from the people, to avoid telling the whole truth openly and directly on important issues. Richard Moore, his White House counsel, pleaded with him to no avail to go before the country and tell the whole story—at least as he knew it—on Watergate.

Other Presidents told the truth after erring and retained the public's confidence. Kennedy survived the Bay of Pigs fiasco by accepting full responsibility for it, while Roosevelt admitted being caught unaware by the Japanese bombing of Pearl Harbor. For Nixon, however, to err may have been human, but it certainly wasn't presidential or even Nixonian. Prime examples of Nixon's grievous shortcoming in this regard was his astounding statement which placed the full responsibility for the Watergate affair on the deceased Martha Mitchell. Speaking to David Frost in a television interview, Nixon, serious-faced and head nodding as if to underline his assertions, said:

> . . . if it hadn't been for Martha Mitchell's mental and emotional problem . . . there'd have been no Watergate. . . . John [Mitchell] wasn't minding the store. He was practically out of his mind about Martha in the spring of 1972. He was letting Magruder and all those boys, these kids, these nuts, run this thing.

Nixon seemed reluctant ever to accept blame for anything throughout his entire career, the sole exception being the self-serving statements made as the Watergate revelations threatened to overwhelm him. At that time he made a series of concessions of error—moving too slowly, believing too much in his aides, being too busy with international affairs, being too compassionate. Finally, such admissions turned out to be clearly calculating, concealing the one almost unimaginable fact—a President leading the way in the cover-up of criminal acts.

More than any President in memory, Nixon seemed determined to bend the truth whenever it conflicted with his purposes as he perceived them. Illustrations abound, but perhaps the technique can be best expressed by a recounting of what took place in Nixon's first election—the victory over incumbent Democratic Congressman Jerry Voorhis.

The campaign began with Nixon fresh from the Navy, identifying Voorhis as someone who "stayed safely behind the front in Washington" while Nixon was, according to his own campaign circulars, "a clean, forthright, young American who fought in the defense of his country in the stinking mud and jungles of the Solomons."

Voorhis came from a wealthy family, was a Phi Beta Kappa at Yale, and had a pleasantly radical streak which led him to try life as a laborer after college graduation. He went to Congress in 1936 as a Roosevelt man and in 1940 demonstrated his conservative tendencies by sponsoring a bill to register Communists. By 1946 he was regarded by the Washington press corps as the "hardest-working Congressman in Washington." He was a markedly liberal man, although he openly opposed the CIO's Political Action Committee (PAC), which he said was Communist-dominated. Despite Voorhis' rejection of it, the PAC chose to support him. Nixon leaped with a vengeance on Voorhis' statement that he did not have PAC support. When Nixon produced a letter on CIO stationery which endorsed Voorhis, the trap was sprung. Garry Wills's review of the campaign follows below:

> He [Voorhis] was off balance, explaining, splitting hairs all the rest of the campaign, which reached a climax in the fifth debate at San Gabriel Mission. Nixon, of course, went to Congress, to the Un-American Activities Committee, to Hiss.

In truth, Voorhis was every inch a responsible Congressman. Earl Mazo recounts that Nixon told him that his first move after securing the Republican Congressional nomination was to decide that Voorhis' "conservative reputation must be blasted." He branded Voorhis as a man who stood for federal controls (hardly popular after the war) and as an enemy of free enterprise. Nixon called his opponent "a lip-service American, a front for un-American elements." When the Republican campaign headquarters came out with a statement accusing Voorhis of voting the "Moscow line" and calling Nixon "a man who will talk American and at the same time vote American in Congress," the election was all but over.

Voorhis was simply not a radical, and Nixon's campaign was palpably dishonest in its effort to suggest that this was a fair statement of the particulars. Moreover, while Voorhis had remained in Washington as a Congressman during the war, this was by no means an unusual circumstance. It was also not true that Nixon was fighting during the same time in the "jungles of the Solomons."

Nixon was commissioned as a Navy officer and was shipped to New Caledonia, where he served as a supply officer in the Naval Air Transport Command. There is no evidence of Nixon seeing combat, and his military career has been described by an old Navy friend, Lester Wroble, as consisting of three essentials: (1) he set up a service where men could buy whiskey, food, and other staples not available as general use, (2) he consistently made money in poker games, and (3) he always knew what he was going to do after the war. Thus, there is some question about whose war service was more meritorious, the hard-working Congressman or the soft-berthed naval officer.

An innately distrustful man himself, it is not surprising that Nixon was so concerned with issues of trust in his relations with others. The Watergate hearings gave ample evidence of the reverence in which the President and his principal aides were held by their subordinates. Witness after witness indicated that they did the strange things they did—and the persistent image remains of lawyers colluding in lawbreaking—because they felt that they were acting in the ultimate interests of the President. And it is here that the Nixon Presidency ran into its most severe trouble. Mr. Nixon believed that the people of the country *should* believe in him and in America—whether he was telling the truth or not—and that those who did not believe lacked faith and suffered from a patriotic deficiency as well.

For a man who could tell an interviewer, "When the President does it, it is not illegal," it requires no great jump to imagine Nixon saying, "When a President lies, it is not lying." Mendacity, too, was at the heart of Nixon's military policy. This was rationalized by his overweening concern for the people he favored. We bombed in North Vietnam to bring peace to our allies in the South, and we propped up a dictator in Saigon to keep the country from falling into the hands of those who were even worse. We lied to the American people about bombing in Cambodia because to have told the truth would betray a private commitment to Prince Sihanouk.

In this light, Garry Wills's argument about Nixon's disengagement with his own past may be something of an oversimplification. Nixon does indeed possess a past, but it is not centrally connected to him; rather in some Horatio Alger way, it is an idealized past, made up of deep-dish apple pie and large bank deposits. Nixon's background was not particularly special; what was special was the way he perceived it. He seemed drawn to something that can be described as an American counterfeit nostalgia, an attachment to the most commonplace features of life.

Thus, the great American fairy tale with its old-fashioned virtues of simplicity, endurance, trust, and the natural life are continuously pitted against the immense complexities of modern life and most importantly come out ahead at every measure. While all this Americana has some appeal, it was in the final analysis rooted in a never-never land near Whittier, California, where red-blooded Americans worked and lived and watched television and stood tall as the flag passed by at the Fourth-of-July parade.

Roger Rosenblatt has written a very provocative evaluation of the revival of interest in nostalgia that has developed in this country. Rosenblatt points out that it takes the comfort and safety of marriage to make us pine for the lost loves of our youth. He argues that there is some connection between the distance we have moved from our past and the guilt we feel about the "ruthlessness and thrust of our forward motion." Nostalgia takes on political colorations in connection to theories of regeneration and absolution, and Rosenblatt concludes that as a cultural aftermath to the war in Southeast Asia, the nation seeks most avidly "instant innocence and painless guilt."

One is compelled to think of the strange contrast between the two Nixons: the Quaker who waved the flag and dropped the bombs,* encouraged political burglaries, and supported capital punishment for "big-time dope peddlers" and the other man who nostalgically pined for

---

*It is hard to know how Nixon could continue to describe himself as a Quaker when he patently disavowed a cardinal religious tenet—pacifism. He was identified as the American President who was responsible for the largest bombing "tonnage" in history in the Vietnam War. Saddest of all was the story in the Washington *Post* on the first day of 1974 that Quaker congregations in such diverse settings as Philadelphia; Plainfield, Vermont; Adelphi, Maryland; and Stamford–Greenwich, Connecticut, were calling for Nixon's resignation. The group's request was especially poignant and began as an appeal:

> . . . to you directly, Richard Nixon, as one who has on various occasions referred to your Quaker background and hence may be expected to respect the traditional testimonies of the Society.
>
> It seems to us that serious discrepancies exist between these testimonies and the actions which you have either taken or for which you may be considered responsible. . . . In view of the above we urge you to resign.

all that was sentimental and golden in our national life. Surely no small part in the Nixon dichotomy was the immense distance that the President moved from the grocery store in Whittier. The strange admixture of Nixon's two sides could be clearly seen when high-school cheerleaders in Pomona, California, were instructed to chant during a Nixon campaign visit, "Nixon is blue-hot," to avoid offending the President's well-known sensitivity to the color red. Here small-town America stands in an uneasy relationship to the complexities of political life.

In Nixon's world, those who were unappreciative of all America offers and those who felt alienated from America's bounties were the real enemies. Truman came from humble stock and refused to be anything else; Acheson and Stevenson were aristocrats who threw in with the common people; the young Kennedys profited much in the new world but chose to align themselves with their historical enemies.

Among this group, the man who was least explicable to Nixon was Dean Acheson, who served under Truman as Secretary of State and who was immutably linked in Nixon's eyes with Alger Hiss. Nixon found it incomprehensible that Acheson, during his Senate confirmation hearing, would say of Hiss in the latter's worst moments, "I should like the committee to understand my friendship was not easily given or easily withdrawn. Alger Hiss had been an officer in the State Department most of the time I served there. We became friends and remained friends."

A few days after Hiss's conviction, Acheson held a routine press conference. The transcript read:

> I should like to make it clear to you that whatever the outcome of any appeal which Mr. Hiss or his lawyers may take in this case I do not intend to turn my back on Alger Hiss. I think every person who has known Alger Hiss or has served with him at any time has upon his conscience the very serious task of deciding what his attitude is and what his conduct should be. That must be done by each person in the light of his own standards and his own principles.

How this story of a highly placed man standing by a friend who, according to Acheson, was "in the greatest trouble a man could be in" could become a *cause célèbre* with Nixon is very revealing of his character. Calling the Acheson statement "disgusting," Nixon could not conceive that the Secretary of State could be motivated by simple sentiment in time of presumed national danger and fired salvo after salvo at Acheson during the trial years and during the national campaign of 1952 that immediately followed. He referred to Acheson as suffering from "color blindness—a form of pink eye toward the Communist threat in the United States."

Even after becoming Vice President, Nixon continued to flail away and enlarged the visual field of enemies by the addition of Truman and Adlai Stevenson, saying that "real Democrats are outraged by the Truman–Acheson–Stevenson gang's defense of Communism in high places" and also that Truman and Acheson were "traitors to the high principles in which many of the nation's Democrats believed." Oddly enough, it has been reported by Earl Mazo that Nixon stated that he and Truman were much alike, that Nixon admired Truman's courage, that he felt that each was his party's best campaigner and that both "had plenty of guts."

All this Nixonian invective could not help but contribute to a scathing denunciation of him that appeared in a Walter Lippmann column about mid-point in Nixon's service as Vice President: "[Nixon is] a ruthless partisan . . . [who] does not have within his conscience those scruples which the country has a right to expect in the President of the United States."

Up to the very end, Nixon's career was a continuous series of ups and downs (the so-called "crises"), and allowing the Watergate affair climate to develop at a time when his political and historical future seemed so clear contributed to the incredulous and absolute disappointment expressed by his longtime supporter, publisher John S. Knight, who wrote in the Detroit *Free Press* in the midst of the Watergate hearings:

> The present plight of President Nixon is an American tragedy such as we have not witnessed in our times. For here was a President almost universally acclaimed for the building of bridges with Russia and China in the cause of peace who has now been toppled from the pedestal of public esteem.

Robert Sherrill, a much less sympathetic auditor of the Nixon career, starts out from a different perspective but ends up sounding the same melancholy note in an article in *The Nation* on May 14, 1973. Sherrill is caught by the relationship between Watergate and Nixon's previous political style, although he remarks that Watergate represented Nixon's exceeding the normal bounds of political dirty work, even by Nixon's own standards. What is key here, contends Sherrill, is that there was no possible excuse since Nixon was really home free in the 1972 election. Sherrill concludes in reference to Nixon and his close associates: "These saboteurs cannot claim, as someone has pointed out, to have acted out of such normal impulses as greed, ambition, or sex; theirs was a sick and hysterical ideological drive."

Here again is an argument made much in the manner of the one advanced earlier by Mark Harris—that Nixon's career was a "mindless one, that he lacked the intellectual underpinnings which attach stronger men to the forces of both morality and reality." This was expressed in another way by Robert Hess, one of Nixon's aides in the 1962 California gubernatorial campaign, who said, "Nixon decides the position he wants to take and we find the facts for him." The truth is that Nixon had no regard for intellectual achievement, that he was suspicious of those who were well educated and was dubious of their loyalty. This attitude was expressed most openly in a paragraph near the end of his chapter on the Hiss case in *Six Crises:*

> At a less rigorous level—somewhere in a vague area that goes by such names as "positivism" or "pragmatism" or "ethical neutralism"—Hiss was clearly the symbol of a considerable number of perfectly loyal citizens whose theaters of operation are the nation's mass media and universities, its scholarly foundations, and its government bureaucracies. This group likes to throw the cloak of liberalism around all its beliefs. Eric Sevareid's term "liberalists" probably describes them most accurately. They are not Communists; they are not even remotely disloyal; and, give or take a normal dose of human fallibility, they are neither dishonest nor dishonorable. But they are of a mind-set, as doctrinaire as those on the extreme right, which makes them singularly vulnerable to the Communist popular front appeal under the banner of social justice. In the time of the Hiss case they were "patsies" for the Communist line.

This retrograde argument, which directly attacks all of the intellectual bastions of the country ("mass media," "universities," "scholarly foundations"), followed on the heels of another strange paragraph in which Nixon undertook to analyze the character flaws of Alger Hiss. It must be remembered that Nixon knew remarkably little of Hiss in any personal sense, and since Hiss denied all Communist complicity, all Nixon had to go on were Chambers' recollections, often changing and unquestionably self-serving in many instances. Yet much in the style of the armchair psychologists who dominated the field of philosophy at the turn of the century, Nixon undertook the following evaluation of Hiss:

> . . . But none of the typical excuses fit Alger Hiss. He did not join the

> Communist Party, accept its rigid discipline, and steal State Department secrets for money, position, or a desire for power, or for psychological reasons stemming from some obscure incident in his early life, or because he had been duped or led astray by his wife. He joined the Communist Party and became a Communist espionage agent because he deeply believed in Communist theory, Communist principles, and the Communist "vision" of the ideal society still to come. He believed in an absolutely materialistic view of the world, in principles of deliberate manipulation by a dedicated elite, and in an ideal world society in which "the party of the workers" replaces God as the prime mover and the sole judge of right and wrong. His morality could be reduced to one perverted rule: anything that advances the goals of Communism is good. Hiss followed his beliefs deliberately and consciously to the utmost logical extreme, and ended up in the area of espionage.

This simplistic evaluation of Hiss's character had to be of the template variety. Despite Nixon's willingness to traffic in his own psychohistory, he was markedly unwilling to accord Hiss the same privilege. Hiss was a Communist without regard to "psychological reasons stemming from some obscure incident in his early life." Instead he became a foreign agent because of his mind, because of his intellectual beliefs, surely small currency in Nixon's world.

The description abounds in unsupported assertions, all of which seem to indict Hiss without regard for his real character. This is not to argue that Hiss did or did not know Chambers or did or did not pass him contraband documents, but rather to point out that nowhere in the lengthy record of the Hiss–Chambers–Nixon trilogy was there any evidence of Hiss's devotion to "an absolutely materialistic view of the world," to "the principles of deliberate manipulation by a dedicated elite," or to "a society which replaces God . . . as the sole judge of right and wrong."

Nixon's description may well fit some Communists, but it hardly does justice to whatever complicated formulation tied Alger Hiss to the movement—if he were indeed tied to the movement. It is the kind of evaluation that one might expect from someone so devoted to his own concept of America that he cannot conceive of another political philosophy. The language is so pious and the argument so hackneyed that the paragraph can well stand as Nixon's voice-print. This is American Legion mentality, hardhat rhetoric coming from "their" own President, the man who telephoned some New York City construction workers to offer congratulations the day after they attacked student demonstrators.

And what about Hiss? Certainly a less partisan and less superficial view of Hiss's character than provided heretofore is required to make a final judgment on the man and his strange case. From Dumbarton Oaks, Yalta, the founding conference of the United Nations in San Francisco to Lewisburg Prison, a cold-water flat on Manhattan's lower East Side, and work as a stationery salesman—this has been the up-and-down road of Hiss's life. The distance fallen is almost immeasurable. Only one other name comes readily to mind in comparison, and that is Richard Nixon, who clearly fell further but whose retirement from the public scene is still graced by many of the perquisites of the good life.

Hiss has commented that his mother paid very little attention to him when he was growing up; she seemed finally to discover him when he was a man, being called a Communist. Hiss grew up in a highly protected environment, sought paternal substitutes, dealt with strong feelings through denial, and often found himself surprised by life's intractable nature. After learning about his father's suicide, his determination to recover for the Hiss family its good name was described by friends to approximate saintliness.* Also, Hiss often seemed to lack—and still does to this day—insight into his own character and that of others.

His sister's suicide was a "sudden irrational act," concluded Hiss, and he felt much the better for this simplistic explanation. He maintains that Chambers' entire story was a fabrication, and yet he cannot—or, possibly because he finds the reason so disturbing, will not—explain the etiology of this attack. "The secret to him [Chambers] was a psychological disturbance," offers Hiss lamely in way of explanation. That is the best he can come up with after thinking about it—probably every day of his life—for the last thirty years. These are the words of a man who is incapable of understanding the nature of the most important relationship in his life.

Let us suppose for a moment that Hiss is, as he maintains, innocent of all charges. What, then, was he hiding in his relationship with Chambers, or whom was he protecting? Was there, as some theorized, a homosexual relationship between him and Chambers? "No," both men say, although Chambers admitted to the FBI that he did have such relations with others. Did Chambers make advances toward Hiss and meet rejec-

---

*A classmate of Hiss's at Johns Hopkins described how Hiss really "dressed him down" for using profanity in ordinary conversation.

tion?* Was it because Hiss was protecting his wife who had had an illegal abortion? Or was it the fact that Priscilla was the real Communist in the family, and Hiss was acting to shelter her? Was he worried that his stepson's homosexuality would become known and ruin the young man's life? Was it true that Tim Hobson had a sexual relationship with Chambers? All of these reasons, and many others as well—both more and less preposterous—have been offered by observers to explain Hiss's evasiveness at the HUAC hearings and later at the two trials.

Although Hiss described himself to the authors as a "positivist," saying "something which is a void is disturbing to me," he here again cannot find an explanation for—or evince interest in—Richard Nixon, who was as much a protagonist in the tangled skein of his life as was Whittaker Chambers. Hiss might well have escaped the grand-jury indictment had not Nixon, microfilm in hand, appeared before the grand jury in its eleventh hour to be certain that "the right man was indicted." Had Nixon not been consulting with Hiss's chief at the Carnegie Foundation, John Foster Dulles, who was at that same time also advising Hiss on how to defend himself, Hiss might well have walked over to *Time* to confront his accuser.

If Hiss had done this, the sorry scenes involving identification at the hearings which hardened the committee's attitude toward him might have been avoided. Hiss's detachment from Nixon is even more disconcerting when one considers that the verdict might have turned out differently had Nixon not made an angry speech before Congress demanding that the fitness of the judge in the first trial be investigated and, in particular, criticized two of his rulings on which (possibly not by coincidence) the judge at the second trial took the opposite position.

When asked for his feelings on Nixon's Watergate difficulties, Hiss replied stiffly, "I have no personal interest in Nixon. I have not attempted to study him." Much in the same vein was an answer he

---

*In 1975 Hiss was reported, in a story on the Associated Press wire, as describing Chambers as a spurned homosexual who testified against him out of jealousy and resentment. Hiss is quoted as saying: "He never made a pass at me, but he had a hostility to the point of jealousy about my wife and a calm, almost paternal attitude toward me which is unexplainable in any other way." Still, it must be remembered that this late statement is just possibly an example of revisionist history. Hiss's review of the FBI files had led to the discovery that Chambers admitted to being a homosexual. Moreover, this statement stands in marked contrast to denials about homosexuality made by Hiss to the authors at much the same time as the Associated Press story.

offered his son, Tony, who asked his father in a *Rolling Stone* interview conducted in 1973, "Why no mention" of Nixon when then people were talking about little else? Hiss responded, "I don't know Nixon. I've never had occasion even to shake hands with him. I'm in no position to do more than say that I faced him across a dais at a number of hearings."

Is this really so surprising coming from the same man who from the first viewed the entire case against him as an effort to discredit Franklin Roosevelt? Hiss said:

> . . . Roosevelt himself had been so popular that it was hardly good politics to attack him directly but he could be attacked indirectly through minor lieutenants like me. So, as I've said, I was sort of a handy compendium of various prejudices and hates. . . . Well, I'd been at Yalta, so that helped make me an inviting target, but if it hadn't been me, they would have found somebody else. They were looking for a target.

In a speech in 1973 in Los Angeles, Hiss asked rhetorically, "Why does a sniper shoot down one man and not another? Why will someone kill a total stranger who resembles his father? Why did Bremmer shoot Wallace? . . . *Motivation baffles me* [authors' italics]. It was always one of the most tormenting aspects of the case. . . . "

Although Hiss's parallel between Bremmer and Chambers is compelling at first blush, it is not quite accurate. Bremmer did not know Governor Wallace personally before he tried to assassinate him; he never ate any meals with him and told him stories of sexual conquests; he never stayed overnight in his apartment, bringing his wife and daughter with him; he never "welshed" on the rent nor, after this transgression, receive an old car from Wallace as a gift a year later.

A statement offered by Hiss's friend, Dean Acheson, is instructive. Acheson wrote in his autobiography that during the Senate hearing in 1949 to discuss his nomination as Secretary of State

> We [Acheson and the members of the Senate committee] tried the experiment of examining the conduct of both [Hiss and Chambers] on the assumption that what Chambers had said was either true or false. On neither assumption was their conduct explicable on any reasonable basis. Some part of the puzzle seemed to be missing.

Today, even with the release of presumably much of the FBI files, "some part of the puzzle" is still missing. Perhaps the missing link lies in the heads and hearts of the two survivors, Hiss and Nixon. Unfortunately, none of the three protagonists extant—the FBI, Alger Hiss, or Richard Nixon—seem willing and/or able to provide the necessary information.

The FBI, which has acted throughout this matter like a sovereign state, has not released its secrets without a fight, and those now available may represent only a small fraction of the entire lot. Nixon went on to bigger and better crises and seems unwilling to talk about anything save for unimaginable sums of money; Hiss tells his story over and over—as one writer recently concluded, "making converts without even trying"—to reporters and college students. It is a polished performance, one which seems orchestrated to meet all contradictions, great and small.

To return to the earlier theme, assuming that Hiss is innocent of the charge of lying, then why the blundering? His naïve plan during the hearings to rely on "unaided memory," which caused him to say "to the best of my recollection" 200 times, was a stratagem that alienated any friends, Republican or Democrat, that he might have made at the time. His request to hear the testimony of the dentist who fixed Chambers' teeth before he could make a positive identification is something more than just the actions of a "cautious man," as Hiss liked to describe himself.

Why did he turn the typewriter and the typed correspondence over to the FBI when he didn't have to? Hiss said he "wanted the record to be complete." Hiss was a well-trained attorney, and somewhere during his three years at Harvard and two and a half years of private practice he must have learned that a good attorney is selective in introducing evidence. As an attorney, Hiss must have also known that innocent people sometimes go to jail. Yet, by his own statements, he had unbounded faith in his ability to win such battles. To Hiss there always was—and is—a higher tribunal, one which would ultimately see things his way. In describing the House committee to his son, he said, "[The committee] wasn't very bright, wasn't very alert, and didn't think the way I did." The snobbery in such a statement is ineffable.

When Hiss was indicted by the grand jury, he couldn't wait to get to court, and as recently as 1976 he still couldn't understand how a jury "could believe that creep after seeing the two of us." As Hiss told the authors, "I couldn't wait to get to court with the case. I felt that I was badly treated by the House committee, but I was absolutely sure that a jury of my peers would deal with me more favorably."

After his conviction, Hiss was certain he would win his appeal, and after losing this round, he was certain he would win the next or the one after that. Today, he has unbounded faith that the process of *coram nobis* will clear his record.

Alger Hiss was either a very brave and honest man or a very foolish and guilty one. Perhaps he indeed had, as his psychoanalyst suggested, a phobia against fear. This might be one reason why he was such a perfect foil for Nixon. A more fearful or a more realistic man would not have appeared as self-destructive as Hiss did.

Recognizing the perfect vision of hindsight, a more prudent man might have done many things to receive better treatment. If Hiss had not asked to appear before the House committee, it is a fair guess he would not have become a symbol of the Cold War era. Hiss, however, disclaimed this possibility. In a comment that was remarkable for its hollow pomposity, he said:

> I believe in "character-typology." I was ordained to do what I did. I could not have lived with myself if I had not come forward and volunteered to testify before the committee.

Not only did Hiss come forward; he volunteered to do this even before the committee required his appearance. He sent the committee the following telegram:

> MY ATTENTION HAS BEEN CALLED BY REPRESENTATIVES OF THE PRESS TO STATEMENTS MADE ABOUT ME BEFORE YOUR COMMITTEE THIS MORNING BY ONE WHITTAKER CHAMBERS. I DO NOT KNOW MR. CHAMBERS AND INSOFAR AS I AM AWARE HAVE NEVER LAID EYES ON HIM. THERE IS NO BASIS FOR THE STATEMENTS MADE ABOUT ME TO YOUR COMMITTEE, I WOULD APPRECIATE IT IF YOU WOULD MAKE THIS TELEGRAM PART OF YOUR COMMITTEE'S RECORD, AND I WOULD FURTHER APPRECIATE THE OPPORTUNITY TO APPEAR BEFORE YOUR COMMITTEE TO MAKE THESE STATEMENTS FORMALLY AND UNDER OATH.

Had Hiss not played games with the name puzzle at the hearings, things might have been different. His refusal to name George Crosley until the bitter end cost him heavily in the court of public opinion. Why not name Crosley? "I wanted to save him any embarrassment," Hiss responded. And forty pounds added and new teeth notwithstanding, Hiss had seen the man's photograph in the newspapers, and his wife had no trouble identifying Chambers/Crosley from the same picture. Hiss's response was far more ambiguous. "Looks a little like the committee chairman,'' he offered in jocular (but peculiarly clumsy) fashion. "Not entirely unfamiliar,'' he added. Was there no resemblance to Crosley in the pictures whatsoever?

The authors' efforts to solve this conundrum brought a very meager return. To the direct question "Why did it take you so long to connect Chambers to Crosley?'' Hiss replied in surprising fashion, as if the inquiry covered an area that was *terra incognita:* "How long did it take?'' Further discussion finally brought a clarifying statement, but it too deals only tangentially with the real issue. Hiss commented:

> When Priscilla first got word of Chambers' testimony, I can't say. I was here [in New York City]. I certainly telephoned her [in Vermont] sometime between the third and the fifth of August when I appeared, and I certainly told her that I intended to testify on the fifth. Now, if she told me then that it might have been Crosley, this would not have affected my testimony on the fifth because I said I wished to confront the man. The next time Mr. Crosley's name came to my attention was when I was coming to the hearing and two friends of mine then told me of leaks from the committee which could point only to Crosley. In other words, it's quite possible that Priscilla made a feminine intuition. That would not have satisfied me. . . . But I was in [Vermont] on the weekend before the sixteenth, and I am sure Priscilla must have mentioned it.

When it was suggested to Hiss that the FBI had mentioned Chambers to him on at least two occasions, that his brother Donald had been informed by a friend, Edward Miller, that Chambers was the source of rampant rumors that Hiss was a member of the Communist underground, and that this accuser had been identified as being an editor at *Time* magazine, Hiss replied:

> But I had never heard *before that* [authors' italics] Chambers worked on the magazine. The first time I heard that was from Miller. When the FBI mentioned it, I ignored the name. I asked them if it could be a fellow named Bob Chambers that I had grown up with who worked for the Bureau. They said no. They told me nothing about Chambers, so I ignored that. The first time I connected Chambers with *Time* magazine was from Ed Miller. And then Eddie reported back, "Forget it; the fellow isn't saying it anymore."

And there it was once again. Feminine intuition couldn't be relied on; Hiss needed "to confront the man''; two friends told him of leaks from the committee "which could only point to Chambers,'' but still Hiss could not allow the accursed name to cross his lips. It was, admittedly, only two weeks from August 3, 1948, when Chambers first accused Hiss before HUAC to August 17, when Hiss finally uttered the name "Crosley," but he had heard the full name "Whittaker Chambers" several times before.

Why not find out more about this strange man from *Time* who was saying these terrible things about him? It was, said Hiss, "the first time I connected Chambers with *Time*. And then Eddie [Miller] reported back: 'Forget it; the fellow isn't saying it anymore.'"

So forget it Hiss did, and he forgot it so well that a brief eight months later the name meant nothing to him. There is clearly something here that defies rational understanding, and it is this very point which scored so heavily against Hiss.

John Chabot Smith finds all of the above quite comprehensible in his biography of Hiss. He wrote:

> He [Hiss] didn't remember that the FBI had asked him six months before if Whittaker Chambers had ever visited his home, and he didn't know that Whittaker Chambers was now the name of the man he had once known as George Crosley. So it meant nothing to him. And when Mrs. Kerr's promise to stop calling him a Communist was relayed back to Hiss through Miller and Donald, he thought it meant Chambers wouldn't be talking about him anymore. Wishful thinking perhaps; but nothing could have been farther from the truth.

The last phrase in the Smith quotation has an ambivalent cast. Could Hiss forget Chambers so completely? Was there justifiable confusion between the name "Crosley," which fit the circumstances, and "Chambers," which did not? The right/wrong Chambers worked just a few blocks away in another tall building in Manhattan. Why not walk over, or at least call? Hiss said he wanted to but was warned by John Foster Dulles not to do so. And Hiss, ever dutiful and law-abiding, accepted this advice.

Bearing directly on the issue of veracity was the matter of Dr. Margaret Nicholson, the Hisses' pediatrician. She became a peripheral government witness at the second trial when the prosecutor was attempting to establish the fact that the Hisses had entertained the Chambers at a cozy New Year's Eve party on December 31, 1936. If the above were indeed true, Hiss's argument that he never saw Chambers in 1937 would be highly suspect. Hiss denied any recollection of the party, maintaining instead that his wife and stepson were visiting relatives in

Chappaqua, New York, at the end of the year and that at this time stepson Tim contracted a case of the chicken pox.

The issue was a stand-off. Hiss was able to prove that his family was away in New York at the end of 1936, but the prosecution required Dr. Nicholson to testify at the second trial that her office records reflected house calls to treat Tim in *Washington* on January 2, 3, and 6 of 1937. Little else was asked of Nicholson, and her inconclusive testimony seemed of little import at the time.

There was, however, another Nicholson story that never surfaced at the trial but was of somewhat greater significance. When Hiss's attorneys learned that the family pediatrician was subpoenaed to appear at the trial, they interviewed her at her own request. At that time, according to attorney Claude Cross, Dr. Nicholson described a bizarre incident. When she left the Hiss household following one of her early January calls, she discovered that she had left her gloves behind. She attempted to return and reclaim them but was refused admission by a stout and gruff man who finally brought the gloves out to her car. The man was later identified by Nicholson as Whittaker Chambers when newspaper pictures of him appeared twelve years later.

The above incident was understandably of much interest to Allen Weinstein, who tried to interview all the available participants. Cross recalled the entire story, as did Hiss. The attorney who interviewed Nicholson—John F. Davis— had no clear recollection, although he told Weinstein that the conversation "probably took place." Nicholson herself refused to be interviewed. Hiss's comments were notable. He said that Chambers had undoubtedly come "to sponge off Priscilla or ask us for help again. He was always doing that."

Troubled by the free-and-easy Hiss recollection that Chambers was "always" around the Hiss household, the present authors asked Hiss for clarification. How could it be that this frequent habitué was so difficult to remember when push came to shove? Hiss's response was evasive: "If it is true [that Nicholson reported she saw Chambers in the Hiss house], she could have mistaken anybody in my house for Chambers. If it was in fact Chambers, it doesn't prove I saw him. On the contrary, she says I was not home. If he had come asking for a handout, after all he knew Priscilla, it is quite possible. It doesn't work against my testimony that I had not seen Chambers after the first of the year. . . . But again, there's never been any secret about the Nicholson possible identification of Chambers. As a matter of fact, I think it's dealt with in the Smith book."

Apart from the fact that Smith never referred to Nicholson in his book, Hiss's reply is very confusing. He begins by suggesting that it may not be true that the pediatrician saw Chambers in the Hiss home, then offers the idea that she could easily be mistaken. This thrust is succeeded by the idea that it might well have happened, but nobody placed Hiss himself in his home at that time. And if it were Chambers and if Hiss were not at home, "It doesn't work against my testimony," for Chambers was always around harassing poor Priscilla and looking for a handout. But then again, continues Hiss, there was nothing "secret" about Nicholson seeing Chambers.

The confusion is rampant. Nicholson never testified to any of the above at the second trial; rather she sought an appointment with Claude Cross to inform him of her fear that she might be asked about the Chamberses, whose part-time pediatrician she also apparently was. Judging from these facts, Nicholson's sympathies were with Hiss, whom she sought to alert to a clear danger. Hiss minimizes her recollections by obfuscation and distortion and chooses never to face the full implications of her testimony. Completely absent is a sense of openness and candor; the circumlocutions are worthy of Nixon himself in the best of his Watergate defenses.

The Bokhara rug incident provides still another view of Hiss's need to dissimulate in *l'affaire* Chambers. The rug was given by Chambers to Hiss from Boris Bykov in gratitude for Hiss's contributions to the Communist underground, said Chambers; it was in partial repayment of money loaned by Hiss to Chambers, said Hiss. Date and purpose notwithstanding, both men acknowledge that a Bokhara rug changed hands.

What happened then soon came to resemble a Mack Sennett film. The rug was kept rolled up in a closet or a small room in the basement of the Hiss home on 30th Street when Chambers last saw it later in 1937. Not so, said Hiss; the rug was put immediately to use on the floor of the Hiss home on P Street in 1935. The dates were of much importance, for if what Chambers says is true, it would prove that they saw each other after Hiss had sworn their episodic relationship had come to an end. Nineteen thirty-five was a much safer year for Hiss for the opposite reasons. And then Hiss said gratuitously and unfortunately to the FBI, "We have used the rug ever since and we have it today."

The implication was clear: There was no need for Hiss to hide a rug given to him in payment for an old debt, and the rug was out in the open for all to see. The Bureau interviewed the Hiss maid, Clydie Catlett, whose memory proved to be variable. The rug was rolled up as Chambers claimed, but it was also in use as Hiss contended. It indeed had been rolled in the 30th Street house, but it was in open use on P Street.

Still undeterred, the FBI scoured Washington and found a storage firm which acknowledged that it kept the rug for the Hisses during 1937 and 1938. "In constant use," declared Hiss. "We have it today," he offered. Not so, said the Bureau.

Looking back over the years, Hiss remains unmoved by these contradictions. In fact, he fails to acknowledge that a contradiction exists, saying, "If it [the stored rug] was the same rug. . . . We had other rugs. I don't say I've always been accurate, but I do say I've always been honest. When I say 'constantly in use,' I mean it was in each house I can remember. I don't know how long the period of storage was; I don't know whether it was the same rug, but if it was the same rug . . . I don't know."

The delay in admitting knowing Chambers and the confusion about the family pediatrician meeting Chambers and the Bokhara rug seem all of a piece. Things are not said when they should be, events which likely happened are not clearly perceived, and information is volunteered (the rug that "was constantly in use") when such an explanation would prove to be an unnecessary complication. Perhaps these are human errors made by a very human man. They all do bear, however, on the most transparent item in the whole Hiss arsenal: Did Hiss "know" Chambers and were they "social friends" as one man claimed and as the other denied?

In the very best sense, Hiss was prosecuted because he dared to stand up to the House committee, which was engaged in a witch hunt. This was surely President Truman's view. On the surface, such forthright behavior is laudable.* Sad to report, Hiss went to the committee hearings with such arrogance, such surety that he would win that he alienated rather than persuaded his audience. When his son, Tony, asked, "When you first went before the committee, did you think that you would be able to dispose of the charges?" Hiss responded:

---

*When the Hiss trial and the McCarthy era which succeeded it were over, some two and a half million federal employees had been reviewed for loyalty. After many hearings, only 270 individuals were dismissed, and of this number sixty-nine were later re-employed. Most important is the fact that not a single case of espionage had been discovered.

> Oh, certainly. . . . So I just thought, you know, a Congressional hearing, familiar territory to me, and it took me quite a while to realize that this was all rigged, that they were not asking tendentious questions, they were trying to turn it all upside down. . . . It was a shock to discover that they were saying one thing and doing another. They were really in a witch hunt.

Was it this naïve belief in his omnipotence that led many people to feel disenchanted with Hiss? David Riesman, the well-respected Harvard social scientist, felt Hiss was guilty and wondered why he continued to do his country a greater disservice than espionage in not frankly telling how an idealistic and successful young lawyer could get involved with the Communist Party. Riesman wrote:

> He would have contributed to clarification instead of mystification, and perhaps partially disentangled the knots of identification binding so many decent people to him hence to the view that he was being victimized. It might have been revealed that his case had special elements (special guilts, special arrogances, special impatiences) and that therefore, despite appearances, it was not a generation on trial but a fringe. Perhaps Hiss thought he could brazen it out. As the square-jawed, clean-cut hero of the two he would have a comic-strip advantage. Perhaps he was ashamed to disillusion his non-Communist friends and preferred to drag them down with him.

Riesman later admitted in personal correspondence to "some uneasiness" about the above judgment, and perhaps its categorical assumption of Hiss's guilt might well have provided this doubt. Nonetheless, in 1973, Riesman still viewed Hiss as a manipulator—that is, someone who saw "ways of getting around the great Philistine American Majority by stratagems, whether this involved persuading a judge or persuading a Congressional committee." Since Riesman believes that in a democracy one "has to persuade people, to bring them along," it seems likely that he would distrust Hiss's sense of omnipotence.

In their own time and in their own way, but in apparently contradictory fashion, both Hiss and Nixon argued for a high personal morality. Each behaved as if a crime had been committed, but neither was willing to admit to any personal responsibility for its commission. Hiss portrayed himself as a beleaguered man being destroyed by an ungrateful acquaintance whose enmity was incomprehensible to him. Nixon throughout his entire career viewed himself as a man simply and completely dedicated to a single issue, that of national security, but who was crossed at every turn by the liberal news media, who treated him unfairly because of the work he did during the Hiss case.

Unfortunately, life does not allow for such simplistic constructions. There was about Hiss a curious blind secretiveness, a purity that was so unreal as to seem opaque. His eternal optimism, his inability to believe the worst was occurring, his unending civility to an interrogator (Nixon) whose hatred for him was manifest are literally without precedent. Hiss's formidable rectitude over the past thirty years reflects the picture of a man determined not to look within himself. Clearly Hiss's "humaneness" had a self-serving quality. It projected a strong sense of innocence which was designed to deny culpability in matters large and small.

To this day very little can be added in a personal sense to the virtuous picture Hiss has so persistently presented to the public. The story still reads as follows: Hiss was completely unprepared for the Chambers attack and never understood its motivation; he quickly became a symbol of anti-New Deal antipathy and was surprised by the perfidy of the House committee; he was dismayed by his conviction by his peers but remains optimistic and confident of his eventual redemption. This was how he saw his case when he wrote his book *In the Court of Public Opinion* in 1957, and this is how he sees it today.

CHAPTER 12

Tis morning: but no morning can restore What we have forfeited. I see no sin: The wrong is mixed. In tragic life, God wot, No villain need be! Passions spin the plot: We are betrayed by what is false within. —GEORGE MEREDITH

The statute of limitations prevented Hiss from being tried for espionage; rather, he was prosecuted on two counts of perjury. One reason such a statute exists is the recognition that memory is an imperfect tool. The Nixon men could not, or perhaps out of convenience would not, remember significant, even historical decisions made or actions taken one or two years removed in time. Still, Hiss was asked to recall fully events that occurred almost a dozen years before.

Ehrlichman said "I can't remember" 125 times at his Watergate Senate hearing. Mitchell's disclaimer was "Not to my recollection." Then there was no talk of a trial for perjury for the President's men. Alger Hiss sounded like a recorded message as he prefaced each and every answer at the HUAC hearings with the caveat "to the best of my knowledge." Perhaps it was his best recollection, but such ambiguity made the committee first distrustful and later antagonistic toward him. The Earl Jowitt wrote of the difficulty of the witnesses at the Hiss trials. "Indeed, it was inevitable that it [memory] should be vague since the witnesses were trying to recall evidence which took place some 12 years before the date of their testimony." Consequently, it might be argued that Hiss was convicted of having a faulty memory—a "crime" most of us are guilty of.

Another reason for the statute of limitations, reads Supreme Court decision 323 U.S. 606, is to safeguard "honest witnesses from hasty and spiteful retaliation in the form of unfounded perjury prosecutions." The federal rules governing perjury trials require that the testimony of a witness be corroborated by another, or by trustworthy evidence. This stricture was reaffirmed by a unanimous decision of the Supreme Court (Weiler v. United States) in 1945.

Justice William O. Douglas was on the bench when the Hiss case was first appealed to the Supreme Court in March 1951. In his autobiography, *Go East Young Man,* the jurist commented on the dubious quality of the evidence presented in the trial. "The inference that Hiss was 'framed' was strong," wrote Douglas. "If either Reed or Frankfurter had not testified at the trial, we would doubtless have had three to grant [the hearing]." What might the Supreme Court decision be if all justices were able to participate? Douglas said, "In my view, no court at any time could possibly have sustained the conviction." It was Douglas' opinion that the Hiss case spoke volumes on the "wisdom of having two witnesses on a perjury charge or if there is only one, as in the Hiss case, that the court ride herd on the nature of the corroborative evidence to make certain it has that 'trustworthy' character which will prevent an accused from being 'framed.'"

Nixon's guilt in the most serious crisis of his life, the Watergate scandal, is a matter of public record. The tape recordings provided immutable ("smoking pistol") evidence of his culpability. Unlike the case of Richard Nixon, no such neat assay of the guilt or innocence of Alger Hiss has been offered since his conviction in 1950. Certainly if the public accepted the premise that he received a fair trial, in fact two fair trials, the corpse of Alger Hiss might long ago have been buried and forgotten. But in the minds of any number of unpersuaded people, Hiss did not receive due process. Supporters claim that Hiss was tried and convicted as a symbol of a generation.

Despite the fact that twenty of twenty-four jurors believed Hiss to be guilty as charged, Hiss and many others continue to argue—and, yes, write—about the verdict. Why should this be so after all these years? The editors of *The Nation* provided something of an answer in 1973:

> The cases that refuse to die are those in which because of the nature of the charges and the political clamor they create, the public itself becomes implicated. It is as if not twelve men, but a whole nation, sits in the jury box. Charge and countercharge, the personalities involved become part of the moral and social history of the times; and even when justice is finally done, the cases are remembered as testament to a people's persistence in the pursuit of justice as a paramount national ideal. All such cases, and the Hiss one followed the pattern, carry with them the clear and appalling implication that if the court verdict was wrong, a government agency, federal or state, might be compromised; and that majority opinion, in a moment of passion, has sanctioned injustice. Inevitably, in such instances, the final verdict—the verdict not of twelve but of 170,000,000—comes slowly. It is only as passions subside that reappraisal of the evidence is possible.

In assessing the guilt or innocence of Alger Hiss, a review of the hard evidence from the trial must be undertaken, as well as an examination of the environment in the courtroom and the mood of the nation. At the first trial, no other witness, save Whittaker Chambers, accused Hiss of espionage; no one even called him a Communist—although one witness, Hede Massing, was waiting in the wings to do just that.

Congressman Nixon found much to criticize in Judge Kaufman's decision not to permit Mrs. Massing's testimony. He attacked Judge Kaufman's fitness and asserted that the judge, who was a Truman appointee, kept Mrs. Massing off the stand for political reasons. "The entire Truman administration was extremely anxious that nothing bad happen to Hiss," said Nixon. Perhaps it was nothing more than coincidence, but Judge Goddard allowed Massing to testify at the second trial, an event that followed the Nixon condemnation.

Julian Wadleigh, never prosecuted although he confessed to the crimes Hiss was convicted of, said he was "genuinely amazed" when Chambers named Hiss as a Communist courier. Wadleigh knew Chambers during the critical period. He also knew Hiss from their work in the State Department and considered him more a closet conservative than a Communist. Wadleigh related how he and "Carl," the name he knew Chambers by during that era, often discussed other sources of contraband material in the government, but the name "Alger Hiss" was never mentioned.

The easiest question to answer in this very murky business is whether Alger Hiss was guilty of Count II, the charge that he saw Chambers after

January 1, 1937. There is less than "reasonable doubt" that Hiss did in fact see Whittaker Chambers after 1937. Hiss received a rug from him which Hiss said was in lieu of the rent but which Chambers said was a gift from their superior in the Party, Colonel Bykov. Dr. Meyer Schapiro's testimony was consistent with Chambers' story. Schapiro said he sent four rugs from New York to Washington to a man named Silverman in the first few days of the year 1937. The FBI was also able to locate Dr. Schapiro's cancelled check and the rug company receipt showing the delivery date to Schapiro as "December 29, 1936."

Hiss testified that the rug was in his house months before the 1937 year began. Concurring with this was the Hiss maid, Clydie Catlett. Certainly Chambers could have bought a rug earlier in the year and presented it to Hiss without the services of Dr. Schapiro, but the Hiss lawyers either didn't consider this possibility or chose purposefully not to pursue any line of questioning, and Schapiro was not cross-examined. What could a jury then or someone reading this today conclude other than that the defense was conceding the point that Hiss saw Chambers after January 1, 1937? There is also the condemning statement of Hiss's pediatrician, Dr. Nicholson, who told a Hiss attorney, but not the jury, that she was refused entry to the Hiss household in early 1937 by a man she later identified as Whittaker Chambers.

Surely it is no heinous crime to see someone in 1937. Alger Hiss would not have spent four years in prison for seeing someone then, perjury or not. It is fair, therefore, to argue, as Stryker did, that Count II is really subsidiary to Count I. It is clearly Count I, the perjury pertaining to espionage, which sent the former State Department official to jail. Did Hiss, in the words from the indictment which were often heard at the trials, "furnish, transmit, and deliver" restricted documents to Chambers? Since there was no other witness to this act save Chambers, there was no corroboration, and one is thereby forced to examine the evidence itself.

Was the evidence used to convict Hiss, in Douglas' word, "trustworthy"? After nearly thirty years of investigation and review, the answer is sadly, "No." There is a fluid quality to the hard evidence in this case: the typewriter, automobiles, the typed and handwritten documents, and microfilm. Time has not treated kindly the evidence that convicted Hiss. Under close scrutiny, these materials assume wholly new and often unconvincing forms.

For example, the typewriter, which became in Richard Nixon's full-blown rhetoric "the key witness in the case," turned out to be a false witness. The evidence surrounding the Hiss Woodstock typewriter is so dubious and so shocking that an entire chapter in this book has been devoted to this matter. Briefly to reprise it here: The FBI knew the machine discovered by the defense could not have been the Hiss typewriter, for it was manufactured at least two years *after* the defense's Woodstock was in use. Moreover, the FBI went to great lengths to keep this particular information secret, an unnecessary effort if the typewriter recovered by Hiss were the correct one. Finally, despite Nixon's certitude that a typewriter is like a fingerprint—unimpeachable—typewriters could be forged. Intelligence operatives had been doing this since World War II, and the FBI—and seemingly Richard Nixon for that matter—was well aware of this capability, for he reportedly told Charles Colson, "We built one [a typewriter] in the Hiss case."

The evidence pertaining to the old and new Ford automobiles does not survive the "reasonable doubt" test of validity either. Hiss said the old car was given to Chambers as part of their rental agreement; Chambers said Hiss wished to give the vehicle for the use of a poor Communist Party organizer and permission was reluctantly given by their Party superior, J. Peters. Neither version holds together under scrutiny. The FBI discovered that the car was sold a year after Chambers left Hiss's apartment—*after* failing to pay the rent. The prosecutor was to use this information well in belittling Hiss's story, saying in a voice brimming with sarcasm, "He [Hiss] gave him the car despite the fact that the guy gypped him a little bit in between."

Neither Chambers' name nor any of his known aliases is found on the District of Columbia title for this car. The automobile was sold to the Cherner Motor Company, and on the same day it was transferred to a William Rosen, whom the FBI was able to prove was a member of the Communist Party. Curiously enough, although there is a District of Columbia title, there is no record of the transaction on the books of the Cherner Motor Company. Later it was discovered by the House Committee on Un-American Activities that the signature on this title was not Rosen's. It was in fact a forgery, and the man who notarized the title, a co-worker of Hiss at the State Department, committed suicide two months after testifying at the hearings.

Although the committee got astonishingly good mileage out of the story of the *old* Ford (making much out of Hiss's faulty memory in relation to the timing of the transaction), they were finally forced to admit that the title matter was ". . . a mystery." As HUAC investigator Robert Stripling put it, ". . . We endeavored then to trace the path of the Ford . . . without too much success. It remains something of a mystery, deepened by the awful fact that the man who notarized the papers on the car's transfer either leaped or fell to a ghastly death from the top floor of the Justice Department in Washington.''

William Rosen, the man whose name appears on the title, was called before the committee to answer questions about the transaction. He sought the protection of the Fifth Amendment and was not called to the stand during the first trial because the judge knew that Rosen would again exercise his right against self-incrimination. The jurist recognized that a witness invoking the Fifth Amendment can adversely reflect on a defendant, and as a consequence he did not allow the prosecution to call Rosen. Once again, Nixon made much of this ruling, citing it as evidence of judicial bias.

At the second trial, Rosen took the stand and, as everyone—including Prosecutor Murphy—expected, invoked the Fifth Amendment. However, before claiming his constitutional privilege, he testified he had never seen Hiss before nor had any dealings with him. It is of interest that shortly after the Hiss conviction the Court of Appeals ruled in another case that there ''might be grounds for reversal if the party who called a witness connected with a challenged transaction knew, or had reasonable cause to know, before putting the witness on the stand that he would claim his privilege.''

There are three different stories about the *new* Ford and the alleged corresponding $400 bank withdrawal by Hiss which preceded the purchase by Chambers. Chambers claimed that at the direction of the same J. Peters, Hiss gave him exactly $400 for the car. (It should be remembered that Chambers told the FBI it was $500, changing it only to the lesser amount after the FBI had an opportunity to examine Hiss's bank records, which noted the $400 withdrawal.) Esther Chambers had already testified that a relative loaned them the money for the car purchase. Hiss claimed the substantial withdrawal had nothing whatsoever to do with this car or any other; it was to buy furniture for the move to the larger house on Volta Place. The government countered this claim by showing Hiss had not even signed the lease for the new place—and wouldn't for two weeks.

There are also discrepancies on the ledger page from the dealership where Esther Chambers bought the new car. Only one page was introduced at the trial, and on that page there are two purchases recorded. One reflects the Chambers' sale, while the other entry shows a transaction that occurred three weeks earlier—an unlikely commingling in a busy automobile agency. The folio numbers for the two purchases are seventy digits apart. There is a difference in the style of the entries, and there is an error in Chambers' address. Certainly such anomalies are not sufficient to conclude that the title was a forgery, but it is markedly suspicious, especially in view of the dubious quality of the other evidence. The Hiss attorneys did not challenge this piece of hard evidence either, and once again it would appear that they made a serious mistake in not doing so.

Ambiguities are apparent as well in the microfilm and in the typed and handwritten documents, which HUAC member Karl Mundt once characterized as "of such startling and significant importance." When the frames of microfilm not introduced at the trial were finally released to the public after more than a quarter of a century of suppression, they turned out to be pictures of Navy documents dealing with such weighty matters as fire extinguishers and chest parachutes. Since Hiss worked in the State Department, it is very unlikely he would have had access to Navy Department documents, and their ultimate importance to espionage was neither "startling nor significant."

The documents that did circulate in the State Department were distributed to some fifteen federal agencies in a period of admittedly lax security. Although records were kept on where they went, none was kept on who returned them. Also, two employees in the State Department testified that many of the documents had never been routed through the office where Hiss worked. Some of these materials were in Hiss's handwriting, and others bore his initials; the defense conceded these two points.

It has been argued by Hiss biographer John Chabot Smith that it would be naïve and foolhardy as well as plain stupid for a spy to sign his name to purloined documents. Although Hiss had the two former qualities in abundance, he was certainly not stupid. Another argument can be offered that such obviousness was precisely what a spy might have hoped for—i.e., a variation on the familiar theme: The best hiding place is the most obvious one. Still, such logic is convoluted at best.

There are other disturbing elements in this regard. The first relates specifically to one of the cables transcribed in its entirety by Hiss, and the second problem stems from some public-relations legerdemain involving the microfilm performed by Hiss in 1975.

To begin with, there were serious disagreements between Hiss and his superior at the State Department, Francis B. Sayre, over Hiss's responsibilities for abstracting cables. Sayre's specific interests were in the area of international economic relations, and some of the cables that Hiss reprised bore only distant connection to economic matters. Moreover, Sayre was clearly uncomfortable with the impression given by Hiss that he routinely required Hiss to prepare "briefing memos" or to "sift cables and digest them and make oral report on [their] content."

Allen Weinstein was much taken by such testimonial discrepancies and laboriously researched Hiss's attorneys' records and found that Sayre yielded very little on his original assertions that (1) he had small interest in matters covered by his subordinate's four handwritten memoranda and (2) that it was not routine for Hiss to perform such duties; Sayre said that " . . . it would not have been part of Alger's duties to prepare such summaries . . . and [he] never knew of Alger dictating such summaries. . . ."

At the heart of this matter was the telegram which was sent to Secretary of State Cordell Hull from Loy Henderson of the U.S. Embassy in Moscow. It contained a warning from the widow of an American undercover agent asking Henderson not to become active in securing the release of an American woman who had been arrested in Russia along with her husband. The case itself was a confusing one. The husband, later identified by Chambers as a Soviet spy, was ordered back to Russia in the late 1930s, and he hoped that the presence of his American-born wife would preclude the possibility of his being imprisoned. The effort was a failure; both were jailed, and the wife was requesting American authorities to secure her release.

Would such an issue concern Sayre? Not according to Sayre, who told the FBI that he was "always disturbed by the Robinson [one of the false passport names used by the couple] cable" and that he wouldn't even know why he was on the distribution list or why Hiss chose to copy it. But copy it Hiss did and in its entirety. And that cable, according to Weinstein, (once again citing Chambers) went directly back to Moscow.

To the present authors Hiss insisted that he firmly believed this cable would be of interest to Sayre. He said, "Sayre in fact testified that one of my jobs was to keep him informed of general matters of interest. . . . Mr. Sayre was constantly consulted on neutrality matters and he liked to know what was going on. The Robinson case was in the papers, nothing secret about it. It was a long brouhaha, and the State Department had been called upon to intervene in behalf of someone who seemed to be an American citizen and who seemed to be mistreated. It was perfectly natural that I would simply say, 'Oh, this is something he'd be interested in. There is something to be learned from the Robinson experience.'"

This, however, was not always Hiss's position on this particular cable, for he denied for many months that the Henderson cable was transcribed in his own handwriting, only conceding the point after both the FBI and Hiss's own handwriting experts agreed that he had indeed copied the cable.

Although Sayre failed to support Hiss's story about his cable responsibilities, he did, according to Hiss, continue to his very death to state his belief in his subordinate's innocence. The facts themselves seem clear. Hiss did transcribe a cable in which his superior professed no interest. That cable did turn up in Chambers' packet. Hiss was covert in his initial recollection, lending belief to the notion that he had something to hide. As an incident by and of itself, it adds little support to the picture of Hiss as honest and open; but it also adds little proof that Chambers received this or other memoranda from Hiss.

This was not Weinstein's conclusion, for he pointed out that Assistant Secretary of State Adolf Berle's notes in 1938 bore the following comment under the entry "Alger Hiss": "When Loy Henderson interviewed Mrs. Rubens, [Rubens was another name used by the Russian-American couple], his report immediately went back to Moscow. Who sent it—such came from Washington."

Weinstein also notes that an article written in November 1938 entitled "The Faking of Americans: the Soviet Passport Racket" describes the Robinson–Rubens couple and contains the Henderson cable, which is recorded in the article "almost verbatim."* But—and it is a very large "but"—the material found in Berle's memo and the article referred to above both came from Whittaker Chambers himself. The Berle notes were taken from an interview with Chambers, and Chambers is also the author of the article. To use Chambers' own assertions to prop up the entire Henderson cable incident is self-serving on Weinstein's part at the very best.

And what of the public-relations ploy referred to above? When the secret documents were finally released to the public by the Justice Department in the spring of 1976, Hiss held a major press conference.

---

*When one searches the voluminous Weinstein footnotes for the journal in which the Chambers article appeared, one is referred to Herbert Solow. Tracking this reference down brings the disturbing information that the Chambers article never was published but that Weinstein has come into possession of a notarized copy of material written by Solow covering some talks in 1938 with Chambers. Both Solow and his presumed literary executor, Sylvia Salmi Solow, are now dead, and there the trail grows very cold indeed.

There was, however, something dishonest in Hiss's performance, which conveyed to the press and spectators that some startlingly new revelations had surfaced. Weinstein is quick to assert that Nixon had offered Hiss's attorney, Edward McLean, full access to the microfilm on January 17, 1949, and even specified what they contained. McLean's memo read: "Nixon says that of the five rolls of microfilm, two were developed and three were not. Of the latter, one was completely blank and illegible. The other two contained partial photographs of documents containing information from the Bureau of Standards for use of the Navy. These rolls have no State Department papers." It is a fact that McLean accepted Nixon's offer. Thus, Hiss's virtuous stand two decades later was simply a well-orchestrated public-relations effort to keep interest alive in his case. Perhaps it is a coincidence, but Hiss's license to practice law was reinstated by the Massachusetts Supreme Court some four days after the press conference.

As recounted before, twenty of twenty-four jurors who heard the case at the two trials concluded that Hiss was guilty. It is now known that three of those jurors, voting for conviction at the second trial when Hiss was convicted, had relatives working for the FBI. This alone is grounds for a mistrial, as they might have harbored some *a priori* prejudice toward the defendant. In the midst of the second trial, Murphy, the prosecutor, learned of this from the FBI. The FBI papers say he thanked the agents for this information and requested " . . . that it be kept quiet." Also Judge Goddard's charge to the jury at the end of the second trial—"It is the contention of the government that this" (indicating the defense Woodstock) "is the typewriter upon which [the] Baltimore Exhibits . . . were typed"—was erroneous. The government's expert said no such thing. Had Hiss's attorneys noticed this error and understood the significance, this also might have provided grounds for a new trial.

Supreme Court Justice Douglas was to conclude of the Hiss case that the "legal issues [were] never conclusively resolved." Its effect on the times "was to exalt the informer, who in Anglo-American history has had an odious history. It gave agencies of the federal government unparalleled power over the private lives of citizens. It initiated the regime of sheep-like conformity by intimidating the curiosity and idealism of our youth. It fashioned a powerful political weapon out of vigilantism."

Even in the well-censored—and self-censored at that—FBI files, the reader is struck by the Bureau's interference in the facts of the Alger

Hiss case. Rather than evaluating the facts—surely the prime responsibility of the nation's largest law-enforcement agency—the Bureau seemingly concluded that Hiss was guilty and then went out to find or even manufacture the facts that would prove this. Therefore, it is not surprising that it was J. Edgar Hoover for whom Hiss reserved his rarely seen spleen. The latter was quoted recently by Fred Cook as saying:

> I never felt bitter or vindictive toward Whittaker Chambers because I considered him of unsound mind, and it was difficult to get steamed up about him. I didn't even feel particularly bitter about Richard Nixon; he was just a cheap, opportunistic politician. But Hoover was supposed to be the head of an impartial investigating agency, and the documents we are getting show that he played a partisan role throughout. He issued specific instructions: "Get evidence to support what Chambers has to say." And again, when there was a chance the grand jury might indict Chambers for perjury instead of me, Hoover issued instructions: "Forget Chambers. Get Hiss."

The entire matter of the release by the Bureau of the materials relating to the Hiss case is illuminating. The first disclosure by the FBI under court order brought forth a literally worthless series of cut-up sheets of papers. Further court action was necessary to force the FBI to produce a larger and more useful set of documents.

The data now available cannot in any way be considered complete, and here the FBI's position is not very far removed from Nixon's familiar one: Those documents withheld from scrutiny are those which reflect "national security" concerns. There is absolutely no reason to believe that the files are truly "open." It is easy to conclude that rather than jeopardizing national security, the material withheld deals with issues that are embarrassing to the agency.*

Beginning in 1975, Hiss's attorneys were in court seeking release of the FBI files in the New York, Boston, and Philadelphia agency offices. Particularly important was the material presumed to be found in the New York office, where there was much pretrial contact with Whittaker Chambers. After first arguing that there were no files in the New York

---

*Athan Theoharis, professor of history at Marquette University, has reviewed extensively the FBI filing system and concludes that the Bureau customarily utilized "separate file systems by which it isolated from a central filing system that pertinent to either illegal, sensitive or embarrassing activities which involved the Bureau." Theoharis' work lays waste to the assumption that there exists a single FBI file on any sensitive matter in which the Bureau was involved. ("The Guilt of Alger Hiss," *Firing Line*, telecast on Public Broadcasting System April 7, 1978)

office, the Bureau finally turned over some 60,000 pages to Hiss. Two important facts reportedly emerge from the first quick assay of the material, and both bear upon the FBI's performance as a law-enforcement agency.

The first dealt with some confusion about Chambers' previous testimony before HUAC and the grand jury as to which of the Hisses (Alger or brother Donald) had belonged to a Washington Communist cell headed by Harold Ware. According to Raymond Murphy, a security officer for the State Department who first interviewed Chambers in 1945, Chambers' initial statement was, "The top leaders of the [Communist] underground were: (1) Harold Ware, (2) Lee Pressman, and (3) Alger Hiss. In the order of their importance." Murphy's notes go on to quote Chambers as saying:

> The heads of the various underground groups in Washington who met with Peters were the Hisses, Kramer (Krevitzky) (sic), Henry Collins, who was either secretary or treasurer of the group, John Abt, Lee Pressman, Nat Perlow (sic), and Nat Witt. These men met regularly at special meetings. With the exception of Donald Hiss, who did not have an organization, they headed parallel organizations. But they did not know the personnel of the different organizations.

Along these lines, the printed record of the House committee on August 3, 1948, contains evidence that Chambers testified that day that the organizer of the Washington Communist group was Harold Ware and that Nathan Witt was its first chairman. The following were listed as members: Lee Pressman, Alger Hiss, Donald Hiss, Victor Perlo, Charles Kramer (alias Krivitsky), John Abt, and Henry Collins.

An FBI telex dated January 14, 1949, one day after Hiss's indictment, reads:

> He [Laughlin] stated that the Director had questioned whether the name of Donald Hiss [Alger's brother] should have been included in the group Chambers claimed was operating under Harold Ware [a Communist cell leader]. I advised him, after checking with S[pecial] A[gent] Thomas Spencer, who was at that moment interviewing Chambers, that the teletype as sent was correct; that Donald Hiss was, according to Chambers, in the group. I also advised that he [Chambers] had not put Alger Hiss in the group because he was not sure Alger Hiss was in the group. Consequently, no statement was made regarding Alger Hiss.

This was surprising in at least two contexts. Chambers had told

Murphy that Donald Hiss had no group organization but was "a head of an underground group." The contradiction and/or ambiguity is never explained: How can a man who "heads" an underground group have no group to head? Moreover, Chambers had told HUAC that both the Hisses belonged to the Ware–Witt group. Five brief months later, Chambers was advising the FBI that Donald Hiss was "a member of the group," but Alger might not have been. This is said, presumedly, without embarrassment, despite the fact that Chambers had under oath testified repeatedly that Alger Hiss had belonged to a Washington Communist cell led by Harold Ware in the 1930s.

The second development to emerge from the New York Bureau's documents is the news that the FBI constructed a pretrial scenario which identified the evidence needed to convict Hiss and how it should be presented at the trial. One fascinating section entitled "Documentary Evidence" treated the typed State Department papers presumably copied by Mrs. Hiss. The Bureau report, dated February 2, 1949, contained a very strange warning:

> It is not clear at this time if Chambers can testify that he received these particular 69 documents from Hiss, but upon establishing the facts of this situation, decision can be thereafter reached as to who is in position to introduce these documents.

How disconcerting it must have been for the Bureau to discover that at that late date, just before the first trial was to begin, Chambers was still recasting his testimony in some of its most important essentials and that he might testify one way or another, depending on his recollection at any particular time.

There should, of course, have been ample warning of Chambers' unreliability. His first conversation with Assistant Secretary of State Adolf Berle simply identified Alger Hiss as a member of the Communist underground. Chambers' complete description of Hiss's activity to Raymond Murphy was "Alger Hiss was never to make converts. His job was to mess up policy." In Chambers' early testimony before the House committee, espionage was never mentioned. In fact, in sworn testimony, Chambers averred that Hiss had never been a spy. Soon he was calmly testifying that Hiss had always been a spy.

Behind all the contradictions described loomed the large figure of the FBI, perhaps producing some evidence, perhaps suppressing some, but clearly counseling, preparing, and defending its "witness." This is in direct conflict with its stated responsibilities. Its charter reads:

> The Federal Bureau of Investigation (FBI) is the principal investigative arm of the United States Department of Justice. It is charged with gathering and reporting facts, locating witnesses, and compiling evidence in matters in which the Federal Government is, or may be, a party in interest. The FBI does not express opinions concerning the guilt or innocence of subjects of its investigations; nor does it otherwise assume the role of accuser, prosecutor, jury, or judge.

In refusing to provide both sides with a full accounting of what it had uncovered, the Bureau was obviously acting illegally. The Schmahl double-agent material alone creates serious question about the agency's willingness to subvert the processes of justice, and the refusal to make clear its suspicions that the Woodstock was not the original Hiss typewriter can only be viewed as an effort to withhold exculpatory evidence from the defense. This was not careful fact-finding, a function the public had been led to expect from the FBI; rather it was an act of partisan prosecution by an executive agency whose charter portrayed it as the major servant of law and order in the land.

Times have clearly changed, and a public bloodied by the Watergate scandals and fast becoming cynical about government operations in general and its investigative agencies in particular* might well take another look at the final consequence of the strange case of Alger Hiss. It must be remembered that Prosecutor Murphy hung his case on the FBI, saying to the jury, "If you think that any bit of evidence in this case, any bit, material, or immaterial, was manufactured, conceived or suborned by the FBI, acquit this man. The FBI ought to be told by you that they can't tamper with witnesses and evidence, if you believe that to be a fact. . . ." No lawyer in this day and age would risk such a challenge, and it is surprising, knowing even what Murphy knew, that he would take such a chance in his summation.

We do argue that the evidence that convicted Alger Hiss was flimsy

---

*In the late fall of 1977 Richard Helms, former director of the CIA, pleaded "no contest" in a federal courtroom in the District of Columbia to a misdemeanor charge that he failed to "answer questions fully and completely as required by law about the CIA's role in opposing the election of Salvador Allende in Chile in 1970." Helms was fined $2,000 and given a suspended sentence.

and controvertible. Yet, Hiss's very demeanor, his persona as it were, aroused suspicion. His immediate and laudable demand for a hearing, his stellar first performance before HUAC, the shaky second act before the same audience, and then the bizarre behavior vis-à-vis Chambers at the Hotel Commodore—these were all different pictures of a very complex man. The strait-jacketed life that followed, the absolute *noblesse oblige,* the pusillanimous public face, and the persistent lack of candor have done little since then to clear up the questions which persist.

Perhaps it is this last quality that is now and always was the most distracting feature of Alger Hiss's personality. He came into the limelight straight and honest, and three decades later his appearance is markedly unchanged, despite prison, dishonor, and poverty. Either Hiss was innocent of any wrongdoing in his entire life and had the strength only associated with the very pure, or he was the world's most accomplished artificer. It is this contrast between the man and the circumstance which divided both friend and foe from the very beginning of the affair in 1948, and this division exists even today.

When at the first trial defense attorney Lloyd Paul Stryker solemnly intoned, "I will take Alger Hiss by hand, and I will lead him before you from the date of his birth down to this hour, even though I would go through the valley of the shadow of death, I will fear no evil, because there is no blot or blemish on him," one can think that Stryker fully believed what he was saying. He took the case for a very small fee and had never met Hiss before he was employed to defend him. It is also possible to conceive that Hiss thoroughly approved the Stryker stratagem (although he probably objected to the flowery language)—in fact, that Hiss thoroughly believed the statement as well. And it is this characteristic that made Hiss's friends the most uncomfortable. It was as if he concluded that purity would win out over all considerations—even sound legal preparation.

Hiss's concern was always for someone else. It was his wife Priscilla who could not sleep when he was indicted; Tim Hobson had to be "spared" from testifying even though the stepson volunteered to do so; Crosley could not be identified until too late in order to spare him (Crosley–Chambers) embarrassment; Hiss worried that Priscilla's abortion, which occurred years earlier, might become known if she testified at the trial. Clearly this man couldn't recognize a crisis if it exploded in his face, in marked contrast to Nixon, who created his own crises and used them for his own advancement and to harden his character.

Close and persistent contact with Alger Hiss reveals something mildly incongruous, something just a bit awry in a carefully orchestrated exterior. While the word has a pejorative ring, it must finally be said. There is a covert, hidden quality about the man. Knowing him is no simple task, and the cardinal question can likely never be fully answered: Is Hiss covert because life has taught him its most bitter lesson, or is he covert because he has much to hide? All of this is by way of saying that Hiss affects one as a man who was guilty of something, but our careful review fails to establish that he was guilty of passing documents to Whittaker Chambers, the crime for which he went to jail.

But was Hiss telling the truth, the "whole truth" as required by law? We are now back again to the familiar "no exit" point. David Riesman, who presumed Hiss was guilty in 1953 but who was listed as being "uncertain" about that fact in 1976 in the Nobile survey, felt that Hiss could have done his country a great service by "telling how a successful and idealistic young lawyer could get involved with the Communist Party." Leslie Fiedler asked plaintively in 1950: "Why did he [Hiss] lie and lying lose the whole point of the case in a maze of irrelevant data: the signature on the transfer of ownership of the car, the date a typewriter was repaired . . . ?"

Both Riesman and Fiedler make the same perplexing point, for Fiedler even proposes what Hiss should have said, "Yes, I did these things—things it is now possible to call treason—not for money or prestige, but out of a higher allegiance than patriotism," but each assumes that Hiss was guilty and that the truthful confession would have saved him. Fiedler is dead certain of this, for he says that if Hiss had told the truth, "then he need not even have gone to prison." Such easy conviction is patently incorrect; the history of that particular era produces a somewhat more ambiguous scoreboard for confessors and deniers. Moreover, most who remained silent, the third alternative, did not seem to do too badly at all.

But what if neither Riesman nor Fiedler, nor anyone else for that matter, has conceived of another option? Is it possible to believe that Hiss lied early and late and persistently about a single issue (perhaps the nature of his relationship with Crosley–Chambers) but still was not proven guilty of espionage, of passing the controverted documents to Chambers? Despite persistent statements by the two judges, by the prosecution and the defense, by Congressmen Nixon and Hebert about truth-telling and its importance in determinations of guilt and innocence, two juries elected decisively to believe a man who freely admitted to lying and to renounce a man who insisted that he told nothing but the truth.

In what may be the most awful irony of all, the crucial lie for Hiss was, in our opinion, his refusal to admit to knowing Whittaker Chambers by that name or any of the eight or nine other names he used over the years. That first denial led unfortunately to the necessity for further denials. When Hiss asserted that a man's name and his picture did not summon up any recollection at all, he was then forced into being logically consistent even in the face of the enlarging evidentiary field. There was at first no relationship recalled, then only a very small one, almost forgettable, really; then just a little more, a car, a flat, a rug, a picture painted by one woman and accepted by another, but still no more than that, still easily forgettable; and then finally, thirty years later, the offhand comment that contained the suggestion of something much closer, a relationship in which Chambers was jealous to the point of cold hostility toward Priscilla Hiss and evinced an "almost paternal attitude" toward Hiss himself.

When Stripling asked Hiss at the first hearing, "You say you have never seen Mr. Chambers?" Hiss's response—"The name means absolutely nothing to me"—was evasive. Hiss had more to go on for this identification than just the man's name. In the next exchange he admitted, "I have looked at all the pictures [of Whittaker Chambers] I was able to get hold of in, I think it was, yesterday's paper which had the pictures." His response was, as Nixon described it, " . . . qualified carefully." In one sense, Hiss admitted being covert. He told his son, Anthony, and others that he remembered hearing the name "Whittaker Chambers" when interviewed by the FBI in 1947.

The Chambers name game became mightily complicated for Hiss at the subsequent hearing when he gave his accuser the name "George Crosley" but was unable—and, as it later turned out, unwilling—to produce any other witnesses to substantiate this recollection. Although the Nye Committee receptionist was by then dead, certainly there were other reporters covering the committee's hearing who could remember this "gregarious"—at least this was Hiss's estimation—writer by this or any other name.

When Hiss found the man who remembered publishing poems by Chambers during the years 1925 and 1926 under the *nom de plume* George Crosley, this witness was not called to the stand. Why? Hiss says his attorneys dissuaded him because the publisher, Samuel Roth, who had been convicted under the obscenity laws, would be viewed by the jury as a man of dubious credibility. Moreover, it was feared that his association with the defense might reflect adversely on their client. These were their conclusions, but another argument could be offered that the testimony could reflect adversely on Chambers, who, after all, had submitted the erotic poems under a false name to a man who would eventually be convicted for obscenity. A Hiss lie or blunder? It is impossible to say, but many have concluded that the decision to keep Roth off the stand was one of the biggest blunders in the defense's case.

While it is often claimed that legal verdicts derive from the relationship between facts and the law, Justice Oliver Wendell Holmes took a somewhat less sanguine view of this balance when he wrote:

> The life of the law has not been logic; it has been experience. The felt necessities of the time, the prevalent moral and political theories, intuitions of public policy, avowed or unconscious, even the prejudices which judges share with their fellow men, have had a good deal more to do than the syllogism in determining the rules by which men should be governed. . . . The substance of the law at any given time pretty nearly corresponds, so far as it goes, with what is then understood to be convenient. . . .

It would seem, therefore, that the more important facts for a jury are the human ones—the man's demeanor, whether he "appears" to be innocent or guilty—in effect, does his story make human sense?

And it was here that Hiss failed badly. No one believed that a man could not recognize another with whom he lunched occasionally, who spent a week or so in his home, whom he loaned or gave a car, and from whom he received a rug. And it was here that Fiedler, who was so annoyingly wrong about some things in the Hiss case, was so right in making the following sobering comparison of the two men:

> Every word he [Chambers] spoke declared him an ex-traitor, a present turncoat and squealer; and Hiss, sensing his inestimable advantage in a society whose values are largely set in boyhood when snitching is the ultimate sin, had traded on his role as the honest man confronted by the "rat." Really, Hiss kept insisting, they'd have to call the Harvard Club, say he'd be a few minutes late to dinner—after taking care of this unpleasantness. For a while it came off quite successfully, coming from one who visibly belonged, whose clothes beautifully fitted, whose manners were adequate to all occasions.

But then finally it no longer came off "successfully" at all. Hiss was lying about knowing Chambers, and all the birds were coming home to roost, for this was the very first issue in the case, whether Hiss knew

Chambers or not, and from then on Hiss's credibility was badly shattered; he was in—nay, he had deliberately put himself in—the position of being an absolutely honest man. Chambers, of course, had no such obstacles to overcome. He had lied much, sometimes too often even to straighten out in his own mind, sometimes only yesterday or the day before, but this time it was the truth, the whole truth, so help him, God. Chambers, the confessed sinner, had now found religion and with it, in inevitable accompaniment, the truth as well. Hiss, the professed saint, was really a Communist, a man who lied about the most simple issue, personal recollection, and thus must be lying about much else as well.

And so Alger Hiss, the man erroneously believed to be named by his father for Horatio Alger,* the earnest and ambitious American youth, may well have come to his sad end because he felt compelled to deny his relationship with the shabby Chambers and, not making that denial stick, could not then be believed by his peers in a number of far more important issues. Remarkably, this was Nixon's conclusion, although he never informed the public of his posture. In a letter Nixon wrote to John Foster Dulles before the first trial, he said of Hiss: "Whether he was guilty of technical perjury or whether it has been established definitely that he was a member of the Communist Party are issues which may still be open to debate, but there is no longer any doubt in my mind that for reasons only he can give, he was trying to keep the [HUAC] committee from learning the truth in regard to his relationship with Chambers."

"In for a dime, in for a dollar" runs an American street refrain; a small lie escalates, and before you look around, your whole fortune is imperiled. So it was with Hiss, the All-American boy laid low for a mistruth by a confessed Communist spy who lied as some men say "good morning" in a scenario master-minded by a young Congressman from California whose own career ended in shambles after being exposed in a monumental series of lies.

Nixon's demeanor in its public and private aspects was quite different from Hiss's. In public Nixon was clearly Hiss's equal in piety. He was sworn to defend the Constitution, and his high-mindedness was clearly visible. His explanations were reasoned and sober, and the occasional signs of irritation or flashes of anger were to be viewed as the signs of humanity in a sorely tried man. Not unlike Hiss, he cried, "Believe me,

---

*Hiss's father's name was Charles Alger Hiss. Charles's father, George M. Hiss, named his son after his friend Russell Alexander Alger, who became a major general during the Civil War.

believe me!" but his voice was strident. Much taken by the trappings of high office, he sought to cloak himself in its protective colors. "When the President does something, it's not illegal" was one of his last public explanations, and it might well be offered in a slightly altered form for all the crimes of commission and omission that dotted his entire career: "When Nixon does it, it's not illegal."

Is it reasonable to think that Watergate was only the last act in a career which began thirty years earlier with a misrepresentation of his own naval service record and an unwarranted assault on the politics of his first political opponent? What Watergate simply did was clearly to expose Nixon's persecutory zeal. There was a "criminal" in the Nixon make-up who was managed by a lifelong assault on the immorality of others. Perhaps it is platitudinous to recall here La Rochefoucauld's maxim, "Our virtues are most frequently vices in disguise."

Still, there was a real surliness in Nixon, a pugnaciousness which verged on ugliness. It was likely this which prompted the uncomplimentary comparison between Truman and Nixon in the Washington *Post* and also led to Walter Lippmann's thumbs-down review of Nixon's capability for high office during the years of his Vice Presidency.

To state it directly, Nixon's piety and unctuous morality rested uneasily atop some very contradictory tendencies. Perhaps this is why, more than most men, he appeared so uneasy, so uncertain, so covert. And perhaps his downfall came when the accouterments of honor, of the nation's highest office, allowed the always inherently flawed personality parts to become unglued, to separate even further. Thus, his faulty perceptions of reality and his exaggeration of the threat from without (always a Nixonian soft spot) led to grievous miscalculations and finally to the loss of public trust. Each new revelation of official misconduct then served only to confirm the growing suspicion that Nixon had an unreliable character, and the final confirmation for many came when that much beleaguered, chalky-faced, and visibly tired man cried defiantly near the end of one of his last news conferences, "I am not a crook! I've earned everything I've got!"

This disclaimer notwithstanding, Nixon knew quite a bit about shady practices. Big business with its own rules and own special pleadings found a ready ear in Nixon's White House. The President knew how to shade the regulations (the ITT case, the Milk Fund, his presidential papers) and not get caught. Big crime also had entry to Nixon, who was familiar with both crime and its suppression from public notice (the White House "plumbers," the Huston Plan, the hush money for the

Watergate defendants). The likes of Jimmy Hoffa, Frank Fitzsimmons, Frank Sinatra, Robert Vesco, and others found here a man who had a feeling for the seamy side of life as well.

Unlike Hiss, who sanctimoniously professed to be uninterested in Nixon at all, the former President could never forget his first major victim. The Watergate tapes made this abundantly clear. Mr. Hiss came in and out of Nixon's closet, and his periodic appearances usually signaled the approach of hard times for the President. The Hiss prosecution was his shining hour, and it came to symbolize Nixon's strong and masterful behavior under extreme stress.

The published versions of the White House tapes present evidence of Nixon's need to reach back into the past to re-establish contact with what he considered his most outstanding success. On February 28, 1973, Nixon instructed John Dean about the Hiss case in advising Dean how to deal with Attorney General Richard Kleindienst. The three passages are instructive:

> When you talk to Kleindienst—because I have raised this in previous things with him on the Hiss case—he got, he'd forgotten, and I said, "Well, go back and read the first chapter of *Six Crises*."
>
> These guys, you know—the informers, look what it did to Chambers. Chambers informed because he didn't give a God damn. . . . But then, one of the most brilliant writers according to Jim Shepley* we've ever seen in this country—and I am not referring to the Communist issue—this greatest single guy in the time of 25 or 30 years ago, probably, probably the best writer [unintelligible] this century. They finished him.
>
> Tell Kleindienst that Kleindienst in talking to Baker and Ervin should emphasize that the way to have a successful hearing and a fair one is to run it like a court: no hearsay, no innuendo. . . . Now you know God damned well they aren't going to. . . . Tell them that is the way Nixon ran the Hiss case. Now, as a matter of fact, some innuendo came out, but there was God damned little hearsay. We really—we, we just got them on the facts and just tore them to pieces. Say "No hearsay; no innuendo." And that, that he, Ervin should sit like a court there.

The personal grandiosity in the above references is literally overwhelming. Nixon describes himself in the third person, holds himself up

---

*James R. Shepley was a former newspaper reporter who became publisher of both *Time* and *Fortune* and who is currently president of Time, Incorporated.

to Kleindienst, Baker, and Ervin as a behavioral model, and describes Whittaker Chambers as this century's "best writer."

This last judgment is so extreme as to make one wonder about Nixon's intelligence. It may well be that Whittaker Chambers was the best thing that happened to Nixon "in this country" or "in this century," but to describe him as Nixon does above so extends conventional wisdom as to render the whole matter completely incomprehensible.

One month later (March 27), Nixon, in speaking to H. R. Haldeman about Jeb Magruder, was heard to say:

> Chambers is a case in point. Chambers told the truth, but he was an informer, obviously because he informed against Hiss. First of all, it wouldn't have made any difference whether the informer [unintelligible]. First of all he was an [unintelligible]. Hiss was destroyed because he lied—perjury. Chambers was destroyed because he was an informer, but Chambers knew he was going to be destroyed.

Nixon was much moved by *Six Crises* in totality; he asked Haldeman, who volunteered that he had already read it, to "Warm up to it, and it makes fascinating reading. . . . I want you to reread it." Charles Colson, no mean presidential lackey, happily admitted that he had read the book at least fourteen times, and Colson commented that the crisis in the book which always drew the President's attention was the Hiss case.

When the President was concerned that the Senate investigation of Watergate was endangering the White House, he told John Dean how to seduce Senator Sam Ervin into granting executive privilege:

> We're making a lot of history. And that's it—we're setting a historic precedent. The President, after all, let's point out that the President [Truman], uh, how he bitched about the Hiss case. Which is true, I raised holy hell about it.

Colson, in a phone call to E. Howard Hunt, described the Ellsberg affair as becoming "another Alger Hiss case where the guy is exposed, other people were operating with him, and this may be the way to really carry it out. We might be able to put this bastard into a helluva situation and discredit the New Left."

Although it didn't appear in the White House tapes, there was another sign of Nixon's continuing interest in Hiss during the President's moments of adversity. As the impeachment hearings were drawing near, Nixon had taken to inviting Congressmen to accompany him on the

Presidential yacht, *Sequoia,* in order to evaluate their degree of support for his Presidency. To one such guest Nixon offered the so-called "true story" of the Hiss case: that the two principals were both homosexuals. This was not idle chatter, as Nixon offered this simplistic (locker-room) version to others in the Nixon entourage.

When Nixon was trying to confuse Henry Peterson, who led the Justice Department investigation of the Watergate matter, by planting the notion that money offered to Watergate prisoners was for lawyers rather than for their silence, the President once again said cryptically without identifying the word "they": "They helped the Scottsboro people. They helped the Berrigans; you remember the Alger Hiss defense fund?" There it was, over and over; the more trouble Nixon was in, the more firmly he held to Alger Hiss and the old days.

In a lengthy article by Seymour Hersch in The New York *Times* on December 10, 1973, the reporter wrote about Nixon's active role in the creation of the White House plumbers. Hersch quoted "authoritative sources" in saying that when Nixon gave John D. Ehrlichman overall responsibility for the activities of the plumbers, he told Ehrlichman to urge them to "read a chapter in his autobiography, *Six Crises,* dealing with Alger Hiss." Hersch noted that Nixon wanted the group to know that when he was a Congressman investigating Hiss in 1948, he "did not trust the Justice Department to prosecute the case with the vigor we thought it deserved."

The news story vividly described Nixon's angry reaction to White House leaks and quoted him as saying, "I want every son-of-a-bitch in the State Department polygraphed until you find the guy." When Egil Krogh, co-director of the plumbers group, protested about the legality of this move, Mr. Nixon allegedly said, "I don't give a good goddamn about that; it is more important to find the source of these leaks rather than worry about the civil rights of some bureaucrats."

When Mr. Krogh had accepted his assignment and reported to San Clemente for his initial meeting with Mr. Nixon, the President did indeed urge him to read the Hiss chapter in *Six Crises,* and both Mr. Ehrlichman and Mr. Krogh are reported to have said that they had no doubt that the President "equated Dr. Ellsberg with Mr. Hiss, a comparison that heightened their sense of urgency."

Hersch went on to recount some other interesting material which evoked comparisons between these two events separated by more than a quarter of a century. One government official has said that the publication of the Pentagon Papers brought a "crosscurrent of purposes with national security considerations mingled with an effort to discredit Ellsberg and left wing elements." Nixon avowedly took an active part in drafting the anti-Ellsberg strategy, and when Assistant U.S. Attorney Whitney North Seymour complained that he did not have enough information to justify the restraining order against publication of the papers on national-security grounds, Mr. Ehrlichman was directed by the President to secure an affidavit from the director of the National Security Agency listing the intelligence operations that would be compromised by their publication.

Unfortunately for Nixon, the U.S. District Court in New York lifted the restraining order, and the case rapidly appeared on the Supreme Court docket. After the presentation to the court and during the justices' deliberations, there were two happenings which made one think of the Hiss affair. The first was the announcement by the Justice Department in Washington of Ellsberg's indictment in Los Angeles on June 28, 1971, on charges of theft and unlawful possession of the Pentagon Papers. Particularly interesting was the information now available that the indictment had been forced by the White House and issued over the protests of Justice Department officials in Los Angeles, who did not feel the government was ready to make the indictment. Here one is reminded of Nixon's efforts in that long-gone era to manipulate the New York grand jury in order to secure the Hiss indictment.

The second instance was the publication the next day by Victor Lasky, a columnist closely linked to the White House (and one identified in the FBI files as "friendly to our [the FBI] cause"), of a report that the Pentagon Papers had been provided by Ellsberg to the Soviet Embassy in Washington. The New York *Times* quoted a government investigator as saying, "It [the story] was definitely a White House leak." This was surely not very far removed from the successful release of privileged information during the Hiss case, a maneuver which kept Hiss and his attorneys constantly off balance.

If this attempt to pressure the court and to utilize press leaks for his own advantage were old Nixon tactics, they failed this time, for the U.S. Supreme Court decided on July 30, 1971, to invalidate the injunction against publication of the papers. The White House saw this as a serious political rebuff, and it likely became a key factor in what one senior government investigator felt was the "subsequent decision by the Nixon Administration to move against Dr. Ellsberg."

The President said publicly, "I told Mr. Krogh that as a matter of first priority, the unit should find out all it could about Dr. Ellsberg's associates and his motives." The break-in to Dr. Fielding's office was a natural extension of Nixon's instructions to Krogh, and public knowledge of the Fielding burglary and of Nixon's and Ehrlichman's intercessions with Judge Matthew Byrne ended the Ellsberg trial and brought the President into his most serious difficulty. Krogh pleaded guilty to criminal charges in the Ellsberg burglary and then proclaimed that in "good conscience" he could not assert national security as a defense. John Ehrlichman, David Young, and G. Gordon Liddy were convicted of three counts of burglary, conspiracy, and perjury in the Ellsberg case, and Mr. Nixon's resignation could be traced directly to this matter.

Throughout it all, Nixon seemed obsessed with his own history. In *Six Crises,* in reference to President Truman sheltering Hiss, Nixon wrote in a statement which seems absolutely prophetic in its personal reference:

> His [Truman] error was sheer stubbornness in refusing to admit a mistake. He viewed the Hiss case only in its political implications and he chose to handle the crisis which faced his Administration with an outworn political rule of thumb: leave the political skeletons hidden in the closet and keep the door locked.

Garry Wills, writing in The New York *Times* magazine on August 25, 1974, suggests that Nixon "dwelt upon the Hiss case in order to nerve himself to a necessary hatred. . . . A new, an evil Nixon was born in the fires of the Hiss case, and he felt a duty to nurture that Nixon even as it was consuming his own substance." Wills's article is an intriguing one, and it strongly underlines the self-destructive quality in Nixon. His conclusion that Nixon would be "undone as a result of the effort to repeat the undoing of Hiss" makes good psychological sense.

Although Nixon's interest in Hiss remained high throughout the former's public career, there was little mention of Hiss after Nixon became a private citizen. He occupied no "special" place in Nixon's biography in comparison to the top billing accorded Hiss in *Six Crises** and

---

*The only new material which appears in what is essentially a rewrite of the 1962 manuscript of *Six Crises* is a curious reference in which Nixon describes what he purports to be Harry Truman's real feelings about Alger Hiss. Attributing the material to Bert Andrews, Nixon reports that Truman said angrily to a Justice Department official, "That son of a bitch, Hiss, he betrayed his country." There is another occasion described where Truman was alleged to say to an aide, "Of course, Hiss was guilty. But that Committee [HUAC] isn't interested in that. All it cares about is politics."

Since both Truman and Andrews are long dead, there is no way of ascertaining the

appeared not at all in the Nixon–Frost interviews. It is surprising, therefore, to discover that Nixon chose to discuss Hiss one final time in a conversation with the New York University professor Sidney Hook, longtime New York *Times* reviewer of books dealing with Hiss and Chambers.

Hook had never been very sympathetic to Hiss's protestations of innocence, and it might well have been this fact that brought Nixon to the telephone at San Clemente on November 24, 1976, to answer Hook's inquiries about John Dean's attribution to Nixon in a conversation with Charles Colson—"Typewriters are always the key. We built one in the Hiss case."

Nixon's comment is interesting not so much for what it says (there are no surprises there) but for what it reveals of the absolute persistence of the man. As Nixon told Hook:

> The statement I was alleged to have made with regard to having built a typewriter in the Hiss case is totally false.
>
> This, as you know, was a thesis Hiss and his lawyers put forward 25 years ago—that Chambers, the FBI or others involved in the Hiss prosecution had built a typewriter. Even his most ardent supporters could not swallow such a ridiculous charge. A typewriter is, as you know, almost the same as a fingerprint. It is impossible, according to experts in the field, to duplicate exactly the characteristics of one typewriter by manufacturing another one.
>
> What I said after Hiss was convicted, and have consistently said since that time whenever I discussed the case, was that "the typewriter evidence was a major factor in leading to the Hiss conviction." I have never in any conversation at any time said or implied that "we built a typewriter in the Hiss case."

Clearly, nothing new emerged, and the former President was engaged in the same exercises that characterized his 1962 account in *Six Crises*.

---

truth in the above references. Nixon's statements are at some remove from Truman's own comments, which were very supportive of Hiss, made to Merle Miller in the latter's authorized biography of the late President. Moreover, Nixon makes markedly little use of footnotes, and although this may well be one of the perquisites of high office, the reader is unable to discover where or when Nixon and Andrews discovered how Truman really felt about Hiss.

Nixon was not an entirely reliable historian, as witness his statement in 1962 to an interviewer from the London *Times* who asked about Hiss's behavior since he had left prison. Nixon replied, "He [Hiss] had gone to bits completely. [To the question] . . . Had Hiss shown any overt sign of being a Communist since he came out of prison? Nixon thought that he [Hiss] had recently attended a Communist meeting, but could not give chapter and verse."

The public is prepared to accept some of the personal peccadilloes of those elected to public office, as the careers of James Curley, Adam Clayton Powell, and Joseph McCarthy clearly evidence. But what is important is that in order to continue to support such men, the electorate must be able in some way to identify with that behavior—i.e., to find it either comprehensible or benign or amusing. Padding Congressional payrolls with relatives may well offend the Justice Department but not necessarily the rest of us, and sexual misbehavior is so endemic a feature of life in high places that only the occasionally more complicated scandals such as Chappaquiddick or the Wayne Hays–Elizabeth Ray liaison bother us.

But Richard Nixon seemed to have pushed the American public too far. It could understand the sly wink and the quick grab, the hustle tactics of a poor boy on the political make, but it could not identify with the need to continue the scary game once on top, and it had come to distrust the judgment of a man who once had it all and appeared to be unable to hold his empire together under the *best* of circumstances.

It is the last matter that occasioned the greatest difficulty. The citizenry could no longer trust Nixon's appraisal of their external reality since he had made such a mess of his own. Watergate, after all, didn't need to take place, since no one, McGovern supporters included, could conceive of a Nixon loss in 1972. And the same analogy held true for the other scandals. The price tag for ITT was a paltry $400,000, for Vesco $200,000, and even for the dairy interests a slim $2,000,000 for a campaign war chest that exceeded $60,000,000.

One thing is irrefutable: Both Nixon and Hiss (and Chambers as well) were men of their times. The Earl Jowitt contends that Hiss could never even be tried in England for the offenses which ruined him in the United States, and there exists a real question in our minds as to whether Chambers would be believed in the United States today. Most political observers now hold, moreover, that Richard Nixon—or, more accurately, men like Nixon—would not be elected to high office given the prevailing public winds in the land as the decade ends.

Nixon's career began with Hiss and espionage and ended with Ellsberg and Watergate. The tactics that succeeded so well in 1948 were centrally related to his ultimate downfall in 1974. This is the Nixon tragedy. And Hiss, whose star burned so brightly, whose early career seemed to epitomize the best and the brightest, fell into disgrace because he either couldn't or wouldn't grasp the enormity of the crisis which faced him and respond to it in an effective manner. In retrospect, Hiss's defense tactics (both in a legal and personal sense) were so opaque as to seem inpenetrable. And his failure to acknowledge even "knowing" Chambers in the simplest sense of that term came finally to be viewed as nothing save the act of an untruthful man. Perhaps Freud said it best about both Nixon and Hiss in writing about Dostoevsky's novel *The Brothers Karamazov:*

> It is a matter of indifference who actually committed crime; psychology is only concerned to know who desired it emotionally and who welcomed it when it was done.

NOTES

CHAPTER 1

The quotations attributed to Alistair Cooke are from *A Generation on Trial* (New York: Alfred A. Knopf, 1951). The attributions elsewhere in the book to a "foreign observer" usually mark Mr. Cooke's contributions as well.

Information about Richard Nixon comes from his biography by Earl Mazo, *Richard Nixon: A Political and Personal Portrait* (New York: Harper & Brothers, 1959).

Nixon's review of the Hiss case comes from the first chapter of Nixon's political autobiography, *Six Crises* (New York: Doubleday and Company, 1962). Much of this material was also to be found earlier in Mazo's book and later in Nixon's autobiography, *RN: The Memoirs of Richard Nixon* (New York: Grosset & Dunlap, 1978).

The material about the Nixon era in American political life credited to James Reston appeared in various columns in the New York *Times*.

CHAPTER 2

The question-answer dialogue and reprises of direct statements by Hiss and Chambers and by various members of the House of Representatives come from the Hearings Regarding Communist Espionage in the United States Government—Part I: Hearings before the Committee on Un-American Activities, House of Representatives, Eightieth Congress, Second Session (August 5, 16, 17, 25, 1948).

Trial material reproduced in this chapter was abstracted from the First Trial (May 31, 1949) and Second Trial (November 17, 1949), United States of America versus Alger Hiss, defendant, the United States District Court, Southern District of New York.

Nixon's recollections of the HUAC hearings come from *Six Crises* (cited earlier) and from his speech before the United States House of Representatives on January 26, 1950.

Hiss's comments on the early phases of the House hearings were made to the authors in interviews.

The unknown relationship between Nixon and Father John Cronin was first described by Garry Wills in *Nixon Agonistes* (Boston: Houghton Mifflin, 1970).

Hiss's description of the House hearings are from his book *In The Court of Public Opinion* (New York: Alfred A. Knopf, 1957).

CHAPTER 3

Whittaker Chambers' view of his relations to Hiss and Nixon and to the world at large comes from his autobiography, *Witness* (New York: Random House, 1952).

Priscilla Hiss's interview with Richard Nixon is abstracted from the Hearings Regarding Communist Espionage in the United States Government—Part I: Hearings before the Committee on Un-American Activities, House of Representatives, Eightieth Congress, Second Session (August 18, 1949).

The other quotations in this chapter came from Nixon's *Six Crises*, Hiss's *In the Court of Public Opinion*, and from personal interviews with Hiss (all cited earlier).

CHAPTER 4

Hiss's, Nixon's, and Chambers' personal recollections of the HUAC hearings come (as cited before) from their autobiographical writings.

The testimony from the HUAC hearings is from the Hearings Regarding

Communist Espionage in the United States Government—Part I: Hearings before the Committee on Un-American Activities, House of Representatives, Eightieth Congress, Second Session (August 25, 1948).

Hiss's comments on various aspects of the period described come from interviews with the authors.

CHAPTER 5

Meyer Zeligs interviewed Hiss over a long period of time about his early life, the HUAC hearings, and the two trials. Zeligs corresponded as well with some members of the grand jury after the hearing was over. His conclusions are found in a psychohistorical review of Hiss and Chambers entitled *Friendship and Fratricide: An Analysis of Whittaker Chambers and Alger Hiss* (New York: Viking Press, 1967).

Allen Weinstein's book *Perjury* (New York: Alfred A. Knopf, 1978) provides an interesting summary of relations between Nixon and HUAC investigator Robert Stripling, both during their early days on the committee as well as later.

Material cited about Hede Massing, U.S. Attorney General Tom C. Clarke, Secretary of State James Byrnes, FBI Director J. Edgar Hoover, and Whittaker Chambers comes from the FBI files.

The material about Alger Hiss's relationship to Noel Field and Lawrence Duggan comes from Flora Lewis' book *Red Pawn: the Story of Noel Field* (New York: Doubleday and Company, 1965), as well as from interviews with Hiss.

The report of Hiss's experience with Alex Campbell on the last day the New York grand jury met comes from a note from Hiss to attorney Edward C. McLean. The communication was dated December 14, 1948, and was entitled "Appearance before Grand Jury Today."

McLean's experience with Chambers on the occasion of the former's visit to the Maryland farm is derived from interviews with Alger Hiss.

All other quotations are from the autobiographical writings of Nixon, Hiss, and Chambers.

CHAPTER 6

Transcripts of trial testimony cited in this chapter come from the minutes of the first and second Hiss trials, dated May 31–July 7, 1949, and November 17, 1949–January 20, 1950 (both previously cited).

Evaluations of the importance of the documents used in the hearings and trials come from The Earl Jowitt's book *The Strange Case of Alger Hiss* (London: Hodder and Stroughton, 1953).

Material on potential conflict of interest in the trial jury, on interviews with Chambers about his homosexuality and other matters, and on Julian Wadleigh were obtained from the FBI files.

Comments on Hiss's behavior during the hearings and trials, as well as an assay of Hiss's role in relation to the trade agreements issue, come from John Chabot Smith's book *Alger Hiss: The True Story* (New York: Holt, Rinehart and Winston, 1976).

Hiss's attitude toward both grand and trial jurors, his anticipation of the trial outcome, and factors relating to his decision to protect his stepson are the product of personal interviews.

CHAPTER 7

Quotations from members of the jury at the Hiss trial, materials relating to the Chambers' maid, Edith Murray, to the date of Chambers' defection from the Communist Party, to Chambers' complaints about colleagues at *Time* magazine, to Chambers' prospective *Newsweek* interview, to contributors to the Hiss Defense Fund, to Bert Andrews as an FBI champion, to Chambers' reaction to Hiss's book and to his own volume, and finally to Chambers' break with the FBI all come from the FBI files.

Hiss's comments about the FBI "priming" Edith Murray, his feelings about the proposed change in trial venue from New York to Vermont, his decision to insist on psychiatric testimony at his trials, and his deep antipathy to J. Edgar Hoover come from personal interviews with him.

Statements regarding Hiss's being "framed" are derived from Fred Rodell's article entitled, "Come Clean, Alger Hiss," *The Progressive*, June 1950.

Material about William Buckley's relationship to Whittaker Chambers has its origin in Charles L. Markmann's book *The Buckleys: A Family Examined* (New York: Morrow, 1973).

CHAPTER 8

Testimony regarding the Hiss typewriter is abstracted from the first and second trials, United States of America versus Alger Hiss, defendant, the United

States District Court, Southern District of New York (May 31, 1949, and November 17, 1949).

Other sources for information concerning typewriter evidence are the FBI files, Hiss's latest legal effort known as Petition for a Writ of Error Coram Nobis, United States District Court, Southern District of New York, July 27, 1978, and Zeligs' book *Friendship and Fratricide* (previously cited).

Alger Hiss's current attitude to the state of the art of typewriter evidence is derived from personal interviews with him by the authors, while Nixon's retrospective assays of the importance of the typewriter in legal proceedings come from *The White House Transcripts*, edited by Irvin Horowitz, *et al.* (New York: Viking Press, 1974).

Details about the conventional use of "forged" typewriters are offered in William Stevenson's *A Man Called Intrepid* (New York: Harcourt Brace Jovanovich, 1976).

Information about double agent Horace Schmahl has come through interviews with Hiss and through examinations of the FBI files as well as a telephone interview with Fred J. Cook.

CHAPTER 9

Testimony relating to the Watergate matter comes directly from the official record: The U.S. Senate Select Committee on Presidential Campaign Activities of 1972 (U.S. Government Printing Office, 1974).

Data about Nixon's life come from a variety of sources. Prime among them are Earl Mazo's *Richard Nixon* (previously cited); Bela Kornitzer's *The Real Nixon* (New York: Rand McNally, 1960); James David Barber's book, which offers a particularly instructive view of the Nixon–Eisenhower relations, *The Presidential Character* (Englewood Cliffs, New Jersey: Prentice-Hall, 1972); Fawn Brodie's presentation to the Friends of the San Francisco Psychoanalytic Society entitled "The Child in Richard Nixon"; Garry Wills's *Nixon Agonistes* (previously cited); Nixon's autobiography *RN* (previously cited); and the Oral History Program: Richard M. Nixon Project at California State University at Fullerton, which provides a wealth of familial reference to the young Nixon.

The view of Nixon as a gubernatorial campaigner is from Mark Harris' *Mark, the Glove Boy* (New York: the Macmillan Company, 1964).

The material about the Nixon Administration credited to Tom Wicker appeared in various columns in the New York *Times*.

E. H. Hunt's comments come from his book entitled *Undercover: The Auto-*

biography of America's Most Famous Secret Agent (London: W. H. Allen, 1975).

Howard F. Stein's evaluation of Nixon and McGovern in the 1972 election entitled "The Silent Complicity at Watergate" appeared in *The American Scholar,* Winter 1973, vol. 43.

CHAPTER 10

Biographical material about Alger Hiss derives from four major sources: John C. Smith's biography *Alger Hiss: The True Story* (previously cited); Meyer Zeligs' *Friendship and Fratricide* (also cited before); Anthony Hiss's *Laughing Last* (Boston: Houghton Mifflin, 1977); and the authors' interviews with Hiss.

Leslie Fiedler's assay of the Hiss case comes from *An End to Innocence* (Boston: Beacon Press, 1955).

The most critical assay of Allen Weinstein's scholarship is offered in *The Nation,* June 17, 1978. Weinstein's defenders had their say in *The New Republic,* April 8 and 29, 1978.

Warren Hinckle's article on Sam Krieger and a number of others cited in Weinstein's *Perjury* appeared in the San Francisco *Chronicle* on May 4, 1978.

CHAPTER 11

Roger Kahn's description of Nixon during the years spent away from politics comes from his book *How the Weather Was* (New York: Harper & Row, 1973).

Garry Wills's opinions about Nixon's political career are from his book *Nixon Agonistes,* previously cited.

Roger Rosenblatt's article on false nostalgia entitled "Look Back in Sentiment" appeared in the New York *Times* on July 28, 1973.

Dean Acheson's relations to Alger Hiss and the details of his testimony about the case before the U.S. Senate at the time of Acheson's confirmation hearing for the post of Secretary of State are taken from the latter's book *Present at the Creation* (New York: Norton, 1969).

Richard Nixon's impression of Alger Hiss's character is derived from the former's book *Six Crises,* previously cited, while Hiss's appraisal of the same issues comes from personal interviews by the authors as well as from an article written by his son, Anthony Hiss, entitled "I Call on Alger," which appeared in *Rolling Stone,* September 13, 1973.

David Riesman's review of the central issues in the Hiss case comes from two sources: an article in Riesman's book of essays, *Individualism Reconsidered* (Glencoe, Illinois: The Free Press, 1954) and from personal correspondence with the authors.

CHAPTER 12

Fred Cook's article entitled "The true grit of Alger Hiss," which appeared in *New Times,* October 14, 1977, contained information about Hiss's attitude toward J. Edgar Hoover and other pertinent material regarding the attitude of the FBI toward the Hiss case.

Information relating to Chambers' confusion regarding the early structure of Communist cells in Washington, as well as information on the FBI pretrial scenario, is derived from the FBI files.

Nixon's instructions to John Dean, Jeb Magruder, and Attorney General Richard Kleindienst come from the *Watergate Tapes,* edited by Gerald Gold (New York: Viking Press, 1973).

Nixon's assumption about homosexuality as the presumed root cause of the difficulty between Hiss and Chambers is discussed in Allen Weinstein's book *Perjury,* previously cited, as well as in an article by Weinstein that appeared in *Esquire* in November 1975 entitled "Nixon versus Hiss." Similarly, information relating to Sidney Hook's phone call to Nixon regarding the typewriter evidence is derived from Weinstein's book.

The Earl Jowitt's opinion of how Hiss might have fared in England comes from his book *The Strange Case of Alger Hiss* (London: Hodder and Stoughton, 1953).

INDEX